Name	Symbol	Atomic number	Atomic weight	Name	Symbol	Atomic number	Atomic weight
Molybdenum	Mo	42	95.94	Samarium	Sm	62	150.35
Neodymium	Nd	60	144.24	Scandium	Sc	21	44.956
Neon	Ne	10	20.183	Selenium	Se	34	78.96
Neptunium	Np	93		Silicon	Si	14	28.086
Nickel	Ni	28	58.71	Silver	Ag	47	107.868
Niobium	Nb	41	92.906	Sodium	Na	11	22.9898
Nitrogen	N	7	14.0067	Strontium	Sr	38	87.62
Nobelium	No	102		Sulfur	S	16	32.064
Osmium	Os	76	190.2	Tantalum	Ta	73	180.948
Oxygen	O	8	15.9994	Technetium	Tc	43	
Palladium	Pd	46	106.4	Tellurium	Te	52	127.60
Phosphorus	P	15	30.9738	Terbium	Tb	65	158.924
Platinum	Pt	78	195.09	Thallium	Tl	81	204.37
Plutonium	Pu	94		Thorium	Th	90	232.038
Polonium	Po	84		Thulium	Tm	69	168.934
Potassium	K	19	39.102	Tin	Sn	50	118.69
Praseodymium	Pr	59	140.907	Titanium	Ti	22	47.90
Promethium	Pm	61		Tungsten	W	74	183.85
Protactinium	Pa	91		Uranium	U	92	238.03
Radium	Ra	88		Vanadium	V	23	50.942
Radon	Rn	86		Xenon	Xe	54	131.30
Rhenium	Re	75	186.2	Ytterbium	Yb	70	173.04
Rhodium	Rh	45	102.905	Yttrium	Y	39	88.905
Rubidium	Rb	37	85.47	Zinc	Zn	30	65.37
Ruthenium	Ru	44	101.07	Zirconium	Zr	40	91.22

A. E. Cameron and Edward Wichers, *J. Am. Chem. Soc.*, 84:4192 (1962). The table includes the revisions of the atomic weights adopted by the International Commission on Atomic Weights in July, 1965.

EXPERIMENTAL
PHYSICAL
CHEMISTRY

FARRINGTON DANIELS

Professor Emeritus of Chemistry
University of Wisconsin

ROBERT A. ALBERTY

Dean of Sciences
Massachusetts Institute of Technology

J. W. WILLIAMS

Professor Emeritus of Chemistry
University of Wisconsin

C. DANIEL CORNWELL

Professor of Chemistry
University of Wisconsin

PAUL BENDER

Professor of Chemistry
University of Wisconsin

JOHN E. HARRIMAN

Associate Professor of Chemistry
University of Wisconsin

EXPERIMENTAL PHYSICAL CHEMISTRY

SEVENTH EDITION

McGRAW-HILL BOOK COMPANY

New York   Sydney
St. Louis   Toronto
San Francisco   Mexico
London   Panama

This book was set in Bodoni Book by Graphic Services, Inc., and printed on permanent paper and bound by Von Hoffman Press, Inc. The designer was Marsha Cohen; the drawings were done by J & R Technical Services, Inc. The editors were James L. Smith and Antonia Stires. Les Kaplan supervised the production.

EXPERIMENTAL PHYSICAL CHEMISTRY

Library of Congress Catalog Card Number: 75-77952

15339

234567890 VHVH 76543210

PREFACE

To illustrate the principles of physical chemistry, to train in careful experimentation, to encourage ability in research—these are the purposes of this book, as stated in the first edition 40 years ago. In each of the six revised editions an attempt has been made both to keep pace with the new developments in physical chemistry and to achieve a text that is representative of the teaching of the laboratory course in physical chemistry at the University of Wisconsin.

There are many more experiments in this book than can be performed by any one student. Selection will be made on the basis of the time and apparatus available and on the capacity and ultimate aims of the student. If an experiment is too short, the student will find interesting projects under Suggestions for Further Work; if it is too long, the instructor may designate parts of the Procedure to be omitted.

The imperative is not used. Procedures are described, but orders are not given. The student studies the experiment first and then plans his work—a method which develops both his power and his interest.

The high cost of laboratory apparatus restricts the choice of experiments, particularly where classes are small. Nevertheless, the authors have not hesitated to introduce advanced apparatus and concepts. If students are not given an opportunity to become familiar with a variety of modern developments and new techniques, they will be handicapped in their later practice of chemistry. Space for additional material has been obtained by abbreviating parts of the sixth edition and omitting older classical experiments, some of which have found their way into first- and second-year chemistry courses. For example, the determination of molecular weights by the Victor Meyer method has been eliminated in the present edition because, in spite of its excellent teaching opportunities, the method is now obsolete.

There is a trend toward the use of expensive equipment which can be bought off the shelf. Better results are obtained and much laboratory time is saved, but there is a danger in going too far in this direction. The student must know what is in the "mysterious box" and how it works. It is the instructor's responsibility to see that the principles and operation of the equipment are fully understood by the student.

Every alert laboratory instructor should consider making full use of the modern teaching equipment now available. It is not practical to explain an operation once, to all the students, and expect them to remember the details until their turn for use of the equipment comes, perhaps months later, but a large amount of the instructor's time can be spent in explaining an operation to each student in turn. If the principles of operation, as well as the practice, are to be explained, the demands of time are even greater. This problem can be greatly alleviated by the use of video tape recordings. Recordings are readily made in each laboratory with the equipment the student will actually be using. The student may view a recording whenever he wishes, and even review it if he desires. At the same time, the instructor is spared the frequent repetition of explanations. Recordings are presently being used in the laboratory of the University of Wisconsin for experiments involving electron spin resonance, infrared measurements, rubberlike elasticity, and others. It is desirable for the student to use a computer for some of the calculations if one is available. Illustrative programs have been included.

All the experiments which have been retained from earlier editions have been reviewed, and changes have been made based on continuing class experience. New experiments include Vacuum Techniques and Molecular-weight Determinations, Dielectric Constants of Gases, Heat of Formation of Polyatomic Molecules, Infrared and Raman Spectra of Triatomic Gases, and Electron Spin Resonance. There are extensive modifications in the experiments on Sedimentation, Nuclear Magnetic Resonance, Rubberlike Elasticity, and others.

The second part of the book describes apparatus and techniques, particularly for more advanced work. It is designed not only to encourage students to undertake special work but to aid them in later years in the solution of their laboratory problems. No claim whatever is made for completeness. In selecting material the authors have been guided by their own experience. The difficulty of selection increases with each new edition because the literature on new apparatus and techniques is expanding so rapidly and because commercial manufacturers have developed so many new and improved instruments. More comprehensive descriptions of experimental methods are available in references cited throughout the book.

This edition, like its predecessors, owes much to many people—colleagues at the University of Wisconsin, students, laboratory assistants, and teachers in other universities and colleges—who, over the years, have offered criticisms and provided many worthwhile suggestions for improvements. The authors greatly appreciate this help and welcome further suggestions for future editions.

The authors of the sixth edition are pleased to welcome John E. Harriman, who has shared with them the responsibilities and labor of writing the seventh edition. Professor Harriman has been active in the operation of the Physical Chemistry Laboratory at the University of Wisconsin for several years and brings to the authorship fresh and younger viewpoints and special competence in the field of theoretical physical chemistry, spectroscopy, and mathematical techniques.

The authors are indebted to Professors Lawrence F. Dahl and John E. Willard and to members of the Physical Chemistry Laboratory Staff and several teaching assistants for suggestions and critical reading of parts of the manuscript. They wish to thank Professor Worth Vaughan for many helpful suggestions, particularly for improving the experiment on sedimentation of particles. They wish to acknowledge the important suggestions of Lawrence Barlow, of the laboratory staff, and the typing of the manuscript by Mrs. Mary Wilson and Mrs. Garnie Mullen.

Farrington Daniels   *Robert A. Alberty*
J. W. Williams   *C. Daniel Cornwell*
Paul Bender   *John E. Harriman*

CONTENTS

PART 2.   APPARATUS AND METHODS

SYMBOLS

A	reciprocal moment of inertia, absorbancy
Å	angstrom
B	rotational constant, reciprocal moment of inertia
C	heat capacity, capacitance, moles per liter
D	diffusion coefficient
E	potential difference, voltage, electric field
E_a	Arrhenius activation energy
F	faraday
G	Gibbs free energy
H	enthalpy, magnetic field strength
I	ionic strength, moment of inertia, intensity of light, current, angular momentum (nuclear spin)
J	angular momentum (generalized), spin-spin coupling constant, rotational quantum number
K	equilibrium constant
K	Kelvin scale
K_b	boiling-point elevation constant
K_f	freezing-point depression constant
L	angular momentum (orbital)
M	molecular weight, molar scale
N_0	Avogadro's number
P	pressure, molar polarization, angular momentum (molecular rotation)
R	gas constant
S	entropy
T	absolute temperature, Kelvin scale
T_+, T_-	transference number
U	internal energy, potential-energy function for diatomic molecule
V	volume
W	energy
X	mole fraction
$\mathfrak{Z}$	compressibility factor
$\mathcal{G}$	electrical conductance
$\mathcal{H}$	hamiltonian operator
$\mathcal{P}$	molar polarizability, driving pressure
$\mathcal{R}$	molar refractivity, Reynolds number, Rydberg constant
a	activity, lattice constant
a_s	absorbancy index

c	concentration in moles per liter (or in grams per specified volume in some macromolecular chemistry experiments), number of components in phase rule, velocity of light
d	distance
e	electronic charge, electron, voltage (time-dependent part)
ev	electron volts
f	fugacity, frequency
g	acceleration of gravity
h	Planck's constant
i	current (time-dependent part), imaginary unit $\sqrt{-1}$
k	Boltzmann constant, reaction-rate constant, cell constant
l	mean free path, ionic conductance
m	mass, molal concentration molality, molal scale
n	number of moles, refractive index
p	partial pressure, number of phases in phase rule, linear momentum
$\mathbf{p}$	dipole moment
pH	measure of hydrogen-ion activity
q	quantity of heat absorbed
r	internuclear distance, particle radius
t	Celsius (centigrade) temperature, time
v	velocity, variance
w	work done
y	activity coefficient on molar scale
α	degree of dissociation, angle of optical rotation, alpha particle
β	beta particle, Bohr magneton
γ	activity coefficient on the molal scale, gamma ray, surface tension, gyro-magnetic ratio
ϵ	dielectric constant, energy of a radiation quantum
η	coefficient of viscosity
θ	freezing-point depression, boiling-point elevation
ϑ	fugacity coefficient
κ	specific conductance
λ	wavelength
Λ	equivalent conductance
μ	dipole moment, chemical potential, micron, reduced mass
$\boldsymbol{\mu}_e$	electron magnetic moment
$\boldsymbol{\mu}_N$	nuclear magnetic moment
ν	frequency
$\tilde{\nu}$	wave number

ρ	density
τ	torque
Φ	quantum yield, phase volume, magnetic flux
χ	susceptibility
ω	angular velocity

A superscript zero on a symbol for a thermodynamic quantity means that the value given corresponds to standard-state conditions.

A symbol Δ, as in ΔH, indicates the increment in the thermodynamic quantity accompanying the change from the initial to the final state.

A bar over the symbol for a thermodynamic quantity designates the partial molal quantity (which for a pure substance is equal to the molar value of the quantity).

The notation used for vectors (quantities having magnitude and direction) is illustrated by the following example:

$\mathbf{A}$ = vector quantity

A = magnitude of $\mathbf{A}$

A_x, A_y, A_z = components of $\mathbf{A}$ along axes x, y, z, respectively

EXPERIMENTAL
PHYSICAL
CHEMISTRY

PART 1

LABORATORY EXPERIMENTS

Chapter
1

Gases

1   *GAS DENSITY*

According to the ideal-gas law,

$$PV = \frac{g}{M} RT \tag{1}$$

where P = pressure
$\quad V$ = volume
$\quad T$ = absolute temperature ($t°C + 273.15°$)
$\quad g$ = weight of gas of molecular weight M

The ideal-gas constant R is determined by finding the weight of a measured volume of an ideal gas at a definite temperature and pressure. The molecular weights M are based on the atomic-weight scale, in which the isotope of carbon $^{12}_{6}C$ is taken as exactly 12. The molecular weight of oxygen, O_2, on this scale is 31.988, which may be taken as practically 32.0. Careful experiments have shown that the pressure-volume product for 31.988 g of oxygen at $0°C$ approaches 22.413 as the pressure approaches zero and the gas becomes ideal in its behavior. This value of 22.413 gives to the gas constant R a value of 0.08205 liter-atm mole^{-1} deg^{-1}.

Equation (1) is used to calculate the volume of gas of known molecular weight at any specified pressure and temperature or to determine the molecular weight of a gas from measurements of pressure, temperature, and volume of a given weight of the gas. For permanent gases and pressures of the order of 1 atm this value of the molecular weight is reasonably accurate, but strictly speaking, Eq. (1) is obeyed exactly only as the pressure approaches zero. This experiment shows how the density of a gas is determined in the laboratory from measurements of pressure, volume, and weight.

The relative density of a gas may be determined more easily than the absolute density using a gas-density balance. The principle of Archimedes is used, according to which the upward force is equal to the weight of the gas displaced. A large glass bulb on a balance beam is counterpoised with a weight. A reference gas, such as oxygen, is admitted and its pressure adjusted until the balance pointer reaches a certain mark on the scale. The gas is pumped out, a second gas is introduced, and the pressure is again adjusted to give the same setting of the balance pointer. When the absolute density of oxygen or other reference gas and the ratio of the densities of the two gases are known, it is easy to calculate the absolute density of the second gas.

The density of gas 1, of known molecular weight M_1, at the specified pressure P_1 is given by the expression

$$\frac{g_1}{V_1} = \frac{P_1 M_1}{RT} \tag{2}$$

Then the pressure P_2 of the second gas, 2, is adjusted to give the same position of the pointer. The density of the second gas g_2/V_2 is the same as that of the first gas at its pressure P_1, and at constant temperature

$$P_1 M_1 = P_2 M_2 \tag{3}$$

Then M_2 is readily calculated.

PROCEDURE. Gases are weighed in a glass bulb of the type shown at A in Fig. 1. The bulb is evacuated and weighed. It is then filled to a measured pressure with a gas the density of which is to be determined, and weighed again. In weighing large glass vessels it is advisable to use a counterpoise to minimize errors due to the adsorption of moisture and to changes in the buoyancy of the air caused by barometric fluctuations. If a two-pan balance is used, another glass bulb of the same type and nearly the same volume is placed on the second pan and the additional weights required to effect a complete balance are recorded. The operation is repeated after the bulb is filled with gas, the difference in the weights required for a balance being the weight of the gas.

The pressure is read to millimeters and estimated to tenths of millimeters on a closed-end manometer. The pressure inside the glass bulb A is equal to the difference in height of the mercury in the two limbs of the manometer. The bottom of the manometer is provided with a section of small diameter in order to slow down any rapid movement of the mercury caused by sudden pressure changes and thus prevent splashing of the mercury.

The bulb is first evacuated to 1 mm or less with a motor-driven oil pump protected by a small flask in the vacuum line which traps any oil spray. In order to obtain a steady pressure it is important that the system be free from leaks, as proved by evacuating the bulb and observing that the mercury levels in the manometer remain constant when the connection to the pump is closed. Leaks at the rubber connections may be stopped by substituting a new rubber tube which fits tightly over the glass tubes and by winding wire around the connection.

The stopcock is closed, and the bulb A is removed from the joint B, wiped with a clean, lintless damp (but not wet) cloth, and allowed to hang in the balance case for 5 or 10 min to come to constant weight. The second bulb, acting as a counterpoise, is wiped in the same way, and the weights necessary to bring the balance pointer to zero are recorded in each of the weighings. The joint B is of ground glass so smooth that stopcock grease (with its uncertain weights) can be eliminated.

The gas bulb is replaced in the apparatus and subjected to a second evacuation, after which it is weighed again. If the two weights do not check, the process is repeated until two successive weighings agree.

figure 1.

Manometer and vacuum system for gas-density measurements.

The stopcocks at A, C, and D are opened, and the system is evacuated. It is then filled to a pressure of about 1 atm with carbon dioxide from a tank through the double stopcock D and drying tube E. A diaphragm regulator connected to the tank is used to regulate the pressure. With the two-way stopcock closed the pressure is measured on the manometer.

If a closed-end manometer is used, the observed pressure in millimeters of mercury is corrected to 0° by use of the equation

$$P_0 = P - P\frac{\alpha t - \beta(t - t_s)}{1 + \alpha t} \tag{4}$$

where P_0, P = corrected and observed pressures

$\quad\quad t$ = temperature† of manometer

$\quad\quad t_s$ = temperature at which scale was calibrated, normally 20°C

$\quad\quad \alpha$ = mean cubical coefficient of expansion of mercury between 0 and 35°

$\quad\quad \beta$ = linear coefficient of expansion of scale material

The value of α is 181.8×10^{-6}, and the value of β is 18.4×10^{-6} for brass. If a wooden scale is used, taking the value of β equal to zero introduces a negligible error, since β for wood is about 5×10^{-6}.

If an open-end manometer is used, the pressure in the bulb is equal to the difference between the corrected barometer pressure (page 652) and the manometer pressure corrected by use of Eq. (4). The temperature of the air in the vicinity of the bulb is recorded. After wiping with a damp cloth and waiting for moisture equilibrium in the balance case, the bulb of gas is weighed again. After weighing, the bulb is evacuated and refilled to about the same pressure in order to obtain a check determination, the pressure and temperature are recorded, and the bulb is weighed.

The weight of carbon dioxide is determined in the same manner at about $\frac{3}{4}$ atm and at $\frac{1}{2}$ atm. It is essential in this work that there be no leaks.

The density of additional gases may be determined by introducing the gases through the drying tube E. Air, city gas, or an unknown gas or mixtures of gases may be used.

Since it is necessary to determine the volume of the bulb C, another weighing is made in which the bulb is filled to the stopcock with water at a known temperature. The density of water is given in the Appendix, page 653. The bulb is filled by evacuating it, closing the stopcock, immersing the end of the tube in a beaker of distilled water, and then opening the stopcock to allow the water to flow in. The bulb is removed and placed with the stopcock on top, and the remaining air space under the stopcock is filled with water using a large hypodermic syringe which passes through the bore of the stopcock. After weighing, the water is removed from the bulb with the help of a water aspirator, and the bulb is placed in a drying oven. It is

---

† Unless otherwise specified, all temperatures are on the Celsius (centigrade) scale.

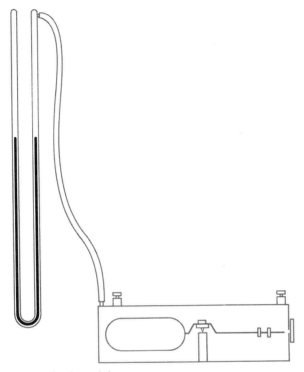

figure 2. Gas-density balance.

evacuated several times while hot to remove the last traces of moisture for further measurements.

In the second part of the experiment, the gas-density balance is used to determine the average molecular weight of dry air. A convenient gas-density balance developed by Edwards[1] is shown in Fig. 2. The outer case of the balance is connected to a closed-end manometer and to a vacuum pump or a tank of pure gas. The window at the end of the balance connected to the counterpoise is provided with a scale so that the position of the pointer may be determined.

Several readings are made on a gas of known molecular weight such as oxygen. The pressure necessary to bring the pointer to the zero point is determined.

The pressure required for a zero balance with air is determined next. The air is purified by passage through a tube of soda lime. When the gas-density balance is filled with a new gas for measurement, it is necessary to sweep out the last traces of the preceding gas by evacuation and filling with the new gas. The evacuation and refilling are repeated, and if the two readings do not agree, the sweeping process is continued until two successive readings agree closely.

CALCULATIONS. If the counterpoise bulb used in the first part of the experiment has practically the same volume as the gas-filled bulb, no correction for the buoyancy of the air is necessary. If a counterpoise bulb is not used, a correction for the buoyancy of the air is applied as described on page 494.

The weight g of the gas is obtained by subtracting the weight of the dry evacuated bulb from the weight obtained when the bulb is filled with gas.

The molecular weight of carbon dioxide is calculated at each pressure. With an imperfect gas, such as carbon dioxide, the apparent molecular weight increases as the pressure is increased and the molecules are crowded closer together, giving a greater intermolecular attraction. To determine the true molecular weight of a gas it is necessary to plot the molecular weight obtained at different pressures and extrapolate to zero pressure.

In the second part of the experiment the mean molecular weight of air is calculated from measurements with the gas-density balance by using Eq. (3) and compared with the average molecular weight obtained from the known composition of air.

Practical applications. The formula of a chemical compound may be calculated from the molecular weight, together with the atomic weights, and the percentage composition found by chemical analysis. In the most accurate work, globes of 8 to 20 liters have been used, and corrections were made for the loss of buoyancy of the globe when it contracted on evacuation.[2] The determination of the density of ammonia gas by Dietrichson and coworkers[3] illustrates the experimental techniques used in accurate work. Birge and Jenkins[4] have discussed the methods for extrapolating to the limiting gas density and the errors involved. The work of Cady and Rarick[5] indicates the high precision with which molecular weights may be determined with a gas-density balance.

The chemical equilibrium between different gases may often be calculated from the density of the equilibrium mixture of gases.

Suggestions for further work. The accurate determination of the molecular weight of hydrogen gas by this method constitutes a real test of a student's care and skill.

The molecular weights of other gases may be determined. Small tanks of methane, ethylene, nitrous oxide, Freon, and other gases may be used.

The percentage composition of a mixture of two gases such as oxygen and carbon dioxide may be determined from the density of the mixture.

More exact values of the molecular weights may be calculated with equations of state such as those of Berthelot, van der Waals, or Beattie and Bridgman. However, the use of these equations requires a knowledge of certain constants characteristic of the gas which will not be available for an unknown gas.

The gas-density balance is well adapted to measuring the density of a mixture of gases. For example, the carbon dioxide content of the exhaled breath may be determined by blowing the breath through a calcium chloride drying tube and a cotton packing into the balance.

In the analysis of a mixture of gases, the density in grams per liter is determined with the balance, and the density of each pure gas at the same pressure is known. A formula is derived that will give the percentage composition corresponding to the observed density at the observed pressure. It is assumed that any interaction among the different gases leading to density changes is negligible in the experiments described here.

It is easier to weigh accurately a small sealed glass bulb of a liquid than to weigh the material in a large glass vessel after it is vaporized by heating. Good measurements of molecular weights can be made by the Victor Meyer method, in which an organic liquid is sealed in a glass bulblet, placed in a gastight steam jacket, and broken. The vaporized material forces out an equal volume of air, which is measured at constant pressure in a gas burette. The calculations are the same as those in which the vapor is weighed in large bulbs. This Victor Meyer method may be used for determining the molecular weight of benzene or carbon tetrachloride vapor. It has been carried out in most laboratory courses for many decades, but the method is not used much now.

References

1. J. D. Edwards, *Ind. Eng. Chem.*, **9:** 790 (1917).
2. G. P. Baxter and H. W. Starkweather, *Proc. Natl. Acad. Sci. U.S.*, **12:** 699 (1926).
3. G. Dietrichson, L. J. Bircher, and J. J. O'Brien, *J. Am. Chem. Soc.*, **55:** 1–21 (1933).
4. R. T. Birge and F. A. Jenkins, *J. Chem. Phys.*, **2:** 167 (1934).
5. H. P. Cady and M. J. Rarick, *J. Am. Chem. Soc.*, **63:** 1357 (1941).

2   *VACUUM TECHNIQUE: DETERMINATION OF MOLECULAR WEIGHT BY GAS-DENSITY MEASUREMENTS*

This experiment illustrates the use of a glass vacuum system of the type commonly used in research for the preparation or purification of samples or for the transfer of a volatile sample from one vessel to another without exposure to air or moisture of the atmosphere.†

THEORY.   The vapor-phase molecular weight is an important property for characterizing a volatile substance. If the empirical formula is uncertain, the gas-phase molecular weight may be helpful, in conjunction with other information, in establishing it. If, as is more often the case, the empirical formula is known, the measured vapor-phase molecular weight can provide reliable information about the molecular state of the vapor.

The molecular weight is calculated from the equation

$$M = \frac{gRT}{PV}$$

† References on vacuum equipment and techniques are given in Experiment 58. A useful guide to simple chemical vacuum techniques is R. T. Sanderson, "Vacuum Manipulation of Volatile Compounds," John Wiley & Sons, Inc., New York, 1948.

At the moderate pressures used, the ideal-gas equation is ordinarily satisfactory to a degree of accuracy consistent with the other measurements made. If there is extensive association or dissociation in the sample vapor, the present method yields a number-average molecular weight for the given temperature and pressure:

$$\frac{gRT}{PV} = \frac{g}{n} = \frac{\sum n_i M_i}{\sum n_i}$$

where n is the total number of moles in the sample and g the total weight and the sum is taken over the various molecular species present.

It is desirable, when several parallel runs are made, to do them at different pressures, as a trend in the apparent molecular weight with pressure can provide an indication of association or dissociation if either is important in the gas phase studied. Certain systematic errors may also be revealed by disagreement among the results for different pressures. Deviations from ideal-gas behavior due to van der Waals interactions are of the order of a few percent or less under the conditions of these measurements for most substances and will usually not be detected with the apparatus described here.

Apparatus. Vacuum manifold and associated flasks and sample tubes (Fig. 3), two-stage mechanical pump, oil-diffusion pump, Pirani gauge and indicating circuit, mercury manometer or precision Bourdon gauge, gas torch, three Dewar flasks, liquid nitrogen, sample, e.g., $CFCl_3$.

PROCEDURE.   The vacuum manifold to be used is shown in Fig. 3. It is a standard type of system constructed of Pyrex glass, with stopcocks used as valves and standard-taper or O-ring connectors used as detachable joints. The use of ordinary stopcocks limits the system to use at pressures below atmospheric, as higher pressures inside could cause the stopcock barrels to pop out. The pumping system for exhaustion of air consists of a mechanical pump and an oil-diffusion pump (pages 403 to 406). The trap, at the left end of the diagram, is surrounded by a liquid-nitrogen bath in a Dewar flask. Its purpose is to collect condensable vapors when the vacuum system is being exhausted, since they would foul the oil in the pumps if permitted to pass through in quantity.

The manometer serves for measurement of pressures in the range 1 to 760 torrs. (1 torr = 1000 mtorr = 1 millimeter of mercury.) The Pirani gauge (page 413) is sensitive in the range 1 to 1000 mtorr. It is used here mostly as a sensitive leak tester, as a slow leak can be detected much more quickly with the Pirani gauge than with the mercury manometer. The closed-end tube attached at port A is used for introduction of the sample as a liquid, and the tube at B is used to hold the middle fraction when the sample is distilled in a purification step. A sample of the vapor is

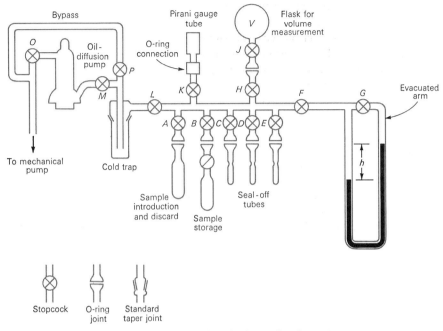

figure 3. Vacuum system for the determination of the molecular weights of gases.

admitted to the flask V, of known volume, at a measured pressure and temperature. This vapor is then transferred quantitatively by condensation into a tube at port C, D, or E. While the sample is kept frozen, the constricted neck is sealed off and the sample is removed and weighed.

As a preliminary step, the student should examine the vacuum system to be sure that he understands its plan and the functions of the components. The stopcocks should all be examined and rotated slightly to see that the lubricant is in good condition. Large stopcocks should be turned slowly, and it is a good idea to support the outer barrel with one hand. Also, if the grease is stiff, even slight warming by the hand may make a noticeable difference. If there are striations (thin channels) in the grease, or if the grease is too stiff, the stopcock should be cleaned and regreased. Standard-taper joints should also be examined for striations. The O-ring joints should be checked visually to see that they are properly seated and clamped.

Pumping Down the Vacuum System and Testing for Leaks. A Dewar flask with liquid nitrogen (initially about half full) is brought up around the cold trap, slowly enough to avoid boil-over of the nitrogen. Since there is some danger from implosion of the Dewar flask or the trap, it is imperative to wear safety glasses.

With the stopcocks open from the mechanical pump through the bypass to the manifold, the mechanical pump is turned on. The Pirani gauge is set on the least sensitive scale and turned on, and the stopcock between it and the manifold is opened. The sound of the mechanical pump changes as the pressure goes through the millimeter region, and the Pirani gauge should then begin to show an indication of decreasing pressure. The mechanical pump should bring the pressure below 100 mtorr within a few minutes, unless something is drastically wrong. With continued pumping, the mechanical pump may bring the system down to a few millitorr, but the rate becomes quite slow.

Next, stopcock F to the manometer can be opened and the left arm pumped out. When the pressure again drops to below 100 mtorr, the mercury level in the two arms should be equal. If the level in the right arm is lower, the stopcock G between the two arms should be opened long enough to pump out accumulated gas in the right arm and then closed again. The stopcock F should be closed. It is not desirable to pump on the manometer for long periods of time as mercury vapor distills slowly into the trap. (The vapor pressure of mercury at room temperature is about 2 mtorr.)

When the pressure has been reduced to about 100 mtorr or less, the diffusion pump is opened to the mechanical pump and to the manifold, the bypass is closed, and the blower and heater for the diffusion pump are turned on.

After the oil-diffusion pump has begun to function, the pressure should drop further and should reach a few mtorr in 10 min or so. If the oil pump is operating properly, there will be a steady ring of oil condensing at the throat and no frothing or bumping of oil. Frothing may occur if the oil is contaminated with air or other volatile substances, and if this occurs, it may be necessary to reduce the heating rate somewhat until the oil has been purged. Bumping may occur if the back pressure against which the pump works is too high. This could mean that the mechanical pump is not performing properly or that there is a bad leak in the system.

If the pressure cannot be brought below 10 or 20 mtorr, it is necessary to test for leaks, and at this stage this can be done most efficiently with the manifold left open to the pumps. Each of the stopcocks and ground joints may be examined and the former rotated back and forth a few times while the gauge is watched. If the leak is not found in this way, the O rings can be tested by applying a wad of cotton soaked with alcohol; this should produce a sudden effect on the reading of the Pirani gauge as the liquid alcohol temporarily blocks the leak. If a bad joint is found, it is taken apart, the O ring cleaned or replaced, and then reassembled. A Tesla coil may be used to search for pinholes in the glass.

When the worst leaks have been identified and corrected, it should be possible to get the pressure down below 10 mtorr. For leak testing at this level, it is best to close the stopcock M between the trap and the pumps and then watch for a rise in reading of the Pirani gauge. A leak rate corresponding to a pressure rise of a few millitorr per minute is tolerable in the present experiment, but if the rate is much greater than this, the stopcocks and O-ring joints should be checked further.

Even if the system is vacuum-tight, a slow rise in pressure will usually result from outgassing, i.e., the gradual evolution of air, water vapor, or other substances adsorbed on the inner walls of the glass or entrapped in the stopcock lubricant.

Outgassing of air should not be a serious problem in the pressure range considered here. Outgassing of water or other vapors can be recognized by noting the effect on the pressure, after a period of outgassing with stopcock L closed, of opening L while leaving M closed. Liquid nitrogen condenses most vapors to such a low vapor pressure (ordinarily far below 1 mtorr) that the trap alone acts very effectively as a pump for "condensables"; the only effect on accumulated air is a small reduction in pressure in proportion to the ratio of volumes with and without the trap included in the system. The removal of adsorbed substances may be accelerated by playing a Tesla coil over the walls of the system or by very prudent use of a hand torch. However, even a slight warming of the stopcocks may cause the grease to flow and cause leakage. Unnecessarily high outgassing rates may result if the flask or a sample tube was connected to the system without having been properly dried, or if the stopcock grease was contaminated.

Introduction of the Sample. After the system is found to be adequately vacuum-tight, stopcock A is closed and the sample tube below this is taken off. About 10 cm³ of the liquid sample to be used is placed in the tube, which is reattached to the vacuum manifold. A Dewar flask with liquid nitrogen is brought up around the sample, and after it is frozen, stopcock A is opened and the system pumped down again.

Since some air is usually entrapped with a frozen liquid sample, it is necessary to outgas the sample. This is accomplished by closing stopcock A, allowing the sample to warm up above the melting point, and then freezing it again after the evolution of bubbles has ceased. Then stopcock A is opened to the pump to exhaust air from the sample tube.

Finally, stopcock L should be closed and the Pirani gauge watched to check the tightness of the O-ring joint at A.

Bulb-to-Bulb Distillation of the Sample. The purpose of this step is to purify the sample and get rid of any remaining occluded air. With stopcock A closed, the sample is allowed to warm up to an intermediate temperature, where it has a vapor pressure of several torr. With stopcock L open, A is carefully opened, and about a quarter of the liquid is allowed to vaporize and be collected in the cold trap. Then L is closed, and a liquid-nitrogen bath is placed around the sample tube at B to condense sample there. The middle half of the sample is transferred in this way into the tube at B. Then stopcock A and the lower one at B are closed, the sample at B is allowed to warm up to room temperature, and the nitrogen bath is placed at A.

From this time on, the tube at A can serve as a container for discarded sample material. There should now be about 4 cm³ of sample in the tube at B.

If the experiment has to be interrupted, the sample can be removed from the system and stored until another laboratory period. To remove the sample, it is only necessary to close the stopcocks on both sides of the joint and then to open the joint by removing the clamp and carefully forcing the joint apart to admit air. When the joint is subsequently reattached, the air between the stopcocks should be pumped out and the sample frozen with liquid nitrogen. Before the stopcock to the sample tube is opened, the stopcock at L should be closed and the manifold tested for leaks. When the stopcock to the sample is opened, the Pirani gauge is watched for indications that air has leaked into the sample tube. If the gauge rises only a few mtorr or less, no further action is required to resume the experiment. If a considerable amount of air has leaked into the sample tube, it may be desirable to outgas the sample.

Preliminary Weighings and Determination of Volume. The sealoff tubes to be used should be cleaned as necessary and weighed. As the sample weights to be determined are rather small, the usual precautions for analytical weighings should be observed. The tubes should not be handled directly before weighing or rubbed with a dry cloth (because of electrostatic effects), and care should be taken to avoid error due to dust or lint on the surface of the glass. The sample tubes may then be attached to the vacuum manifold.

The volume of the flask, if not already known, may be found most accurately by weighing it empty and then filled with water. The stopcock is removed and the lubricant cleaned off with carbon tetrachloride. Then the water is removed, and the flask is rinsed with alcohol and dried in a warm oven or under a heat lamp. The stopcock is then regreased and the flask connected to the manifold.

P, V, T Measurements and Sealing Off of Samples. Final checks are now made for leaks in the system including the flask at J, the manometer, and the sealoff tubes. With the sample at B at room temperature, the stopcock below B is opened slightly and the pressure allowed to build up to the value desired for the first measurement, say about 100 torr, and then the stopcock is closed again. After a minute or so, the pressure and temperature are noted, and F and J are closed. Next, the vapor in the manifold is removed by condensation in the discard tube at A. Then liquid nitrogen is placed around the tube at C, and the sample in the flask V is now transferred completely to C.

Samples may be adsorbed in the stopcock grease and come out slowly, giving an apparent vapor pressure of 10 to 100 mtorr. Since one does not want to collect and weigh adsorbed material, sample collection at C should be stopped as soon as the pressure reaches this region.

With stopcock C closed and the sample kept frozen with liquid nitrogen, the flame of a hand torch is applied to the constricted neck of the tube. The flame should be moved around quickly to soften the glass on all sides. As the glass collapses, the tube is twisted by rolling it between the thumb and fingers and at the same time lowered gently in order to effect a leaktight seal. The trick is to heat the glass to a proper working temperature without having the atmosphere burst through the softened glass walls. A student who has never done this before should ask an instructor to demonstrate the technique.

After the glass has hardened, the tube is removed from the Dewar flask and set aside, along with the corresponding upper section, for weighing.

A similar procedure is used to collect the two other samples, with pressures ranging up to about 400 torrs.

Weighing the Samples. Each set of sample tubes is weighed, after they have come to room temperature, with care taken, as before, to avoid errors due to moisture or other surface effects. It is a good idea to reweigh the samples at a later time as a check against leaks.

When the weighings have been completed, the O-ring connections should be returned to the stockroom, as they can be reused.

Closing Down the Vacuum System. The system, including the manometer, is pumped down to remove any residue of sample vapor. Then, stopcocks F and M are closed, and the trap is allowed to warm up to room temperature. The stopcocks at A, B, H, and L are closed, and the two tubes are removed. Stopcocks M and O are then closed, and the manifold is left evacuated.

The diffusion-pump heater is turned off, and after it has cooled, the blower is turned off. Finally, P is opened, air is vented into the system above the mechanical pump, and the latter is turned off. (Failure to vent air into the system before turning off the pump can cause oil to back up into the system.) The trap can be removed for cleaning by admitting air through the bypass.

The trap can be washed with carbon tetrachloride, *thoroughly* dried with a stream of air, and then replaced on the vacuum line. (It must be supported until the pressure inside it has been lowered.) The air in the trap should be pumped out by the mechanical pump through the bypass. If the system is to be used again in the near future, it is a good idea to leave the mechanical pump on, open to the trap and manifold, in order to remove adsorbed sample material. Note that in these final operations, air is *not* admitted to the diffusion pump, the stopcocks on each side of it being left *closed*.

CALCULATIONS. The apparent molecular weight is calculated from the data and tabulated as a function of pressure.

If the vacuum system was not completely tight, an estimate should be made of the possible error from this source.

Practical applications. The techniques of manipulation of volatile samples in a vacuum manifold illustrated here are used in many other research applications, especially where exclusion of air or moisture is essential.

Suggestions for further work. With addition of provision for keeping a liquid sample at various known temperatures below room temperature, the system can be used for measurement of vapor pressure as a function of temperature. Trichlorofluoromethane or other Freons are suitable samples.

The second virial coefficient of a gas can be measured if the precision of measurement is improved. The sample should be immersed in a bath to hold the temperature constant to within $\pm 0.1°$. The pressure measurements can be improved as required by using wide-diameter glass tubing for the manometer (to reduce error from surface-tension effects) and by using a cathetometer for measurement of the mercury level. Manometers of sufficient accuracy for this work are available from commercial sources.

References (see page 420)

Chapter
2

Thermochemistry

Nearly all chemical reactions and physical transformations are accompanied by the evolution or absorption of heat, which can be measured quantitatively. The heats of reaction are important in thermodynamical calculations and in many practical applications. This chapter includes four experiments: Experiment 3, heat of combustion at constant volume, under a high pressure of oxygen to ensure rapid and complete reaction into simple, final products such as carbon dioxide and water; Experiment 4, heat of solution, showing the simplicity of measuring endothermic reactions and illustrating the mathematical treatment of data as differential and integral heats of solution; Experiment 5, heat of reaction at constant pressure with considerations of the influence of temperature on the heat of reaction; and Experiment 6, heat of formation of a metallic halide.

In Experiment 3 the reaction is carried out adiabatically, and the water jacket around the calorimeter is heated as necessary to keep pace with the rise in temperature of the combustion bomb. In this way there is no loss or gain of heat.

It is difficult to keep the two temperatures exactly equal, and for precision work an isothermal jacket is used and accurate cooling corrections are made. Cooling corrections are utilized in Experiment 5.

3 HEAT OF COMBUSTION: BOMB CALORIMETER

The enthalpy of combustion is determined in this experiment by means of a bomb calorimeter.

THEORY.[1-3a] The standard enthalpy of combustion for a substance is defined as the enthalpy change ΔH_T° which accompanies a process in which the given substance undergoes reaction with oxygen gas to form specified combustion products [such as $CO_2(g)$, $H_2O(l)$, $N_2(g)$, $SO_2(g)$], all reactants and products being in their respective standard states at the given temperature T. Thus the standard enthalpy of combustion of benzoic acid at $298.15°K$ is $\Delta H_{298.15^\circ}$ for the process

$$C_6H_5CO_2H(s) + \tfrac{15}{2}O_2(g) = 7CO_2(g) + 3H_2O(l) \tag{1}$$

with reactants and products in their standard states for this temperature.

As will be shown below, the enthalpy of combustion can be calculated from the temperature rise which results when the combustion reaction occurs under adiabatic conditions in a calorimeter. It is important that the reaction in the calorimeter take place rapidly and completely. To this end, the material is burned in a steel bomb with oxygen under a pressure of about 25 atm. A special acid-resistant alloy is used for the construction of the bomb because water and acids are produced in the reaction.

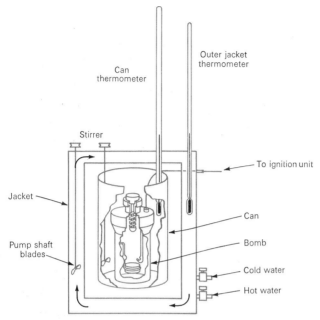

Outer jacket
thermometer

Can
thermometer

Stirrer

Jacket

Pump shaft
blades

To ignition unit

Can

Bomb

Cold water

Hot water

figure 4. Adiabatic combustion calorimeter.

In the adiabatic-jacket bomb calorimeter (Fig. 4) the bomb is immersed in a can of water fitted with a precise thermometer. This assembly is placed within an outer water-filled jacket. Both before and after the combustion occurs, the jacket temperature is maintained (by external means) at the same value as that of the water in the can. If the temperatures are matched with sufficient accuracy, the can and contents do not gain or lose energy by radiation or conduction and the process is therefore adiabatic.

This method affords convenience in work of moderate accuracy, but there is inevitably some error due to time lag in adjustment of the outer-jacket temperature. For the most exacting research measurements, an isothermal jacket is used, and accurate cooling corrections are made.[1-3a,4]

It should be recognized that the process which actually takes place in the calorimeter does not correspond exactly to one of the type of Eq. (1). In the actual calorimeter process, the final and initial temperatures are not equal, and the reactants and products are not in their standard states.

The relationship of the calorimeter process to the isothermal standard-state process may be clarified by reference to Fig. 5. The initial and final temperatures in the calorimeter process are T_1 and T_2, respectively. The various states of interest

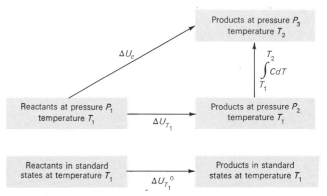

figure 5. Relationship between pertinent states of the calorimeter system (comprising the can and its entire contents). The energy change in going from one state to another is shown near the arrow connecting the states.

are shown, and on each arrow is written the energy change for the *can and contents* in going from one state to another in the direction indicated. Here ΔU_c is the energy change for the actual calorimeter process, while ΔU_{T_1} is the energy change for an imagined process in which the final state is at T_1 rather than T_2. The heat capacity C is that for the can and its contents under the conditions of the experiment. The work of expansion of water in the can is entirely negligible.

The first law of thermodynamics,

$$\Delta U = q - w \tag{2}$$

where $\Delta U =$ internal energy change for system

$q =$ energy transfer into system by heat flow

$w =$ work done by system

may be applied to the actual calorimeter process, which is assumed to be adiabatic ($q = 0$). In the present experiment, w, which consists mainly of the work of stirring, can be neglected,† and Eq. (2) then becomes

$$\Delta U_c = 0 \tag{3}$$

Since the energy change is independent of path, one has

$$\Delta U_c = \Delta U_{T_1} + \int_{T_1}^{T_2} C \, dT \tag{4}$$

---

† In the isothermal-jacket method, mentioned above, the stirring term is not neglected but rather is effectively eliminated along with the heat transfer by making a correction to the observed temperature rise.[1,3a,4]

Since the temperature change is small, it is usually valid to consider C to be constant, so that the integral becomes equal to $C(T_2 - T_1)$. One then obtains

$$\Delta U_{T_1} = -C(T_2 - T_1) \tag{5}$$

It may be observed that a temperature rise corresponds to a negative ΔU_{T_1}, that is, to a decrease in energy for the imagined isothermal process.

The next step is to calculate $\Delta U_{T_1}^{\circ}$ from ΔU_{T_1}. Although the energy is not sensitive to changes in pressure, the correction to standard states, called the *Washburn correction*, may amount to several tenths of 1 percent and is important in work of high accuracy.[2b,3b] The principal Washburn correction terms allow for the changes in U associated with (*a*) changes in pressure, (*b*) mixing of reactant gases and separating product gases, and (*c*) dissolving reactant gases in, and extracting product gases from, the water in the bomb.

The standard enthalpy change $\Delta H_{T_1}^{\circ}$ may then be calculated. The definition of H leads directly to

$$\Delta H_{T_1}^{\circ} = \Delta U_{T_1}^{\circ} + \Delta(PV) \tag{6}$$

Since the standard enthalpy and energy for a real gas are so defined as to be the same, respectively, as the enthalpy and energy of the gas in the zero-pressure limit, the ideal-gas equation may be used to evaluate the contribution of gases to $\Delta(PV)$ in Eq. (6). The result is

$$\Delta(PV) = (n_2 - n_1)RT \tag{7}$$

where n_2 = number of moles of gaseous products

n_1 = number of moles of gaseous reactants

The contribution to $\Delta(PV)$ from the net change in PV of solids and liquids in going from reactants to products is generally negligible.

Apparatus. Parr, Emerson, or other adiabatic calorimeter; pellet press; two thermometers graduated to 0.01°; fuse wire having a known heat of combustion per unit length; benzoic acid, naphthalene, or other samples.

PROCEDURE.   A Parr type calorimeter† is shown in Fig. 4 and a suitable ignition circuit in Fig. 6. The outer-jacket temperature is adjusted by adding hot or cold water.

The two thermometers read in the range 18 to 30°C or thereabouts and should be graduated to 0.01°. It is convenient to use a matched pair, although, as an alter-

---

† A calorimeter with semimicro bomb (22 ml) suitable for samples up to about 0.2 g is available from Parr Instrument Co., Moline, Ill., at a cost appreciably below that of the standard-size bomb calorimeter described above.

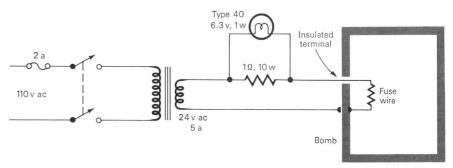

figure 6. Ignition circuit for bomb calorimeter. The fuse wire inside the bomb becomes hot and initiates combustion of the sample. The wire itself is oxidized, and the circuit thereby is opened. When this happens, the lamp ceases to glow. The switch is of the momentary-contact type. The circuit as shown is suitable for use with a 10-cm length of Parr 45C10 fuse wire.

native, one may be calibrated relative to the other; the correction may be assumed to be constant.

The following precautions must be followed if the danger of explosion is to be avoided.[4]

1. The amount of sample must not exceed 1 g.
2. The oxygen pressure must not exceed 30 atm.
3. The bomb must not be fired if gas bubbles are leaking from it when submerged in water.
4. The operator should stand back for at least 15 sec after igniting the sample and should keep clear of the top of the calorimeter. An explosion would be most likely to drive the top upward.
5. Much less than 1 g of sample should be used for testing materials of unknown combustion characteristics.
6. The use of high-voltage ignition systems is to be avoided. Arcing between electrodes may cause the electrode seals to fail and permit the escape of hot gases with explosive force.

A little less than 1 g of the sample is formed into a pellet by means of a pellet press; this is done to prevent scattering of material during the combustion, with consequent incompleteness of reaction. The pellet is weighed and placed in the sample pan. The fuse wire, of measured length about 10 cm and known heat of combustion per unit length, is attached to the two terminals and adjusted to give firm contact with the pellet. It is important to avoid getting kinks in the fuse wire since fusion may occur at such points before the portion of wire in contact with the pellet becomes hot enough to initiate combustion.

The surfaces at which closure of the bomb is to be effected must be kept scrupulously clean and every precaution taken to avoid marring them. The parts of the dismantled bomb should be placed on a clean, folded towel.

The cover is carefully assembled with the bomb and tightened. The bomb is then connected to the oxygen tank, and oxygen is admitted *slowly* until the pressure is 25 atm. The valves are then closed, the pressure in the line is relieved, and the bomb is removed.

About 2000 ml of water, the temperature of which has been adjusted so as to be at least several degrees below the upper limit of the thermometer range and preferably close to room temperature, is weighed in the calorimeter can; the latter is then placed within the adiabatic jacket. The ignition leads are connected, and the bomb is immersed in the water.

The water in the can must cover the bomb. If gas bubbles escape, the assembly ring may require tightening, or the gaskets may need to be replaced.

The cover of the adiabatic jacket is set in place and the thermometers lowered into position. The stirrer is started, and the jacket temperature is then adjusted to within $0.03°$ of that of the water in the can. The thermometer in the can is read for a few minutes to be sure that equilibrium has been attained. This temperature is recorded as the initial temperature T_1. The ignition switch is then closed until fusion of the wire is indicated by extinction of the lamp (Fig. 6). However, the switch should not be held closed for more than about 5 sec because damage to the ignition unit or undue heating by passage of current through the water may result.

If combustion has occurred, the temperature of the water in the can will be seen to rise within a few seconds. Otherwise the leads should be examined, the voltage output of the ignition circuit checked, or the bomb opened and examined for possible sources of trouble.

After a successful ignition, the temperature of the calorimeter rises quickly. The jacket temperature should be kept as close as possible to that of the can. After several minutes the rate of change of the temperature becomes small enough to permit reduction of the difference between the can and jacket temperatures to a few hundredths of a degree. The final steady temperature of the can is then recorded as T_2.

Next, the bomb is removed, the pressure relieved by opening the valve, and the cover removed. If the sample contained nitrogen, the acid residue in the bomb is washed quantitatively with water into a flask and titrated with $0.1 \ N$ NaOH. However, the nitric acid titration may be omitted in work of moderate accuracy if the only source of nitrogen is the air initially present in the bomb. In any case, the length of the residue of unoxidized fuse wire is measured. The bomb and calorimeter are carefully cleaned and dried after each experiment.

Two runs are made with benzoic acid for determination of the heat capacity of the calorimeter, and two with naphthalene (or other sample) for determination of the enthalpy of combusion.

CALCULATIONS.   The heat capacity C may be written as

$$C = mC_{\mathrm{H_2O}} + C_0 \tag{8}$$

where m = mass of water in can

$C_{\mathrm{H_2O}}$ = heat capacity of water per gram = 0.999 cal deg^{-1} g^{-1} at room temperature

Here C_0 represents the heat capacity of the calorimeter (bomb and contents, can, immersed portion of thermometer, etc.). The value of C_0 may be assumed to be the same for all four runs.

For the benzoic acid runs, ΔU_{T_1} is considered to be known, and C_0 may be calculated from the temperature rise. The value of ΔU_{T_1} is calculated by allowing[1] -6318 cal per gram of benzoic acid burned and the value specified by the manufacturer† for the wire burned. The contribution from nitric acid formed may be calculated when necessary as $-13,800$ cal per mole of HNO_3 produced.

The data for the naphthalene runs are used to calculate $\Delta U_{T_1}^{\circ}$ and the standard heat of combustion $\Delta H_{T_1}^{\circ}$; for student work, the difference between ΔU_{T_1} and $\Delta U_{T_1}^{\circ}$ may be considered negligible. The values of $\Delta U_{T_1}^{\circ}$ and $\Delta H_{T_1}^{\circ}$ should be reported for 1 mole of sample.

The standard enthalpy of formation of naphthalene is then calculated from the values -94.05 and -68.32 kcal mole^{-1} for the standard enthalpies of formation of $CO_2(g)$ and $H_2O(l)$, respectively, at 25°. If the experimental values of T_1 are within a few degrees of 25°, the correction of the enthalpy of combustion from T_1 to 25° may be omitted.

Practical applications. For many modern technological developments, such as rocket-propulsion systems, it is obviously necessary to have good thermochemical data.

Suggestions for further work. The enthalpy of combustion of a liquid sample may be determined. The sample is sealed off in a thin-walled glass bulb.[3a] A weighed quantity of naphthalene or paraffin oil is placed around the bulb. When ignited, this will break the glass and start combustion of the liquid.

Solid samples which cannot be formed easily into pellets are placed in the sample pan around the fuse wire. The oxygen must be admitted slowly to prevent scattering of the powder.

References

1. R. S. Jessup, Precise Measurement of Heat of Combustion with a Bomb Calorimeter, *Natl. Bur. Std. U.S. Monograph* 7, 1960.
2. F. D. Rossini (ed.), "Experimental Thermochemistry," Interscience Publishers, Inc., New York, 1956: (*a*) chap. 3, by J. Coops, R. S. Jessup, and K. van Nes; (*b*) chap. 5, by W. N. Hubbard, D. W. Scott, and G. Waddington; (*c*) chap. 6, by E. J. Prosen.

† For example, -2.3 cal cm^{-1} for Parr 45C10 (No. 34 B & S gauge Chromel C).

3. J. M. Sturtevant in A. Weissberger (ed.), "Technique of Organic Chemistry," vol. 1, pt. 1, chap. 10; Interscience Publishers, Inc., New York, 1959: (*a*) "Physical Methods of Organic Chemistry," 3d ed.; (*b*) pp. 597–598.
4. Oxygen Bomb Calorimetry and Combustion Methods, Parr Instrument Co. (Moline, Ill.) *Tech. Manual* 130, 1960.
5. International Critical Tables, vol. V, p. 162, McGraw-Hill Book Company, New York, 1929.
6. Selected Values of Chemical Thermodynamic Properties, *Natl. Bur. Std. U.S. Circ.* 500, 1952.
7. Selected Values of Physical and Thermodynamic Properties of Hydrocarbons and Related Compounds, *Am. Inst. Res. Proj.* 44 Rept., 1955.

4   HEAT OF SOLUTION

The integral heat of solution of potassium nitrate in water is determined as a function of concentration. From these data differential heats of solution and integral heats of dilution are calculated.

THEORY.[1]   The quantitative study of the thermal effects which accompany the solution of a solute in a pure solvent or a solution has been systematized through the introduction of the concepts of the integral and differential heats of solution.

The *integral heat of solution* ΔH_{IS} at a particular concentration is the heat of reaction† at a specified temperature and pressure when *one mole of solute* is dissolved in enough pure solvent to produce a solution of the given concentration.

Thus ΔH_{IS} for KNO_3 in water equals the enthalpy change for the process§

$$KNO_3(s) + nH_2O(l) = [KNO_3, nH_2O]$$

For example, if 1 mole of solute is dissolved in 500 g of water at constant T and P, the heat of reaction gives the value of the integral heat of solution at the concentration 2 m, corresponding to $n = 27.75$.

For a solution of given concentration, the *differential heat of solution* of the solute, $\Delta H_{DS}(T,P,n_2/n_1)$, is the heat of solution per mole of solute added under conditions in which the concentration of the solute is changed only differentially. Thus, if δq represents the enthalpy change when dn_2 moles of solute is added (at constant T, P) to a solution phase already containing n_2 moles of solute and n_1 moles of solvent,

$$\Delta H_{DS}(T,P,n_2/n_1) = \frac{\delta q}{dn_2}$$

† Because the processes considered here take place at constant pressure and only pressure-volume work is involved, the heats of reaction are given by the corresponding enthalpy changes.

§ The symbol on the right-hand side of this equation refers to a solution consisting of 1 mole of KNO_3 in n moles of water.

Alternatively, the differential heat of solution may be identified with the value of Q_p for the addition of 1 mole of solute to an infinite quantity of solution of the concentration, i.e., mole ratio n_2/n_1, concerned; in this case also the concentration of the solution would be changed only differentially.

It will be obvious that a direct measurement of the differential heat of solution is not possible; its value must be calculated from results of integral-heat-of-solution measurements. Let $\Delta H_S(n_2)$ represent the enthalpy change for the dissolving, at constant T, P, of n_2 moles of solute in a fixed number of moles of solvent. The differential $d\Delta H_S = \Delta H_S(n_2 + dn_2) - \Delta H_S(n_2)$ then can be identified with the enthalpy change for the addition of dn_2 moles of solute to a solution already containing n_2 moles of solute and the fixed number of moles n_1 of solvent. From the definition of the differential heat of solution,

$$\Delta H_{DS}(T,P,n_2/n_1) = \left(\frac{\partial \Delta H_S}{\partial n_2} \right)_{n_1,T,P} = \left[\frac{\partial (n_2 \, \Delta H_{IS})}{\partial n_2} \right]_{n_1,T,P} \tag{1}$$

since the definition of the integral heat of solution requires $\Delta H_S = n_2 \, \Delta H_{IS}(T,P,n_2/n_1)$.

If the constant number of moles of solvent n_1 is specified as $1000/M_1$ ($M_1 =$ gram formula weight of solvent), then n_2 can be replaced by the molality m of the solute, and

$$\Delta H_{DS}(T,P,m) = \left[\frac{\partial (m \, \Delta H_{IS}(T,P,m))}{\partial m} \right]_{T,P} \tag{2a}$$

$$= \Delta H_{IS}(T,P,m) + m \left[\frac{\partial \Delta H_{IS}(T,P,m)}{\partial m} \right]_{T,P} \tag{2b}$$

Equations (2a) and (2b) are exactly equivalent mathematically, but the practical problem involved in the accurate determination of the slope of a curve will often permit better calculations to be made by use of Eq. (2b) instead of (2a).

The magnitudes of these heats of solution depend specifically on the solute and solvent involved. The value of the heat of solution at high dilutions is determined by the properties of the pure solute and by the interactions of the solute with the solvent. As the concentration of the solution increases, the corresponding changes in the differential and integral heats of solution reflect the changing solute-solvent and solute-solute interaction effects.

For the interpretation of heat-of-solution data for some systems, it is instructive to compare the results with the behavior predicted for ideal solutions. An ideal solution may be defined as one which obeys Raoult's law over the entire range of concentrations being considered. It can be shown[1] that the mixing, at constant T and P, of pure liquid solvent with such a solution produces no change in enthalpy. From this it follows that for the case of a *liquid* solute, dissolving to form an ideal solution, $\Delta H_{IS} = 0$. For the case of a *solid* solute which dissolves to form an ideal solution, ΔH_{IS} equals the molar heat of fusion to give the (supercooled) liquid at the temperature of the solution. Such behavior is approximated in some actual cases, which involve nonelectrolyte solutes and nonpolar solvents, e.g., naphthalene in benzene. For electrolyte solutes the actual behavior is very different from the ideal case, because of marked solute-solvent and solute-solute interactions.

The *integral heat of dilution* $\Delta H_{D,m_1 \to m_2}$ between two molalities m_1 and m_2 is defined as the heat effect, at constant temperature and pressure, accompanying the addition of enough solvent to a quantity of solution of molality m_1 containing *one mole of solute* to reduce the molality to the lower value m_2. The process to which the integral heat of solution at molality m_2 refers is equivalent to the initial formation of the more concentrated solution of molality m_1 followed by its dilution to the lower molality m_2; the integral heat of dilution is thus equal to the difference of the integral heats of solution at the two concentrations involved:

$$\Delta H_{D,m_1 \to m_2} = \Delta H_{IS}(m_2) - \Delta H_{IS}(m_1) \tag{3}$$

Apparatus. Heat-of-solution calorimeter; sensitive mercury thermometer (18 to 31° in 0.01° divisions); potentiometer and potential divider; thermistor thermometer with Wheatstone bridge and lamp and scale galvanometer; calibrated ammeter with 1-amp scale; 6-volt dc power supply; stopwatch or electric timer; six weighing bottles or 10-ml Erlenmeyer flasks with stoppers; potassium nitrate; switch and wire.

PROCEDURE.[2,3]   This experiment illustrates the special advantage that endothermic reactions offer for calorimetric measurements. When the reaction absorbs heat, the cooling effect may be balanced with electrical heating to prevent any change of temperature. It thus becomes unnecessary to know the heat capacity of the calorimeter or of the solution being studied. No cooling correction is necessary, and the method is simpler than the ordinary adiabatic method.

The essential features of a suitable calorimeter for work of moderate precision are shown in Fig. 7. The vacuum bottle minimizes heat exchange between the solution and the surroundings. A mechanical stirrer is used to provide the efficient and uniform stirring essential to the rapid solution of the solute. The rate of stirring, however, should be kept as low as efficiency permits to minimize the energy introduced by stirring. The stirrer shaft should be of a poor heat conductor, and proper bearings must be provided to eliminate (as far as possible) heat generation by friction. (The shaft bearings should be located above the calorimeter proper.) A belt-and-pulley drive is used to keep the heat transfer by conduction and radiation from the motor at a minimum.

The calorimeter heating element should have a low temperature coefficient of resistance and a low time lag in transferring heat to the solution and must be elec-

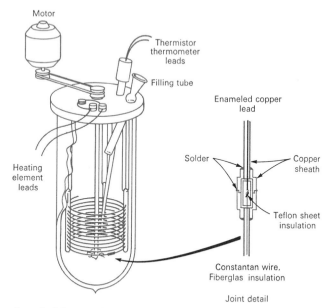

Motor
Thermistor thermometer leads
Filling tube
Enameled copper lead
Solder
Copper sheath
Heating element leads
Teflon sheet insulation
Constantan wire, Fiberglas insulation
Joint detail

figure 7. Calorimeter for measuring heats of solution.

trically insulated from the solution. These requirements can conveniently be met by use of copper-sheathed Fiberglas-insulated constantan wire.† The copper sheath, soldered to the resistance wire at one end, serves as one electric lead. At the other end, the sheath is cut back to expose a short section of the wire, to which an enameled copper lead wire is then soldered. A section of copper sheath from which the resistance wire has been removed is slipped over the lead wire, and a watertight joint§ is made between the two sections, as shown, for example, in Fig. 7. The heater-coil assembly should remain immersed for some distance past the joint to ensure its temperature equilibration with the solution. A heater resistance of about 7 ohms, as given by about 5 ft of 24-gauge constantan wire, is recommended because the current obtained by use of a 6-volt storage battery or regulated dc supply¶ can be determined accurately with a 1-amp ammeter.

† This wire assembly is made by the Precision Tube Company, Inc., North Wales, Pa., under the name Precision Coaxitube.

§ The brass fittings can be prepared at any machine shop. Insulation is readily accomplished with heat-shrinkable Teflon tubing, available from electronic component suppliers.

¶ Solid-state power supplies, such as those available from the Harrison Division of the Hewlett-Packard Co., Palo Alto, Calif., have been found very convenient.

As temperature indicator, a sensitive mercury thermometer may be used, but a thermistor (page 609) is recommended because of its rapid and sensitive response to temperature changes. A conventional student dc Wheatstone bridge with lamp and scale galvanometer readily makes available a thermometric sensitivity of the order of a millidegree.

A schematic diagram of the electrical circuits involved is given in Fig. 8. The timer is actuated by a ganged switch so that the time that current flows through the heating coil will be measured. Temperature indication is achieved by means of a thermistor immersed in the solution. The resistance R_t of the thermistor varies rapidly with temperature (about 80 ohms deg^{-1} near room temperature with R_t about 2000 ohms).

The resistance of the heating coil R_h must be measured with the current flowing since its resistance increases with increasing temperature. This is accomplished by measuring the potential drop across R_h by means of a potentiometer while a known current is passing. The value of R_h is then calculated by using Ohm's law. Since the potential drop across R_h is about 7 volts, it is necessary to use a potential divider if the ordinary type of laboratory potentiometer is to be used. The potential-

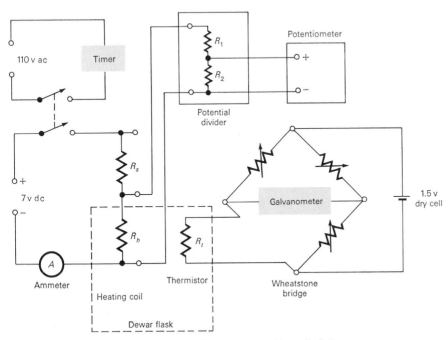

figure 8. Schematic diagram of electric circuits for measurement of heat of solution.

divider ratio $(R_1 + R_2)/R_2$ is chosen to be $10:1$. Since the current may change somewhat during the balancing process, the current should be noted at a time when the potentiometer is in balance. For the calculation of R_h, it is to be noted that the current indicated by the ammeter includes a current through the potential divider as well as the current through the heating coil.

The calibration of the ammeter is checked by measuring the potential drop across a standard resistance R_s in series with the heater.

Six samples of finely pulverized potassium nitrate, three of about 8 g and one each of 3, 4, and 5 g, are transferred to numbered weighing bottles, which are then stoppered and weighed. About 200 ml of distilled water, which has been carefully adjusted to room temperature, is weighed into the vacuum bottle, and the calorimeter assembled. The thermistor thermometer is connected to the Wheatstone bridge, and the stirrer is started. The time that the stirrer is on must be recorded. The thermometer resistance is checked by means of the bridge; if the water temperature was properly adjusted, the resistance will soon show only a very slow decrease due to the temperature rise caused by stirring.

When this condition has been reached, the position of the galvanometer light balance is noted as a reference point for later use, and the first determination is started by adding the 3-g sample of potassium nitrate through the sample tube. The emptied weighing bottle is set aside to be reweighed later. The heating-circuit switch is then closed. The heating current is recorded, and any salt adhering to the surface of the filling tube is pushed down with a blunt glass rod or a camel's-hair brush.

The galvanometer deflection is checked at frequent intervals; *the bridge dial settings are not changed.* When the unbalance has been reduced far enough for the galvanometer light spot to remain on the scale, the galvanometer circuit switch is closed and the heating current and timer are turned off. The number n of scale divisions traversed thereafter by the spot due to the lag in the heater and thermometer is noted. The heating current and timer are switched on again and turned off when the light spot has reached the point n scale divisions short of the initial balance reference point. The spot will then come to rest very close to the latter, and in this way the final temperature of the solution is matched to the initial temperature. The total heating time is recorded.

The thermometer bridge balance is checked, and the second determination is made as above with the addition of the 4-g sample. Since heat exchange with the surroundings is influenced by the magnitude of the temperature differential between the calorimeter and the room, the solute may profitably be added gradually during the heating period rather than all at once. The remaining samples are used in turn to extend the concentration range studied to near 2 m. The empty weighing bottles are then reweighed.

To determine the work of stirring per second, the solution is stirred for a time of the order of 15 min and the resulting change in the thermistor resistance ΔR_t is noted. Next, a measured current is passed through the heating coil for about 1 min, and the change ΔR_t again noted. From these data, the work of stirring per second can be calculated if it is assumed that the resistance of the thermistor R_t varies linearly with temperature. This assumption is entirely satisfactory over the very small temperature intervals involved.

It is wise to make one or two test experiments to learn the proper technique.

CALCULATIONS.   For this experiment the first law of thermodynamics leads to

$$\Delta H = q - w_s \tag{4}$$

The work of stirring w_s is negative because work is done *on* the system. The heat q absorbed by the solution is equal to the electrical energy dissipated in the resistor of resistance R due to passage of current I for time t. Hence

$$\Delta H = I^2 Rt - w_s \tag{5}$$

If I is expressed in amperes, R in ohms, and t in seconds, the energy dissipated is given in joules; to convert to calories the energy in joules is divided by 4.184.

For each solution the total number of moles of solute present and the corresponding total enthalpy change ΔH are calculated. The molality of the solution and the integral heat of solution at that concentration are then calculated. The integral heat of solution ΔH_{IS} is plotted against the molality, and the differential heat of solution is evaluated at 0.5 and 1.5 m by use of Eq. (2). This method is employed, rather than direct use of Eq. (1), to minimize the uncertainty in the calculated values due to the difficulty of determining accurately the slope of a curve.

The integral heats of solution at 0.5, 1, and 1.5 m are obtained by interpolation, and the integral heats of dilution from 1.5 to 1 m and 1 to 0.5 m are evaluated. The various experimental results are compared with accepted values.[4] For comparison with literature data it should be remembered that the enthalpy of formation of a solute is defined as the enthalpy change for formation of the solution from the solvent and the elements of the solute, each in its standard state.

Practical applications. Integral-heat-of-solution data are often required in energy-balance calculations for chemical processes for engineering purposes. They are also used in the indirect evaluation of standard heats of formation of compounds for which heats of reaction in solution must be utilized. Measurements of the integral heat of solution may be used for the calculation of integral heats of dilution when no direct determinations of the latter are available. The differential heat of solution at saturation determines the temperature coefficient of solubility of the solute[5] (see Experiment 21).

Suggestions for further work. The method here described is suitable for measurements on most endothermic reactions. The apparatus may be used for exothermic reactions by determining the temperature rise due to the reaction, then cooling the system to the initial temperature and reheating it through the identical temperature range by means of the electrical heating coil.

The heat of solution of urea may be determined as typical of a nonelectrolyte. The individual samples used should be larger (about 15 g) because of the smaller heat of solution. The heats of solution of urea, phenol, and the compound $(NH_2)_2CO \cdot 2C_6H_5OH$ are measured separately, and the heat of formation of the compound calculated.[6] The compound is prepared by fusing 9.40 g of phenol with 3 g of urea in a test tube immersed in boiling water.

The heat of hydration of calcium chloride may be determined indirectly from measurements of the endothermic heat of solution of $CaCl_2 \cdot 6H_2O$ and the exothermic heat of solution of $CaCl_2$. A test of the equipment and operating technique may be made by measuring the relatively small heat of solution of sodium chloride. Comparison data for concentrations up to 1.3 m have been given by Benson and Benson.[7]

The heat of mixing of organic liquids may be determined with the same apparatus.[8] The results for chloroform-acetone and carbon tetrachloride–acetone may be discussed on the basis of hydrogen bonding.

References

1. G. N. Lewis and M. Randall, "Thermodynamics," 2d ed., rev. by K. Pitzer and L. Brewer, McGraw-Hill Book Company, New York, 1961.
2. J. M. Sturtevant in A. Weissberger (ed.), "Technique of Organic Chemistry," vol. 1, "Physical Methods of Organic Chemistry," 3d ed., pt. 1, chap. 10, Interscience Publishers, Inc., New York, 1959.
3. H. A. Skinner, J. M. Sturtevant, and S. Sunner, in H. A. Skinner (ed.), "Experimental Thermochemistry," vol. II, chap. 9, Interscience Publishers, Inc., New York, 1962.
4. Selected Values of Chemical Thermodynamic Properties, *Natl. Bur. Std. U.S. Circ.* 500, 1952.
5. A. T. Williamson, *Trans. Faraday Soc.*, **40:** 421, (1944).
6. A. N. Campbell and A. J. R. Campbell, *J. Am. Chem. Soc.*, **62:** 291, (1940).
7. G. C. Benson and G. W. Benson, *Rev. Sci. Instr.*, **26:** 477 (1955).
8. B. Zaslow, *J. Chem. Educ.*, **37:** 578 (1960).
9. M. L. McGlashan in H. A. Skinner (ed.), "Experimental Thermochemistry," vol. II, chap. 15, Interscience Publishers, Inc., New York, 1962.

5   HEAT OF REACTION IN SOLUTION: CONSTANT-PRESSURE CALORIMETER

The enthalpy change for a reaction in solution at constant pressure is found from the measured temperature change which occurs when the reaction takes place in a thermally insulated vessel.

THEORY.[1]   The present experiment is concerned with the determination of the enthalpy change for a chemical reaction in solution. As an example, the following reaction may be considered

$$PbI_2(s) = Pb^{++}(aq) + 2I^-(aq) \tag{1}$$

where the symbol $Pb^{++}(aq)$ means 1 mole of Pb^{++} in an infinitely dilute solution in water solvent. Thus, for the usual choice of standard states for solution components, the enthalpy change for reaction (1) equals $\Delta H°$, the standard state enthalpy change for the given temperature.

As the measurement cannot actually be performed at infinite dilution, the procedure is to measure ΔH for the process

$$2[KI,50H_2O] + [Pb(NO_3)_2,400H_2O] = PbI_2(s) + 2[KNO_3,250H_2O] \tag{2}$$

The notation $[KI,50H_2O]$ represents a solution containing 1 mole of KI dissolved in 50 moles of H_2O. Then ΔH for reaction (2) can be combined with heat-of-dilution data to obtain ΔH for reaction (1). The relationship of (2) to (1) may be seen clearly from the following diagram:

$$PbI_2(s) + 2K^+(aq) + 2NO_3^-(aq) \xrightarrow{\Delta H_1} Pb^{++}(aq) + 2I^-(aq) + 2K^+(aq) + 2NO_3^-(aq)$$

$$\downarrow \Delta H_2 \qquad\qquad\qquad\qquad\qquad\qquad \uparrow \Delta H_4$$

$$PbI_2(s) + 2[KNO_3,250H_2O] \xrightarrow{\Delta H_3} 2[KI,50H_2O] + [Pb(NO_3)_2,400H_2O]$$

Thus,

$$\Delta H_1 = \Delta H_2 + \Delta H_3 + \Delta H_4$$

Here ΔH_3 is the negative of the measured enthalpy change for the process in Eq. (2), and ΔH_2 and ΔH_4 can be obtained from heat-of-dilution measurements (Experiment 4) or from tabulated heat-of-formation data for the solutes at various concentrations. ΔH_1 is, of course, equal to ΔH for the process in Eq. (1), since at infinite dilution the solutes are independent and the $K^+(aq)$ and $NO_3^-(aq)$ terms cancel.

The principles involved in the measurement of ΔH for a process at finite concentrations of the type

$$\text{Reactants } (T) \longrightarrow \text{ products } (T) \tag{3}$$

may now be considered. It is to be noted that in Eq. (3), reactants and products are specified to be at the same temperature; the enthalpy change for a reaction process of this type will be designated ΔH_T.

The method of measurement involves a simple application of two important properties of the enthalpy function:

1. The energy transfer to the system by heat flow q is given by

$$q = \Delta H \tag{4}$$

for any process carried out in such a manner that the conditions $P = $ constant and $w = \int P\,dV$ hold.

2. The enthalpy is a function of state; hence ΔH is independent of path, while q in general is not.

The present method is particularly well suited to the determination of ΔH_T for reactions involving solutions. The reaction is carried out under a constant (atmospheric) pressure in a vacuum bottle, which is an excellent thermal insulator. Therefore the actual calorimeter process is adiabatic ($q = 0$) rather than isothermal ($T =$ constant). The relationship between the calorimeter process and an isothermal process of the type of Eq. (2) is shown in Fig. 9. Clearly,

$$\Delta H_c = \Delta H_{T_1} + \int_{T_1}^{T_2} C_P \, dT \tag{5}$$

where $\Delta H_c =$ enthalpy change for calorimeter process
$\quad\quad T_1 =$ initial temperature
$\quad\quad T_2 =$ final temperature
$\quad\quad C_P =$ heat capacity, at constant pressure, of contents and inner wall of vacuum flask, after reaction has occurred

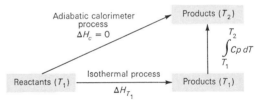

figure 9. Relationship between the actual adiabatic calorimeter process and the imagined isothermal process for which ΔH_T is to be determined. Near each arrow is written the enthalpy change in going from one state to another in the direction indicated.

The temperature interval is generally small enough for C_P to be considered constant, and the integral in Eq. (5) then becomes $C_P(T_2 - T_1)$. Furthermore, one has from Eq. (4), applied to the calorimeter process,

$$\Delta H_c = q = 0 \tag{6}$$

and combining this with Eq. (5) yields

$$\Delta H_{T_1} = -C_P(T_2 - T_1) \tag{7}$$

It may be noted that if the temperature *rises* in the adiabatic calorimeter process, then ΔH_{T_1} is *negative*; in this case, heat would be evolved, according to Eq. (4), were the same reaction to be carried out isothermally at temperature T_1.

In order that Eq. (7) may be used to calculate ΔH_{T_1} from the measured temperature change, it is necessary to determine C_P for the product system. This is best done by an electrical method. Shortly after the reaction has occurred, a current is passed through the coil of wire immersed in the solution and the resulting temperature rise $T_4 - T_3$ is measured. The enthalpy change for the system in this process is $\Delta H = C_P(T_4 - T_3)$. Since electrical work $I^2 R_h t$ is done on the system, Eq. (4) does not apply and must be replaced by

$$q - w_{el} = \Delta H \tag{8}$$

with $q = 0$ and $w_{el} = -I^2 R_h t$. Here R_h stands for the resistance of the heater coil and t for the time of current flow. Hence, C_P may be calculated from

$$C_P(T_4 - T_3) = \Delta H = I^2 R_h t \tag{9}$$

For the calculation of ΔH for the dilution steps, enthalpy-of-formation tables[2] may be used. The true enthalpy of formation $\Delta H_{f,T}$ for a *solution* would be ΔH for the formation of the specified solution from the *elements of all components* in their standard states at the given temperature. However, when one component is designated as solvent, it is conventional to define an enthalpy of formation for a *solute* as ΔH for the formation of the specified solution from the pure *solvent* and *elements of the solute* in their standard states at the given temperature. An enthalpy of formation for a solute in the latter sense will be denoted by $\Delta H_{f',T}$. Figure 10 may serve to clarify the distinction between $\Delta H_{f,T}$ and $\Delta H_{f',T}$.

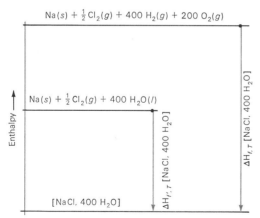

figure 10. The distinction between $\Delta H_{f,T}$ and $\Delta H_{f',T}$ for the solution [NaCl, 400H_2O].

From the definition it is easily shown that ΔH for a dilution process such as

$$[\text{NaCl},300\text{H}_2\text{O}] + 100\text{H}_2\text{O} \longrightarrow [\text{NaCl},400\text{H}_2\text{O}] \tag{10}$$

is found by subtracting the heats of formation $\Delta H_{f,T}$ for the solute at the two concentrations. If ΔH_T for a reaction is to be calculated from solute enthalpies of formation by the equation

$$\Delta H_T = \Sigma\ \Delta H_{f,T}(\text{products}) - \Sigma\ \Delta H_{f,T}(\text{reactants}) \tag{11}$$

an explicit term involving $\Delta H_{f,T}(H_2O)$ must be included if the number of moles of solvent, here considered to be water, changes in the reaction; e.g., for the reaction

$$[\text{HCl},100H_2O] + [\text{NaOH},100H_2O] = [\text{NaCl},201H_2O] \tag{12}$$

ΔH_T is given by

$$\Delta H_T = \Delta H_{f,T}[\text{NaCl},201H_2O] + \Delta H_{f,T}(H_2O) \\ - \Delta H_{f,T}[\text{NaOH},100H_2O] - \Delta H_{f,T}[\text{HCl},100H_2O] \tag{13}$$

It may be desired to calculate $\Delta H_{T_0}^{\circ}$ for a temperature T_0 different from that at which the measurement was made. This calculation is most efficiently accomplished through introduction of partial molar quantities (page 95). For the kth component of a solution, the partial molar enthalpy $\overline{H}_k$ is defined by

$$\overline{H}_k = \left(\frac{\partial H}{\partial n_k}\right)_{T,P,n_l,\,\ldots} \tag{14}$$

where H = enthalpy of solution

$\quad\quad n_k$ = number of moles of component k

$\quad n_l,\ldots$ = numbers of moles of all other components

For a reaction,

$$\Delta H_T = \sum_j \nu_j \overline{H}_j - \sum_i \nu_i \overline{H}_i \tag{15}$$

where ν_i is the coefficient of the ith reactant, and ν_j that of the jth product, in the chemical equation for the reaction as written; Σ_j and Σ_i denote summation over products and reactants, respectively. The validity of Eq. (15) may be verified by considering the occurrence of an infinitesimal amount of reaction by removal of differential quantities of reactants and introduction of differential quantities of products in amounts consistent with the stoichiometry of the reaction.

To differentiate Eq. (15) with respect to T, we use

$$\frac{\partial \overline{H}_k}{\partial T} = \frac{\partial}{\partial T}\left(\frac{\partial H}{\partial n_k}\right) = \frac{\partial}{\partial n_k}\left(\frac{\partial H}{\partial T}\right) = \frac{\partial}{\partial n_k}\ C_P \equiv \overline{C}_{P_k} \tag{16}$$

and obtain

$$\frac{\partial \Delta H_T}{\partial T} = \sum_j \nu_j \overline{C}_{P_j} - \sum_i \nu_i \overline{C}_{P_i} \tag{17}$$

Integration of Eq. (17) from T_1 to T_0 yields

$$\Delta H_{T_0} - \Delta H_{T_1} = \int_{T_1}^{T_0} \left(\sum_j \nu_j \overline{C}_{P_j} - \sum_i \nu_i \overline{C}_{P_i} \right) dT \tag{18}$$

If the temperature interval $T_0 - T_1$ is only a few degrees, so that the partial-molar-heat capacities may be considered to be constant, and if specification is made to the standard-state case, Eq. (18) becomes

$$\Delta H_{T_0}^\circ = \Delta H_{T_1}^\circ + \left(\sum_j \nu_j \overline{C}_{P_j}^\circ - \sum_i \nu_i \overline{C}_{P_i}^\circ \right) (T_0 - T_1) \tag{19}$$

Values of $\overline{C}_P^\circ$ for a number of solutes in water are available from tables[2,3] or can be calculated from specific-heat data for solutions given as a function of concentration.

Clearly, the solvent terms in Eq. (19) will cancel unless the number of moles of solvent changes in the reaction. If the standard state for the solvent is taken to be the pure solvent at temperature T, as is generally the case, then $\overline{C}_P^\circ$ for the solvent is just its molar heat capacity. For pure water, this is

$$\overline{C}_P^\circ = (0.9989)(18.016) = 17.996 \text{ cal deg}^{-1} \text{ mole}^{-1} \text{ at } 25^\circ\text{C}$$

····▸ ***Apparatus.*** Calorimeter (Fig. 11) consisting of 500-ml vacuum bottle, sensitive (0.01°) thermometer, stirrer, special inner flask, and 15-watt heater coil; source of water-saturated, pressurized air; 6 or 12 volts dc supply for heater; heater; heater timer; double-pole double-throw switch controlling both heater and heater timer; additional timer or clock with sweep second hand; calibrated ammeter; potentiometer; precision potential divider; reactant solutions, 500 ml of $[\text{Pb(NO}_3)_2, 400\text{H}_2\text{O}]$, 150 ml of $[\text{KI}, 50\text{H}_2\text{O}]$.

PROCEDURE.   The calorimeter is shown in Fig. 11. A vacuum bottle provides a simple yet effective means of keeping the rate of heat exchange with the environment to a very low value. The cover is fitted with a thermometer† sensitive to 0.01° and a polystyrene stirrer. The inner Pyrex vessel serves to hold one reactant separate from the other while permitting the two solutions to come to the same temperature before the reaction takes place. At the proper time, air pressure is used to force the reactant out of the inner flask. Water-saturated air should be used to minimize the cooling effect of evaporation. If the reactant in the inner flask is to be the limiting reactant, the open end of the spout should be set low enough for the solution after delivery to be well below the surface of the liquid. Then, after mixing, a few cubic centimeters of the liquid can be drawn back into the inner flask and

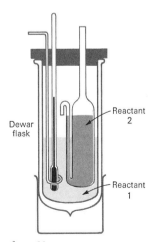

figure 11.

Vacuum-flask calorimeter for solution reactions.

† A thermistor may be used in place of the mercury-in-glass type of thermometer shown in Fig. 11. The thermistor offers the advantages of smaller size, lower heat capacity, and more rapid response and is therefore particularly appropriate if it is desired to use small quantities of reactants. Suitable apparatus and calibration techniques for use with a thermistor are described in Experiments 4 and 13, respectively.

forced out again, to ensure complete reaction of the contents of the inner flask. A suction bulb should be used for this operation.

A simpler method, which yields satisfactory though somewhat less accurate results, is to place the reactants initially in separate Dewar flasks, in one of which a heating coil is immersed, and to mix the two solutions at the proper time by pouring one reactant into the other. The product mixture should then be in the Dewar flask with the heater coil. An advantage of this procedure is that it allows greater flexibility in the choice of relative volumes of the two reactant solutions.

The heater coil, used in the determination of heat capacity, may be of the same design as that described in Experiment 4. The power required is of the order of 15 watts. A 6- or 12-volt dc power supply is convenient. A regulated supply is preferable to a storage battery because the voltage of the latter will drop slightly even in the course of a single run. A calibrated ammeter serves to indicate the heater current, and a timer actuated by the switch which controls the heater current (Fig. 8) measures the period of current flow. Procedures for calibration of the ammeter and measurement of R_h are described in Experiment 4. If the heater wire is Advance or constantan, which has a very low temperature coefficient of resistance (~ 2 ppm $°C^{-1}$), the resistance can be measured cold with an ordinary Wheatstone bridge.

In addition to the heater timer, a timing device is needed which will operate continuously over the duration of the run. A clock with sweep second hand will serve for this purpose if a second laboratory timer is not available.

Effective stirring is essential. For the outer solution, a motor-driven stirrer is to be recommended. This must be positioned with care if the thermometer is to survive the experiment. A simple glass or plastic rod stirrer may be used for the inner flask.

Several factors are involved in selecting the reactant concentrations to be used. The concentrations should be large enough to give an accurately measurable temperature rise, of the order of several degrees, and the relative volumes of the two reactants should be appropriate for the calorimeter. For the case of reaction (2), the concentrations indicated correspond to relative reactant volumes of about 4:1. If a different ratio is desired, it is only necessary to change the numbers of moles of water in the solutions in Eq. (2).

The reactant solutions may be prepared by adding known quantities of water to weighed amounts of the solid salts. One run requires about 0.056 mole of *KI* (MW = 166.0) dissolved in 2.8 moles of water (about 50 ml) and an amount of $Pb(NO_3)_2$ (MW = 331.2) dissolved in 200 ml of water, about 2 percent less than that required to react with the KI. Thus, the $Pb(NO_3)_2$ is the limiting reagent. In general, the amount of solute which reacts must be accurately known; the proportions of solvent to solute should correspond nominally to the chemical reaction equation as written.

When the apparatus is ready for a run, it is assembled with reactant solutions in place. The solutions are stirred until the system has reached thermal equilibrium. Then the temperature is noted, the second reactant is forced out of its flask to mix with the first, and the timer is started (or the time noted if a clock is used). Temperature readings are taken at about 1-min intervals until a constant rate of change has been reached. (This slow change is due to heat exchange with the surroundings and to work of stirring.)

Next, the heater is turned on for a measured interval of time, such as to produce a rise of the order of 1 or 2° in temperature. (With a 15-watt heater, the time required for this would be of the order of several minutes.) After the current has stopped, time-temperature readings are again taken until the rate of change becomes constant, as shown in Fig. 12.

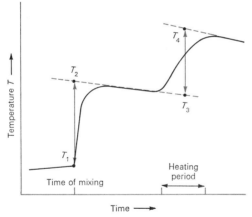

figure 12. Typical time-temperature graph.

CALCULATIONS. Time-temperature curves are plotted (Fig. 12). The nearly straight line obtained a short time after mixing is extended back to the time of mixing to obtain T_2. One seeks with this procedure to find the temperature which would have been reached if the mixing of reactants and the subsequent thermal equilibration of solution and thermometer had occurred instantaneously, i.e., in too short a time for heat exchange with the environment to have occurred.

In this calculation it is implicitly assumed that the rate of heat transfer between system and surroundings is constant from the time of mixing until the linear rate of change is reached. This assumption would not be acceptable if the solution temperature actually changed as slowly after mixing as the graph would seem to indicate. For typical ionic reactions in solution, however, the temperature rise of the solution itself probably occurs almost as quickly as the reactants can be mixed, and the slower rise in observed readings is then due mainly to lag in the thermometer response. When this is the case, the method given above for determining T_2 will be valid.

The temperature rise $T_4 - T_3$ produced by electrical heating is found by drawing, on the time-temperature graph, straight lines through the linear portions which precede and follow the passage of the current and extending these forward and backward, respectively, beyond the middle of the heating period. The temperature rise produced by the electrical heating alone is the vertical distance between these nearly parallel lines, measured at the midpoint of the electrical-heating period.

The enthalpy change due to electrical heating is $\Delta H = I^2 R_h t$, where I is the current, t the time, and R_h the resistance of the heating coil. In the calculation of R_h, one must take into account the fact that the current indicated by the ammeter includes not only the current through the heater coil but also that through the potential divider used for the voltage measurement.

The calculation of ΔH_{T_1} is straightforward. The value reported should be for the amounts shown in the chemical equation as written, and the temperature should be specified.

The next step is the calculation of $\Delta H^\circ_{T_1}$ corresponding to the reaction of Eq. (1). The required heats of dilution can easily be obtained as differences in tabulated heats of formation.[2] For KNO_3, the heat-of-formation data do not go to a sufficiently high concentration, and the following additional heat-of-dilution data,[4a] which are for 25°C, bridge the gap:

$$[KNO_3,25H_2O] + 225H_2O = [KNO_3,250H_2O] \qquad \Delta H = 1219 \text{ cal}$$
$$[KNO_3,25H_2O] + 475H_2O = [KNO_3,500H_2O] \qquad \Delta H = 1309 \text{ cal}$$

Finally, the value of $\Delta H^\circ_{298.15}$ may be calculated from $\Delta H^\circ_{T_1}$ by using the heat-capacity data of Table 1. The solution values are partial molar heat capacities for the solutes in an infinitely dilute solution. These can be considered to be constant in the neighborhood of room temperature, since the entire heat-capacity term in the present calculation is not large.

Table 1. Partial Molar Heat-capacity Data

Substance	$\bar{C}^\circ_P$, cal mole^{-1} deg^{-1}	Source	t, °C
$PbI_2(s)$	19.5	Ref. 4b	25
$Pb(NO)_3(aq)$	5.2	Ref. 4c	18
$KI(aq)$	−29.0	Ref. 3	25
$KNO_3(aq)$	−15.5	Ref. 3	25

Practical applications. The study of heats of reaction in solution has contributed to an understanding of the behavior of some systems, particularly electrolytes. Enthalpy data for reactions are used extensively for calculating the rate of change of equilibrium constant with temperature, through the relation $\partial \ln K / \partial T = \Delta H° / RT^2$. Enthalpy data may also be combined with entropy data for the calculation of K at a single temperature by use of the relation $-RT \ln K = \Delta G° = \Delta H° - T \Delta S°$.

Suggestions for further work. It is interesting to study heats of neutralization of a series of weak acids by NaOH. Miller, Lowell, and Lucasse[5] suggest the acids sulfamic, acetic, monochloroacetic, oxalic, and tartaric. The results may be combined with the accurately known enthalpy change for the process

$$H^+(aq) + OH^-(aq) = H_2O$$

to obtain the enthalpy changes for dissociation of the acids at infinite dilution. While the association of ions to form the weak acids could be studied directly by the use of reactions of the sodium or potassium salts with a strong acid, the neutralization reactions offer the advantage of not involving ternary (three-component) solutions, for which heat-capacity and heat-of-dilution data are usually not available.

Pattison, Miller, and Lucasse[6] describe procedures for determining heats of several types of reactions. A few examples are:

1. Gas evolution: decomposition of H_2O_2
2. Oxidation-reduction: reaction of $KBrO_3$ with HBr
3. Precipitation: reaction of $MgSO_4$ with NaOH
4. Dilution: H_2SO_4 with water; ethyl alcohol with water
5. Organic reaction: reaction of hydroxylamine with acetone

Approximate values can be obtained in favorable cases for the heat of formation of a complex ion from its constituents even though the stability of the complex ion is not high. An illustration is the determination of ΔH for the reaction

$$Ni^{++} + 6NH_3 \longrightarrow Ni(NH_3)_6^{++}$$

by Yatsimirskiĭ and Grafova.[7] A large excess of NH_3 is used, and the heat of reaction is measured as a function of the concentration of NH_3 in the final solution. Care must be taken in interpreting such data if there is any possibility that several species of complex ions are present in significant amounts.

References

1. J. M. Sturtevant in A. Weissberger (ed.): "Technique of Organic Chemistry," vol. 1, "Physical Methods of Organic Chemistry," 3d ed., pt. 1, chap. 10, Interscience Publishers, Inc., New York, 1959.
2. Selected Values of Chemical Thermodynamic Properties, *Natl. Bur. Std. U.S. Circ.* 500, 1952.
3. G. N. Lewis and M. Randall, "Thermodynamics," 2d ed., p. 652, rev. by K. S. Pitzer and L. Brewer, McGraw-Hill Book Company, New York, 1961.
4. "International Critical Tables," vol. V, McGraw-Hill Book Company, New York, 1929: (*a*) p. 162; (*b*) p. 96; (*c*) p. 122.
5. J. G. Miller, A. I. Lowell, and W. W. Lucasse, *J. Chem. Educ.*, **24:** 121–122 (1947).

6. D. B. Pattison, J. G. Miller, and W. W. Lucasse, *J. Chem. Educ.,* **20:** 319–326 (1943).
7. K. B. Yatsimirskii and Z. M. Grafova, *Zhr. Obshch. Khim.,* **22:** 1726 (1952); **23:** 717 (1953); see also *Chem. Abstr.,* **47:** 2030*c*, 10327*e* (1953).

6   ENTHALPY OF FORMATION OF TITANIUM TETRACHLORIDE

Experience is gained in the application of calorimetric techniques and in the practical use of elementary procedures in glass blowing.

THEORY.   For a given temperature, the standard enthalpy of formation of a compound is the standard enthalpy change for the reaction in which 1 formula weight of this compound is formed from the stoichiometrically equivalent elements in their conventional reference forms. Such a standard enthalpy of formation can be calculated from a knowledge of the standard enthalpy change for any reaction in which the compound participates and for which the standard enthalpy of formation is known for all other reactants and products. This process must start with a measurement of the heat of reaction for practical experimental conditions which permit accurate characterization of the initial and final states involved; the effects of all differences between experimental and standard-state conditions must then be accounted for.

The simplest possible case appears to be that of a compound whose synthesis by direct reaction of the elements proceeds in clear-cut fashion and sufficiently rapidly for adequately accurate calorimetric measurements to be possible. These conditions can be satisfied in the case of titanium tetrachloride, but complications are still encountered. Two alternatives are available: (*a*) study of the reaction under conditions in which the product $TiCl_4$ is in the gas phase,[1] with a result for liquid $TiCl_4$ then derived through use of the heat of vaporization, or (*b*) measurements at room temperature,[2-4] with more complex calculations required since the final state then involves a heterogeneous system with the liquid phase a solution of titanium chloride in chlorine.

····➤ *Apparatus.* Calorimeter with stirrer, calibration heater, and accessories; reaction vessel; break-seal ampoules; gas-handling manifold; titanium metal; chlorine tank with reducing valve; liquid nitrogen; Dry Ice; trichloroethylene; vacuum pump.

Caution: This experiment should not be attempted before the necessary background has been obtained by successful completion of Experiments 5 and 57. It must be emphasized that particular care must be exercised in the use of pressurized glassware and such reagents as chlorine and titanium tetrachloride. Safety glasses or a

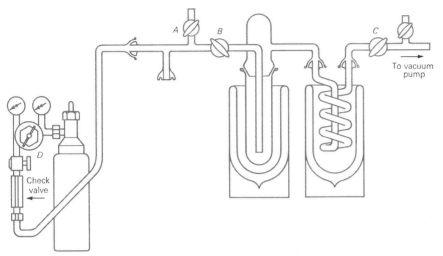

figure 13. Gas manifold for filling chlorine ampoules.

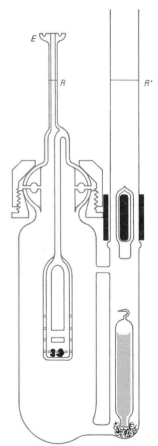

figure 14.

*Glass reaction
vessel for calorimeter.*

face shield should be worn at all times, and due consideration must be given to the potential hazard to other workers in the laboratory.

PROCEDURE.   The reaction vessel, of about 200 to 240 cm³ volume, is shown in Fig. 14; the actual volume between the two reference marks indicated should be determined for use in the later calculations. The lower section of the clamp for the O-ring joint is of two-piece construction, with the sections pinned together for assembly. A standard *screw*-type joint clamp can be used instead, with some sacrifice of compactness. The ampoule-filling assemblies are constructed as shown in Fig. 15, and should be tested before use, by means of nitrogen pressure, to ensure that they can withstand an internal pressure of about 10 atm, since the vapor pressure of liquid chlorine at room temperature is nominally 8 atm.

Caution: All glassware to be used in this experiment must be carefully annealed, and precautions must be taken to prevent the surfaces from being scratched.

When this test is made, the assembly should be wrapped in a towel or other precautions should be taken against flying glass fragments if the test fails. The internal diameter of the ampoule tubing should be known so that the mass of chlorine introduced can be estimated from the nominal density value of 1.5 g cm⁻³.

The ampoule-assembly unit is weighed accurately and connected to the gas manifold (Fig. 13), which is then evacuated. The cold traps are cooled in liquid nitro-

gen to protect the vacuum pump. Stopcocks *B* and *C* are then closed. The valve on the chlorine cylinder proper is opened, and the reducing valve set so that the outlet pressure gauge reads slightly above zero.

Note: Such a gauge indicates the *difference* between outlet line pressure and the ambient atmospheric pressure.

The use of stopcocks which can withstand a moderate internal positive gas pressure (Ace Glass Type 8115A or equivalent) is recommended for positions *A* and *B*. The metering valve *D* is opened to fill the manifold with chlorine gas, then closed. First valve *B* and then valve *C* are opened, and the flushing gas is pumped out of the manifold, after which *B* and *C* are again closed. Valve *D* is reopened, and a Dry Ice–trichloroethylene bath is placed about the ampoule. Chlorine is allowed to condense in desired amount (about 2 cm³), after which valve *D* is closed and the ampoule carefully sealed off† at the constriction. A pair of forceps whose tips have been shaped to fit around the ampoule will be found very useful here. The ampoule is placed on a clean towel in the hood and allowed to warm up to room temperature.

Stopcock *B* is next opened to remove the residual chlorine by condensation in the cold trap, then closed. The manifold is vented to the atmosphere through stopcock *A*. The upper section of the ampoule-assembly unit can now be removed from the manifold. This section and the filled ampoule are now weighed to permit calculation of the mass of chlorine in the ampoule. Since accuracy only to within a few milligrams is required here, and since there is an element of hazard to the balance, it is recommended that an old balance be used.

This sequence of operations can be repeated to prepare several additional ampoules. It should be emphasized that for efficient transfer of chlorine to the ampoule a good vacuum must be obtained initially in the manifold before the chlorine is introduced.

A small pad of glass wool or Teflon is placed at the bottom of the side tube of the reaction vessel, which is clamped in a nearly horizontal position. The filled ampoule, with the break seal toward the top, is slipped into place at the end of the tube. The iron-core glass striker is inserted appropriately next, and the retaining magnet is put in place around the tube. The side tube is then sealed off carefully and symmetrically, about 1 in. above reference mark *R'* (Fig. 14), with a suitable trap being used to keep water vapor from entering the reaction vessel through the blowing tube connected at *E* (Fig. 14). The basket assembly is then removed from the reaction vessel, and the weighed titanium sample (about 0.5 g) is added to it. The reaction vessel is then raised *carefully* to the vertical position and the basket

figure 15.
Ampoule-loading
assembly.

† The Little Torch, available from Tescom Corporation, Instrument Division, Minneapolis, Minn. 55414, is very convenient in this application.

unit replaced. The vessel is then evacuated, after being connected to the gas manifold at the joint B, and is sealed off under vacuum at a point 1 in. above reference mark R. The volumes above R, R' are estimated as additions to the volume between the reference marks.

The reaction vessel is now transferred to its holder in the calorimeter proper. In the further calorimetric work the conventional procedures described in Experiments 4 and 5 are followed. To initiate the reaction, the magnet is raised sufficiently to release the striker, whose fall will result in fracture of the break seal and release of the chlorine. There will usually be a moderate induction period before the rate of reaction becomes appreciable. Subsequent to this determination, an electrical calibration of the calorimeter is made.

The reaction vessel is then removed from the calorimeter and opened *in the hood*. It is rinsed out first with dilute HCl then with distilled water. The upper ends of the side tube and the evacuation tube are then cut off. The glassware is dried and new sections added as necessary to restore the original configuration. The main valve of the chlorine cylinder is closed, and the manifold cold traps are removed and transferred immediately to a hood to warm up.

Note: Some samples of titanium do not ignite readily under the conditions of this experiment. Satisfactory results have been obtained by use of granular titanium lump, CP.† The quality of the tank chlorine should be checked; air and carbon dioxide in appreciable amounts may be encountered as contaminants. For accurate results, it is appropriate to purify the chlorine used by fractional distillation.

Table 1. *Vapor-pressure and Latent-heat-of-vaporization Data for Chlorine*[5]

t, °F	P, atm	L_v, Btu lb^{-1}
0	1.940	119.7
20	2.895	116.8
40	4.170	113.8
60	5.825	110.7
80	7.925	107.3

Table 2. *Vapor Pressure of Titanium Tetrachloride*[6]

$$\log P_{mm} \approx 8.001 - \frac{2{,}077}{T}$$

t, °C	P, mm Hg
40.3	23.5
52.7	42.0
63.5	67.5
75.0	108.0

CALCULATIONS.   The experimental heat of reaction is calculated as usual (Experiments 4 and 5). The initial state of the reaction system is readily identified. To characterize the final state it may be assumed that the liquid and gaseous solutions present are ideal. From the known volume of the system, the initial masses of the

† Supplied by A. D. Mackay, Inc., New York, 10038.

titanium and chlorine, and vapor-pressure data for $TiCl_4$ and Cl_2 as given in Tables 1 and 2, the masses and compositions of the two phases may be calculated. The additional calculation steps required to obtain results referred to standard-state conditions are then identified, and values are obtained for the standard enthalpy of formation of $TiCl_4(l)$ and $TiCl_4(g)$ at 25°C, where the latent heat of vaporization of $TiCl_4$ may be taken as 9.9 kcal mole^{-1} (Ref. 7). These values are compared with results given in the JANAF Thermochemical Tables,[8] which also will provide auxiliary data required for the calculations.

Practical applications. Calorimetric determinations of heats of reaction and other thermochemical data are of basic importance in the generation of such tabulations as the JANAF Thermochemical Tables,[8] which greatly facilitate the practical application of thermodynamic principles.

Suggestions for further work. The experimental procedure described by Skinner and Ruehrwein[4] may be tried.

References

1. W. H. Johnson, R. A. Nelson, and E. J. Prosen, *J. Res. Natl. Bur. Std. U.S.*, **62:** 49 (1959).
2. P. Gross, C. Hayman, and D. L. Levi, *Trans. Faraday Soc.*, **51:** 626 (1955); **53:** 1601 (1957).
3. W. F. Krieve, S. P. Vango, and D. M. Mason, *J. Chem. Phys.*, **25:** 519 (1956).
4. G. B. Skinner and R. A. Ruehrwein, *J. Phys. Chem.*, **59:** 113 (1955).
5. E. J. Laubusch in J. S. Scone (ed.), "Chlorine: Its Manufacture, Properties and Uses," Reinhold Publishing Corporation, New York, 1962.
6. H. Schafer and F. Zeppenick, *Z. Anorg. Chem.*, **272:** 274 (1953).
7. H. Schafer, G. Breil, and G. Pfeffer, *Z. Anorg. Chem.*, **276:** 325 (1954).
8. D. R. Stull et al., JANAF Thermochemical Tables, *Dept. Commerce, Clearinghouse Sci. Tech. Inform.*, Doc. PB-168-370 and addenda PB-168-370-1, PB-168-370-2.

Chapter 3

Vapor Pressure of Pure Substances

7 VAPOR PRESSURE OF A PURE LIQUID

In this experiment the vapor pressure of a liquid is measured at several temperatures. The enthalpy of vaporization is calculated using the Clausius-Clapeyron equation.

THEORY.   When the temperature is raised, the vapor pressure of a liquid increases, because more molecules gain sufficient kinetic energy to break away from the surface of the liquid. When the vapor pressure becomes equal to the pressure of the gas space, the liquid boils. The temperature at which the vapor pressure reaches 760 mm Hg is the *standard* boiling point.

   According to the Clapeyron equation, the temperature coefficient of the vapor pressure of a liquid is given by

$$\frac{dP}{dT} = \frac{\Delta \overline{H}_{\text{vap}}}{T(\overline{V}_v - \overline{V}_l)} \tag{1}$$

where $\Delta \overline{H}_{\text{vap}}$ = enthalpy of vaporization at temperature T

   $\overline{V}_v, \overline{V}_l$ = molar volumes of vapor and liquid

The Clausius-Clapeyron equation

$$\ln P = -\frac{\Delta \overline{H}_{\text{vap}}}{RT} + \text{constant} \tag{2}$$

is derived from this exact equation with the following three assumptions: (*a*) the volume of a mole of liquid may be neglected in comparison with a mole of vapor at its saturation pressure; (*b*) the vapor behaves as an ideal gas; and (*c*) the enthalpy of vaporization is independent of temperature. Although Eq. (2) leads to a very simple interpretation of experimental data, the values of $\Delta \overline{H}_{\text{vap}}$ calculated in this way may disagree significantly with the directly determined calorimetric values. Better values may be obtained by use of a more complete equation derived in the following way.[1]

   The volume factor in the Clapeyron equation may be written

$$\overline{V}_v - \overline{V}_l = \overline{V}_v \left(1 - \frac{\overline{V}_l}{\overline{V}_v}\right) = \frac{ZRT(1 - \overline{V}_l/\overline{V}_v)}{P} \tag{3}$$

where Z is the compressibility factor for the vapor. The expression on the right is introduced in the Clapeyron equation, which can then be rearranged to yield

$$\frac{d \ln P}{d(1/T)} = -\frac{\Delta \overline{H}_{\text{vap}}}{ZR(1 - \overline{V}_l/\overline{V}_v)} \tag{4}$$

or, approximately,

$$\ln P = \frac{-\Delta \overline{H}_{\text{vap}}}{ZR(1 - \overline{V}_l/\overline{V}_v)T} + \text{constant} \tag{5}$$

This is a better equation, but it involves three quantities which are functions of the temperature, $\Delta \overline{H}_{vap}$, Z, and $1 - \overline{V}_l/\overline{V}_v$. When PVT data on the compound being studied are available, the enthalpy of vaporization may be calculated from the slope of the plot of ln P versus $1/T$ using Eq. (4). If PVT data are not available, a good estimate of the required quantities may often be made using the Berthelot equation if the critical constants are known.

Apparatus. Vacuum system consisting of water aspirator, ballast tank, mercury manometer, and connections; Ramsay-Young vapor-pressure tube; one or more liquids chosen from carbon tetrachloride, acetone, chloroform, benzene, or other liquid boiling below 100°; autotransformer and heating mantle.

PROCEDURE.[2,3]   This experiment may be carried out using the classical Ramsay-Young[4] apparatus or, preferably, the Tobey[5] modification. In the usual Ramsay-Young apparatus the organic liquid to be studied is allowed to drip from a dropping funnel and down the thermometer to a hygrometer wick or layer of muslin cloth around the thermometer bulb. The temperature measured is that of the liquid in equilibrium with the vapor at the pressure indicated by the manometer. In the Tobey modification shown in Fig. 16, the dropping funnel is replaced by a cold finger, and the liquid is vaporized at the lower end of the long column and recondensed at the upper end. Thus there is no loss of material during the experiment as with the Ramsay-Young apparatus, and the difficulties of adjusting the flow rate, which are characteristic of that apparatus, are avoided.

A long Pyrex tube A, about 25 mm in diameter, is sealed to a 200-ml round-bottom flask, and a 29/42 standard-taper joint is fastened to the top. The flash guard at the bottom of the column prevents liquid from bumping up into the column, with consequent superheating of the thermometer bulb. The cold finger should be sufficiently long for the amount of organic liquid escaping into the vacuum system to be negligible. The dimensions in the diagram are recommended. The thermometer C is a 0.0 to 110.0°C complete-immersion precision type graduated to 0.1°. The bulb is wrapped with a hygrometer wick or layer of muslin cloth.

The vacuum system consists of an aspirator connected through a three-way T stopcock B to a ballast tank of 10 liters or more capacity.† This tank prevents sudden pressure surges from disturbing the equilibrium. Connections between the various parts of the apparatus can be made with rubber vacuum hose, but Tygon tubing is better. The stopcocks and the glass joints are greased, if this operation has not already been done.

The bulb at the bottom of the column is filled approximately half full with the liquid to be studied. If a liquid different from the one being studied has been used

---

† Stainless-steel tanks, which were used by the Army air forces for breathing oxygen, make excellent ballast tanks. These may be available at surplus-property supply houses. If an ordinary glass bottle is used, it should be wrapped in heavy wire mesh or placed in a wooden box to eliminate hazards due to flying glass in case of breakage when evacuated.

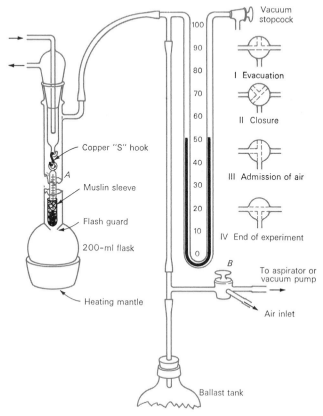

100
90
80
70
60
50
40
30
20
10
0

Vacuum stopcock

I Evacuation

II Closure

III Admission of air

IV End of experiment

Copper "S" hook

Muslin sleeve

Flash guard

200-ml flask

Heating mantle

To aspirator or vacuum pump

Air inlet

Ballast tank

figure 16. Vapor-pressure apparatus.

previously in the apparatus, the system must be cleaned out and the liquid completely removed.

The experiment may be performed using either a closed-end or open-end manometer, but closed-end manometers have the advantage that they read the pressure rather than the difference from atmospheric pressure.

If a manometer of the closed-end type is used, the manometer is connected to the atmosphere by use of stopcock B to see if the indicated pressure agrees with that given by the precision barometer in the laboratory. If the difference is greater than 1 mm, the instructor should be consulted.

The cooling water for the reflux head is turned on. With stopcock B in position I, the system is pumped down until the manometer indicates approximately

100 mm pressure in the system or until the liquid in the bulb begins to boil. To test for leaks, stopcock *B* is closed (position II) to disconnect the aspirator, and the mercury meniscus in the manometer is watched. If it rises continuously, there is a leak in the system. If there is leakage, the source must be located and eliminated.

An autotransformer connected to the heating mantle is turned on, the dial is set at about 30 volts, and the heater is allowed to come to temperature. When steady-state conditions within the system are reached, there will be a steady flow of liquid back into the pot through the hole in the flash guard and a drop of liquid should drop off the thermometer bulb about every 5 to 10 sec. The thermometer bulb is heated by condensation of vapor and by radiation from the flask and cooled by evaporation until a steady temperature is registered. This is the boiling temperature of the liquid at the pressure registered on the manometer, unaffected by superheating. When the temperature registered by the thermometer reaches a steady value, the temperature and pressure are recorded.

The pressure in the system is then raised slightly by turning stopcock *B* gradually toward position III. Sufficient air is introduced to increase the pressure in the system so that the temperature indicated by the thermometer rises about 5°C, or the pressure is increased about 100 mm. The stopcock is then closed, and the system allowed to come to a steady state as before. This process is continued until the pressure in the system is returned to atmospheric. As the pressure is raised, it will be necessary to advance the setting of the Variac to obtain a suitable rate of reflux. If suitable clamps are provided for the ground joints and stopcocks, the measurements may be extended by use of compressed air to raise the pressure in the system *moderately* above ambient atmospheric pressure.

Caution: Glass systems are hazardous when evacuated and can be even more so when pressurized.

The pressure is now successively reduced in steps, and another series of data points is taken on a descending run. To lower the pressure, stopcock *B* is opened briefly so that air is pumped out of the system. The Variac is turned back as required to maintain a suitable rate of reflux.

At the close of the experiment, stopcock *B* is turned to position IV, the Variac is turned off, and the cooling water is shut off.

CALCULATIONS. Two types of graphs are plotted. In one the vapor pressures are plotted against the temperatures, and in the second the logarithms of the vapor pressures are plotted against the reciprocals of the *absolute* temperatures. Values taken from the literature are plotted also.

The values of the constants A and B in the equation

$$\log P = \frac{A}{T} + B \tag{6}$$

are determined by two methods.

First, the best straight line is drawn through the points by eye, and the constants A and B are calculated for this straight line. This may be done by using two points on the line which are far apart and solving the two simultaneous equations for A and B. Alternatively, the slope A may be calculated, and then B calculated, using $\log P$ at some particular temperature.

Second, more objective values of A and B are calculated using the method of least squares. As explained in Chapter 20, this method yields the values of A and B such that the sum of the squares of the deviations of the experimental points from the theoretical line is a minimum. The details of the method are given on page 446. The first step is to tabulate $\log P$, $1/T$, $(1/T)^2$, and $(1/T) \log P$. In carrying out these calculations a desk calculator is required since the calculations must be carried out with several more figures than represented by the number of significant figures in the experimental results.

The heat of vaporization ΔH_{vap} is then calculated from the values of A obtained by using the two methods. By comparison of Eqs. (2) and (6) it can be seen that

$$A = \frac{-\Delta H_{\text{vap}}}{2.303R} \tag{7}$$

The assumptions made in the derivation of the Clausius-Clapeyron equation limit the accuracy of the heats of vaporization calculated in this way.

Better values may be calculated using Eq. (4). The compressibility factor may be obtained directly from PVT data for the compound under study if they are available. When PVT data are not available, the compressibility factor may be estimated using the Berthelot equation, which may be written

$$Z = 1 + \frac{9}{128} \frac{P}{P_c} \frac{T_c}{T} \left(1 - 6 \frac{T_c^2}{T^2}\right) \tag{8}$$

where T_c = critical temperature
$\quad\quad P_c$ = critical pressure
The compressibility factor is calculated for a particular point on the vapor-pressure curve, and the slope of the plot of $\ln P$ versus $1/T$ at that point is used. The ideal-gas law is used to calculate $\overline{V}_v$ in the factor $1 - \overline{V}_l/\overline{V}_v$.

It should also be noted that an excellent correlation of vapor-pressure-temperature data may be obtained by use of the Antoine equation:[6]

$$\log P = \frac{-A}{t + C} + B$$

where A, B, C = constants empirically determined from experimental data

t = temperature, °C

Practical applications. Vapor-pressure measurements are important in all distillation problems and in the calculation of certain other physical properties. They are used in the correction of boiling points and in the recovery of solvents. The concentration of vapor in a gas space may be regulated nicely by controlling the temperature of the evaporating liquid. Humidity conditions, which are so important in many manufacturing processes, depend largely on the vapor pressure of water.

Suggestions for further work. The vapor pressures of other liquids may be determined, using, if possible, liquids whose vapor pressures have not yet been recorded in tables. The sublimation temperature of a solid may be obtained by covering the thermometer bulb with a thin layer of the solid.

   The vapor pressure may be determined by an entirely different method, evaluating the amount of liquid evaporated by a measured volume of air, as described in Chapter 23.

References

1. O. L. I. Brown, *J. Chem. Educ.*, **28:** 428 (1951).
2. G. W. Thomson in A. Weissberger (ed.), "Technique of Organic Chemistry," vol. 1, "Physical Methods of Organic Chemistry," 3d ed., pt. 1, chap. 9, Interscience Publishers, Inc., New York, 1959.
3. C. B. Willingham, W. J. Taylor, J. M. Pignocco, and F. D. Rossini, *J. Res. Natl. Bur. Std. U.S.*, **35:** 219 (1945).
4. W. Ramsay and S. Young, *J. Chem. Soc.*, **47:** 42 (1885).
5. S. W. Tobey, *J. Chem. Educ.*, **35:** 352 (1958).
6. G. W. Thomson, *Chem. Rev.*, **38:** 1 (1946).

8   KNUDSEN SUBLIMATION-PRESSURE MEASUREMENT

This experiment illustrates a method for determining the sublimation pressure of a solid and provides experience in the use of high-vacuum equipment.

THEORY.   The rate of escape of molecules of a gas through a small hole into a vacuum is directly proportional to the pressure under certain conditions, and the proportionality constant may be calculated from kinetic theory.[1-3] In order to obtain a simple relation it is necessary that the pressure outside of the hole be sufficiently low for the mean free path of the molecules, the average distance traversed by a molecule between collisions, to be long compared with the diameter of the hole and the thickness of the foil through which the hole is punched. For an ideal gas of rigid spherical molecules, the mean free path l is given by

$$l = \frac{1}{\sqrt{2}\,\pi\sigma^2 n} = \frac{kT}{\sqrt{2}\,\pi\sigma^2 P} \tag{1}$$

where σ = collision diameter for gas molecule, for example, 3.75×10^{-8} cm for N_2 and 4.63×10^{-8} cm for CO_2

n = number of molecules per cubic centimeter

P = pressure

k = Boltzmann constant = R/N_0

When the mean free path on the low-pressure side is large compared with the diameter of the hole, the number of molecules passing through the hole is equal to the number of molecules that would collide with the corresponding area of wall. The kinetic-theory result for the number of gas molecules ν colliding with a square centimeter of wall in a second is

$$\nu = n \left(\frac{RT}{2\pi M} \right)^{\frac{1}{2}} \tag{2}$$

where R = ideal-gas constant

M = molecular weight of gas

Since the rate of escape depends upon the number n of molecules per cubic centimeter, it may be expressed in terms of the pressure by introducing $P = nRT/N_0$, where N_0 is Avogadro's number. The effusion rate is conveniently expressed in terms of the mass m of gas passing through the hole per square centimeter area per second.

$$m = \frac{M\nu}{N_0} \tag{3}$$

Eliminating ν between Eqs. (2) and (3) and introducing the ideal-gas law yields

$$P = m \left(\frac{2\pi RT}{M} \right)^{\frac{1}{2}} \tag{4}$$

If m is expressed in grams per square centimeter per second and R is written as 8.314×10^7 ergs deg^{-1} mole^{-1}, P comes out in dynes per square centimeter. The pressure in atmospheres may then be obtained by dividing by

$$(76 \text{ cm})(13.595 \text{ g cm}^{-3})(980.7 \text{ cm sec}^{-2}) = 1.013 \times 10^6 \text{ dynes cm}^{-2} \text{ atm}^{-1}$$

Equation (4) is often referred to as the *Herz-Knudsen equation*.

By measuring the rate of escape of gas from a saturated vapor through a small hole into a vacuum, the vapor pressure may be obtained. This method was developed by Knudsen[1,2] and is useful for the measurement of the sublimation pressures of

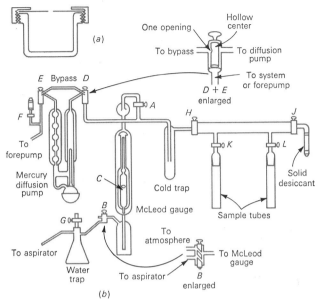

figure 17. (a) Diagram of Knudsen cell; (b) vacuum system for measurements of vapor pressure by Knudsen method. The vacuum line should be at least 76 cm above the mercury level.

relatively nonvolatile solids at high temperatures. The range of sublimation pressures which may be studied is about 10^{-9} to 10^{-3} atm. A number of refinements of this method have been developed.[4]

Apparatus. Mechanical vacuum pump; mercury diffusion pump to reduce the pressure to below $1\ \mu$ ($= 10^{-3}$ mm) of mercury; McLeod gauge; Knudsen cell; brass shim stock (0.002 in. thick); cold trap and chamber for Knudsen cell; thermostat, 10 to 50°; naphthalene.

PROCEDURE.   The Knudsen cell is shown in Fig. 17*a*. In order to get accurate results with the Knudsen method, the pressure outside the Knudsen cell should be maintained at about 10^{-3} mm of mercury or less. In the laboratory this is accomplished with a mechanical vacuum pump and a mercury diffusion pump. The pressure is measured with a McLeod gauge.

Caution: Before starting this experiment, the student should study the principles of operation of the rotary vacuum pump, mercury diffusion pump, and McLeod gauge (Experiment 58).

The vacuum system to be used is shown in Fig. 17*b*.

Part A. Turning On the Vacuum System

The vacuum system is turned on, and the pressure measured by the following procedure:

1. The water to the condenser of the mercury diffusion pump is turned on.
2. Stopcock *G* is opened, the water aspirator is turned on full force, and then stopcock *G* is closed.
3. A slush of finely powdered Dry Ice in trichloroethylene is prepared in the Dewar flask. The Dry Ice as obtained from a mechanical crusher is too coarse for this purpose and should be reduced to fine powder by placing a cupful at a time in a towel and beating it with a hammer. The final mixture should contain enough liquid so that it can be placed around the cold trap without danger of breakage. The Dewar flask should be almost half filled. When the bath has been prepared, it is placed very carefully around the cold trap.
4. Initially stopcocks *A*, *D*, *E*, *F*, *H*, *J*, *K*, and *L* should be closed. The forepump is started. *E* and then *D* are opened to bypass the diffusion pump. Pumping is continued for about 1 min, until the pump is relatively quiet. It is well to listen to the pump while opening stopcock *H*. If the right-hand portion of the system was full of air initially, the characteristic sound of large quantities of air passing through the forepump will be heard. This sound should be remembered.
5. A test is carried out to see what the pressure is after the forepump has been running about 10 to 15 min. Stopcock *A* to the McLeod gauge is opened cautiously.
   a. If the mercury begins rising in the McLeod gauge, this stopcock is closed and an instructor is consulted.
   b. If the mercury does not begin rising into the gauge, stopcock *B* is slowly opened to the atmosphere. The mercury will then begin to rise in the McLeod. The stopcock is left open until the mercury is about 2 cm above the midpoint of the large glass tube marked *C* on the diagram. Stopcock *B* is closed. The mercury will continue to rise until it reaches the capillaries. The pressure can now be read. To pull the mercury down, stopcock *B* is opened to the aspirator. (The aspirator is left on throughout the experiment.) Then stopcock *A* is closed. When the mercury is down, stopcock *B* is closed.
6. When the pressure is below 100 μ, stopcock *E* is turned to open the diffusion pump to the forepump. By listening to the forepump the presence of air in the diffusion pump is checked. If air is present, it should be pumped out before turning stopcock *D* to open the diffusion pump to the rest of the system. The diffusion pump is turned on. The Variac which controls the diffusion-pump heater should be set at the optimum value for operation of the pump.
7. When the mercury in the diffusion pump has been distilling for about 10 min, the pressure is checked again as above. If the pressure is not below 1 μ,

an instructor should be consulted; the difficulty could result from a leak, from faulty operation of the pumps, or from outgassing. If the pressure is below 1 μ, the experiment is continued. The mercury level is lowered, and the McLeod gauge is closed off by shutting stopcocks *A* and *B*. These are *not* opened again during the experiment.

Part B. *Preparing the Knudsen Cells*

Two Knudsen cells are used. A sketch showing the design of a Knudsen cell appears in Fig. 17*a*. The top plate is a thin disk of brass (about 0.002 in. thick) with an orifice in the center. The orifice can best be fabricated by means of a drill press, with the top plate sandwiched between two $\frac{1}{16}$-in. plates of brass the upper of which has a pilot hole to guide the drill. Suggested bore diameters are 1.3, 0.8, and 0.4 mm.

Because of a tendency for the orifice to collect obstructing particles of dust, it should be washed with acetone and carefully blown out with filtered compressed air before the plate is inserted into the effusion cell. Furthermore, just before introducing each cell into the vacuum system, its orifice should be cleaned by inserting a fine wire and *very* lightly rubbing it against the edges.

The Knudsen cells are carefully cleaned and dried and are half filled with finely powdered naphthalene. Top plates having the desired orifice sizes (see below) are placed on top of the Knudsen cells, and the caps are screwed on. Fingerprints are wiped off the cells, and the filled cells are weighed to 0.1 mg.

Part C. *Placing the Knudsen Cell in the Vacuum System*

Observations are to be made at four different temperatures, for which four different top plates (three orifice sizes) are provided. The largest orifice size is used at 0°C, the medium size at 15°, and the two smallest sizes at 30 and 45°. Diameters are given in Part B.

The two higher temperatures may be attained in the regulated water baths, the two lower in Dewar flasks containing water to which ice may be added from time to time as required. Two runs may be made on each day.

The following procedure is suitable for introduction of the Knudsen cell:

1. Stopcock *H* is closed.
2. Stopcocks *K* and *L* are opened. Then *J* is opened slowly to admit dry air.
3. The sample tubes are removed.
4. The Knudsen cells are placed in the sample tubes. A long spring clamp may be used to facilitate this operation. The glass sample tubes should be clean because

any dirt or grease transferred to the Knudsen cell will cause an error in the weight loss.

5. The standard-taper connections are regreased if necessary, and the sample tubes reconnected to the vacuum system. Small springs or rubber bands may be used to support the sample tubes.
6. The water baths are put into place.
7. Stopcock *J* is closed.
8. Stopcocks *K* and *L* are opened.
9. Stopcock *H* is opened. The time is recorded, as the vaporization begins at this time.

Part D. Closing Down the Vacuum System

After 2 or 3 hr, the cells are removed and the vacuum system closed down according to the following procedure.

1. Stopcock *H* is closed.
2. Stopcock *J* is opened to the atmosphere. The time at which this is done is recorded.
3. Stopcocks *K* and *L* to the sample tubes are closed; *then* the samples are removed. (*K* and *L* are closed to prevent moist laboratory air from entering the manifold.)
4. Stopcocks *D* and *E* are closed to prevent the vacuum from being lost in the diffusion pump.
5. Stopcock *F* is opened to the atmosphere. Then the forepump is promptly turned off.
6. Stopcock *G* on the water trap leading to the aspirator is opened. Then the water aspirator is turned off.
7. The diffusion pump heater is turned off.
8. The cooling water to the diffusion pump is turned off.
9. Stopcocks *F* and *J* are closed.

Part E. Weighing the Knudsen Cells

If the cell has been at a temperature appreciably different from room temperature, it must be allowed to come to the temperature of the balance case before weighing.

CALCULATIONS.   The vapor pressure of naphthalene is calculated from the rate of weight loss, using Eq. (4). The reliability of the calculated vapor pressure is estimated.

The experimental vapor pressures are compared with literature values[5] by means of a plot of log P versus $1/T$. The heat of sublimation per mole $\Delta \overline{H}$ is calculated from the slope of this plot, assuming $\Delta \overline{H}$ is independent of temperature.

Suggestions for further work. Other compounds which may be studied are benzoic acid, *d*-camphor, acetamide, and hexachlorobenzene.

If the vapor pressure is determined over a sufficiently wide range of temperature, it will be found that the plot of log P versus $1/T$ is not straight. This is primarily a result of the change in heat of sublimation with temperature. If the temperature range is not too large, the change in heat of sublimation with temperature is given by

$$\Delta \overline{H}_{T_2} = \Delta \overline{H}_{T_1} + \Delta \overline{C}_P(T_2 - T_1) \tag{5}$$

where $\Delta \overline{C}_P$ is the difference in molar heat capacity of the vapor and solid at constant pressure. Actually, the total pressure varies over the range of temperature, but this causes little error. The enthalpies of sublimation are calculated at two widely separated temperatures from the slope of tangents to the plot of log P versus $1/T$. Then $\Delta \overline{C}_P$ is calculated by using Eq. (5).

This method is especially useful for the determination of sublimation pressures at high temperatures. The measurement of the sublimation pressures of tungsten[6] and beryllium[7] are examples.

The Knudsen cell has been used in connection with the mass spectrometer to determine such quantities as vapor pressures and equilibrium constants after electron-impact ionization reactions.[8]

References

1. M. Knudsen, *Ann. Physik,* **28:** 75, 299 (1909); **29:** 179 (1909).
2. M. Knudsen, "Kinetic Theory of Gases," Methuen & Co., Ltd., London, 1934.
3. R. D. Present, "Kinetic Theory of Gases," McGraw-Hill Book Company, New York, 1958.
4. J. L. Margrave in J. O'M. Bockris, J. L. White, and J. D. MacKenzie, "Physico-chemical Measurements at High Temperatures," Academic Press Inc., New York, 1959.
5. R. W. Bradley and T. G. Gleasby, *J. Chem. Soc.,* **1953:** 1690.
6. I. Langmuir, *Phys. Rev.,* **2:** 329 (1913).
7. R. B. Holden, R. Speiser, and H. L. Johnston, *J. Am. Chem. Soc.,* **70:** 3897 (1948).
8. J. Berkowitz, W. A. Chupka, and M. G. Ingraham, *J. Chem. Phys.,* **27:** 85 (1957).

Chapter 4

Solutions

9 LIQUID-VAPOR EQUILIBRIUM IN BINARY SYSTEMS

Boiling-point and vapor-composition data for a binary solution system at constant pressure may be correlated in a graph of temperature versus composition. Data for such a plot are obtained in this experiment, in which the liquid and vapor compositions are determined refractometrically. The calculation of the activity coefficients for the components in the liquid phase and their representation by the van Laar equations are considered.

THEORY.[1,2]   The relation between the composition of a liquid solution (phase l) of two volatile liquids and that of the vapor (phase v) in equilibrium with it at a given temperature and pressure may be established by use of the thermodynamic requirement that the chemical potential μ_i for component i have a common value for the two equilibrium phases

$$\mu_{i,l} = \mu_{i,v}$$

The fugacity f_i for component i, irrespective of the phase in which it is present, is defined by the relation

$$\mu_i = \mu_i^* + RT \ln f_i \tag{1}$$

where μ_i^* corresponds to the chemical potential that component i would have as an ideal gas at 1 atm pressure at the temperature T. This definition makes the fugacity of a component of an ideal-gas mixture identical with its formal partial pressure $P_i = X_{i,v}P$, where $X_{i,v}$ is the mole fraction of constituent i present and P the total pressure of the gas phase. For a real-gas mixture, $f_{i,v} = \vartheta_i P_i = \vartheta_i X_{i,v} P$, where the *fugacity coefficient* ϑ_i is a function of temperature, pressure, and composition which can be calculated from the equation of state of the gas phase. For a condensed phase, the fugacity of a constituent can be found by determining its value for the equilibrium vapor phase. Condensed phases are of interest, however, under conditions in which no such calculation is possible, as, for example, a solid of unmeasurably low vapor pressure or a solution of a nonvolatile solute such as sodium chloride in aqueous solution. It is thus convenient to introduce the *thermodynamic activity a_i* for a constituent of a given phase by the relation

$$\mu_i = \mu_i^\circ + RT \ln a_i = \mu_i^\circ + RT \ln \frac{f_i}{f_i^\circ} \tag{2}$$

where f_i° is the fugacity for a selected standard state for which the chemical potential is μ_i°. The activity a_i is a ratio of two fugacities and may readily be determined even when the individual fugacities involved cannot be. The standard state employed may be selected arbitrarily on a basis of practical convenience but will normally be so chosen as to provide the simplest possible relation between the activity and the concentration of the constituent in the phase concerned. It thus becomes common to select a different standard state for a component for each phase in which it is present, so that the activity, unlike the fugacity, usually does not have a common value for different equilibrium phases.

For nonelectrolytic solutions the standard state for each component is normally taken to be the pure liquid at the temperature and pressure of the solution, and the activity is correlated with the concentration on the mole-fraction scale by means of the *activity coefficient* γ.

$$a_i = \frac{f_{i,l}}{f^{\circ}_{i,l}} = \gamma_i X_{i,l} \tag{3}$$

For what is called an *ideal* solution, γ_i as defined above is identically equal to unity for any component at any concentration. For real solutions the activity coefficients must be determined by experiment.

The fugacity $f^{\circ}_{i,l}$ for this standard state is calculated as follows. The vapor pressure $P^{*}_i(T)$ of pure liquid i at the given temperature is multiplied by the fugacity coefficient $\vartheta^{*}_i(P^{*}_i,T)$ of the vapor as calculated from the equation of state of the vapor to obtain the fugacity of the saturated vapor at the temperature T. This then gives the fugacity of pure liquid i for temperature T and pressure P^{*}_i. The fugacity $f^{\circ}_{i,l} = f_{i,l}(P,T)$ may then be calculated by taking into account the difference between P^{*}_i and P, using the thermodynamic relation

$$\ln f^{\circ}_{i,l} = \ln f_{i,l}(P,T) = \ln \vartheta^{*}_i P^{*}_i + \int_{P^{*}_i}^{P} \frac{\overline{V}^{*}_i}{RT} dP \tag{4}$$

where $\overline{V}^{*}_i$ is the molar volume of pure *liquid i*. The integral in Eq. (4) will be negligible if $P - P^{*}_i$ is not large.

For liquid-vapor equilibrium, since the fugacity of each constituent must have a common value for both phases,

$$f_{i,l} = \gamma_i f^{\circ}_{i,l} X_{i,l} = f_{i,v} = \vartheta_i X_{i,v} P \tag{5}$$

Assuming the effect of pressure on the fugacity of the pure liquid to be negligible, $f^{\circ}_{i,l} = \vartheta^{*}_i P^{*}_i$ and

$$X_{i,l} = \frac{1}{\gamma_i} \frac{\vartheta_i}{\vartheta^{*}_i} \frac{P}{P^{*}_i} X_{i,v} \tag{6}$$

If the gas phases involved are considered to behave ideally,

$$X_{i,l} = \frac{1}{\gamma_i} \frac{P}{P^{*}_i} X_{i,v} \tag{7}$$

For ideal liquid solutions, this desired relation between the liquid and vapor compositions further simplifies to

$$X_{i,l} = \frac{P}{P_i^*} X_{i,v} \tag{8}$$

For real solutions the activity coefficients are functions of concentration, temperature, and pressure. For a binary nonelectrolytic solution system the concentration dependence may often be represented to a good degree of approximation by the van Laar equations, which have been written as follows by Carlson and Colburn:[3]

$$\log \gamma_1 = \frac{A_1}{[1 + (A_1 X_1 / A_2 X_2)]^2} \qquad \log \gamma_2 = \frac{A_2}{[1 + (A_2 X_2 / A_1 X_1)]^2} \tag{9}$$

The van Laar coefficients A_1, A_2 are functions of temperature and pressure. Even substantial changes in pressure have only a small effect. The dependence on temperature is more important, but over a range of 10 or 20° the resultant change in a typical activity coefficient will usually be only a few percent.

Real solution systems vary widely in their degree of departure from the ideal-solution rule, for which the boiling points of the solutions are always intermediate between those of the pure liquids. In many cases, however, the deviation from ideality becomes so great that a minimum or maximum results in the plot of boiling point versus liquid or vapor composition. At such a maximum or minimum, the equilibrium vapor and liquid compositions are identical. Such solutions are called *azeotropes*. A comprehensive description of methods for the experimental study of vapor-liquid equilibria and for the correlation of the results has been given by Hala et al.[1] An extensive table of azeotropes has been prepared by Horsley et al.,[4] and data for many binary solution systems have been summarized by Timmermans.[5]

Apparatus. Distilling apparatus as illustrated in Fig. 18; pipette of about 1 ml; resistance wire for electric heater; step-down transformer (110 to 6 volts); thermometer graduated to 0.1°; refractometer with thermostated prism; weighing bottle; benzene; ethanol.

PROCEDURE.   The apparatus which is shown in Fig. 18 may be readily constructed from a 50-ml distilling flask. Superheating is avoided by internal electrical heating with a resistance coil. An alternative apparatus is described by Rogers, Knight, and Choppin.[6]

The heating coil of No. 26 nichrome wire about 14 cm long is wound in the form of a helix about 3 mm in diameter. It is brazed to No. 14 copper wire leads set into the cork. The coil should touch the bottom. A small step-down transformer capable of at least 25 watts output is used.

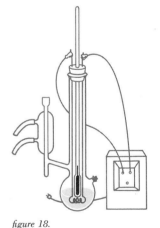

figure 18.

Apparatus for determining liquid and vapor compositions of binary systems as a function of temperature.

Other types and sizes of resistance wire may be used, but the current should be such that the wire is heated to a dull red heat when out in the open air. A heater of 2 ohms operating at 6 volts is satisfactory.

A thermometer graduated to 0.1° and reading from 50 to 100° serves very well, but any accurate thermometer with large 1° divisions will do. A short length of glass tubing surrounds the bulb of the thermometer; this enables the boiling liquid to circulate over the entire thermometer bulb. The bulb must not touch the heating coil.

The arm of the distilling flask is bent upward to act as a reflux condenser; at the bottom of the bend is a bulb of about 1 ml capacity to act as a pocket for retaining condensed distillate as it flows down from the short condenser.

The transformer is adjusted so that the liquid boils vigorously at a constant rate, and the vapor condenses in the reflux condenser. Additional regulation may be accomplished with a Variac if necessary. The boiling is continued until the pocket below the reflux condenser has been thoroughly rinsed out with condensed liquid and the thermometer reading has become constant. The approach to equilibrium is hastened by stirring the liquid in the pocket with a long glass rod. The current is then turned off, and samples of about 1 ml are taken with a small tube or pipette from the distillate in the pocket and then from the residue in the flask through the sidearm. The sample of distillate is removed by inserting the end of the pipette through the open end of the reflux condenser directly into the pocket below. A dry pipette should be used for taking the samples. The refractive indices of the samples are determined with a refractometer. Samples for this determination may be preserved for a short time in small *stoppered* vials or test tubes, but errors caused by partial evaporation of the samples must be considered. It is important to close the jaws of the refractometer quickly to avoid evaporation from the liquid film on the prism.

The Abbe refractometer, illustrated in Fig. 19, makes use of the principle of the grazing angle. The field in the telescope will show a light region and a dark region, the sharp line of demarcation between which corresponds to the grazing angle.

White light from a frosted electric light bulb is used for convenience, and if it were not for the compensating Amici prism of different kinds of glass in the telescope, the line of demarcation between the dark and light fields would be colored and indistinct because the refraction of light is different for different wavelengths. The light of different wavelengths is dispersed by the refractometer prism, by the first compensating prism A', and by the sample of liquid. Since the extent of the dispersion differs for each liquid, the second compensating prism A is adjusted manually so that its dispersion is exactly equal and opposite to the dispersion produced by the refractometer and the liquid. A knurled ring in the middle of the telescope

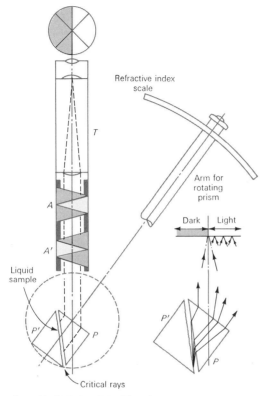

figure 19. Optical path in Abbe refractometer.

barrel is turned until the compensation is complete and the color fringes disappear, leaving a sharp line of demarcation between the two parts of the field.

Although white light is used, the refractive index measured, n_D, is for the D line, 5893 Å, because the Amici compensating prisms are constructed with special glasses so that light of this wavelength is not deviated but all other light is deviated.

The Abbe refractometer has two prisms, the first of which, P', has a ground-glass face. It is used to confine the thin sample of liquid and to illuminate it with scattered light. The upper prism P is the refracting prism. The prisms are jacketed so that the temperature may be controlled to $0.2°$ with water from a thermostat. The refractometer prism is rotated by a protruding arm so as to set the edge of the shadow directly on the intersection of the cross hairs as shown in Fig. 19.

The prisms are opened like jaws after turning the lock nut, and they are wiped with lens tissue paper, care being taken not to scratch the prism surfaces. A few drops of liquid are placed on the face of the lower prism, and the prism jaws are then closed and locked. The compensating ring is turned to eliminate color fringes. The telescope is set in a convenient position, and the mirror is adjusted to reflect the light from a frosted electric lamp into the refractometer. The prism is rotated by means of the arm until the border between the dark and light fields passes exactly through the intersection of the cross hairs. The telescope eyepiece is adjusted until the cross hairs are in good focus, and the eyepiece on the movable arm is adjusted to give a sharp focus on the scale. The scale is graduated directly in terms of refractive index calculated for the glass used in the prism as shown in Fig. 19. The reproducibility of the individual readings on the scale is ±0.0002 in refractive index. Accurate temperature control is important because the refractive indices of many organic liquids change 0.0004 per degree. After a liquid is used, it is absorbed with lens paper or rinsed off with a volatile liquid in which it is soluble.

About 25 ml of benzene is measured into the flask, and its boiling point is determined. Boiling points and refractive indices of the residue and distillate are then determined after successive additions of 0.2, 0.5, 1, 5, 5, and 5 ml of ethanol. The refractive indices are used to obtain the mole fractions of ethanol in these solutions.

In order to construct a plot of refractive index versus mole percent ethanol, the refractive indices are determined for the pure benzene and ethanol and for a series of solutions containing accurately known weights of benzene and ethanol. Mixtures about 5 ml in volume containing approximately 1 volume of ethanol to 1, 3, and 6 volumes of benzene are convenient.

The boiling flask is drained and dried, and about 25 ml of ethanol is introduced for a boiling-point determination. Boiling points and compositions of the residue and distillate are then determined after successive additions of benzene, for example, 2, 4, 5, 7, and 10 ml.

The barometer should be read occasionally. In case the atmospheric pressure changes considerably, it is necessary to estimate a correction for the boiling point, taking an average correction for the two liquids as an approximation. Such a correction may usually be avoided by performing all the distillation experiments within a few hours.

CALCULATIONS.   The refractive indices of the weighed samples and the pure liquids are plotted against the compositions of the solutions expressed in mole fractions of ethanol. The composition of each sample of distillate and residue may then be determined by interpolation on this graph. In a second graph three sets of curves are plotted:

The Boiling-point Diagram (I) for the System as Determined Experimentally. Two curves are plotted, one in which boiling temperature is plotted against the mole fraction of ethanol in the residue; in the other, the same boiling temperatures are plotted against the mole fraction of ethanol in the distillate. The composition as mole-fraction ethanol is plotted along the horizontal axis. Different symbols should be used for the two sets of points. The significance of this graph is discussed with respect to the feasibility of separating benzene and ethanol by fractional distillation.

The Boiling-point Diagram (II) for the System as Predicted by the Ideal-solution Rule. Points for the two curves involved may be calculated as follows for a given pressure P. A temperature T is selected between the boiling points of the two pure liquids as *calculated* from accurate relations such as those given below. The terms $P_1^*(T)$, $P_2^*(T)$ represent the calculated vapor pressures of the pure liquids at this temperature.

$$P = P_1 + P_2 = X_{1,l}P_1^*(T) + X_{2,l}P_2^*(T) \overset{\text{use } X_1 + X_2 = 1}{\nwarrow}$$
$$= P_1^*(T) + X_{2,l}[P_2^*(T) - P_1^*(T)] \quad (10)$$

From Eq. (10) there is then calculated the mole fraction $X_{2,l}$ of component 2 in the solution having vapor pressure P at temperature T. Then the mole fraction $X_{2,v}$ for the equilibrium vapor phase is given by

$$X_{2,v} = \frac{P_2}{P} = \frac{X_{2,l}P_2^*(T)}{P} \quad (11)$$

The Boiling-point Diagram (III) for the System as Predicted by the van Laar Equations for Values of A_1, A_2 Consistent with the Experimentally Determined Azeotrope Temperature and Composition. In this calculation it will be necessary to assume that the activity coefficients are functions of composition only; that is, A_1, A_2 = constants, an approximation justified by the small temperature range involved.

The activity coefficient γ_i is given by relation (7) (for the *ideal-gas* approximation)

$$\gamma_i = \frac{X_{i,v}}{X_{i,l}} \frac{P}{P_i^*(T)}$$

For the azeotropic solution, the mole fraction of each component has the same value for the liquid and vapor phases; hence the activity coefficients $\gamma_{1,az}$ and $\gamma_{2,az}$ for the azeotropic solution are given by

$$\gamma_{1,az} = \frac{P}{P_1^*(T)} \qquad \gamma_{2,az} = \frac{P}{P_2^*(T)} \quad (12)$$

From the pair of activity coefficients so calculated and the composition of the azeotrope, the van Laar coefficients may be calculated. It is convenient first to calculate the ratio A_2/A_1.

$$\frac{A_2}{A_1} = \frac{X_{1,l}^2}{X_{2,l}^2} \frac{\log \gamma_1}{\log \gamma_2} \tag{13}$$

Then

$$A_2 = \left(1 + \frac{A_2}{A_1} \frac{X_{2,l}}{X_{1,l}}\right)^2 \log \gamma_2 \qquad A_1 = \frac{A_2}{A_2/A_1} \tag{14}$$

Now select some concentration $X_{2,l}$, $X_{1,l} = 1 - X_{2,l}$ and calculate from the van Laar equations γ_2 and γ_1.

$$P = P_1 + P_2 = \gamma_1 X_{1,l} P_1^*(T) + \gamma_2 X_{2,l} P_2^*(T) \tag{15}$$

By successive approximation, determine the temperature T for which this equation is satisfied for the known pressure P. Then

$$X_{2,v} = \frac{P_2}{P} = \frac{\gamma_2 X_{2,l} P_2^*(T)}{P} \tag{16}$$

Several sets of points should be calculated in this way for compositions on each side of the azeotropic composition. Such calculations as these may be greatly expedited by use of modern computing methods. A typical computer program for this purpose

Table 1. *Vapor-pressure Data for Benzene, Standard Millimeters of Mercury*

$\log P = 6.90522 - \dfrac{1211.215}{t + 220.87}$ Ref. 7

t, °C	0.0	0.1	0.2	0.3	0.4	0.5	0.6	0.7	0.8	0.9
65	465.9	467.5	469.1	470.7	472.3	473.9	475.5	477.2	478.8	480.4
66	482.1	483.7	485.3	487.0	488.6	490.3	491.9	493.6	495.3	496.9
67	498.6	500.3	502.0	503.7	505.4	507.1	508.8	510.5	512.2	513.9
68	515.6	517.3	519.1	520.8	522.6	524.3	526.0	527.8	529.6	531.3
69	533.1	534.8	536.6	538.4	540.2	542.0	543.8	545.6	547.4	549.2
70	551.0	552.8	554.6	556.5	558.3	560.2	562.0	563.8	565.7	567.5
71	569.4	571.3	573.1	575.0	576.9	578.8	580.7	582.6	584.5	586.4
72	588.3	590.2	592.1	594.1	596.0	597.9	599.9	601.8	603.8	605.7
73	607.7	609.6	611.6	613.6	615.6	617.5	619.5	621.5	623.5	625.5
74	627.6	629.6	631.6	633.6	635.6	637.7	639.7	641.8	643.8	645.9
75	647.9	650.0	652.1	654.1	656.2	658.3	660.4	662.5	664.6	666.7
76	668.8	670.9	673.1	675.2	677.3	679.5	681.6	683.8	685.9	688.1
77	690.3	692.4	694.6	696.8	699.0	701.2	703.4	705.6	707.8	710.0
78	712.2	714.4	716.7	718.9	721.2	723.4	725.7	727.9	730.2	732.5
79	734.7	737.0	739.3	741.6	743.9	746.2	748.5	750.8	753.1	755.5
80	757.8	760.1	762.5	764.8	767.2	769.6	771.9	774.3	776.7	779.1

Table 2. Vapor-pressure Data for Ethanol, Standard Millimeters of Mercury

$$\log P = 8.11576 - \frac{1596.76}{t + 226.5} \qquad \text{Ref. 8}$$

$t, °C$	0.0	0.1	0.2	0.3	0.4	0.5	0.6	0.7	0.8	0.9
65	437.9	439.8	441.7	443.6	445.5	447.5	449.4	451.3	453.3	455.2
66	457.2	459.2	461.1	463.1	465.1	467.1	469.1	471.1	473.1	475.1
67	477.2	479.2	481.3	483.3	485.4	487.5	489.5	491.6	493.7	495.8
68	497.9	500.0	502.1	504.3	506.4	508.5	510.7	512.9	515.0	517.2
69	519.4	521.6	523.8	526.0	528.2	530.4	532.6	534.9	537.1	539.4
70	541.6	543.9	546.2	548.5	550.7	553.0	555.4	557.7	560.0	562.3
71	564.7	567.0	569.4	571.7	574.1	576.5	578.9	581.3	583.7	586.1
72	588.5	590.9	593.4	595.8	598.3	600.8	603.2	605.7	608.2	610.7
73	613.2	615.7	618.2	620.8	623.3	625.9	628.4	631.0	633.6	636.2
74	638.8	641.4	644.0	646.6	649.2	651.9	654.5	657.2	659.8	662.5
75	665.2	667.9	670.6	673.3	676.0	678.8	681.5	684.3	687.0	689.8
76	692.5	695.3	698.1	700.9	703.7	706.6	709.4	712.2	715.1	718.0
77	720.8	723.7	726.6	729.5	732.4	735.3	738.3	741.2	744.1	747.1
78	750.1	753.0	756.0	759.0	762.0	765.1	768.1	771.1	774.2	777.2
79	780.3	783.4	786.4	789.5	792.7	795.8	798.9	802.0	805.2	808.3
80	811.5	814.7	817.9	821.1	824.3	827.5	830.7	834.0	837.3	840.5

is given in the section in Part Two on Treatment of Experimental Data (page 452).

The required data for the vapor pressures of benzene and ethanol will be found in Tables 1 and 2.

Practical applications. Vapor-composition curves are necessary for the efficient separation of liquids by distillation. Fractional distillation under controlled conditions is essential in the purification of liquids and in many industries, such as the petroleum industry and solvent industries.[7-9]

Suggestions for further work. Solutions of chloroform and acetone, giving a maximum in the boiling-point curve, may be studied in exactly the same manner as described for ethanol and benzene.

The maximum in the boiling-point curve of hydrochloric acid and water occurs at 108.5° and a composition of 20.2 percent hydrochloric acid at a pressure of 760 mm. The distillate at the maximum boiling point is so reproducible in composition at a given pressure and so easily obtained that it may be used to prepare solutions of HCl for volumetric analysis. A solution of hydrochloric acid is made up roughly to approximate the constant-boiling composition, and after boiling off the first third, the remaining distillate is retained. The barometer is read accurately, and the corresponding composition is obtained from the literature.[10,11]

Solutions of chloroform and methanol, giving a minimum in the boiling-point curve, may be studied by using a Westphal density balance for determining the compositions instead of a refractometer. A density–mole-fraction curve is plotted, and the compositions of the samples are determined by interpolation. Since larger samples are needed for the density measurements, more material and a larger flask are required.

The gas-saturation method for vapor-pressure measurements may be used in studying binary liquids. Using this technique, Smyth and Engel[12] have determined vapor-pressure–composition curves for a number of ideal and nonideal types.

Vapor-liquid equilibria at different total pressures provide an interesting study. The acetonitrile-water system has an azeotrope which varies considerably in composition as the pressure is reduced.[13] Othmer and Morley[14] describe an apparatus for the study of vapor-liquid compositions at pressures up to 500 psi. The earlier papers of Othmer may be consulted for a number of binary vapor-liquid equilibria.

References

1. E. Hala, J. Pick, V. Fried, and O. Vilim, "Vapor-Liquid Equilibrium," Pergamon Press, New York, 1958.
2. K. G. Denbigh, "The Principles of Chemical Equilibrium," 2d ed., Cambridge University Press, New York, 1966.
3. H. C. Carlson and A. P. Colburn, *Ind. Eng. Chem.*, **34:** 581 (1942).
4. L. H. Horsley et al., Azeotropic Data, *Advan. Chem. Ser.*, **6** (1952), **35** (1962); Tables of Azeotropes and Non-azeotropes, *Anal. Chem.*, **19:** 508–600 (1947).
5. J. Timmermans, "Physico-chemical Constants of Binary Systems in Concentrated Solutions," Interscience Publishers, Inc., New York, 1959.
6. J. W. Rogers, J. K. Knight, and A. R. Choppin, *J. Chem. Educ.*, **24:** 491 (1947).
7. E. R. Smith, *J. Res. Natl. Bur. Std. U.S.*, **26:** 129 (1941).
8. C. B. Kretschmer and R. Wiebe, *J. Am. Chem. Soc.*, **71:** 1793 (1949).
9. J. M. Prausnitz, C. A. Eckert, R. V. Orye, and J. P. O'Connell, "Computer Calculations for Multicomponent Vapor-Liquid Equilibria," Prentice-Hall, Inc., Englewood Cliffs, N.J., 1967.
10. W. D. Bonner and B. F. Branting, *J. Am. Chem. Soc.*, **48:** 3093 (1926).
11. C. W. Foulk and M. Hollingsworth, *J. Am. Chem. Soc.*, **45:** 1220 (1923).
12. C. P. Smyth and E. W. Engel, *J. Am. Chem. Soc.*, **51:** 2646, 2660 (1929).
13. D. F. Othmer and S. Josefowitz, *Ind. Eng. Chem.*, **39:** 1175 (1947).
14. D. F. Othmer and F. R. Morley, *Ind. Eng. Chem.*, **38:** 751 (1946).

10   FRACTIONAL DISTILLATION

In this experiment the efficiencies of packed and unpacked columns are compared at total reflux. The separation of a binary mixture by fractional distillation is studied by using refractive-index measurements to analyze the distillate.

THEORY.   The separation of liquids by distillation is one of the oldest operations of chemistry, but considerable improvement has been made in the understanding of the process and in the design of apparatus for separating materials by fractional distillation.[1-6] The developments in petroleum refining and the need for increased efficiency in laboratory operations and purification have been largely responsible for these improvements. The concentration of isotopes has made still greater demands on fractional distillation.

The separation of two liquids which is obtained by a simple vaporization and condensation is not effective except in the case of liquids with widely differing boiling points. Greater separation may be achieved by a series of simple distillations, but this is laborious. The same result is obtained by using a fractionating column through which the vapor is passed and brought into contact with part of the condensate flowing down the column. The less volatile components in the ascending vapor are condensed in such a column, and the more volatile components are boiled out of the descending liquid phase, so that distillation through the column is equivalent to a number of successive simple distillations. Greater separation is obtained in a fractionating column if most of the vapor condensed at the top of the column is returned as reflux, flowing back into the top of the column. The reflux ratio is defined as the ratio of the volume of liquid flowing back into the column to the volume of liquid removed as distillate.

It is necessary to insulate or heat a fractionating column so that the net condensation in the column will not be too great. The temperature will be lower at the top of the column, where the more volatile component is concentrated, than at the bottom. The purpose of the packing is to provide good contact between the vapor and liquid phases in the column, but it is undesirable for the packing to hold a large fraction of the batch being distilled because of the resulting decrease in sharpness of separation.

The effectiveness of a distilling column is expressed in terms of the theoretical plate. The *theoretical plate* is a hypothetical section of column which produces a separation of components such that the vapor leaving the top of the section has the composition of the vapor which is in equilibrium with the liquid leaving the bottom of the section. A column consisting of a simple 1-cm tube 1 m long might be equivalent to only 1 theoretical plate, whereas the same tube filled with adequate packing can give the equivalent of 20 or more theoretical plates. A column with 12 theoretical plates is adequate for the practical separation of benzene (bp 80.1°) and toluene (bp 110.8°). The number of theoretical plates required for a given separation increases when the reflux ratio is decreased.

The number of theoretical plates cannot be determined from the geometry of the distilling column; it is calculated from the separation effected with a liquid mixture for which the liquid-vapor equilibrium data are fully known. As an example, the determination of the number of theoretical plates in a column by distilling a mixture of carbon tetrachloride and benzene is illustrated in Fig. 20.

The ordinates of the upper curve give the experimentally determined mole percent of carbon tetrachloride in the vapor which is in equilibrium with the liquid having the composition given on the abscissa. Temperatures are not involved in this

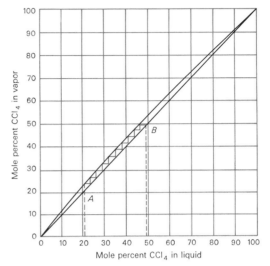

figure 20. Vapor-liquid graph for calculating the number of theoretical plates.

diagram. They vary from the boiling point of pure benzene to that of pure carbon tetrachloride.

Suppose, for example, that a sample of the distillate obtained under conditions of practically total reflux had composition *B* and the residue in the distilling flask had composition *A*, as indicated by the dotted lines. A series of vertical and horizontal lines is drawn stepwise from *A* as shown until the composition of the distillate is reached. Each vertical line represents an ideal distillation step in which the intersection with the upper line gives the composition of the vapor which is in equilibrium with the liquid indicated by the intersection of the vertical line with the lower line. Each horizontal line represents complete condensation of all the vapor to give a liquid of the same composition as the vapor. The number of these vertical-line steps minus 1 is equal to the number *n* of theoretical plates in the fractionating column. The liquid-vapor surface in the distillation pot acts as one theoretical plate.

Depending upon the construction of the distillation column, the number of theoretical plates may vary somewhat with the rate of entry of vapor into the bottom of the column and the rate of return of liquid from the top of the column. In the case of small-scale laboratory columns, it is found that the actual separations obtainable at finite reflux ratios are, in general, lower than would be predicted from the number of theoretical plates determined at total reflux.

The number of theoretical plates in a column under actual operating conditions may be determined by the method of McCabe and Thiele.[7,8] In this method the number of theoretical plates is obtained by plotting a curve representing mole fraction of the more volatile component in the binary mixture in the vapor phase versus its mole fraction in the liquid phase and counting steps between this curve and the *operating line,* rather than the 45° line, which is used for the calculation of theoretical plates at total reflux. The operating line is a straight line drawn through the 45° line at the composition of the distillate with a slope equal to $R/(1 + R)$, where R is the reflux ratio. The student is referred to chemical engineering texts for a complete description of the use of this important concept in actual separations by fractional distillation.

····→ ***Apparatus.*** Vigreux column; vacuum-jacketed packed columns; still head with cold thumb and thermometer; glass-cloth-covered heating mantle; variable autotransformer; distilling flask; 100-ml graduated cylinder; small test tubes and corks; Abbe refractometer; carbon tetrachloride; chloroform; benzene.

PROCEDURE.   Two types of fractionating columns are used: (*a*) an unpacked column of the Vigreux type with small inwardly protruding cones, sloping downward, punched in the wall, and (*b*) a vacuum-insulated column of the bubble-plate type or a column packed with double cones of metal screen (Stedman), stainless-steel saddles, or glass helices. A distillation apparatus with a vacuum-jacketed fractionating column is illustrated in Fig. 21. The unpacked column and one of the packed columns are compared by determining the number of theoretical plates for each at total reflux, using solutions of benzene and carbon tetrachloride. Samples from the distilling pot and distillate are analyzed, using measurements of refractive index obtained with an Abbe refractometer.

About 10 ml of carbon tetrachloride and 40 ml of benzene are mixed and placed in the distillation pot, which is attached to the Vigreux column. The distilling pot is heated by an electric heating mantle controlled with a variable autotransformer. The liquid is boiled vigorously until condensation takes place in the top of the distilling column, and then the heating is decreased so that the column is no longer flooded. The cold-finger condenser should be in such a position that all the condensate is returned to the column.

After equilibrium is attained, as indicated by the fact that the thermometer reading is constant, a sample of the material is taken from the top of the column. The cold-finger condenser is withdrawn to a position such that the condensed liquid drips into the graduated cylinder and is collected. The first two or three drops are discarded, and then a 1-ml sample is collected for analysis with the Abbe refractometer (page 64). The electric heater is then turned off, and a sample of the liquid (about 1 ml) is removed through the sidearm of the distillation pot with a pipette. It is to be emphasized that the efficiency of the column as determined by use of these two samples is the efficiency essentially at total reflux, since the column is brought to

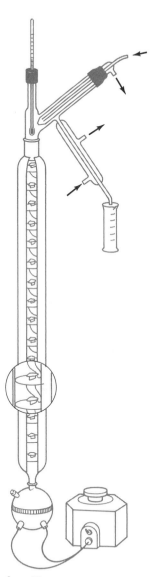

figure 21.

Bubble-plate fractionating column.

equilibrium at total reflux and only a small sample is withdrawn from the column.

The above procedure is repeated using a packed fractionating column or a bubble-plate column (page 73) rather than the Vigreux column. The packed column requires a longer time to come to equilibrium.

While equilibrium is being established in the fractionating column, refractive-index measurements with an Abbe refractometer are made on pure benzene, pure carbon tetrachloride, and two or three mixtures of known composition. These measurements are used for the determination of the compositions of samples of distillate and residue.

The packed column is used next to demonstrate the separation of two liquids. The column and distilling pot are emptied and dried, and 30 ml of chloroform and 30 ml of benzene are introduced. The reflux condenser is set for a reflux ratio of from 5:1 to 10:1, as estimated from the rates of dripping from the reflux condenser and from the distillation tube. The distillate is collected in the graduated cylinder. A 1-ml sample of the distillate is collected in a small stoppered bottle after every 3 ml of distillate, and the samples are then used for refractive-index measurements. A record is kept of the total volume of liquid distilled and the thermometer reading at the time of taking each sample. The final total volume of the distillate is also recorded.

CALCULATIONS.   The refractive indices of benzene, of carbon tetrachloride, and of the known mixtures of the two are plotted on coordinate paper against the mole percent of carbon tetrachloride. A smooth curve is drawn to represent these data and used to determine the composition of an unknown liquid mixture by interpolation of the refractive index.

The numbers of theoretical plates effective in the Vigreux column and in the packed column at total reflux are calculated with the help of a large graph in which the mole percent of the more volatile component in the vapor is plotted against the mole fraction of this component in the liquid, as indicated in Fig. 20. The data required for the benzene–carbon tetrachloride system are given in Table 1.

The efficiency of various types of packing and construction of the fractionating column may be compared by calculating the length equivalent to one theoretical plate. The length of column per theoretical plate is called the *height equivalent per theoretical plate* (HETP). This value is calculated for the various columns used.

The effectiveness of the chloroform-benzene distillation is illustrated by plotting the temperatures of the condensing distillate against the percentage of the total volume of the mixture distilled. The refractive indices of the samples of distillate are also plotted on the same graph against the percentage of the total volume of liquid distilled. For a column with a very large number of theoretical plates operated at a high reflux ratio, the refractive index of the distillate changes abruptly from that of the more volatile to that of the less volatile component when the volatile component has all distilled out. Likewise, the temperature of distillation rises abruptly when the more volatile component is distilled out.

Table 1. *Liquid and Vapor Composition of Mixtures of Carbon Tetrachloride and Benzene at 760 mm and at Temperatures between the Boiling Points*[8]

Mole percent CCl₄ in liquid	0	13.64	21.57	25.73	29.44	36.34	40.57	52.69	62.02	72.2
Mole percent CCl₄ in vapor	0	15.82	24.15	28.80	32.15	39.15	43.50	54.80	63.80	73.3

Suggestions for further work. Additional pairs of liquids may be separated by fractionation with an efficient column. A mixture of carbon tetrachloride and toluene may be used to determine the number of theoretical plates. The data for this system are given in Table 2.

Table 2. *Liquid and Vapor Compositions of Mixtures of Carbon Tetrachloride and Toluene*[8]

Mole percent CCl₄ in liquid	0	5.75	16.25	28.85	42.60	56.05	64.25	78.20	94.55
Mole percent CCl₄ in vapor	0	12.65	31.05	49.35	64.25	75.50	81.22	89.95	97.35

Some of the various types of packing referred to in Chapter 23 may be compared by determining the HETP for each.

The value of a fractionating column depends not only on the number of theoretical plates but also on the amount of liquid held up by the packing.[5,9] Equilibrium conditions are attained more rapidly if the holdup of the column is small. The amount of liquid held up may be determined at the end of an experiment by removing the heating bath, taking out the column and blowing dry air through it, and condensing the material in a weighed U tube surrounded by a freezing bath of Dry Ice. When the packing is completely dry, the increase in weight of the U tube gives the weight of the liquid held up in the column.

References

1. W. L. Badger and J. T. Banchero, "Introduction to Chemical Engineering," McGraw-Hill Book Company, New York, 1955.
2. W. A. Gruse and D. R. Stevens, "Chemical Technology of Petroleum," 3d ed., chap. 6, pp. 255–282, McGraw-Hill Book Company, New York, 1960.
3. T. P. Carney, "Laboratory Fractional Distillation," The Macmillan Company, New York, 1949.
4. A. A. Morton, "Laboratory Techniques in Organic Chemistry," McGraw-Hill Book Company, New York, 1938.
5. C. S. Robinson and E. R. Gilliland, "Elements of Fractional Distillation," 4th ed., McGraw-Hill Book Company, New York, 1950.
6. T. E. Williams and A. L. Glazebrook in A. Wiessberger (ed.), "Technique of Organic Chemistry," vol. 4, E. S. Perry and A. Weissberger (eds.), "Distillation," 2d ed., chap. 2, Interscience Publishers, Inc., New York, 1965.
7. W. L. McCabe and E. W. Thiele, *Ind. Eng. Chem.,* **17:** 605 (1925).

8. R. H. Perry, C. H. Chilton, and S. D. Kirkpatrick (eds.), "Chemical Engineers Handbook," 4th ed., McGraw-Hill Book Company, New York, 1963.
9. W. L. Collins and E. W. Lantz, *Ind. Eng. Chem. Anal. Ed*, **18:** 673 (1946).

11   *VARIATION OF AZEOTROPE COMPOSITION WITH PRESSURE*

The variation of azeotrope composition with pressure is determined for a binary system and the results compared with predictions based on thermodynamic principles. Measurements of heats of mixing of the two pure liquids are made, and their relation to the effect of temperature on the activity coefficients is considered.

THEORY.[1-3]   The Gibbs free energy G of a single homogeneous phase containing two components is a function of temperature, pressure, and the numbers of moles n_1, n_2 of the two components. For a change of state the differential change in Gibbs free energy is given by

$$dG = -S \, dT + V \, dP + \mu_1 \, dn_1 + \mu_2 \, dn_2 \tag{1}$$

where S is the entropy of the phase, V its volume, and μ_1, μ_2 the chemical potentials of its two components. Since at any temperature and pressure the free energy of the phase is given by

$$G = n_1\mu_1 + n_2\mu_2 \tag{2}$$

the differential dG is also given by

$$dG = n_1 \, d\mu_1 + n_2 \, d\mu_2 + \mu_1 \, dn_1 + \mu_2 \, dn_2 \tag{3}$$

Combination of Eqs. (1) and (3) yields the Gibbs-Duhem equation,

$$S \, dT - V \, dP + n_1 \, d\mu_1 + n_2 \, d\mu_2 \equiv 0 \tag{4}$$

which interrelates the changes in the temperature, the pressure, and the chemical potentials of the components for the given change of state. This relation may be written in terms of intensive quantities by dividing by $n_1 + n_2$, the total number of moles, to obtain

$$\overline{S} \, dT - \overline{V} \, dP + X_1 \, d\mu_1 + X_2 \, d\mu_2 \equiv 0 \tag{5}$$

where $\overline{S}, \overline{V}$ are entropy and volume per mole of solution, respectively, and X_i is the mole fraction of component i.

Consider now the simultaneous application of this relation to a liquid phase (subscript l) and the vapor phase (subscript v) in equilibrium with it:

$$\overline{S}_l \, dT - \overline{V}_l \, dP + X_{1,l} \, d\mu_{1,l} + X_{2,l} \, d\mu_{2,l} \equiv 0 \tag{6}$$

$$\bar{S}_v \, dT - \bar{V}_v \, dP + X_{1,v} \, d\mu_{1,v} + X_{2,v} \, d\mu_{2,v} \equiv 0 \tag{7}$$

Since for equilibrium the chemical potential for each component must have a common value for both phases,

$$\mu_{1,v} = \mu_{1,l} = \mu_1 \qquad \mu_{2,l} = \mu_{2,v} = \mu_2$$

and Eqs. (6) and (7) may be combined to give

$$(\bar{S}_v - \bar{S}_l) \, dT - (\bar{V}_v - \bar{V}_l) \, dP + (X_{2,l} - X_{2,v})(d\mu_1 - d\mu_2) \equiv 0 \tag{8}$$

This equation determines the relation between the changes in temperature, in pressure, and in chemical potentials in going from one vapor-liquid equilibrium condition to another. Now consider a variation in composition at constant temperature $(dT = 0)$ away from a state for which the equilibrium vapor and liquid compositions are identical (the azeotropic condition), so that $X_{2,l} - X_{2,v} = 0$. Since in general $\bar{V}_v - \bar{V}_l$ is not zero, it follows that $dP = 0$ for the indicated differential change, and hence there must be a minimum or maximum in the plot of equilibrium pressure versus composition, at constant temperature, at the azeotrope composition. Similarly, for a shift in composition at constant pressure, since $\bar{S}_v - \bar{S}_l \neq 0$, there must be a maximum or minimum at the azeotrope composition in the plot of equilibrium temperature versus composition at constant pressure (the boiling-point diagram).

The azeotrope composition changes as the equilibrium temperature and pressure are changed. If the composition is known for one pressure, the corresponding azeotrope composition for another pressure can be estimated in the following way. As shown in Experiment 9, to the ideal-gas approximation for the vapors involved is added the activity coefficient for a component in the liquid phase, which is given by

$$\gamma_i = \frac{X_{i,v}P}{X_{i,l}P_i^*(T)} \tag{9}$$

where P = pressure of equilibrium vapor phase
$P_i^*(T)$ = vapor pressure of pure-liquid component i at temperature involved
$X_{i,v}, X_{i,l}$ = mole fractions of component i in vapor and liquid phases, respectively
Let P_{az} represent the vapor pressure of the azeotrope (this of course corresponds to the constant pressure P for a boiling-point diagram), for which $X_{i,l} = X_{i,v}$. Then for the azeotropic solution

$$\gamma_{1,az} = \frac{P_{az}}{P_1^*(T)} \qquad \gamma_{2,az} = \frac{P_{az}}{P_2^*(T)} \tag{10}$$

and hence

$$\frac{\gamma_{1,az}}{\gamma_{2,az}} = \frac{P_2^*(T)}{P_1^*(T)} \tag{11}$$

The azeotrope composition thus is that composition for which the ratio of the activity coefficients of the two components is equal to the inverse ratio of the vapor pressure of the two pure liquids at the temperature concerned. If the azeotrope composition and pressure are known for one temperature, the coefficients A_1, A_2 can be calculated for the given T, P, for the van Laar equations:

$$\log \gamma_1 = \frac{A_1}{[1 + (A_1 X_{1,l} / A_2 X_{2,l})]^2} \qquad \log \gamma_2 = \frac{A_2}{[1 + (A_2 X_{2,l} / A_1 X_{1,l})]^2} \tag{12}$$

If A_1 and A_2 as thus determined are assumed to be independent of temperature and pressure, i.e., if the activity coefficients are considered to be functions of composition only, then it is possible to calculate the mole fraction of component 1 for the azeotrope at a different temperature as the value of $X_{1,l}$ for which Eq. (11) is satisfied at this new temperature. The new azeotrope pressure P_{az} can be calculated by use of the relation

$$\gamma_{1,az} = \frac{P_{az}}{P_1^*(T)}$$

The accuracy of the prediction so made necessarily depends on how well the van Laar equations express the concentration dependence of the activity coefficients at a given temperature and pressure; for satisfactory results the azeotropes involved should be reasonably intermediate in composition.

There remains the question of the validity of the assumption that the activity coefficients are insensitive to changes in temperature and pressure at constant composition. From Eq. (12) it is seen that

$$A_1 = \lim_{X_{1,l} \to 0} \log \gamma_1 \qquad A_2 = \lim_{X_{2,l} \to 0} \log \gamma_2 \tag{13}$$

Hence

$$\left(\frac{\partial A_1}{\partial T} \right)_P = \left[\left(\frac{\partial \log \gamma_1}{\partial T} \right)_P \right]_{X_{1,l} \to 0} \tag{14}$$

From the basic relation between the activity coefficient, concentration, and chemical potential,

$$RT \ln \gamma_1 X_{1,l} = \mu_{1,l} - \mu_{1,l}^{\circ}$$

and thus

$$\left(\frac{\partial \log \gamma_1}{\partial T} \right)_{P,X_{1,l}} = -\frac{1}{2.303R} \frac{\overline{H}_1 - \overline{H}_1^{\circ}}{T^2} \tag{15}$$

where the term $\overline{H}_1 - \overline{H}_1^{\circ}$ actually represents the differential heat of solution (cf. Experiment 4) of pure-liquid component 1 in the solution of concentration $X_{1,l}$ at the given temperature and pressure. The temperature coefficient of A_1 is thus determined by the limiting value of the differential heat of solution of component 1 at infinite dilution in component 2 as solvent.

$$\left(\frac{\partial A_1}{\partial T}\right)_P = \frac{-1}{2.303RT^2} \, [\overline{H}_1(X_{1,l}) - \overline{H}_1^\circ]_{X_{1,l}\to 0} \tag{16}$$

The bracketed enthalpy term can be calculated from integral-heat-of-solution data; it can be approximated by measuring the heat of solution per mole of component 1 in a sufficiently large excess of component 2 to give a very dilute product solution. Its value may be as large as several kilocalories per mole or more, particularly when hydroxylic compounds, such as alcohols, are involved. Similarly, the pressure coefficient of A_1 is found as follows:

$$\left(\frac{\partial \log \gamma_1}{\partial P}\right)_T = \frac{1}{2.303RT} \left[\left(\frac{\partial \mu_{1,l}}{\partial P}\right)_{T,X_{1,l}} - \left(\frac{\partial \mu_{1,l}^\circ}{\partial P}\right)_{T,X_{1,l}}\right] \tag{17}$$

$$= \frac{1}{2.303RT} \, [\overline{V}_1(X_1) - \overline{V}_1^\circ] = \frac{1}{2.303RT} \, [\overline{V}_1(X_{1,l}) - V_1^*] \tag{18}$$

and

$$\left(\frac{\partial A_1}{\partial P}\right)_T = \frac{\overline{V}_1(X_{1,l}\to 0) - \overline{V}_1^*}{2.303RT} \tag{19}$$

where $\overline{V}_1^\circ = \overline{V}_1^*$, the molar volume of pure liquid 1 at the temperature and pressure of the solution. The partial molar volume of component 1 at infinite dilution in component 2 as solvent, $\overline{V}_1(X_{1,l}\to 0)$, can be calculated from density measurements on the solutions (cf. Experiment 14). Since for solutions of this type $\overline{V}_1$ changes but slightly with concentration and is usually not much different from $\overline{V}_1^*$, the pressure coefficient of A_1 will be very small.

····→ *Apparatus.* Fractional distillation column equipped for vacuum distillations; distilling flask; heating mantle and autotransformer; pressure-control system; pycnometers or Westphal balance; sample vials; Dewar calorimeter, thermistor thermometer, or sensitive mercury thermometer; distilled water; acetonitrile, ethanol.

PROCEDURE. An aqueous solution containing acetonitrile at a mole fraction of approximately 0.75 is prepared. The quantity of solution required will depend upon the holdup of the fractionating column used but should not exceed 150 ml. The solution is placed in the distilling flask, which is then connected to the fractionating column. The water valve is opened *slowly* to start the flow of cooling water through the distilling head, which is set for total reflux. The autotransformer switch is turned on, and the control knob advanced from zero as required to raise the temperature of the solution to the boiling point. The boil-up rate is increased until condensation takes place in the distilling head, with a steady but not excessive stream of condensate returned to the column. When the temperature shown by the thermometer immersed in the vapor at the top of the column has become constant, indicating that equilibrium has been reached, a preliminary sample of condensate is drained into the first receiver (and subsequently discarded) and the same collector is rotated to obtain a sample for analysis in the second receiver. The barometric pressure and condensation temperature are recorded. Analysis of this sample gives the azeotrope composition for the measured temperature and pressure, since the fractionating column separates out the minimum-boiling azeotrope due to its high volatility. If the analysis is not made immediately, the sample container should be tightly stoppered.

The head is again set for total reflux, and the autotransformer is turned off. The sample-collecting unit is removed, and the container holding the azeotrope sample is tightly stoppered and set aside for later analysis. The sample collector is cleaned, dried, reassembled, and put back in place on the column, which is then connected to the pressure-control system. The pressure in the column is then *gradually* reduced to about 500 mm and maintained there. The liquid boil-up rate is increased, and the column is brought to equilibrium as before. A sample for analysis is then obtained as described above; the pressure and condensation temperature are recorded. The column is reset to total reflux, and the heating mantle turned off. The pressure in the system is now raised slowly to atmospheric. The second azeotrope sample is set aside for analysis, and the apparatus prepared for the next run. Further determinations are then made for pressures of approximately 350 and 200 mm. Care should be taken to record the condensation temperature and pressure for each azeotropic solution collected.

The densities of the samples are measured by means of the Westphal balance (page 495) or by the pycnometric method (page 98), and their concentrations are determined by reference to a working curve of density versus mole fraction of acetonitrile as found for samples of known composition. The data for this curve may be obtained during the periods in which the column is coming to equilibrium.

A measurement is made of the integral heat of solution of liquid acetonitrile in water to give a product solution in which the mole fraction of acetonitrile is 0.05 or less. A Dewar calorimeter of the type used in Experiment 4 or 5 may be used. Fifty milliliters of distilled water (or other accurately measured quantity, as required to cover the thermometer bulb completely) is placed in the calorimeter. The thermometer reading is noted, and a sample of acetonitrile is adjusted to the same temperature level by warming or cooling in a water bath. Ten milliliters of acetonitrile is then pipetted into the calorimeter, the solution is stirred to make it homogeneous, and the change in temperature is determined as described under Experiment 5. In this case the temperature change will be small. The temperature change observed and the quantities of reactants used are recorded.

In a second experiment the same procedure is followed with 100 ml of acetonitrile as solvent and 1 ml of water added as solute.

The heat capacity of the calorimeter used may be determined as described under Experiment 5 if the value is not given for the unit provided.

CALCULATIONS.   From the experimentally determined azeotrope composition at atmospheric pressure, the van Laar coefficients are calculated (cf. Experiment 9). The necessary vapor-pressure data for acetonitrile may be taken from the work of Heim as summarized by Timmermans,[4] and for water from handbooks or the "International Critical Tables."[5] The ratio γ_1/γ_2 is calculated for various values of $X_{1,l}$, the mole fraction of acetonitrile, from 0.6 to 0.9. The ratio $P_2^*(t)/P_1^*(t)$ of the vapor pressures of the two pure liquids is calculated for a series of temperatures in the range from 30 to 90°C. On the same piece of graph paper two curves are plotted, with the same vertical scale in each case:

1. γ_1/γ_2 as ordinate versus $X_{1,l}$ as abscissa
2. $P_2^*(t)/P_1^*(t)$ as ordinate versus t as abscissa

For a selected temperature t' the predicted azeotrope composition is found by reading off the value of $X_{1,l}$ for which $\gamma_1/\gamma_2 = P_2^*(t)/P_1^*(t)$. For this composition the activity coefficient $\gamma_{1,az}$ for acetonitrile in the azeotropic solution is calculated from the van Laar equation, and the azeotrope pressure $P_{az}(t')$ is predicted from the relation

$$\gamma_{1,az} = \frac{P_{az}(t')}{P_1^*(t')}$$

In this way, results are obtained for a comparison with the experimentally determined combinations of azeotrope pressure, temperature, and composition, which is prepared in tabular form.

The heat of solution, for the common temperature of the reactants, of the species added as solute in a given case is calculated from the relation (cf. Experiment 5)

$$\Delta H = -C_P \, \Delta T$$

where $\Delta T =$ observed change in temperature
$C_P =$ total heat capacity of system

Here C_P is the sum of the heat capacities of the product solution and calorimeter. The former is to be calculated as the sum of the heat capacities[4] of the pure liquids, an approximation which is adequate for the present purposes. The latter will be furnished for the unit provided. The integral heat of solution is then calculated by dividing the observed ΔH by the number of moles of solute added. It is now assumed that the product solution is dilute enough for the integral heat of solution and differential heat of solution to differ negligibly and that their common value will not change much for work at lower concentrations. The temperature coefficient of the van Laar parameter A_1 is then calculated by use of Eq. (16). In a similar fashion the temperature coefficient for the quantity A_2 is calculated. The implications of these results (concerning the assumption earlier made that the activity coefficients are functions of composition only) are considered. It should be recognized that the heat-of-solution data, here obtained at room temperature, are subject to some change with temperature.

A summary of experimental results for the acetonitrile-water system has been given by Timmermans.[6] Of particular pertinence are the papers of Othmer and Josefowitz,[7] Maslan and Stoddard,[8] and Vierk.[9] The heat-of-mixing data given in the latter reference were calculated from temperature-pressure-composition data for the system and do not result from direct calorimetric measurements.

Practical applications. The prediction of the behavior of real-solution systems from a minimum of experimental data is a common problem in modern technology.

Suggestions for further work. The heat-of-solution measurements may be made for a range of concentrations to permit a more accurate estimation of the differential heats of solution at infinite dilution. Similar measurements may be made for the system acetonitrile-ethanol and the results compared with those calculated from the heats of mixing given by Thacker and Rowlinson.[10] The term *heat of mixing* ΔH_m, as regularly employed in solution thermochemistry, refers to the increase in enthalpy in the formation of 1 mole of *solution* from the pure-liquid components at the temperature and pressure of the solution.

References

1. E. A. Guggenheim, "Thermodynamics," pp. 188–190, Interscience Publishers, Inc., New York, 1949.
2. E. Hala, J. Pick, V. Fried, and O. Vilm, "Vapor-Liquid Equilibrium," Pergamon Press, New York, 1958.
3. O. A. Hougen and K. M. Watson, "Chemical Process Principles," chap. 15, John Wiley & Sons, Inc., New York, 1949.
4. J. Timmermans, "Physico-chemical Constants of Pure Organic Compounds," Elsevier Press, Inc., Houston, Texas, 1950.
5. "International Critical Tables," vol. III, p. 212, McGraw-Hill Book Company, New York, 1928.
6. J. Timmermans, "Physico-chemical Constants of Binary Systems," vol. IV, Interscience Publishers, Inc., New York, 1960.
7. D. F. Othmer and S. Josefowitz, *Ind. Eng. Chem.*, **39:** 1175 (1947).
8. F. D. Maslan and E. A. Stoddard, *J. Phys. Chem.*, **60:** 1147 (1956).
9. A. Vierk, *Z. Anorg. Chem.*, **261:** 283 (1950).
10. R. Thacker and J. S. Rowlinson, *Trans. Faraday Soc.*, **50:** 1036 (1954).

12   ELEVATION OF THE BOILING POINT

The boiling points of a solution and of the pure solvent are determined and used for calculating the molecular weight of a nonvolatile solute.

THEORY.   When a *nonvolatile* solute is dissolved in a solvent, the vapor pressure of the latter is decreased; as a consequence, the boiling point of the solution is higher than that of the pure solvent. The extent of the elevation θ depends upon the concentration of the solute, and for *dilute, ideal* solutions it may be shown that

$$T_b - T_0 = \theta = K_b m \tag{1}$$

where

$$K_b = \frac{RT_0^2}{1000\lambda_v} \tag{2}$$

and T_0 = boiling point of pure solvent

   T_b = boiling point of solution of molality m, at same pressure

   λ_v = enthalpy of vaporization of pure solvent *per gram* at temperature T_0

   K_b = molal boiling-point elevation constant

K_b is a constant characteristic of the solvent. Relation (1) permits calculation of the molecular weight of the solute, since it may be transformed into the equivalent form

$$M = \frac{1000 K_b g}{G\theta} \tag{3}$$

where θ is the elevation of the boiling point for a solution containing g g of solute of molecular weight M in G g of solvent of boiling-point elevation constant K_b.

It should be noted that even for ideal solutions the foregoing relations are valid only if the solution is also dilute, i.e., if the mole fraction of solute is small.

One of the more interesting applications of this equation is in the study of solutes which can form dimers, trimers, etc. The position of equilibria of the type

$$nA \rightleftharpoons A_n$$

may be strongly dependent on the solvent. Examination of the derivation which leads to Eqs. (1) to (3) will show that if several solute species are present, the value of M (now the *number average molecular weight*) found by naïve application of Eq. (3) is actually the average for the species present; that is,

$$M = \sum_i X_i M_i \tag{4}$$

where X_i is the mole fraction of species i and M_i its molecular weight. Although the position of equilibrium cannot be found with accuracy by this method, it is possible to characterize systems in which the solute is mainly of a single species by comparing the apparent molecular weight with the known formula weight.

More rigorous thermodynamic equations are required for the analysis of boiling-point elevation data on solutions which deviate markedly from ideal behavior. The effort is rewarded, however, because one gains additional information in such cases, pertaining to the nonideal properties of these solutions.[1,2] For example, activity coefficients for sodium chloride in aqueous solution have been determined in this way by Smith[3] over the range 60 to 100°C. Thus boiling-point data represent potentially a valuable source of information concerning electrolyte solutions.

For the determination of the boiling temperature of a solution, the thermometer must be in contact with solution rather than with condensed vapor, and the elimination of superheating of the liquid phase is therefore particularly important. Electric heating may be used to reduce the superheating. Often, however, solvent vapor is passed into the solution, where it condenses, and raises the solution to the boiling point. In this way there is no radiation or conduction of heat from a body at a higher temperature, and superheating is eliminated.

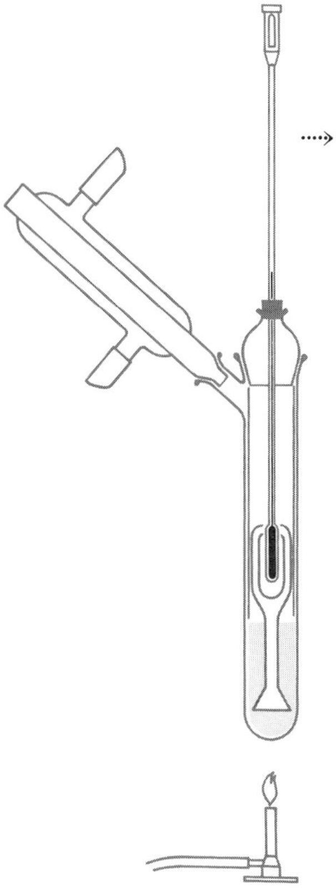

····> **Apparatus.** Boiling-point apparatus of the Cottrell type; Beckmann thermometer or other thermometer graduated to 0.01°; carbon tetrachloride; benzoic acid.

PROCEDURE.   The elevation of the boiling point of carbon tetrachloride produced by dissolved benzoic acid is to be measured.

A commercially available boiling-point apparatus of the Cottrell type[4-6] is shown in Fig. 22. A known quantity of solvent is placed in the tube and a Beckmann thermometer (page 475) or other thermometer graduated to 0.01° inserted. The liquid level must be below the lower end of the glass thermometer. The apparatus is clamped in a vertical position and heated with a small gas flame or, preferably, an electric heater. A hinged vertical shield at the back and sides reduces fluctuations in temperature caused by drafts of air. The purpose of the small inverted funnel, which is raised above the bottom on small projections, is to catch the bubbles of vapor and direct them through the center tube and three vertical spouts. As the bubbles discharge through these outlets, they direct three sprays of liquid and vapor against the thermometer; any superheated solution comes to equilibrium with the vapor by the time it gets to the thermometer bulb.

An inner glass shield, concentric with the outer tube but fastened at the top, improves the efficiency of the pumping system and also serves to shield the thermometer from the cold solvent returning from the condenser. A somewhat more rugged design is one in which the funnel unit is firmly attached by three short pieces of glass rod to the lower end of the inner glass shield.†

figure 22.
Cottrell boiling-point apparatus.

† An apparatus of this design is obtainable on special order from the Scientific Glass Apparatus Co., Bloomfield, N.J., as Model J-2098-1 (modified, Print 577996).

If the liquid does not pump steadily over the thermometer bulb, the rate of heating is changed. The rate should be adjusted so that ebullition takes place primarily within the funnel in order to produce the most efficient pumping action. The rate of heating should be steady and should not be so great as to drive the liquid condensate film too close to the upper end of the condenser, since this may result in loss of solvent and also cause superheating. A metal chimney placed around the burner helps to reduce fluctuations in the rate of heating.

An absolutely constant boiling-point reading cannot be expected, but when equilibrium has been reached, the observed temperature will fluctuate slightly around a mean value and in particular will not show a slow drift, except when there is a corresponding drift in barometric pressure. The thermometer, which must be handled carefully, is tapped gently before a reading is taken. Since the boiling point is sensitive to changes in pressure, the barometer should be read just after the temperature reading is recorded.

After the boiling point of the pure solvent has been determined, the liquid is allowed to cool. The condenser is then removed, and a weighed quantity of benzoic acid sufficient to produce a 2 to 3 percent solution is added. To prevent loss, the benzoic acid is made up into a pellet in a pellet machine or is placed in a short glass tube and rammed tight with a central rod acting as a plunger. Alternatively, if it is feasible to do so, the pellets may be dropped in through the condenser. The steady boiling point of the solution is then determined in the manner previously described. Additional determinations are made by adding more pellets.

When possible, a second series of measurements should be made, starting again with pure solvent and covering the same general range of concentrations as before. In this way, a valuable check is obtained on reproducibility.

Serious error can result from failure to wait for equilibrium to be attained. Other experimental errors may be caused by fluctuations in atmospheric pressure or by appreciable holdup of solvent in the condenser or escape of solvent vapor. The first difficulty may be met by making corrections for pressure changes, by employing a manostat to maintain a constant pressure, or by using two sets of apparatus at the same time, one for solvent and one for solution. Errors from loss or holdup of solvent are tolerable in student work provided reasonable care is taken to minimize these effects. For work of highest accuracy, one may use an apparatus† in which provision has been made for withdrawing a sample of the liquid just after the temperature measurement has been made. The molality is then found by weighing this sample, evaporating the solvent, and then weighing the residue.

CALCULATIONS.   The molecular weight is calculated by means of Eq. (3), from

---

† The Washburn and Read[6] modification of the Cottrell boiling-point apparatus is available from several commercial sources.

Table 1. Molal Boiling-point Elevation Constants

Solvent	Boiling point, °C at 760 mm	K_b at 760 mm	$\dfrac{\Delta K_b}{\Delta P_{mm}}$
Acetone	56.0	1.71	0.0004
Benzene	80.2	2.53	0.0007
Bromobenzene	155.8	6.20	0.0016
Carbon tetrachloride	76.7	5.03	0.0013
Chloroform	60.2	3.63	0.0009
Ethanol	78.3	1.22	0.0003
Ethyl ether	34.4	2.02	0.0005
Methanol	64.7	0.83	0.0002
Water	100.0	0.51	0.0001

the values for K_b shown in Table 1, corrected by use of the pressure coefficient given in the last column, unless the correction is negligible.

The necessary correction to θ required by a difference in the barometric pressures at the times the boiling points of the solvent and solution were recorded may be made by use of Eq. (1) of Experiment 7. For this purpose, it is assumed that dp/dT may be set equal to $\Delta p/\Delta T$. The value of ΔH_{vap} or λ_v may be taken from tables.

The calculated apparent molecular-weight values are graphed against molality and compared with the formula weight. Any discrepancies among these values should be discussed in the light of the estimated experimental error.

Practical applications. Many materials cannot be vaporized for direct determinations of the vapor density without decomposition. In such cases the material is dissolved in a suitable solvent, and the elevation of the boiling point furnishes a rapid and convenient method for determining the molecular weight. Molecular weights of some substances in solution, however, may be different from the values found from vapor-density measurements.

Suggestions for further work. A more elaborate and accurate method may be used in which a thermocouple gives directly the difference in boiling point between the solvent and solution in two different vessels. This method has been described by Mair.[7]

The molecular weight of benzoic acid in a polar solvent such as ethanol may be determined, and the result compared with that obtained with the nonpolar solvent.

Caution is required when an inflammable solvent is used.

A complete treatment of the experimental determination of the boiling point of solutions is given by Swietoslawski and Anderson.[8]

References

1. H. S. Harned and B. B. Owen, "Physical Chemistry of Electrolyte Solutions," 3d ed., Reinhold Publishing Corporation, New York, 1958.
2. R. A. Robinson and R. H. Stokes, "Electrolyte Solutions," 2d ed., Academic Press Inc., New York, 1959.

3. R. P. Smith, *J. Am. Chem. Soc.*, **61:** 497, 500 (1939).
4. F. G. Cottrell, *J. Am. Chem. Soc.*, **41:** 721 (1919).
5. H. L. Davis, *J. Chem. Educ.*, **10:** 47 (1933).
6. E. W. Washburn and J. W. Read, *J. Am. Chem. Soc.*, **41:** 729 (1919).
7. B. J. Mair, *J. Res. Natl. Bur. Std.* U.S., **14:** 345 (1935).
8. W. Swietoslawski and J. R. Anderson in A. Weissberger (ed.), "Technique of Organic Chemistry," vol. 1, "Physical Methods of Organic Chemistry," 3d ed., pt. 1, chap. 8, Interscience Publishers, Inc., New York, 1959.

13 ACTIVITIES FROM FREEZING-POINT DEPRESSION DATA

The freezing point of a solution is lower than that of the pure solvent. Freezing-point depression data are of considerable value in the thermodynamic study of solutions. In particular, activity coefficients of both solvent and solute can be determined as a function of concentration to a high degree of accuracy.

THEORY. The activity a_1 of solvent† in a solution at temperature T is defined by

$$\overline{G}_1 = \overline{G}_1^\circ + RT \ln a_1 \tag{1}$$

where $\overline{G}_1$ = partial molar Gibbs free energy of solvent in solution at temperature T

$\overline{G}_1^\circ$ = molar Gibbs free energy of pure liquid solvent at temperature T

For a solution at the temperature T_0 of freezing of the pure solvent, Eq. (1) is equivalent to

$$\overline{G}_1 = \overline{G}_1^c + RT_0 \ln a_1 \tag{2}$$

where $\overline{G}_1^c$ is the molar free energy of pure crystalline solvent at temperature T_0, because at T_0 the equilibrium condition

$$\overline{G}_1^\circ = \overline{G}_1^c \tag{3}$$

is satisfied. Therefore a_1 for a solution at temperature T_0 can be calculated from

$$R \ln a_1 = \left(\frac{\overline{G}_1}{T_0} - \frac{\overline{G}_1^c}{T_0} \right) = \Delta \frac{\overline{G}_1}{T_0} \tag{4}$$

Here, and throughout this discussion, the increment symbol Δ will refer to a change in a function of state for solvent in going from the pure crystalline solid to a solution at the *same temperature* T; for example, for water solvent,

$$\begin{array}{ccc} H_2O(c, \text{ pure}) & \longrightarrow & H_2O(\text{liquid solution}) \\ (\overline{G}_1^c / T) & & (\overline{G}_1 / T) \end{array} \tag{5}$$

† Solvent properties are designated throughout by subscript 1 and solute properties by subscript 2.

For a series of solutions it is desired to calculate, by Eq. (4), the activity of the solvent at the temperature T_0. To evaluate the right side of Eq. (4), consider the function

$$\Delta \frac{G_1}{T}$$

for general T; this may be expanded in a Taylor series about T_0 to obtain

$$\Delta \frac{\overline{G}_1}{T} = \left[\Delta \frac{\overline{G}_1}{T} \right]_0 + \left[\frac{\partial(\Delta \overline{G}_1/T)}{\partial T} \right]_0 (T - T_0)$$
$$+ \frac{1}{2} \left[\frac{\partial^2(\Delta \overline{G}_1/T)}{\partial T^2} \right]_0 (T - T_0)^2 + \cdots \quad (6)$$

The functions in square brackets in Eq. (6) are to be evaluated at $T = T_0$. The series-expansion coefficients are

$$\left[\Delta \frac{\overline{G}_1}{T} \right]_0 = R[\ln a_1]_0 \tag{7a}$$

$$\left[\frac{\partial(\Delta \overline{G}_1/T)}{\partial T} \right]_0 = - \left[\frac{\Delta \overline{H}_1}{T_0^2} \right]_0 \tag{7b}$$

$$\frac{1}{2} \left[\frac{\partial^2(\Delta \overline{G}_1/T)}{\partial T^2} \right]_0 = - \frac{1}{2} \left[\frac{\partial}{\partial T} \frac{\Delta \overline{H}_1}{T^2} \right]_0 = \left[\frac{\Delta \overline{H}_1}{T_0^3} - \frac{\Delta \overline{C}_P}{2T_0^2} \right]_0 \tag{7c}$$

where again all quantities in square brackets are to be evaluated at $T = T_0$.

Now if T in Eq. (6) is specified to be the freezing point of the solution, T_f, the left side of Eq. (6) becomes zero because of the basic requirement for equilibrium, namely, that $\overline{G}_1$ have the same value for both phases. Combining Eqs. (7) and (6) for $T = T_f$ then yields

$$0 = R[\ln a_1]_0 - \left[\frac{\Delta \overline{H}_1}{T_0^2} \right]_0 (T_f - T_0) + \left[\frac{\Delta \overline{H}_1}{T_0^3} - \frac{\Delta \overline{C}_P}{2T_0^2} \right]_0 (T_f - T_0)^2 + \cdots \tag{8}$$

With the freezing-point depression designated as $\theta = T_0 - T_f$, Eq. (8) leads to

$$-R[\ln a_1]_0 = \left[\frac{\Delta \overline{H}_1}{T_0^2} \right]_0 \theta + \left[\frac{\Delta \overline{H}_1}{T_0^3} - \frac{\Delta \overline{C}_P}{2T_0^2} \right]_0 \theta^2 + \cdots \tag{9}$$

Since $\Delta \overline{H}_1 = \overline{H}_1 - \overline{H}_1^c$ will often be close to $\Delta \overline{H}_f^\circ \equiv \overline{H}_1^\circ - \overline{H}_1^c$, the heat of fusion of the pure solvent, it is convenient to write

$$\Delta \overline{H}_1 = \Delta \overline{H}_f^\circ + \overline{L}_1 \tag{10}$$

where $\overline{L}_1$ is defined by $\overline{L}_1 = \overline{H}_1 - \overline{H}_1^\circ$ at any given temperature. Introduction of

Eq. (10) into (9) then yields

$$-R[\ln a_1]_0 = \left[\frac{\Delta \overline{H}_f^\circ}{T_0^2}\right]_0 \left\{\left[1 + \frac{\overline{L}_1}{\Delta \overline{H}_f^\circ}\right]_0 \theta + \left[\frac{1}{T_0} - \frac{\Delta \overline{C}_P^\circ}{2\Delta \overline{H}_f^\circ}\right]_0 \theta^2 + \cdots\right\} \tag{11}$$

where $\overline{L}_1$ and $\partial \overline{L}_1/\partial T$ are neglected in comparison with $\Delta \overline{H}_f^\circ$ and $\Delta \overline{C}_P^\circ$, respectively, in the θ^2 term.

Thus from Eq. (11) the activity coefficient of solvent in a solution at temperature T_0—the freezing point of the pure solvent—may be found from a measurement of the freezing-point depression θ for the given solution.†

For solutions of moderate or low concentration, the simple equation

$$-R[\ln a_1]_0 = \left[\frac{\Delta \overline{H}_f^\circ}{T_0^2}\right]_0 \theta \tag{12}$$

will often be sufficiently accurate.§ However, one cannot really use Eq. (12) with confidence unless the higher-order terms in Eq. (11) have been estimated for the most concentrated solution being considered and found to be less than experimental error.

For water as solvent at $0°C$, the factor in braces in Eq. (11) becomes

$$\left\{\left[1 + \frac{\overline{L}_1}{1436}\right]_0 \theta + [4.9 \times 10^{-4}]\theta^2 + \cdots\right\} \tag{13}$$

with $\overline{L}_1$ in calories per mole and θ in degrees. The quantity $\overline{L}_1$ may be evaluated experimentally from heat-of-dilution data or calculated from

$$\overline{L}_1 = \frac{\partial \Delta H_{IS}}{\partial n_1} \tag{14}$$

where ΔH_{IS} is the integral heat of solution (page 25) for 1 mole of solute in n_1 moles of solvent. It may also be calculated from solute heats of formation, ΔH_f (pages 35 and 36), which are tabulated for many substances.

A familiar equation for the freezing-point depression for an *ideal, dilute* solution is obtained by introducing into Eq. (12) the approximations

$$a_1 \approx X_1 \qquad\qquad\qquad\qquad \textit{ideal solution} \tag{15}$$

$$\ln X_1 = \ln (1 - X_2) \approx -X_2 \approx -\frac{mM_1}{1000} \quad \textit{dilute solution} \tag{16}$$

† An equation for $\ln a_1$ is sometimes given in a somewhat simpler form, with the small but concentration-dependent $\overline{L}_1$ term absent; in this case, however, a_1 is specified to be the activity at temperature T_f (rather than at T_0); for a series of measurements on solutions of various concentrations, the calculated activities then refer to different temperatures. Equation (11) is given above in the same form as in Ref. 2, although a slightly different method of derivation has been used.

§ Physically, Eq. (12) corresponds to the case $\Delta \overline{H}_1 = \Delta \overline{H}_f^\circ$, $\Delta \overline{C}_P^\circ = 0$, $\theta \ll T_f$.

where m = molality = number of moles of solute in 1000 g of solvent

M_1 = molecular weight of solvent

X_1, X_2 = mole fractions of solvent and solute, respectively

The result is

$$\theta = \frac{RT_0^2 M_1}{1000\Delta \overline{H}_f^\circ} m = K_f m \tag{17}$$

where

$$K_f = \frac{RT_0^2 M_1}{1000\Delta \overline{H}_f^\circ} = \frac{RT_0^2}{1000\lambda_f} \tag{18}$$

where λ_f = heat of fusion of pure solvent, *per gram*, at temperature T_0

K_f = molal freezing-point depression constant

Observe that K_f is a property of the pure solvent and as such does not depend on the nature of the solute.

If the solute is associated or dissociated in solution, the ideal equations can be expected to serve as useful approximations only if the association or dissociation is recognized explicitly. Thus, consider a solution containing n_1 formula-weight units of solvent and n_2 formula-weight units of solute; n_1 and n_2 are the weights divided by the respective formula weights. Now suppose that, in solution, n_2 formula-weight units of solute actually form νn_2 moles of solute, with $\nu > 1$ for dissociation and $\nu < 1$ for association. Then the actual mole fraction of solvent, X_1', calculated with the association or dissociation explicitly recognized, is

$$X_1' = \frac{n_1}{n_1 + \nu n_2} = 1 - \frac{\nu n_2}{n_1 + \nu n_2} \approx 1 - \nu \frac{n_2}{n_1} \tag{19}$$

where the last form is valid for dilute solutions. For a solution of stoichiometric molality m, that is, with m formula-weight units of solute per 1000 g of solvent, Eq. (16) becomes

$$\ln X_1' \approx - \left(\nu \frac{n_2}{n_1} \right) = - \frac{\nu m M_1}{1000} \tag{20}$$

and Eq. (17) becomes

$$\theta = K_f \nu m \tag{21}$$

with K_f defined as before.

An important use of Eq. (21) is the determination of ν. Since the equation holds only for the ideal, dilute case, it may be more appropriately written

$$\nu = \lim_{m \to 0} \frac{\theta}{K_f m} = \frac{1}{K_f} \lim_{m \to 0} \frac{\theta}{m} \tag{22}$$

It should be recognized that the limit in Eq. (22) is to be evaluated from experimental data for a number of solutions at finite concentrations; the value of ν obtained therefore reflects the behavior of the given solute at the lower end of the concentration range studied.

If the data are accurate enough, *solute* activities may also be calculated from the freezing-point data. The only case to be considered in detail here will be that of a solute which is neither associated nor dissociated at the lower end of the concentration range studied. For this case, the activity a_2 and activity coefficient γ_2 of the solute are defined by†

$$\bar{G}_2 = \bar{G}_2^\circ + RT \ln a_2 \tag{23}$$
$$= \bar{G}_2^\circ + RT \ln \gamma_2 m \tag{24}$$

with $\bar{G}_2^\circ$ defined by specifying $\gamma_2 \longrightarrow 1$ as $m \longrightarrow 0$ or, equivalently, by

$$\bar{G}_2^\circ = \lim_{m \to 0} (\bar{G}_2 - RT \ln m) \tag{25}$$

The equation for calculating γ_2 for a given solution of molality m at temperature T_0, based on Eqs. (24) and (25), is

$$\ln \gamma_2 = (\phi_m - 1) + \int_0^m \frac{\phi_m - 1}{m} \, dm \tag{26}$$

where ϕ_m is the osmotic coefficient at molality m, defined by

$$\phi_m = -\frac{X_1}{X_2} \ln a_1 = -\frac{1000}{m M_1} \ln a_1 \tag{27}$$

The derivation of Eq. (26) may be found in a number of texts on chemical thermodynamics.[1-4] When Eq. (12) is applicable, ϕ_m and $\phi_m - 1$ may be calculated from

$$\phi_m = \frac{\theta}{K_f m} \tag{28}$$

$$\phi_m - 1 = \frac{(\theta/m) - K_f}{K_f} \tag{29}$$

A suitable procedure is first to prepare a graph of θ/m versus m to see whether the data indicate either association or dissociation at the low-concentration end; if, to within experimental error, $\nu = 1$, then Eqs. (26) and (29) may be used to find γ_2 for any m in the range studied. The integral is evaluated graphically from a plot of $(\phi_m - 1)/m$ versus m.

If an attempt is made to use Eq. (26) for a solute which is associated or

---

† The molality scale for γ_2 is used here. Alternative scales[1-4] are sometimes used and lead to different equations and different numerical values for γ_2.

dissociated at the lower end of the concentration range studied, the integrand will seem to increase without limit as m approaches zero; the integral will therefore be indeterminate. This observation reflects the fact that the limit in Eq. (25) either does not exist (dissociation case) or is not being approached even at the lowest concentrations studied (association case). One is therefore compelled to modify the definition of $\overline{G}_2^\circ$, and hence also that of γ_2, in order to be able to obtain meaningful results for such systems.[1-4]

····> **Apparatus.** Vacuum bottle with cork and stirrer; precision (calorimeter-type) thermometer ($+0.5°$ to $-5.0°$, graduated to $0.01°$), or differential thermometer with $5°$ range, or thermistor and bridge (Experiment 4); five 6-in. test tubes with tight stoppers; density balance precise to ±0.0002 g/ml with a 10-ml sample, or a set of 5-ml pycnometers and an analytical balance; 100 ml of a suitable solute sample such as isopropyl alcohol, n-propyl alcohol, or acetonitrile.

PROCEDURE.   A suitable apparatus, shown in Fig. 23, consists of a vacuum bottle (1 pt) fitted with a cork, polystyrene† stirrer, and thermometer. A convenient type of thermometer for this experiment is one having a range of 0.5 to $-5.0°$C with $0.01°$ graduations. If a Beckmann thermometer is used, it is adjusted so that the mercury comes near the top of the scale at the freezing point of the pure solvent.

A thermistor§ offers definite advantages as an alternative to a mercury thermometer: the thermistor has a relatively low time lag and, being physically smaller, does not conduct heat from the room as rapidly into the freezing solution. Thus thermal equilibration can be achieved more efficiently. If a thermistor is used, the resistance measurement may be made by means of a dc Wheatstone bridge and sensitive galvanometer, as outlined in Experiment 4. The temperature dependence of the thermistor resistance may be assumed to have the form

$$R = Be^{\beta/T} \tag{30}$$

where R is the resistance of the thermistor at the temperature T. The parameters B and β may be evaluated by measurement of R at several known temperatures, which for convenience may be chosen in the range 0 to 25°C.

To the vacuum bottle is added distilled water and an equal volume of clean cracked ice. The water used should be prechilled, to minimize the amount of impurities set free by melting of ice. The freezing mixture should cover the thermometer bulb or thermistor without being too close to the cork. If the ice is not of adequate purity, the observed freezing point will slowly decrease with time as occluded impurities are freed by melting; in this event, it may be necessary to prepare ice by freezing distilled water.

figure 23.

Freezing-point apparatus.

† Polystyrene is thermoplastic and can be bent when heated in a drying oven or above a bunsen-burner flame.

§ A suitable thermistor and bridge assembly is available as Model S-81601 from E. H. Sargent & Co., Chicago, Ill.

The ice and water are stirred vigorously until a steady temperature is attained. This value is then recorded. If a mercury thermometer is used, it should be gently tapped before being read.

The water is drained off and replaced by a chilled solution of the specified solute in distilled water at a concentration of about 3 *m*. The solution and ice are stirred vigorously until a constant temperature is reached, whereupon the temperature is recorded and a sample is withdrawn quickly with a 10-ml pipette; the tip of the pipette should be held near the bottom of the flask to avoid getting pieces of ice. The pipette is used here only as a matter of convenience, as the sample volume need not be accurately known. To obtain a sample with minimal disturbance of the equilibrium system, the sample should be taken rapidly. The sample is discharged into a 6-in. test tube, which is then stoppered and set aside.

The temperature will appear to rise slightly when the stirring is interrupted for sample-taking, but this apparent rise can be ignored because it results mainly from the warming of the thermometer or thermistor itself and does not indicate a change in temperature of the solution. After the sample has been withdrawn, vigorous stirring may be resumed and a second temperature reading taken. The temperatures recorded for the stirred system immediately before and after sampling should agree to within about 0.01°C, and these values may be averaged for the subsequent calculations.

In a similar fashion, additional samples are taken from equilibrium freezing systems with progressively more dilute solutions, obtained by adding each time an appropriate amount of ice (to replace ice melted in the flask) and ice-cold distilled water. Some allowance should be made for dilution by melting of ice. In all, measurements should be made for about six different concentrations, ranging downward from 3 *m* in roughly equal steps of about 0.5 *m* each. A convenient way of obtaining well-spaced points covering the desired range of molalities is to adjust the concentration each time to bring the freezing point to the desired region.

The molalities of the samples are determined by careful measurement of density. The precision required is of the order of ± 0.0002 g ml^{-1}. If a density balance is used, the measurements can be made with the samples still in the original test tubes if a device is provided to support them in the balance. The temperatures of the samples should be noted at the time of the density measurements, and the density values corrected to a common temperature by assuming that the thermal coefficient of expansion of a dilute aqueous solution is about the same as that of pure water, for which $\dfrac{1}{V}\left(\dfrac{\partial V}{\partial T}\right)_P$ is 2.6×10^{-4} deg^{-1} at 25°C. Alternatively, 5-ml pycnometers may be filled with the samples at a known temperature and subsequently weighed.

The data needed for preparing a calibration graph of density versus molality are obtained by measuring densities of pure solvent and of four or five solutions of known composition made up by mixing known volumes of the two components.

CALCULATIONS.   If a thermistor is used for temperature measurement, the constant β may be found from a plot of log R versus $1/T$. As neither this graph nor Eq. (30) is particularly well suited to the accurate calculation of small temperature increments, it is preferable, provided the range of freezing points does not exceed a few degrees, to calculate the freezing-point depressions from the expansion

$$\theta = \frac{T_0^2}{\beta}\left[\frac{R_f - R_0}{R_0} - \left(\frac{1}{2} + \frac{T_0}{\beta}\right)\left(\frac{R_f - R_0}{R_0}\right)^2 + \cdots \right] \tag{31}$$

where R_0 = resistance at freezing point T_0 of pure solvent

   R_f = resistance at freezing point T_f of solution

Equation (31) is obtained by expanding T as a function of R in a Taylor series about T_0; the derivatives required are most easily calculated from

$$\frac{dR}{dT} = -\frac{R\beta}{T^2} \tag{32}$$

etc. The first neglected term inside the brackets is of the order of $[(R_f - R_0)/R_0]^3$.

   The molalities and freezing-point depressions for the various samples are then tabulated. A graph of θ/m versus m is prepared to illustrate the determination of ν from Eq. (22). The solvent activity and the osmotic coefficient are calculated for each molality and graphed as functions of m. For this calculation it is convenient to express the coefficient in Eq. (12) in terms of K_f. Values of K_f and of several other pertinent quantities appear in Table 1. For most nonelectrolyte solutions in the concentration range considered here, the $\bar{L}_1$ term in Eq. (13) is quite negligible.

   The calculation of solute activities from these data is outlined below.

Table 1.   *Freezing-point Depression Constants and Related Data*

Solvent	Freezing point t, $°C$	λ_f, cal g^{-1}	K_f	$\Delta \bar{C}_P^\circ$, cal deg^{-1} mole^{-1}
Acetic acid[a]	16.61	46.6	3.58	9.4
Benzene[b]	5.5	30.1	5.12	
t-Butanol[c]	25.1	21.88	8.37	
Carbon tetrachloride[a]	−22.9	3.9	32	1.1
Cyclohexane[b]	6.5	7.4	20.0	
Cyclohexanol[d]	25.1	4.27	37.7	
Water[a]	0.00	79.72	1.860	8.911

[a] Selected Values of Chemical Thermodynamic Properties, *Natl. Bur. Std. U.S. Circ.* 500, 1952.
[b] N. A. Lange, "Handbook of Chemistry," 8th ed., McGraw-Hill Book Company, New York, 1952.
[c] F. H. Getman, *J. Am. Chem. Soc.,* **62:** 2179 (1940).
[d] H. N. Wilson and A. E. Heron, *J. Soc. Chem. Ind. London,* **60:** 168 (1941).

Practical applications. Equation (17) is often used for the determination of molecular weights of solutes, since it may be written in the equivalent form

$$M_2 = \frac{1000 K_f g}{G\theta} \tag{33}$$

where θ is the freezing-point depression for a solution containing g g of solute of molecular weight M_2 in G g of solvent of molal freezing-point depression constant K_f. Association or dissociation, if complete, leads to values of M_2 which are multiples or submultiples of the formula weight. By an extension of Eq. (33), ionization constants of weak acids can be found from freezing-point data, since the apparent molecular weight M_2 can be expressed in terms of the fraction dissociated.

Suggestions for further work. If the data are sufficiently accurate, *solute* activities may be calculated from the data obtained in this experiment; Eq. (26) is applicable provided the graph of θ/m versus m indicates $\nu = 1$ to within experimental error. It may be desirable to draw a smooth curve, representing the data, on the θ/m-versus-m graph and to use selected points from this curve (rather than raw data points) for the calculation of solute activities. In any case, the quantities $\phi_m - 1$ and $(\phi_m - 1)/m$ are tabulated and graphed against m. For selected values of m, the integral in Eq. (26) is evaluated from the graph, and γ_2 is calculated from Eq. (26).

A good example of a dissociated solute which is readily studied is KCl. In this case, the samples, after being withdrawn into a 5-ml pipette, should be discharged into previously weighed bottles and then weighed. The weight of KCl in each sample is subsequently found by titrating the samples with 0.1 N AgNO$_3$, with about 0.2 ml of 5 percent potassium chromate solution added as indicator. The AgNO$_3$ solution is standardized with weighed samples of KCl.

References

1. H. S. Harned and B. B. Owen, "Physical Chemistry of Electrolyte Solutions," 3d ed., Reinhold Publishing Corporation, New York, 1958.
2. G. N. Lewis and M. Randall, "Thermodynamics," 2d ed., pp. 404–409, 412–413, rev. by K. S. Pitzer and L. Brewer, McGraw-Hill Book Company, New York, 1961.
3. R. A. Robinson and R. H. Stokes, "Electrolyte Solutions," 2d ed., Academic Press Inc., New York, 1959.
4. F. T. Wall, "Chemical Thermodynamics," 2d ed., pp. 384–385, 387–388, W. H. Freeman and Company, San Francisco, 1965.

14   *PARTIAL MOLAL PROPERTIES OF SOLUTIONS*

The accurate determination of the density of a liquid and the precise mathematical treatment of the properties of solutions are studied.

THEORY.[1-5]   The quantitative study of solutions has been greatly advanced by the introduction of the concept of partial molal quantities. A property of a solution, e.g., the volume of a mixture of alcohol and water, changes continuously as the composition is changed, and considerable confusion existed formerly in expressing these properties as a function of composition. A *partial molal property* of a component of a solution is defined as follows. Let Y represent any extensive property of a binary solution; at constant temperature and pressure, Y then will be a function of the two independent variables n_1 and n_2, which represent the numbers of moles of the two components present. The partial molal property of component 1 is then defined by the relation

$$\overline{Y}_1 = \left(\frac{\partial Y}{\partial n_1}\right)_{n_2, T, P} \tag{1a}$$

Similarly for component 2,

$$\overline{Y}_2 = \left(\frac{\partial Y}{\partial n_2}\right)_{n_1, T, P} \tag{1b}$$

The partial molal quantity may be designated by a bar above the letter representing the property, and by a subscript, which indicates the component to which the value refers. The usefulness of the concept of partial molal quantities lies in the fact that it may be shown mathematically[1,2,4] that

$$Y(n_1, n_2) = n_1 \overline{Y}_1 + n_2 \overline{Y}_2 \qquad T, P \text{ constant} \tag{2}$$

Any extensive property of the solution may be expressed in this manner in terms of partial molal properties, which themselves are functions of the concentration of the solution, the temperature, and the pressure and must be evaluated by means of experimental measurements. The activity a_i of a component of a solution is defined in terms of its relative partial molal free energy $\overline{G}_i - \overline{G}_i^\circ$, and the calculation of heats of reaction for solution systems requires a knowledge of the relative partial molal enthalpies $\overline{H}_i - \overline{H}_i^\circ$ for all the components. The superscript $^\circ$ refers to the standard state.

In the case of the volume of the solution, Eq. (2) gives directly

$$V = n_1 \overline{V}_1 + n_2 \overline{V}_2 \qquad T, P \text{ constant}$$

The partial molal volumes $\overline{V}_1$ and $\overline{V}_2$ may be evaluated from density measurements on the solutions. The graphical methods described fully by Lewis and Randall[2] may be used in the treatment of the data; of these, the use of the apparent molal volume ϕ_V is particularly convenient for binary solutions.

The apparent molal volume is defined by the relation

$$\phi_V = \frac{V - n_1 \overline{V}_1^\circ}{n_2} \qquad T, P \text{ constant} \tag{3}$$

where V = volume of solution containing n_1 moles of component 1 and n_2 moles of component 2

$\bar{V}_1^\circ$ = molar volume of component 1 at given T, P

Since $V = n_2\phi_V + n_1\bar{V}_1^\circ$,

$$\bar{V}_2 = \left(\frac{\partial V}{\partial n_2}\right)_{n_1,T,P} = \phi_V + n_2\left(\frac{\partial \phi_V}{\partial n_2}\right)_{n_1,T,P} \tag{4a}$$

and

$$\bar{V}_1 = \frac{V - n_2\bar{V}_2}{n_1} = \frac{1}{n_1}\left[n_1\bar{V}_1^\circ - n_2^2\left(\frac{\partial \phi_V}{\partial n_2}\right)_{n_1,T,P}\right] \tag{4b}$$

In terms of the experimentally measured density ρ and the molecular weights M_1 and M_2 of the two components,

$$\phi_V = \frac{1}{n_2}\left(\frac{n_1M_1 + n_2M_2}{\rho} - n_1\bar{V}_1^\circ\right) \tag{5}$$

When the molal concentration scale is used, $n_2 = m$, the molality, and n_1 is equal to the number of moles of component 1 in 1000 g of solvent, so that

$$\phi_V = \frac{1}{m}\left(\frac{1000 + mM_2}{\rho} - \frac{1000}{\rho_1}\right) = \frac{1000}{m\rho\rho_1}(\rho_1 - \rho) + \frac{M_2}{\rho} \tag{6}$$

where ρ_1 = density of pure component 1

ρ = density of solution of molality m of component 2 having molecular weight M_2

The second expression is particularly convenient for actual calculations.

The use of the apparent molal volume in this determination is advantageous because the error involved in the graphical determination of the derivative of a function is encountered only in the evaluation of the term giving the difference between the partial molal volume and the apparent molal volume.

The partial molal volume may be visualized by considering a large reservoir of a solution of given composition, so large that the addition of one more mole of a component will not appreciably alter the concentration. If now 1 mole of component 1 is added to this large reservoir of solution, the increase in the volume of the solution is equal to the partial molal volume of component 1 at the indicated concentration, temperature, and pressure. The magnitude of the partial molal volume depends upon the nature of the interactions between the components of the solution under the given conditions; the effects of these interactions are difficult to predict theoretically, but the overall result is readily expressed mathematically, as already shown. In the special case of an ideal solution, the partial molal volume of any component at any concentration is equal to the molar volume of the pure liquid component at the temperature and pressure of the solution.

The concept of the partial molal quantity may, of course, be applied to solutions containing more than two components. The extension of Eqs. (1) and (2) to the general case is discussed in detail elsewhere.[1-5]

Apparatus. Pycnometers; thermostat; rapid analytical balance; six small glass-stoppered bottles or flasks; sodium chloride.

PROCEDURE.[6]   Solutions of sodium chloride in water containing approximately 2, 4, 8, 12, and 16 percent sodium chloride by weight are prepared. The salt and water are weighed out accurately into a weighing bottle or glass-stoppered flask, care being taken to prevent evaporation of the volatile solvent. A total volume of about 75 ml of each solution is required for the execution of duplicate determinations.

The density of each solution is determined accurately at 25.0°C. A pycnometer of the Weld or Ostwald-Sprengel type shown in Fig. 24 may be used. The pycnometer is dried carefully, weighed, then filled with distilled water, and placed in the thermostat for 10 to 15 min.

The Weld pycnometer is initially filled to bring the liquid level about halfway up the throat T of the reservoir R. The pycnometer is placed in the thermostat with the cap C in position to prevent evaporation from the exposed liquid surface. When temperature equilibrium has been reached, the cap C is removed and the plug P is inserted. A moderate pressure is sufficient to seat the plug firmly. Any excess liquid

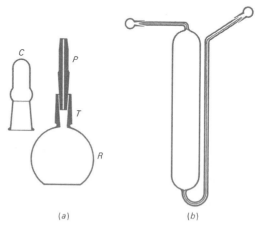

(a)   (b)

*figure 24. (a) Weld pycnometer; (b) Ostwald-Sprengel pyc-
nometer.*

on the tip of the plug is wiped off with a piece of filter paper, care being taken to avoid removing liquid from the plug capillary in the process. The pycnometer is then removed from the thermostat, wiped dry with a lintless cloth, and the (dried) cap C put in place. It is allowed to stand in the balance case for a few minutes before being weighed.

With the Ostwald-Sprengel pycnometer, the quantity of liquid is adjusted so that the liquid meniscus is at the mark on the horizontal capillary when the other capillary arm is filled. This adjustment may be made by tilting the completely filled unit slightly and withdrawing liquid slowly from the other capillary by touching a piece of filter paper to it. The pycnometer is removed from the thermostat and wiped dry with a lintless cloth, and the caps placed on the capillary arms. It is allowed to stand in the balance case for a few minutes before being weighed.

For the best results, it is suggested that the temperature of the balance room be not much above that of the thermostat. Also, there is often difficulty with "creeping" of the salt solutions during storage. It is preferable to make up the solutions and measure the densities on the same day.

In this fashion, duplicate determinations are made of the weight of liquid required to fill the pycnometer at the thermostat temperature, for water and for each of the solutions previously prepared. Two pycnometers may be used to advantage, so that one may be weighed while the other is in the thermostat.

CALCULATIONS.   The weights of the water and of the various salt solutions held by the pycnometers are corrected to vacuum as described in Chapter 23. The density of water at 25°C is taken as 0.99707 g ml^{-1} for the calculation of the volumes of the pycnometers. The density of each solution is then calculated by dividing its vacuum weight by the appropriate pycnometer volume.

The concentration of each solution is expressed in terms of the molal concentration scale, and the apparent molal volume is determined at each concentration. The uncertainty in the apparent molal volume introduced by an uncertainty of 0.02 percent in the density is computed for each solution.

By means of Eqs. (4a) and (4b) the partial molal volumes of solute and solvent are evaluated at each concentration. In this case $n_2 = m$, the molality, and n_1, the number of moles of solvent associated with n_2 moles of solute, is equal to 55.51, that is, 1000/18.016. It is convenient in the case of an electrolytic solution to plot ϕ_V against $m^{\frac{1}{2}}$ instead of against m and to utilize the relationship

$$\left(\frac{\partial \phi_V}{\partial m}\right)_{n_1,T,P} = \frac{1}{2m^{\frac{1}{2}}}\left[\frac{\partial \phi_V}{\partial (m^{\frac{1}{2}})}\right]_{n_1,T,P} \tag{7}$$

A second method is also used for the evaluation of $\overline{V}_2$. The volume of solution containing 1000 g of solvent is plotted against the molality m. The slope of the tangent to the curve at any chosen concentration gives directly the value of $\overline{V}_2$ (Ref. 2). The values given by this method are compared with those obtained by the preceding method.

Practical applications. The use of partial molal quantities is fundamental in the application of thermodynamics to solution systems. This is well illustrated in an article of unusual interest in which their application to the physical chemistry of solutions of substances of biological interest is described.[5]

Suggestions for further work. Other solutions may be investigated. The system benzene–carbon tetrachloride[7] exhibits nearly ideal behavior. The system ethanol-water[7] provides an example of nonideal behavior and is particularly interesting in the region from 0 to 15 mole percent ethanol.

References

1. I. M. Klotz, "Chemical Thermodynamics," Prentice-Hall, Inc., Englewood Cliffs, N.J., 1950; "Introduction to Chemical Thermodynamics," W. A. Benjamin, Inc., New York, 1964.
2. G. N. Lewis and M. Randall, "Thermodynamics," 2d ed., rev. by K. S. Pitzer and L. Brewer, McGraw-Hill Book Company, New York, 1961.
3. F. T. Wall, "Classical Thermodynamics," 2d ed., W. H. Freeman and Company, San Francisco, 1965.
4. K. Denbigh, "The Principles of Chemical Equilibrium," 2d ed., Cambridge University Press, London, 1966.
5. E. F. Casassa and H. Eisenberg, Thermodynamic Analysis of Multicomponent Solutions, *Advan. Protein Chem.*, **19:** 287 (1964).
6. N. Bauer and S. Z. Lewin in A. Weissberger (ed.), "Technique of Organic Chemistry," vol. 1, "Physical Methods of Organic Chemistry," 3d ed., pt. 1, chap. 4, Interscience Publishers, Inc., New York, 1959.
7. S. E. Wood and J. P. Brusie, *J. Am. Chem. Soc.*, **65:** 1891 (1943).

Chapter 5

Homogeneous Equilibria

15   *DISSOCIATION OF NITROGEN TETROXIDE*

The equilibrium constant for a reaction is determined as a function of temperature, and the corresponding heat of reaction is calculated.

THEORY.   Nitrogen tetroxide dissociates in accordance with the reaction

$$N_2O_4 \longrightarrow 2NO_2$$

If the equilibrium degree of dissociation is represented by α, an initial 1 mole of N_2O_4 gives 2α moles of NO_2 and $1 - \alpha$ mole of N_2O_4 at equilibrium. The total number of moles is thus $1 + \alpha$, and the mole fraction of NO_2 is $2\alpha/(1 + \alpha)$, and that of N_2O_4 is $(1 - \alpha)/(1 + \alpha)$. When the partial pressure p of each constituent is set equal to the product of its mole fraction and the total pressure P, the equilibrium constant for the reaction takes the form

$$K_p = \frac{p_{NO_2}^2}{p_{N_2O_4}} = \frac{[2P\alpha/(1 + \alpha)]^2}{[P(1 - \alpha)/(1 + \alpha)]} = \frac{4\alpha^2 P}{1 - \alpha^2} \tag{1}$$

Experimentally, α is found by measuring M, the average molecular weight of the equilibrium gas mixture. One mole of undissociated N_2O_4, of molecular weight $M_0 = 92.06$, dissociates to form $1 + \alpha$ moles in the equilibrium mixture. Since the total weight is unchanged, the average molecular weight is

$$M = \frac{M_0}{1 + \alpha} \qquad \text{or} \qquad \alpha = \frac{M_0 - M}{M} \tag{2}$$

Equation (2) is used to calculate α from M. Equation (1) applies only when the gas mixture is considered to be an ideal mixture of perfect gases.

The standard Gibbs free-energy change for a reaction can be calculated from the thermodynamic equilibrium constant K, which for the ideal case equals K_p.

$$\Delta G° = -RT \ln K_p = -2.303RT \log K_p \tag{3}$$

The determination of the equilibrium constant at a series of temperatures permits the evaluation of the standard enthalpy change by application of the Gibbs-Helmholtz equation in the form

$$\Delta H° = -T^2 \left(\frac{\partial \Delta G°/T}{\partial T}\right)_P = 2.303RT^2 \frac{d \log K_p}{dT} = -2.303R \frac{d \log K_p}{d(1/T)} \tag{4}$$

When the plot of K_p versus $1/T$ is a straight line, the standard enthalpy change is constant over the temperature range involved and may be calculated from the slope of the line. When the line is not straight, the value of $\Delta H°$ depends on the temperature, but the slope of a tangent drawn at a point corresponding to any temperature will give the standard enthalpy change at that temperature.

⋯⋯→ ***Apparatus.*** Two gas-density bulbs; counterpoise bulb of same capacity; small adjustable thermostat; cylinder of nitrogen tetroxide; gas-filling apparatus with stopcocks.

PROCEDURE.   Pyrex bulbs of about 200 ml capacity with capillary glass stopcocks are used for determining the density of the dissociation mixture. Nitrogen tetroxide attacks ordinary stopcock greases, but its effect on silicone grease is negligible during the period of the experiment if a minimum amount is used in lubrication.

Two bulbs are evacuated to 1 mm or less and weighed to 0.1 mg, using the weighing procedure described in Experiment 1.

The weighed bulbs are filled from a small cylinder of nitrogen tetroxide.

Caution: Nitrogen tetroxide is corrosive and very poisonous. All operations should be carried out in a well-ventilated hood.

The arrangement of stopcocks and traps shown in Fig. 25 is used for filling the bulbs with dry nitrogen tetroxide (in equilibrium with nitrogen dioxide). The tank

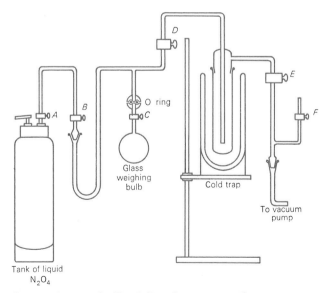

figure 25. Apparatus for filling bulbs with nitrogen tetroxide.

of liquid N_2O_4 and the train of glass tubes are set up permanently on a rack in a ventilated hood.

The steel valve A on the nitrogen tetroxide tank is connected with a small, slightly flexible stainless-steel tube and stopcock B. The outlet from this stopcock is attached with inert cement to the inner half of a standard-taper ground-glass joint. The outer part of the joint is connected to the glass system of tubes and stopcocks including a glass O-ring joint which makes connection with the weighed bulb for holding the N_2O_4. The surfaces at the joint are so accurately formed that a vacuumtight connection is made at the rubber O ring without the use of stopcock grease. Such grease on a replaceable surface makes accurate weighing impossible. The bulb is provided with stopcock C and lubricated with a trace of silicone grease.

The tube from the tank extends to stopcock D of large diameter and then to a cold trap containing trichloroethylene and Dry Ice in a vacuum bottle. The Dry Ice is added in small pieces, slowly at first, to prevent frothing of the liquid over the top of the vacuum bottle. The bottle is partially filled and then raised on a ring stand and holder to immerse the cold trap and freeze out the nitrogen tetroxide and thus protect the pump and pump oil from nitrogen tetroxide and prevent back diffusion of atmospheric moisture. Stopcock E of large bore permits the pump to be closed off after the system has been evacuated, and stopcock F allows the access of air at atmospheric pressure.

The filling operation is carried out, in steps I to V, as follows:

Stopcocks	A	B	C	D	E	F
I	Closed	Closed	Closed	Closed	Closed	Open
II	Closed	Closed	Open	Open	Open	Closed
III	Closed	Open	Open	Open	Open	Closed
IV	Open	Open	Open	Closed	Open	Closed
V	Closed	Closed	Closed	Closed	Open	Closed

The vacuum oil pump is started with stopcocks arranged as indicated in step I.

The system is evacuated as shown at step II.

Evacuation is continued with stopcocks set as shown in step III.

Then the stopcocks are turned as shown at step IV, and the bulb is filled with nitrogen tetroxide at a pressure equal to its vapor pressure. The temperature of the tank must be higher than $21.5°$, the boiling point, in order to have the gas in the bulb slightly above atmospheric pressure.

Steps III and IV are then repeated to eliminate residual air.

The stopcocks are then closed as indicated at step V, and the bulb is removed for weighing.

A second bulb is filled using the same sequence of operations. When there are no more bulbs to be filled, stopcock *F* is opened and the pump is turned off. Then the freezing bath is lowered, the trichloroethylene is stoppered and saved, and the lower part of the trap is removed. The frozen nitrogen tetroxide is flushed out with an excess of water.

A small thermostat which permits rapid setting is used for measurements at successively higher temperatures.

The bulb is placed in the small thermostat, which is first set at the lowest temperature to be used (about 30°). The stopcock is opened momentarily at intervals of 2 or 3 min during the period of thermostating, and the gas is allowed to escape until no more brown fumes are seen to issue from the opening. The bulb is closed and again weighed;† care must be taken to ensure uniform weighing technique. Time should be allowed for moisture equilibrium between the bulb and the air in the balance case to be attained and for the bulb to cool to the temperature of the balance.

The bulb is then placed in a thermostat at a temperature approximately 10° higher, and the above procedure repeated. The same measurements are made every 10° up to 60°.

The volume of the bulb may be determined from the weight of water it can hold. The bulb is evacuated two or three times with a water aspirator in order to remove all corrosive nitrogen tetroxide. It is then evacuated with the aspirator, the stopcock is closed, and the end of the tube is immersed in a beaker of distilled water. The stopcock is then opened to permit the bulb to fill. A hypodermic syringe with a long needle, passing through the stopcock bore, may be used to complete the filling of the bulb. The bulb is then weighed, and from the weight of water it holds the volume is calculated by use of the data on the density of water given in the Appendix.

The bulb is then emptied for later experiments, with the help of the water aspirator, and placed in a drying oven. It is evacuated several times, while hot, to assist in the drying process.

The barometric pressure must be taken at the time of the experiment for use in subsequent calculations.

CALCULATIONS. The average molecular weight of the gas at each temperature is computed by use of the ideal-gas law (cf. Experiment 1). The corresponding values of the degree of dissociation and the equilibrium constant are calculated by application of Eqs. (2) and (1). A plot is made of log K_p versus $1/T$, and the equation is found for the line considered to best represent the set of points. The standard en-

---

† After removing the bulb from the thermostat, the temperature regulator should be reset to a point approximately 10° higher. The thermostat will then be approaching the new temperature while weighings are being made.

thalpy change, the standard free-energy change, and the entropy change for the reaction are calculated for 25°C.[1,2]

Practical applications. The determination of equilibrium constants is of fundamental importance in industrial work.

Suggestions for further work. A glass-diaphragm manometer may be used for studying this equilibrium at various pressures.[3]

A simple photometer may be used for determining the partial pressure of NO_2 in the mixture.[4]

The dissociation of N_2O_4 in carbon tetrachloride solution may be studied.[5]

Other rapid reversible dissociations such as that of phosphorus pentachloride and ammonium chloride may be studied at higher temperatures.

References

1. G. N. Lewis and M. Randall, "Thermodynamics," 2d ed., p. 561, rev. by K. S. Pitzer and L. Brewer, McGraw-Hill Book Company, New York, 1961.
2. W. F. Giauque and J. D. Kemp, *J. Chem. Phys.*, **6:** 40 (1939).
3. F. H. Verhoek and F. Daniels, *J. Am. Chem. Soc.*, **53:** 1250 (1931).
4. L. Harris and B. M. Siegel, *Ind. Eng. Chem. Anal. Ed.*, **14:** 258 (1942).
5. K. Atwood and G. K. Rollefson, *J. Chem. Phys.*, **9:** 506 (1941).

16   SPECTROPHOTOMETRIC DETERMINATION OF AN EQUILIBRIUM CONSTANT

The equilibrium constant for a reaction in solution is determined from a study of the concentration dependence on the intensity of an absorption band in the spectrum of the solution.

THEORY.   The present experiment illustrates an important method for the study of chemical equilibrium in solution. The method utilizes differences in the light-absorbing properties of reactants and products and is particularly appropriate for systems for which classical methods of chemical analysis cannot be used to find the concentrations of the various species present.

The optical absorption spectrum, i.e., the percentage transmission of light as a function of wavelength, has been investigated for iodine in a variety of solvents. Associated with the color of the solutions is a strong absorption in the neighborhood of 500 mμ.† In certain solvents, especially aromatic compounds, a new absorption band appears in the violet or ultraviolet region. The new band has been attributed to a complex (a molecular combination of iodine with solvent) existing in equilibrium with uncomplexed iodine and solvent. By means of a quantitative study of the

† The wavelength unit millimicron (mμ) is equivalent to 10 Å, or 10^{-7} cm.

intensity of absorption at the peak of this band, as a function of concentration, it is possible to test this interpretation and to obtain a value for the equilibrium constant for the formation of the complex.[1,2]

The study is best carried out with iodine and the complexing organic substance, mesitylene in this experiment, both present as solutes in dilute solution in an inert solvent such as CCl_4. (The inertness of CCl_4, expected on chemical grounds, is verified by the absence of new absorption bands in solutions of I_2 in CCl_4.) The initial step involves the measurement of the percentage transmission, over the appropriate wavelength range, of three solutions: one containing I_2 solute only, one containing mesitylene solute only, the third containing both I_2 and mesitylene as solutes. An absorption band present only in the spectrum of the third solution is attributed to a 1:1 complex, $M \cdot I_2$, existing in equilibrium with free mesitylene (M) and I_2,

$$M + I_2 = M \cdot I_2$$

$$K = \frac{x}{(c_1 - x)(c_2 - x)} \tag{1}$$

where c_1 = total concentration of mesitylene
  c_2 = total concentration of I_2
  x = concentration of complex at equilibrium
  K = equilibrium constant

It remains to verify that an equilibrium condition of the form of Eq. (1) is consistent with the concentration dependence of the absorption intensity and to evaluate K. The new absorption band occurs in a wavelength region in which absorption by uncomplexed iodine and mesitylene is very slight; the investigation can therefore proceed without serious interference from absorption due to uncomplexed solutes.

Let I and I_0 be intensities of light of a specified wavelength transmitted, respectively, by solution and by pure solvent. Then the optical absorbancy A, defined by

$$A = -\log \frac{I}{I_0} \tag{2}$$

is given by the Beer-Lambert law, for the case in which only one solute absorbs at the given wavelength, by

$$A = abc$$

where a = molar absorbancy index of absorbing solute
  b = length of light path in cell
  c = concentration of absorbing solute

The molar absorbancy index depends on the wavelength, temperature, and solvent.

If several solutes absorb independently, the absorbances are additive. Thus

$$A = A_1 + A_2 + A_3 \tag{3}$$
$$A_1 = a_1 b(c_1 - x) \tag{4}$$
$$A_2 = a_2 b(c_2 - x) \tag{5}$$
$$A_3 = a_3 b x \tag{6}$$

where a_1 = molar absorbancy index of mesitylene

a_2 = molar absorbancy index of I_2

a_3 = molar absorbancy index of complex

Hence, in the present case, measurement of absorbancy does not lead directly to values for the concentration of complex. A_1 and A_2, though smaller than A_3 at the wavelength of the peak of the new complex band, may not always be negligible; also, a_3 is initially unknown. However, the following indirect procedure yields values for a_3 as well as for K.

A considerable simplification in the work results from restricting the investigation to solutions in which mesitylene is present in large excess. Thus

$$c_1 \gg c_2 > x \tag{7}$$

Accordingly, for the calculation of A_3 from A, the approximations

$$A_1 = a_1 b c_1$$
$$A_2 = 0$$

may be used, with a_1 calculated from the absorbancy measured at the given wavelength for a solution containing only mesitylene as solute at a known concentration.

The equilibrium condition (1), with the approximation (7), simplifies to

$$K = \frac{x}{c_1(c_2 - x)} \tag{8}$$

Upon replacement of x by $A_3/a_3 b$ and rearrangement, this becomes

$$\frac{A_3}{c_1 c_2} = b a_3 K - \frac{A_3}{c_2} K \tag{9}$$

Absorbancies are measured for a series of solutions made up with various known values of c_1 and c_2, and values of $A_3/c_1 c_2$ plotted against A_3/c_2. If the data are well represented by a straight line, the slope and intercept lead to values for a_3 and K.

The inclusion of a correction for absorption due to uncomplexed I_2 may be warranted if the data are of sufficient accuracy. A procedure for carrying out this refinement is outlined under Suggestions for Further Work.

A word of caution is in order with regard to the interpretation of the results.

While it is possible by the procedure outlined here to learn whether or not the absorbancies are consistent with the postulated reaction equilibrium, linearity of the graph does not of itself constitute proof that Eq. (1) represents the true state of affairs. It has in fact been shown[3,4] that a functional relationship among c_1, c_2, and A_3 identical with that of Eq. (9) will exist for a system described by two stages of complex formation,

$$M + X = MX \qquad K_1 = \frac{(MX)}{(M)(X)}$$

$$MX + X = MX_2 \qquad K_2 = \frac{(MX_2)}{(MX)(X)}$$

if the ratio of the molar absorbancies of MX and MX_2 happens to satisfy a certain condition. (In this case A_3, found experimentally as before, stands for the total absorbancy due to complexed M.) The best procedure, therefore, when it can be used, is to make measurements on more than one of the absorption bands due to the complex. The necessary condition on the ratio of the absorbancies is likely not to be satisfied at each of several widely different wavelengths, so that it will then become obvious if the system is not obeying the equilibrium equations corresponding to Eq. (1).

Finally, a few comments will be made about the principle and operation of the spectrophotometer.[5] The function of a spectrophotometer is to produce monochromatic light of a selected wavelength and to measure the intensity of light transmitted by a solution relative to that transmitted by a sample of the pure solvent. The essential elements of a typical spectrophotometer are shown in Fig. 26. The source is a

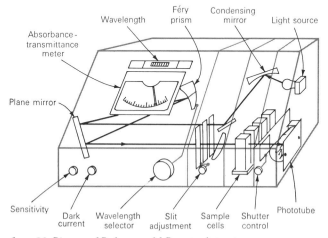

figure 26. Diagram of Beckman model B spectrophotometer.

tungsten lamp. The wavelength to be used is selected by turning a knob which moves the prism until light of the desired wavelength is directed toward the slit. The wavelength scale is mechanically coupled to the same shaft. The slit, an aperture of adjustable width, serves to admit only light of a narrow band of wavelengths; at the same time, it necessarily determines the intensity of light passed. As the slit is widened to provide greater intensity of light for the measurement, a wider range of wavelengths is permitted to pass through. At wavelengths where the transmission of a sample varies rapidly with wavelength, the slit width should be kept as narrow as possible, consistent with the need for adequate intensity.

In the unit shown, any one of four samples may be placed in the light path by manually shifting an external shaft. With the shutter closed, the so-called dark current of the phototube is balanced out so as to give a meter deflection corresponding to zero percent transmission. With the shutter open, and the pure solvent sample in the light path, the sensitivity and slit-width controls are adjusted to give a deflection corresponding to 100 percent transmission. Finally, with the solution sample in the light path, the deflection indicates directly the percent transmission. The output meter scale is also calibrated directly in absorbancy.

If the solution transmission is low, better accuracy may be attained by increasing the sensitivity control by one or more steps; if this is done, the scale is altered: for the Beckman model B spectrophotometer, 0.5 must be added to the absorbancy scale reading for each step by which the sensitivity control has been advanced above that at which the reference setting was made with the pure solvent.

For accurate work it is necessary to take into account differences among spectrophotometer sample cells. The first to be considered is that resulting from imperfections in the cell windows. To determine the correction for this effect, the cells are cleaned and filled with nonabsorbing solvent. It is convenient to select the cell with the highest transmission as the reference cell. The slit width is adjusted to give a deflection corresponding to 100 percent transmission with this cell in the light path. The other cells are then placed in turn in the light path, and the absorbancies, designated A_c, noted. The value of A_c for a particular cell is called the *cell correction;* A_c may vary with wavelength. In subsequent measurements with solutions, the true sample absorbancy is then found by subtracting the value of A_c for the cell used from the measured absorbancy of the cell filled with the solution sample.

Second, the optical path lengths may be different for two cells of nominally the same thickness. The path lengths for two cells can most easily be compared by measuring the transmission for both filled with a series of absorbing solutions. The manufacturing tolerances on cell thickness are close enough to permit omission of the path-length correction in most student work.

····> ***Apparatus.*** Spectrophotometer; set of four cells, preferably matched; special mixing cell; reagent grade CCl_4; 1 ml of 0.04 M I_2 in CCl_4; 8 ml mesitylene (1,3,5-trimethylbenzene); 25-ml volumetric flask.

PROCEDURE.   Two solutions are prepared: (*a*) 100 ml of a 0.0004 M solution of I_2 in CCl_4 and (*b*) 25 ml of a 2 M solution of mesitylene in CCl_4. The first is prepared by dilution of a more concentrated solution, which in turn is made up by dissolving a weighed quantity of I_2 in a measured volume of solution. The second is made up by weighing approximately 8 ml of mesitylene in a 25-ml volumetric flask and filling to the mark with solvent. The concentration of the I_2 solution may be determined by standard titration methods or may be calculated from the measured absorbancy at the peak of the band near 500 mμ. The molar absorbancy index for I_2 in CCl_4 solvent at the wavelength of the peak of this band is 900 liters mole^{-1} cm^{-1} (Ref. 6).

The spectra of three solutions are to be obtained:

1. 0.0002 M I_2 in CCl_4
2. 1 M mesitylene in CCl_4
3. 0.0002 M I_2, 1 M mesitylene, in CCl_4

Each of these is placed in a spectrophotometer cell, and pure solvent is placed in a fourth cell to serve as reference. The percentage transmission is measured at convenient intervals from 700 to 320 mμ. The manufacturer's instructions for the operation of the spectrophotometer should be consulted, but it is to be noted especially that the slit width must be adjusted at each wavelength for a meter deflection of 100 percent for the pure solvent and that, where recommended, a suitable filter should be used. The data may be plotted while the measurements are being made in order that sufficiently closely spaced points may be taken to assure adequate delineation of the spectral curves.

The three cells used for the solutions may then be filled with solvent, and their transmission measured relative to that of the reference cell at several wavelengths. A cell correction can be applied to the data previously obtained if there is an appreciable difference between any cell and the reference cell.

Absorbancies of a number of solutions are to be measured at a wavelength at which the complex absorbs strongly while uncomplexed I_2 and mesitylene do not. For these measurements, a special spectrophotometer cell† (Fig. 27) is employed, together with an ordinary cell containing pure solvent as reference. The entire series of measurements is to be taken without altering the wavelength setting.

† Available commercially from Norman D. Erway, Oregon, Wis.

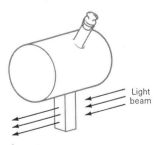

Light beam

figure 27.

Special mixing cell for spectrophotometer. The lower portion is a spectrophotometer cell.

First, pure solvent is placed in the special cell and the cell correction relative to the reference cell determined. Next, a sample of the 0.0004 M I_2 solution is placed in the special cell and its absorbance noted. For best results in the measurements to follow, the absorbancy of the iodine solution should not exceed 0.02 with the proper cell correction; a larger absorbancy may indicate the presence of impurities due to aging of the solution or to the use of impure solvent in its preparation.

After careful rinsing and drying of the special cell, a 5-ml sample of the 2 M mesitylene solution is placed in it and the absorbancy measured. To this are added one 1-ml portion and then in succession five 5-ml portions of the I_2 solution, the absorbancy being measured after each addition. Care must be taken to avoid error due to evaporation of solvent. The solutions should be mixed by gently rocking the cell, and not by shaking it, because loss of solution at the top will cause error.

Finally, the temperature of the solution at the time of the measurement is noted, since K is a function of temperature.

CALCULATIONS.   Percentage transmission is plotted versus wavelength for each of the three solutions and the significance of the curves briefly discussed.

The absorbancy data obtained in the final series of measurements may be treated by the graphical procedure outlined in the theory section to determine a_3 and K for the temperature of the run.

The validity of the approximation $c_1 \gg x$ should be checked by calculating c_1 and x for the least favorable case.

In making comparison of the value of K obtained in this experiment with values in the literature, attention must be paid to the way in which K is defined, as the value is different if mole fractions are used instead of concentrations.

Suggestions for further work. Combination of Eq. (8) with (5) gives an expression which may be used to calculate values of A_2 for each solution after the first approximation to K has been obtained by the procedure given above. Improved values of A_3 may then be calculated, and a new series of points plotted on the original graph. The cycle may be repeated if the new value for K leads to appreciably different estimates for values of A_2. A value for a_2 at the appropriate wavelength may be obtained from measurements on the stock I_2 solution.

References

1. H. A. Benesi and J. H. Hildebrand, *J. Am. Chem. Soc.*, **71:** 2703 (1949).
2. L. J. Andrews and R. M. Keefer, *J. Am. Chem. Soc.*, **74:** 4500 (1952).
3. R. Kruh, *J. Am. Chem. Soc.*, **76:** 4865 (1954).
4. T. W. Newton and F. B. Baker, *J. Phys. Chem.*, **61:** 934 (1957).
5. W. West in A. Weissberger (ed.), "Technique of Organic Chemistry," vol. 1, "Physical Methods of Organic Chemistry," 3d ed., pt. 3, chap. 28, Interscience Publishers, Inc., New York, 1960.
6. O. J. Walker, *Trans. Faraday Soc.*, **31:** 1432 (1935).

17 ACID DISSOCIATION CONSTANT OF METHYL RED

The spectrophotometric determination of the acid dissociation constant of a dye is illustrated.

THEORY. In aqueous solution methyl red is a zwitterion and has a resonance structure somewhere between the two extreme forms shown in Fig. 28. This is the red form HMR in which methyl red exists in acid solutions. When base is added, a proton is lost and the yellow anion MR⁻ of methyl red has the structure shown at the bottom of the figure. The basic form is yellow because it absorbs blue and violet light. The equilibrium constant for the ionization of methyl red is

$$K_c = \frac{(H^+)(MR^-)}{(HMR)} \tag{1}$$

It is convenient to use this equation in the form

$$pK_c = pH - \log\frac{(MR^-)}{(HMR)} \tag{2}$$

The ionization constant may be calculated from measurements of the ratio $(MR^-)/(HMR)$ at known pH values.

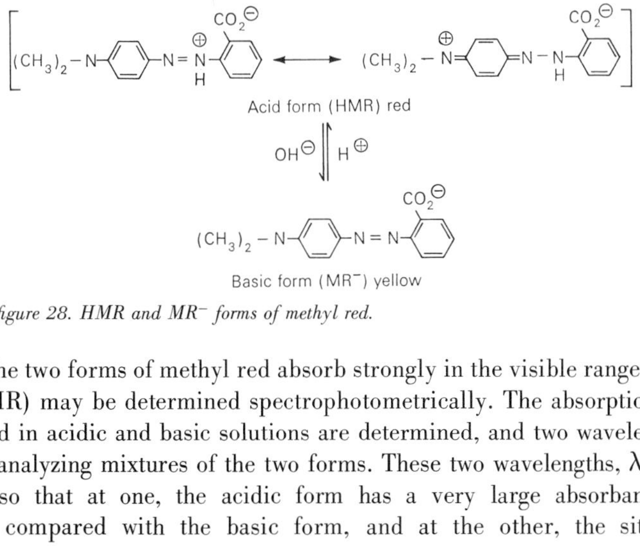

figure 28. HMR and MR⁻ forms of methyl red.

Since the two forms of methyl red absorb strongly in the visible range, the ratio $(MR^-)/(HMR)$ may be determined spectrophotometrically. The absorption spectra of methyl red in acidic and basic solutions are determined, and two wavelengths are selected for analyzing mixtures of the two forms. These two wavelengths, λ_1 and λ_2, are chosen so that at one, the acidic form has a very large absorbancy index (page 107) compared with the basic form, and at the other, the situation is reversed. The absorbancy indices of HMR and MR⁻ are determined at both of these wavelengths, using several concentrations to determine whether Beer's law is obeyed.

The composition of a mixture of HMR and MR^- may be calculated from the absorbancies A_1 and A_2 at wavelengths λ_1 and λ_2 using, at unit cell thickness,

$$A_1 = a_{1,\text{HMR}}(\text{HMR}) + a_{1,\text{MR}^-}(\text{MR}^-) \tag{3}$$
$$A_2 = a_{2,\text{HMR}}(\text{HMR}) + a_{2,\text{MR}^-}(\text{MR}^-) \tag{4}$$

····> **_Apparatus._** Spectrophotometer for measuring absorbancies in the visible range; pH meter; methyl red; sodium acetate; acetic acid; hydrochloric acid; 95 percent ethanol; volumetric flasks and pipettes for preparing solutions.

PROCEDURE.   The procedure for this experiment has been described by Tobey.[1] The methyl red is conveniently supplied as a stock solution made by dissolving 1 g of crystalline methyl red in 300 ml of 95 percent ethanol and diluting to 500 ml with distilled water. The standard solution of methyl red for use in this experiment is made by adding 5 ml of the stock solution to 50 ml of 95 percent ethanol and diluting to 100 ml with water.

The absorption spectrum of methyl red is determined in hydrochloric acid solution as solvent to obtain the spectrum of HMR and in sodium acetate solution as solvent to obtain the spectrum of MR^-. Distilled water is used in the reference cell. The procedure for using the spectrophotometer is described in Experiment 16. Since the equilibrium to be studied is affected by temperature, it is important that all the spectrophotometric and pH measurements be made at the same temperature. If the cell compartment of the spectrophotometer is slightly above room temperature, the filled cells should be placed in the spectrophotometer just before making the measurements. In order to obtain the best results, the cell compartment should be thermostated. The acidic solution is conveniently prepared by diluting a mixture of 10 ml of the standard methyl red solution and 10 ml of 0.1 M hydrochloric acid to 100 ml. The basic solution is conveniently prepared by diluting a mixture of 10 ml of the standard methyl red solution and 25 ml of 0.04 M sodium acetate to 100 ml.

From the plots of absorbancy versus wavelength, two wavelengths are selected for analyzing mixtures of the acidic and basic forms of methyl red. Further spectrophotometric measurements over a range of concentration are made at these two wavelengths with both acidic and basic solutions to check whether Beer's law is obeyed. The solutions are diluted with 0.01 N hydrochloric acid or 0.01 N sodium acetate, respectively, so that the medium is held constant.

In order to determine the ionization constant of the dye, spectrophotometric analyses are carried out on solutions containing 0.01 N sodium acetate, a constant total concentration of dye, and various concentrations of acetic acid. The pH values of these solutions are measured at the same temperature as the spectrophotometric measurements. For methyl red it is convenient to use acetic acid concentrations ranging from 0.001 to 0.05 N.

CALCULATIONS.   Plots are prepared of absorbancy versus wavelength and absorbancy versus concentration of dye in acidic and basic solutions at λ_1 and λ_2. The values of the various absorbancy indices are calculated.

The concentrations of the acidic and basic forms of the dye in the various buffer solutions are calculated by using Eqs. (3) and (4).

Equation (2) is used to calculate the pK value for the dye. As a means of testing and averaging the data log $[(MR^-)/(HMR)]$ may be plotted versus the pH. An average value from the literature[2] is 5.05 ± 0.05 for the 25 to 30° temperature range.

Practical applications.   This method is useful for studying dyes for use as indicators in acid-base titrations, or by an analogous procedure for indicators for oxidation-reduction titrations.

Suggestions for further work.   The pK values for other common dyes may be determined. General references are cited.[3]

References

1. S. W. Tobey, *J. Chem. Educ.,* **35:** 514 (1958).
2. I. M. Kolthoff, "Acid-Base Indicators," The Macmillan Company, New York, 1953.
3. R. P. Bell, "Acids and Bases: Their Quantitative Behavior," Methuen & Co., Ltd., London, 1952; "The Proton in Chemistry," Cornell University Press, Ithaca, N.Y., 1959.

Chapter 6

Heterogeneous Equilibria

18 DISTRIBUTION OF A SOLUTE BETWEEN IMMISCIBLE SOLVENTS

Studies are made of the equilibrium distributions of a solute between two immiscible solvents. Such experiments give evidence of association or dissociation of the solute in one of the phases. They provide information as to the nature of complex ions and their dissociation constants. Distribution-coefficient data are of value in the design and operation of solvent-extraction equipment.

THEORY.[1] When two liquid phases are in equilibrium with each other, a dissolved substance will distribute itself between the two according to a definite equilibrium. If we represent the two solvents in contact as being the α and β phases and the solute which is present in each layer as i, then at equilibrium the chemical potentials μ of i in the two phases will be equal; thus $\mu_i^\alpha = \mu_i^\beta$. In general, and on the C, or volume-based, concentration scale, we then have

$$(\mu_i^\alpha)^\circ + RT \ln y_i^\alpha C_i^\alpha = (\mu_i^\beta)^\circ + RT \ln y_i^\beta C_i^\beta$$

where the activity coefficients are represented by y_i. Thus

$$\ln \frac{y_i^\beta C_i^\beta}{y_i^\alpha C_i^\alpha} = \frac{(\mu_i^\alpha)^\circ - (\mu_i^\beta)^\circ}{RT} = \text{constant at constant } T, P \tag{1}$$

Hence the distribution coefficient K_c defined by

$$\frac{C_i^\beta}{C_i^\alpha} = K_c \tag{2}$$

is a function of solute concentrations only to the extent that the activity coefficients are concentration-dependent.

In this simplest form the behavior of a neutral solute molecule has been considered. However, the relationship $\mu_i^\alpha = \mu_i^\beta$ is a perfectly general one, and it is necessary only to use proper descriptions of these chemical potentials when the solute ionizes or associates in either of the two solvent mediums. For a substance which ionizes, the logarithmic relation of chemical potential to concentration applies in the limit of zero concentration to each ionic species, so that the chemical potential of a salt is obtained by adding the chemical potentials of the ions. If, for example, the equilibrium distribution of hydrochloric acid between benzene and water were being studied,[1] one would write for the chemical potential of the acid in the aqueous phase α

$$\mu_{\text{HCl}} = (\mu_{\text{H}^+})^\circ + (\mu_{\text{Cl}^-})^\circ + 2RT \ln C_{\text{HCl}} + 2RT \ln \chi_\pm \dagger$$

† When a salt (or strong acid or base) is dissolved in water, a system of three chemically different species is formed. The logarithmic relation of chemical potential to concentration applies to each species, so that in this instance the chemical potentials of hydrogen and chloride ions are to be added to give the chemical potential of the acid. Perhaps it should be noted that this is a system of two components rather than of three, because the concentration of one of the ions is known from that of the other, for the system as a whole is electrically neutral.

Equating this to the chemical potential of the undissociated acid in the benzene phase β yields

$$\frac{C_i^\beta}{(C_i^\alpha)^2} = K_c \tag{3a}$$

If, on the other hand, the solute is associated to form an n-mer in the organic, or β, phase according to the reaction

$$nA \rightleftharpoons A_n$$

it can be shown in the same way that

$$\frac{C_i^\beta}{(C_i^\alpha)^n} = K_c \tag{3b}$$

If the association is not complete, the value of n computed from Eq. (3b) will not be an integer and it may vary with concentration.

The carboxylic acids are suitable for a study of an association because they generally form double molecules in nonpolar solvents or in the gas phase but exist as single molecules in polar solvents such as water. Some carboxylic acids, e.g., acetic acid, are so weak that their ionization in water can be practically neglected, but others, e.g., trichloroacetic acid, are almost entirely ionized.

It follows directly from the distribution law that extraction with several portions of solvent is more efficient than with a single portion of the same total volume. In the case of phase distributions which satisfy Eq. (2), it is possible to derive a generalized formula which will show the amount remaining unextracted after a given number of operations. If V ml of solution initially containing x_0 g (or equivalents) of a substance is repeatedly extracted with v ml of an immiscible solvent, we may calculate the number of grams x_n remaining unextracted after n extractions as follows. After one extraction, the concentration in the original solution will be x_1/V and in the extracting phase $(x_0 - x_1)/v$. Therefore the distribution coefficient is

$$\frac{x_1/V}{(x_0 - x_1)/v} = K_c \tag{4}$$

and

$$x_1 = x_0 \frac{K_c V}{K_c V + v} \tag{5}$$

After a second extraction, x_2 g (or equivalents) remains in the original solvent. Thus

$$x_2 = x_1 \frac{K_c V}{K_c V + v} = x_0 \left(\frac{K_c V}{K_c V + v}\right)^2 \tag{6}$$

The last term is obtained by substituting for x_1 its equivalent as given by Eq. (5).

After n extractions, the quantity remaining in the original solvent is

$$x_n = x_0 \left(\frac{K_c V}{K_c V + v} \right)^n \tag{7}$$

Craig and others[2-4] have designed various pieces of apparatus for carrying out multiple extractions conveniently. A countercurrent-distribution apparatus is analogous to a distillation column in that separation is achieved by many two-phase distributions. Such countercurrent-distribution experiments may be used not only for the purpose of fractionation but also for the characterization of unknown organic compounds.

Since the distribution coefficients for organic acids and bases between aqueous solutions and immiscible organic solvents depend markedly on the pH of the aqueous phase, multiple extractions with various buffer solutions may be used to separate various organic acids, such as the penicillins.

In still another application, distribution experiments may be used to investigate the nature of a complex ion and to make estimates of its dissociation constant. We consider the triiodide ion. With pure water and carbon tetrachloride as the immiscible solvents, the distribution coefficient K_c for iodine, I_2, between them is determined. Then, if instead of pure water, solutions of potassium iodide are used as one of the phases, there will be an apparent increase in the solubility of the iodine which may be attributed to the rapidly reversible reaction

$$I^- + I_2 = I_3^-$$

for which the apparent equilibrium constant is

$$K_c = \frac{C_{I_3^-}}{C_{I_2} C_{I^-}} \tag{8}$$

The amount of free iodine in aqueous iodide solution cannot be determined by direct titration with standard thiosulfate solution, because as soon as the iodine is removed, the complex triiodide ion which is present dissociates to give more iodine. However, the concentration of uncombined iodine in this aqueous phase can be obtained by titrating the carbon tetrachloride layer and making use of the known distribution coefficient for iodine between the two pure liquids.

The total iodine present, I_2 and I_3^-, may be obtained by the thiosulfate titration, so that it becomes possible to compute by difference the triiodide-ion concentration. The total iodide-ion concentration is known from the manner in which the experiment is designed.

Part A

→ **Apparatus.** Three 100-ml separatory funnels; three 100-ml Erlenmeyer flasks; 100-ml volumetric flasks; 25-ml pipette; 10-ml pipette; 1.0 N acetic acid or trichloroacetic acid or other acid; glacial acetic acid; carbon tetrachloride; ether; 0.5 N sodium hydroxide; 0.01 N sodium hydroxide.

PROCEDURE.   One hundred milliliters each of approximately 0.50 N, 1 N, and 2 N solutions of acetic acid in water is prepared. Twenty-five milliliters of each of the three solutions is pipetted into closed 100-ml separatory funnels, and to each is added 25 ml of diethyl ether. Closed rubber tubes are put over the outlets to keep out the water of the thermostat, and the separatory funnels are set in a thermostat at 25° for 20 min or more, with frequent shaking.

After the solutions have come to equilibrium, the separatory funnels are removed from the thermostat and the lower layers run out into beakers, care being taken to let none of the upper layer go through. Ten-milliliter samples of each of the lower layers are taken rapidly and drained into Erlenmeyer flasks for titration with sodium hydroxide, using phenolphthalein as an indicator. Samples of the upper layers are removed from the separatory funnels with 10-ml pipettes, care being taken to avoid taking up any of the lower layer. The aqueous and ether solutions are titrated with 0.5 N NaOH. Check titrations should be made in each case.

A second set of experiments is carried out in the same way, using 25-ml aliquots of carbon tetrachloride instead of ether. Ten-milliliter aliquots of the CCl_4 phases are measured rapidly in order to prevent loss due to evaporation. The CCl_4 phases are titrated with 0.01 N sodium hydroxide.

In titrating the acid dissolved in carbon tetrachloride or ether, an equal volume of water is added initially, and it is necessary to shake vigorously to accelerate the passage of dissolved acid across the surface and into the water layer.

CALCULATIONS.   Concentrations of acid C_A in the two layers in the distribution experiments are calculated in moles per liter, and the distribution coefficients are calculated with Eq. (2). Plots of log C_A^β versus log C_A^α are also constructed, and n of Eq. (3b) calculated. If the value of n is not equal to unity, a hypothesis is suggested to explain the results.

As an illustration of the application of the distribution coefficient, the concentration of acetic acid remaining in the ether phase is calculated for the following cases:

1. 50 ml of 1.0 N acetic acid in ether is extracted with 50 ml of water.
2. 50 ml of 1.0 N acetic acid in ether is extracted with five 10-ml portions of water.

Part B

····→ ***Apparatus.*** Glass-stoppered bottles of 250 ml capacity; beakers, burettes, and pipettes; starch indicator; 0.02 and 0.1 N thiosulfate solutions; 0.1 N potassium iodide; carbon tetrachloride saturated with iodine.

PROCEDURE.   In this section we investigate the triiodide equilibrium. To do so, the distribution of iodine between carbon tetrachloride and three different aqueous phases is studied. These aqueous phases are (*a*) pure water, (*b*) 0.01 N potassium iodide solution, and (*c*) 0.1 N potassium iodide solution. To each of six of the glass-stoppered bottles is added 35 ml of the carbon tetrachloride saturated with iodine. The bottles are then segregated in pairs to provide duplicate experiments. To each of the first pair is added 100 ml pure water; to each of the second pair, 75 ml of 0.01 N potassium iodide solution; and to each of the third pair, 35 ml of 0.1 N potassium iodide solution. The bottles are thoroughly shaken from time to time and at frequent intervals for half an hour; otherwise they are to remain immersed to the neck in a 25°C thermostat.

   After the several systems have come to equilibrium, time is permitted for complete phase separation. Then samples may be removed by pipetting directly from the bottles in the thermostat. In pipetting out samples of the carbon tetrachloride phase, the inside of the pipette may be kept dry as it passes through the aqueous phase by placing a finger tightly over the upper end of the pipette. The concentration of iodine in each of the solutions is determined by titration with one of the thiosulfate solutions, using the starch indicator.

   In order to achieve the necessary accuracy, it is suggested that the dilute thiosulfate solution be used for the titration of measured volumes of the aqueous phases, 75, 50, and 25 ml being used in order of increasing potassium iodide concentration. For the carbon tetrachloride layers, 25 ml is a suitable volume, and the more concentrated thiosulfate solution is used as the analytical reagent.

   The titration of the carbon tetrachloride layer presents some problems. In the first place, it is slow, because the iodine must be extracted into the aqueous solution which is being added. The starch-indicator solution is used in small volume only as the end point is approached, as judged by the near disappearance of the iodine color from the carbon tetrachloride phase. It is difficult to avoid overtitration. The addition of a small volume of 10 percent KI solution is of aid in the titration.

CALCULATIONS.   The concentrations of iodine in the two pure-solvent layers, water and carbon tetrachloride, are calculated in moles per liter, and the distribution coefficient is computed with the aid of Eq. (2). Now, with the use of this coefficient, the concentration of molecular iodine C_{I_2} in the two potassium iodide solutions is calculated, it being assumed that

$$\frac{C_{I_2}^{\beta}}{C_{I_2}^{\alpha}} = \text{constant}$$

These values, when subtracted from the total iodine concentration of the solution, provide data for the computation of the concentrations of the I_3^- in the two aqueous potassium iodide systems. The iodide-ion concentrations, C_{I^-}, are obtained by subtracting the values for $C_{I_3^-}$ from the original potassium iodide concentrations.

With these data, values for the equilibrium constant for the reaction $I^- + I_2 = I_3^-$, as defined by Eq. (8), are computed for the two different potassium iodide systems. They are compared with data in the literature.[5]

Practical applications. Extraction of a solute by shaking a solution with an immiscible solvent is an operation that is used extensively in organic chemistry. The efficiency of any such operation may be calculated when the distribution coefficient is known.

The chemical potential of a solute determines its distribution into a second solvent; thus the activity of the solute may be calculated from distribution data.

In some cases, the degree of hydrolysis of a salt may be determined by measuring the distribution ratio of the acid or base when shaken with an immiscible solvent.

The extraction of uranium from its ores and the separation of uranium from its fission products in atomic-energy operations often make use of solvent extraction with an organic solvent which is immiscible with water.

The coloring of polyethylene terephthalate fibers by nonionic dyes is another example of the practical use of a distribution coefficient.[6]

Suggestions for further work. Other distribution systems may be studied, such as the following:

Hydrochloric acid between water and benzene.[1]

Salicylic acid or picric acid between water and benzene and between water and chloroform as a function of the pH of the aqueous phase.

Uranyl nitrate between water and ether or between water and tributyl phosphate. The extraction of uranyl nitrate into the organic phase is improved by the addition of nitric acid.

As an example of the distribution of an inorganic salt between immiscible solvents, the distribution of ferric chloride between ether and water may be studied. The salting-out effect is illustrated by the addition of excess sodium chloride or calcium chloride to the aqueous phase.

References

1. R. W. Knight and C. N. Hinshelwood, *J. Chem. Soc.*, **1927**: 466.
2. L. C. Craig, *J. Biol. Chem.*, **155**: 519 (1944).
3. L. C. Craig and D. Craig in A. Weissberger (ed.), "Technique of Organic Chemistry," 2d ed., vol. 3, pt. 1, "Separation and Purification," Interscience Publishers, Inc., New York, 1956.
4. L. C. Craig in P. Alexander and R. J. Block (eds.), "Analytical Methods of Protein Chemistry," vol. 1, Pergamon Press, New York, 1960.
5. G. Jones and B. B. Kaplan, *J. Am. Chem. Soc.*, **50**: 1845 (1928).
6. M. J. Schuler and W. R. Remington, *Discussions Faraday Soc.*, **16**: 201 (1954).

19 PHASE DIAGRAM OF A BINARY SOLID-LIQUID SYSTEM

This experiment illustrates the use of cooling curves to establish the phase diagram for a binary system. It illustrates also the use of the thermocouple.

THEORY.[1] The purpose of the experiment is to obtain data by thermal analysis for constructing a phase diagram which indicates the solid and liquid phases that are present at each temperature and composition. The temperatures at which solid phases appear upon cooling various solutions of the two components are detected by observation of the changes in slope of the plot of temperature versus time. A slower rate of cooling is obtained while a solid phase is separating out because the heat evolved by solidification partly offsets the heat lost by radiation and conduction to the colder surroundings.

Part A. An Alloy System

Apparatus. Six Pyrex test tubes fitted with thermocouple wells and spacers; tin, lead, and their binary mixtures; benzoic acid; bismuth; chromel-alumel thermocouple and potentiometer; electric furnace or Méker burner; large test tube, spacer, and rack; aluminum foil; watch or electric clock; vacuum bottle. (Recording potentiometer is recommended.)

PROCEDURE. A suggested sample tube and cooling jacket are shown in Fig. 29. A brass spacer centers the thermocouple well in the sample tube, and a second spacer centers the sample tube in the outer Pyrex tube. Bright aluminum foil is rolled into a cylinder and placed between the sample tube and the cooling jacket to reduce radiation losses and extend the time of cooling. The metal to be melted is placed in a steel cup made by drilling out a short piece of steel rod. It is set in the Pyrex tube on a wad of glass wool. If the metals are melted and frozen directly in the glass sample tube, the tube is apt to break because of the differences in thermal expansion. The tube containing a particular sample is heated† until the material has just melted completely. If the tube is markedly overheated, oxidation may be more serious, and furthermore a large number of readings of no practical value will be required at the beginning of the measurements. The Pyrex tube containing the metal cup is then transferred to the outer cooling jacket, and the time-temperature curve determined.

One junction of the thermocouple is placed in the thermocouple well of the sample tube; the other is placed in a tube in an ice bath contained in a vacuum bot-

† The Fieldner volatile-matter furnace, obtainable at supply houses, is recommended for this experiment. The furnace temperature may be controlled conveniently by use of a Variac or other variable autotransformer. A bunsen burner may be used.

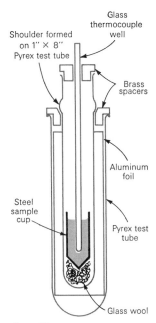

figure 29.

Sample tube and cooling jacket.

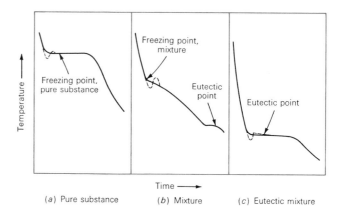

figure 30. Typical cooling curves (dashed portions illustrate supercooling effects).

tle. The thermocouple leads are connected to a potentiometer. The experiment is greatly facilitated by use of a recording potentiometer, which automatically graphs the voltage of the thermocouple against time. The appropriate range for the present application is about 0 to 20 mv.

If a manual potentiometer is used, the commonly available 0-to-16-mv range is appropriate. The potentiometer is adjusted as required to balance the changing electromotive force of the thermocouple as the sample cools. Time and voltage (temperature) readings are recorded approximately every quarter minute, until the results indicate that the freezing point (or, for a mixture, the eutectic temperature) has been reached as illustrated in Fig. 30.

The chromel-alumel thermocouple is calibrated by taking cooling curves with solids of known melting points: benzoic acid (121.7°), tin (232°), bismuth (271°), and lead (327°).

Cooling curves are then determined for the following mixtures:

1. Lead, 90 percent; tin, 10 percent
2. Lead, 80 percent; tin, 20 percent
3. Lead, 60 percent; tin, 40 percent
4. Lead, 40 percent; tin, 60 percent
5. Lead, 20 percent; tin, 80 percent

It is convenient to have each of the samples in a separate sample holder with thermocouple well. This collection of samples is kept in a metal test-tube rack, and the thermocouple is transferred to a new, heated well when one cooling curve is completed. A thin layer of graphite is placed over the samples to minimize air oxidation. When a sample becomes extensively oxidized, it is replaced with fresh material. Cooling curves obtained with impure or oxidized samples will often fail to exhibit the distinct breaks which are sought.

If the recording potentiometer used does not contain an accurate voltage reference, the true voltages corresponding to several scale deflections, spanning the range used, should be determined by using a standard manual potentiometer as an accurate voltage source. Actual voltages (rather than just readings on an arbitrary scale) are required in order to use handbook thermocouple tables as an aid in the temperature calibration.

With standard chromel and alumel thermocouple wire manufactured to standard tolerances, the temperature for a given voltage is within about $\pm 2°$C of that given by the tables. For better accuracy, a correction curve may be set up by plotting the difference between the measured and tabulated emf values at the freezing points of the calibration standards used.[2,3]

Part B. An Organic Compound System

Apparatus. Ten Pyrex tubes with varying compositions of phenol and *p*-toluidine; larger tube for slow cooling; recording potentiometer or 0 to 50° thermometer with 0.1 or 0.2° divisions.

PROCEDURE.   Phase relationships can be illustrated by use of mixtures of organic compounds as well as metal-alloy systems. For example, compound formation between the components is shown by the system phenol-*p*-toluidine.

Caution: These compounds must be handled with great care to avoid contact with the skin.

For the system assigned, a set of about 10 freezing-point tubes should be prepared to cover the composition range from one pure component to the other. The compounds are weighed out carefully into $\frac{1}{2}$- by 6-in. test tubes; about 6 g total mixture weight is sufficient. One of the tubes is heated with hot water or a bunsen burner until the mixture is barely, but completely, melted. It is then inserted in a larger test tube, with the help of a cork ring, to reduce the rate of cooling. A small glass tube, closed at the bottom, is fitted into a cork and set so that it reaches near the bottom of the melted material. The thermocouple junction fits into this well, and a little oil is added to improve the thermal conductance. The alumel-chromel thermocouple and recorder used in Part A may be used here, or at these lower temperatures a more sensitive couple of copper-constantan may be used. The thermocouple is calibrated in an ice bath and a steam bath (with corrections for barometric pressure) and at intermediate temperatures with one or two other materials with sharp melting points.

If a recording potentiometer is not available, a 0 to 50° thermometer graduated to 0.1° may be used in place of the thermocouple and well.

A ring stirrer is used to keep the cooling mixture at a uniform temperature. The stirrer may be made from thin glass rod or nichrome wire or other chemically inert wire.

A time-temperature graph is obtained as described under the procedure of Part A, and the readings are continued until the sample has completely solidified.

The mixture may be melted again, and a check determination made. The thermocouple well is then removed *carefully* and thoroughly cleaned before inserting in the next sample. The procedure is repeated with each of the mixtures. It is important to use pure materials for preparing the mixtures.

For mixtures with low freezing points the test tube and its outer jacket may be immersed in an ice bath or an ice-salt mixture. If the compounds are subject to air oxidation or tend to absorb moisture or carbon dioxide from the air, they may be sealed off in an all-glass tube. The thermometer well is sealed into the top of the tube and extends nearly to the bottom of the tube. The sample is introduced through a side tube, which is then sealed off. It is more difficult to avoid supercooling in this apparatus in which shaking is the means of stirring.

A major experimental problem in all this work is supercooling, i.e., failure of crystallization to take place at the proper temperature. Actually, a *small* extent of supercooling is useful, since then, when crystallization does start, the crystals formed are dispersed widely through the liquid, and equilibrium between the solid and liquid phases is more easily maintained. If supercooling seems too great, the experiment is repeated, with more vigorous stirring at the appropriate stages. Supercooling may usually be avoided by dropping in a "seed crystal" of the solid material.

CALCULATIONS.   If a manual potentiometer was used, the various experimental cooling curves are drawn as plots of potentiometer readings versus times of observation. If a recording potentiometer was used, the recorder charts may be used directly, in terms of arbitrary scale units. The voltages (or scale readings) corresponding to the observed freezing points of the pure substances used as calibration standards are determined from these curves (Fig. 30). A thermocouple calibration curve is constructed from the points thus obtained, and voltages (or scale readings) are then converted into temperatures by interpolation on this curve. As a check, several points based on handbook data for the type of thermocouple used are also entered on the calibration graph. The various sources of error in the calibration graph should be considered.

For each mixture studied, the cooling curve is examined to determine the temperatures at which abrupt changes in slope or complete arrests occur. The former signify changes in the number of phases present, and the latter indicate systems which are invariant under the condition of constant pressure.

A phase diagram is then prepared by plotting these temperatures against the compositions of the mixtures. For each mixture, *all* the data points corresponding to changes in the number of phases present should be put on the graph. Lines are drawn through the points to complete a phase diagram, and each area labeled according to the phases present. The various types of one-, two-, and three-phase systems possible for the particular system studied are listed by identifying the phases present in each case, and the properties of each are discussed in terms of the variance calculated from the phase rule under the assumption of constant pressure.

The limiting slopes of the observed freezing-point curves can be calculated theoretically on the assumption that the solid phases are the pure substances. Thus, for a two-component system, with mole fractions X_1 and X_2, the freezing point depression near $X_1 = 1$ is given by

$$T_f - T_0 = - \frac{RT_0^2}{\Delta H^\circ_{\text{fus}}} X_2 \tag{1}$$

where T_f = melting point of solution of mole fraction X_2

T_0 = melting point of pure component 1

$\Delta H^\circ_{\text{fus}}$ = standard molar enthalpy of fusion for pure component 1 at T_0

Then

$$\left(\frac{\partial T_f}{\partial X_2} \right)_{X_2 \to 0} = - \frac{RT_0^2}{\Delta H^\circ_{\text{fus}}} \tag{2}$$

The limiting slopes are estimated from the phase diagram, and the corresponding heats of fusion are then calculated and compared with values obtained from the literature.

Practical applications. The method of thermal analysis illustrated in this experiment is a basic procedure in the study of phase relationships. A maximum in the freezing-point–composition curve indicates the existence of an intermediate compound, and the composition of the compound is given by the highest point on the composition-temperature curve, for this represents the melting point of the pure compound.

Temperature-composition curves and other phase diagrams are of great value in the technical study of alloys and ceramics and in the recovery of a salt by crystallization from a mixture of salts.

Fractional crystallization is an effective method of purification.

The constancy of the freezing point through the whole solidification from start to finish is one of the best criteria for purity. If the substance is impure, the impurities become concentrated in the mother liquor as the liquid freezes out, and the freezing point is lowered more and more by the impurities.

Suggestions for further work. The following pairs of organic compounds are suitable for study: urea, phenol; naphthalene, nitrophenol; acetamide, β-naphthol; β-naphthol, p-toluidine; phenol, α-naphthylamine; diphenylamine, naphthalene. A number of phase diagrams in organic systems are discussed by Kofler and Kofler[4] and by Skau and Wakeham.[5]

References

1. A. Findlay, A. N. Campbell, and N. O. Smith, "The Phase Rule and Its Applications," Dover Publications, Inc., New York, 1951.
2. Reference Tables for Thermocouples, *Natl. Bur. Std. U.S. Circ.* 561, 1955.
3. Methods of Testing Thermocouples and Thermocouple Materials, *Natl. Bur. Std. U.S. Circ.* 590, 1958.
4. L. Kofler and A. Kofler, "Thermo-Mikro-Methoden," Verlag Chemie GmbH, Weinheim, Germany, 1954.
5. E. L. Skau and H. Wakeham in A. Weissberger (ed.), "Technique of Organic Chemistry," vol. 1, "Physical Methods of Organic Chemistry," 3d ed., pt. 1, chap. 3, Interscience Publishers, Inc., New York, 1959.

20 THREE-COMPONENT SYSTEMS

Solubility and tie-line data for a three-component system are plotted in a triangular graph.

THEORY.[1,2] According to the phase rule of Gibbs, the variance v (number of degrees of freedom) of a system at equilibrium is equal to the number of components c minus the number of phases p plus 2, provided that the equilibrium is influenced only by temperature, pressure, and concentration.

$$v = c - p + 2 \tag{1}$$

The variance v is the smallest number of independent variables required to completely fix the state of the system.

This experiment is carried out at constant temperature and pressure so that these are not variables. Thus Eq. (1) becomes

$$v = c - p \tag{2}$$

and for a three-component system $v = 3 - p$. Thus in a three-component system if only one liquid phase is present, $v = 2$ and the concentration of two of the chemical substances must be specified in order to describe the system completely. If two immiscible liquid phases are present in equilibrium, $v = 1$ and the concentration of only one component needs to be specified—the concentrations of the others may be ascertained by reference to the phase diagram.

It is convenient to represent a three-component system on a triangular diagram such as that illustrated in Fig. 31.

In an equilateral triangle, the sum of the perpendiculars from a given point to the three sides is a constant. The perpendicular distance from each apex, representing a pure compound, to the opposite side is divided into 100 equal parts, corre-

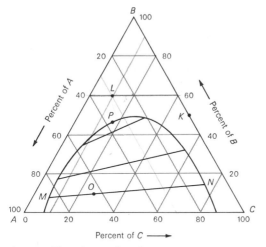

figure 31. Phase diagram for a three-component system.

sponding to percent, and labeled along the side at the right of the perpendicular. A point situated on one of the sides of the triangle indicates that there are two components with the percentage composition indicated. The composition corresponding to any point within the triangle is obtained by measuring on these coordinates the distance toward apex A, the distance toward B, and the distance toward C. These three distances representing percentages always add up to 100. For example, point K indicates 50 percent B, 50 percent C, and 0 percent A; and point L represents the composition 30 percent A, 60 percent B, and 10 percent C.

Several different types of ternary systems are possible, depending upon whether one, two, or three pairs of the liquids are partially miscible in each other.

In the system illustrated in Fig. 31 pair A and B and pair B and C are completely miscible at all concentrations, while pair A and C are miscible only at high concentrations of A or C and at intermediate concentrations form two liquid phases which do not mix with each other. All mixtures of the three liquids having compositions lying below the curve MPN will separate into two separate liquid phases, while all mixtures having compositions lying above the curve will give one homogeneous liquid phase.

For example, mixture O will separate into two liquid phases, one having the composition M and the other the composition N, and the line connecting these two immiscible ternary solutions in equilibrium with each other is called a *tie line*. The relative amount of phase M is given by the ratio ON/MN, and the relative amount of phase N by the ratio MO/MN. These tie lines slope upward to the right, indicating that component B is more soluble in the phase N, which is rich in C, than it is in the phase M, which is rich in A.

As the amount of component B is increased, the compositions of the conjugate solutions approach each other. At point P the two conjugate solutions have the same composition, so that the two layers have become one: this is called the *plait point*.

An excellent book by Francis[3] discusses the various possible types of liquid mixtures with more than one phase and gives data and references on many hundreds of multicomponent liquid systems.

> **Apparatus.** Thermostat (25°); three burettes (25 ml); six glass-stoppered bottles (50 ml); one 5-ml pipette; small separatory funnel with Teflon stopcock; Westphal density balance; chloroform; glacial acetic acid; 0.5 N sodium hydroxide.

PROCEDURE.   The solubility relations of the three-component system chloroform–acetic acid–water are studied in this experiment. The tie lines may be determined conveniently by titrating the acetic acid in the two separate liquid phases, which are separated after they have come into equilibrium with each other.

Three burettes are set up containing water, chloroform, and glacial acetic acid. Nonaq† grease should be used for the burette containing chloroform.

The density of each solution is determined with a Westphal balance (page 495) in order to convert volumes into weights for calculation of the percent composition of the mixtures; or the densities are obtained from tables in the literature. Solutions of accurately known composition are prepared containing approximately 10, 25, 40, and 60 percent by weight acetic acid in water. About 20 g of each is sufficient. Glacial acetic acid must be used quickly because of its objectionable odor. These four solutions are placed in 50-ml glass-stoppered bottles and set in a thermostat at 25°. After coming to temperature, they are removed as needed and titrated with chloroform. During the titration, the bottle is shaken vigorously after each addition of chloroform, and the end point is taken as the first perceptible permanent cloudiness. This cloudiness is produced by the formation of a second liquid phase with a different refractive index.

Samples of approximately 10, 25, 40, and 60 percent acetic acid in chloroform are then prepared and titrated to slight cloudiness with water at the thermostat temperature. The percentage by weight of each component present at the first appearance of the second phase is calculated, and the compositions are plotted on the ternary diagram.

The tie lines are determined by preparing about 40 ml each of mixtures of accurately known concentrations containing approximately 10, 20, 30, or 40 percent

† Eimer and Amend, New York.

acetic acid with 45 percent chloroform in each case, the remainder being water. These two-phase mixtures are prepared in the glass bottles and are allowed to equilibrate in the 25° thermostat. After equilibrium has been reached, the phases are separated by means of a separatory funnel provided with a Teflon stopcock to avoid the use of stopcock grease. The density of each phase is determined with a Westphal balance. Five-milliliter aliquots are titrated with 0.2 N sodium hydroxide, using phenolphthalein as indicator.

CALCULATIONS.　The percentage by weight of chloroform, acetic acid, and water for each of the mixtures that showed the first indication of turbidity is plotted on triangular graph paper.

The determination of the acetic acid concentrations in the two-phase mixtures separated with a separatory funnel allows the compositions of the conjugate phases to be located on the two-phase curve. It may be assumed that the denser phase is the one rich in chloroform. The gross compositions of the two-phase mixtures are also plotted on the triangular graph, and the tie lines should pass through these points.

The phases present at each area and line are recorded, and the effect of adding more of the components at significant points is described.

Practical applications. The increase in mutual solubility of two liquids due to the addition of a third is of practical as well as theoretical importance. Calculations in two-phase extraction processes may be carried out, using triangular diagrams.

Suggestions for further work. The solubility and tie-line determinations may be repeated at a different temperature. As the temperature is raised, the area under the curve corresponding to the region of two liquid phases becomes smaller, because the liquids become more soluble in each other. Several isothermal lines may be drawn on the same diagram, or a space model may be constructed with temperature as the vertical axis and the triangular diagrams lying in horizontal planes. A triangular-prism space model may be made for the liquid-solid phases in the system biphenyl-diphenylamine-benzophenone.[4]

Liquid systems in equilibrium may be selected for experimentation and interpretation from the wealth of material summarized by Francis.[3]

The systems benzene-water-alcohol and cyclohexane-water-alcohol studied by Washburn and his associates illustrate the effect of increasing the hydrocarbon-chain length in the alcohol homologous series.[5]

The system water–ether–succinic nitrile furnishes an example of a three-component system containing three pairs of partially miscible liquids. The system water–methyl (or ethyl) acetate–*n*-butyl alcohol is an example of a system in which two of the pairs of liquids are partially miscible while the third pair is completely miscible.

A three-component system involving two solids and a liquid which is suitable for a laboratory experiment[6] is provided by lead nitrate–sodium nitrate–water. Data are obtained by evaporating the water from saturated solutions of known weight and concentrations of salts.

References

1. A. Findlay, A. N. Campbell, and N. O. Smith, "The Phase Rule and Its Applications," Dover Publications, Inc., New York, 1951.
2. J. E. W. Rhodes, "Phase Rule Studies," chaps. 5 and 6, Oxford University Press, Fair Lawn, N.J., 1933.
3. A. W. Francis, "Liquid-Liquid Equilibriums," Interscience Publishers, Inc., New York, 1963.
4. H. H. Lee and J. C. Warner, *J. Am. Chem. Soc.*, **55:** 4474 (1933).
5. D. R. Simonsen and E. R. Washburn, *J. Am. Chem. Soc.*, **68:** 235 (1946); and earlier contributions by E. R. Washburn et al.
6. E. L. Heric, *J. Chem. Educ.*, **35:** 510 (1958).

21   SOLUBILITY AS A FUNCTION OF TEMPERATURE

The determination of solubility[1-5] and the calculation of the differential heat of solution at saturation are illustrated in this experiment.

THEORY.   One of the simplest cases of equilibrium is that of a saturated solution in contact with excess solute; molecules leave the solid and pass into solution at the same rate at which molecules from the solution are deposited on the solid. The term *solubility* refers to a measure, on some arbitrarily selected scale, of the concentration of the solute in the saturated solution. Here the molal concentration scale will be used, and the solubility then becomes equal to the molality m_s of the solute in the saturated solution.

An equilibrium-constant relation may be written for the equilibrium considered:

$$K = \frac{a_2}{a_2^*} \tag{1}$$

Here a_2 represents the activity of the solute in the saturated solution and a_2^* the activity of the pure solid solute. The conventional choice of standard state for the latter is the pure solute itself at the temperature and pressure involved, making a_2^* identically equal to unity. The activity a_2 is related to the molality m of the solute by means of the activity coefficient γ, a function of T, P, and composition which approaches unity as m approaches zero. Then

$$K = [a_2]_{m=m_s} = \gamma_s m_s \tag{2}$$

where the subscript s indicates that the relation applies to the saturated solution. The symbol $[a_2]_{m=m_s}$ denotes the value of the activity a_2 for the saturated solution.

The change in K with temperature at constant pressure reflects a change in m_s, and also the change in γ_s, which is affected by both the variations in temperature and concentration of the solution. The van't Hoff equation requires that

$$\left(\frac{\partial \ln K}{\partial T}\right)_P = \frac{\Delta H^\circ}{RT^2} \tag{3}$$

where ΔH° is the standard enthalpy change for the solution process. This quantity should not be confused with any actual experimentally measurable heat of solution; it can be determined indirectly, however.

Taking into account the effects of temperature and concentration on γ_s (Ref. 6), there results for constant pressure

$$\left[1 + \left(\frac{\partial \ln \gamma}{\partial \ln m}\right)_{T,P,m=m_s}\right] \frac{d \ln m_s}{dT} = \frac{[\Delta H_{DS}]_{m=m_s}}{RT^2} \tag{4}$$

Here $[\Delta H_{DS}]_{m=m_s}$ is the *differential heat of solution* at saturation at the given temperature and pressure (see Experiment 4). For cases in which the activity coefficient γ for the solute changes only slightly with concentration in the immediate neighborhood of saturation, the bracketed term on the left in Eq. (4) becomes unity, and

$$\frac{d \ln m_s}{dT} = \frac{[\Delta H_{DS}]_{m=m_s}}{RT^2} \tag{5}$$

or

$$\frac{d \log m_s}{d(1/T)} = -\frac{[\Delta H_{DS}]_{m=m_s}}{2.303R} \tag{6}$$

In this approximation, then, the differential heat of solution at saturation may be calculated at a given temperature T by multiplying by $-2.303R$ the slope at this temperature of the plot of $\log m_s$ versus $1/T$.

If it is assumed in addition that $[\Delta H_{DS}]_{m=m_s}$ is independent of T, an assumption which in general is better for nonelectrolytic solutes than for electrolytic types, then integration of Eq. (5) leads to

$$\log \frac{m_s(T_2)}{m_s(T_1)} = \frac{[\Delta H_{DS}]_{m=m_s}(T_2 - T_1)}{2.303RT_2T_1} \tag{7}$$

The heat of solution with which we are concerned here is the heat absorbed when 1 mole of the solid is dissolved in a solution that is already practically saturated. It differs from the heat of solution at infinite dilution, which is the heat of solution often given in tables, by an amount equivalent to the heat of dilution from saturation to infinite dilution.

···→ ***Apparatus.*** Water bath with adjustable temperature; ice bath; one large test tube for a jacket and four medium-sized test tubes; ring stirrer; 0.1° thermometer; 10-ml pipette; two 10-ml weighing bottles: oxalic acid; standardized sodium hydroxide solution, approximately 0.5 *N*.

PROCEDURE.   The solubility of oxalic acid is determined at 25, 20, 15 and 0° by analyzing for the number of moles of oxalic acid per 1000 g of water in the saturated solution. A thermostat is generally available for 25°, and an ice bath is used for 0°. Two intermediate temperatures are obtained with an improvised water bath made of a large beaker of water provided with a small motor stirrer. For temperatures below that of the room, ice or cold water is added as needed.

The saturated solution of oxalic acid is placed in a medium-sized test tube, which in turn is surrounded by a still larger test tube to provide an insulating air jacket and reduce the fluctuations in temperature. The 0.1° thermometer is immersed in the solution, which is stirred vigorously by the hand operation of a vertical ring stirrer which fits closely in the inner test tube so that the projecting ring cannot break the thermometer at the center of the tube. The ring stirrer is made of a plastic or glass rod with a right-angle ring at the bottom.

The distilled water in large test tubes is saturated with oxalic acid by shaking with an excess of crystals at a higher temperature and cooling the solution down to the thermostat temperature so that some of the dissolved material is crystallized out. The solubility is less at the lower temperatures. When the equilibrium is approached in this way from a supersaturated solution, it is achieved rapidly, whereas it may be achieved slowly if heated from a lower temperature so that crystals must be dissolved to give an equilibrium concentration.

Two 10-ml samples are removed at each temperature with a pipette, drained into weighing bottles, and weighed to 0.01 g. To prevent drawing small crystals into the pipette along with the saturated solution a filter is provided as indicated in Fig. 32. The filter is removed before the pipette is drained. Alternatively the filtering operation can be carried out with a small piece of filter paper slipped over the bottom of the pipette and fastened with a rubber band. The solution is then titrated with the

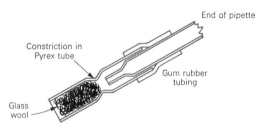

figure 32. Filter for use in solubility determinations.

standardized sodium hydroxide using phenolphthalein as an indicator. Duplicate measurements are made at each temperature.

CALCULATIONS. The solubility in moles per 1000 g of solvent is calculated at each of the four temperatures and compared with the accepted values. It is interesting to compare these values with the solubilities calculated in moles per liter.

The logarithm of the solubility in moles per 1000 g of solvent is plotted against the reciprocal of the absolute temperature, and a smooth curve is drawn through the four points. If the value of $\Delta H°$ were constant, the line would be straight. Tangents are drawn at 25 and 0°, and the heat of solution is determined at the two temperatures with the help of Eq. (6).

Practical applications. The solubility of a substance may be calculated at other temperatures when it has been determined at two different temperatures. The results are more accurate when the heat of solution is not affected by temperature or when the temperature range is small.

Suggestions for further work. The solubility of other materials may be determined in a similar way, e.g., benzoic acid or succinic acid or other solids having low solubility and easy methods of analysis. Nonaqueous solvents may be used also.

Boric acid may be used instead of oxalic acid for this experiment. It is titrated by using phenolphthalein as an indicator and adding 10 to 20 ml of neutral glycerin to give a sharp end point.

The solubility-product rule and the effect of other salts on solubility may be illustrated by determinations of the solubility of a slightly soluble salt such as silver bromate in the presence of common ions, ammonium hydroxide, and other salts.

The influence of salts in reducing the solubility of benzoic acid may be determined.[6] The salting-out constant thus obtained can be used for calculating activity coefficients.

In Eq. (2) it is assumed that the heat of solution is independent of temperature, but this assumption is not often justified. The equation may be made exact by introducing terms for the heat capacity of the solute and solvent and for the solution.

References

1. J. H. Hildebrand and R. L. Scott, "Solubilities of Nonelectrolytes," Reinhold Publishing Corporation, New York, 1950.
2. A. Seidell, "Solubilities of Inorganic and Metal Organic Compounds," 3d ed., vol. I, D. Van Nostrand Company, Inc., Princeton, N.J., 1940.
3. A. Seidell, "Solubilities of Organic Compounds," 3d ed., vol. II, D. Van Nostrand Company, Inc., Princeton, N.J., 1941.
4. A. Seidell and W. F. Linke, "Supplement to Solubilities of Inorganic and Organic Compounds," 3d ed., vol. III, D. Van Nostrand Company, Inc., Princeton, N.J., 1952.
5. W. J. Mader, R. D. Vold, and M. J. Vold in A. Weissberger (ed.), "Technique of Organic Chemistry," vol. 1, "Physical Methods of Organic Chemistry," 3d ed., pt. 1, chap. 11, Interscience Publishers, Inc., New York, 1959.
6. A. T. Williamson, *Trans. Faraday Soc.,* **40:** 421 (1944).

Chapter
7

Chemical
Kinetics

Except in the case of very rapid reactions the rate of a chemical reaction is often as important as the thermodynamics of the reaction. If there is a decrease in free energy when the reaction takes place, at constant T, P it may go spontaneously but will be useful only if it takes place in a reasonably short time. Moreover, if several different reactions are thermodynamically possible, the one which is fastest will use up the reactants first and result in a larger yield of the product. Application of the principles of thermodynamics and chemical kinetics makes possible the prediction and control of chemical reactions, but the overall reaction becomes complicated when several different reactions are taking place together.

In this chapter, the kinetics of four different reactions are studied experimentally. They are chosen because they give results which can be easily described in simple mathematical terms and because they illustrate a first-order reaction, a second-order reaction, a catalyzed reaction, and a complex reaction.

In studies of chemical kinetics (Refs. 1 to 6, pages 143 to 144) it is important to determine the rate expression which will give the concentration of one or more of the reactants or products as a function of time and to obtain the numerical value for the *specific rate constant k.*

Although chemical reactions which accurately fit these formulas are chosen for illustration, the student must realize that a great many chemical reactions involve so many simultaneous competing successive and reverse reactions that the mathematical analysis in simple terms has not been possible. The development of electronic computers is now making possible the mathematical analysis of many of these complicated reactions.

Unimolecular reactions are those which involve the breakdown or rearrangement of one type of molecule such as

$$AB \longrightarrow A + B \qquad or \qquad ABA \longrightarrow BAA$$

Bimolecular reactions involve a collision between two molecules such as

$$A + B \longrightarrow AB \qquad or \qquad AB + CD \longrightarrow AC + BD$$

Termolecular reactions involve a collision between three molecules. But the rate-determining step in the reaction usually does not involve a mechanism of a simple uni-, bi-, or termolecular reaction. The *order of the reaction, n,* which must be evaluated experimentally, is important in determining the mechanism by which the reaction takes place. It is defined by the equation

$$\frac{dc}{dt} = kc^n \tag{1}$$

where n is evaluated from the rate of change of concentration of reactant c with time. If n is 1, the reaction is first order, if it is 2, the reaction is second order, and if it is 3, the reaction is third order. If (as is usually the case) n is found to have other values that are not integers, the reaction is complex and involves more than one uni-, bi-, or termolecular reaction. Fortunately the rates of many unimolecular or bimolecular reactions can be estimated from molecular structure or other properties, and often a complex reaction may be broken up into a series of predictable unimolecular and bimolecular reactions.

The first-order reaction equation

$$-\frac{dc}{dt} = kc^1 \tag{2}$$

is integrated to give

$$-\ln c = -2.303 \log c = kt + \text{constant} \tag{3}$$

or

$$k = \frac{2.303}{t_2 - t_1} \log \frac{c_1}{c_2} \tag{4}$$

where c_1 and c_2 are the concentrations at times t_1 and t_2.

For first-order reactions k is numerically equal to the fraction of the substance which reacts per unit time, usually expressed in reciprocal seconds (or minutes). In such reactions it is not necessary to know the initial concentration of the reactants or the absolute concentrations at various times. The concentrations may be determined directly by experiment using chemical or physical measurements; or any property, e.g., volume, electrical conductance, or light absorption, which is proportional to the concentration may be measured and substituted for c in formulas (3), (4), or (5).

The kinetics of a second-order reaction is described by the equation

$$-\frac{dc}{dt} = kc_A^2 \tag{5}$$

where c_A is the concentration of the reactant A, or

$$-\frac{dc_A}{dt} = kc_A c_B \tag{6}$$

where c_A and c_B are the concentrations of two reactants A and B.

The numerical value of the rate constant k for a second-order reaction depends on the units in which the concentrations are expressed, such as moles per liter, moles per cubic centimeter, or atmospheres. In a first-order reaction these units cancel out, but in a second-order reaction they do not. In a second-order reaction, if one reactant is present in sufficiently large excess, its concentration remains essentially constant and so the second-order reaction then appears to be of the first order.

22 *HYDROLYSIS OF METHYL ACETATE*

THEORY.[1-6]   The hydrolysis of methyl acetate presents several interesting aspects. The reaction, which is extremely slow in pure water, is catalyzed by hydrogen ion:

$$CH_3COOCH_3 + H_2O + H^+ \underset{k_2}{\overset{k_1'}{\rightleftharpoons}} CH_3COOH + CH_3OH + H^+$$

The reaction is reversible, so that the net rate of hydrolysis at any time is the difference between the rates of the forward and reverse reactions, each of which follows the simple rate law given by Eq. (6). Thus

$$-\frac{dc_{CH_3COOCH_3}}{dt} = k_1' c_{H_2O} c_{CH_3COOCH_3} - k_2 c_{CH_3COOH} c_{CH_3OH} \tag{7}$$

where k_1' is the rate constant for the forward reaction and k_2 for the reverse reaction. For dilute solutions, water is present in such large excess that its concentration undergoes a negligible proportional change while that of the methyl acetate is changed considerably. For this case Eq. (7) may be written

$$-\frac{dc_{CH_3COOCH_3}}{dt} = k_1 c_{CH_3COOCH_3} - k_2 c_{CH_3COOH} c_{CH_3OH} \tag{8}$$

In the early stages of the hydrolysis, the concentrations of acetic acid and methanol remain small enough for the term involving them to be negligible, and the reaction appears to be of first order:

$$-\frac{dc_{CH_3COOCH_3}}{dt} = k_1 c_{CH_3COOCH_3} \tag{9}$$

The value of k_1 can then be determined by one of the methods conventional for first-order reactions.

Evaluation of k_1 at two different temperatures permits the calculation of the *Arrhenius heat of activation* ΔH_a for the forward reaction:

$$\frac{d \ln k_1}{dT} = \frac{\Delta H_a}{RT^2} \tag{10}$$

$$\log \frac{k_{1,T_2}}{k_{1,T_1}} = \frac{\Delta H_a}{2.303R} \frac{T_2 - T_1}{T_2 T_1} \tag{11}$$

In obtaining the integrated form, it is assumed that ΔH_a is a constant. The heat of activation is usually expressed in calories per mole and is interpreted as the amount of energy the molecules must have in order to be able to react.

A more accurate calculation of the influence of temperature may be made on the basis of the Eyring equation,

$$k = \kappa \frac{RT}{N_0 h} e^{\Delta S^{\ddagger}/R} e^{-\Delta H^{\ddagger}/RT} \tag{12}$$

where N_0 is Avogadro's number, h is Planck's constant, and $\Delta S^{\ddagger}$ and $\Delta H^{\ddagger}$ are the standard entropy and enthalpy changes for formation of the activated complex from the reactants

$$CH_3COOCH_3 + H_2O + H^+ \longrightarrow \text{[activated complex]}$$

and κ is a constant, of the order of $\frac{1}{2}$, defined as the probability that an activated complex will decompose to form product species (rather than regenerating reactant species). Thus $\Delta H^{\ddagger}$ may be determined from measurements of k at two or more temperatures, on the assumption $\Delta S^{\ddagger}$, $\Delta H^{\ddagger}$, and κ are independent of temperature.

$$\frac{k_{1,T_2}}{k_{1,T_1}} = \frac{T_2}{T_1} \exp\left[-\frac{\Delta H^{\ddagger}}{R}\left(\frac{1}{T_2} - \frac{1}{T_1} \right) \right] \tag{13}$$

$$\Delta H^{\ddagger} = \frac{RT_1 T_2}{T_2 - T_1} 2.303 \log \frac{(k_{1,T_2}) T_1}{(k_{1,T_1}) T_2} \tag{14}$$

Although $\Delta S^{\ddagger}$ cannot be determined from these data, for lack of knowledge of the value of κ, it is sometimes possible to gain some information about the magnitude of $\Delta S^{\ddagger}$ by making a guess as to the value of κ. In ordinary cases, a value of $\frac{1}{2}$ to 1 is considered a reasonable estimate, but under certain circumstances κ may be very small. The value of $\Delta H^{\ddagger}$ can be used, of course, to calculate the value of k_1 at any temperature (over the range in which $\Delta H^{\ddagger}$ and $\Delta S^{\ddagger}$ remain constant) from a knowledge of k_1 at one temperature.

An explicit solution to the kinetic equation may also be written for the case where the reverse reaction cannot be ignored. If the concentration of methyl acetate is a moles per liter initially, and $a - x$ moles per liter at time t, then Eq. (8) can be written as

$$-\frac{d(a - x)}{dt} = \frac{dx}{dt} = k_1(a - x) - k_2 x^2 \tag{15}$$

since for each mole of methyl acetate hydrolyzed a mole of acetic acid and a mole of methanol are produced. Integration of this relation gives

$$t = \frac{1}{k_1(4ak_2/k_1 + 1)^{\frac{1}{2}}} \ln \frac{2a + x[(4ak_2/k_1 + 1)^{\frac{1}{2}} - 1]}{2a - x[(4ak_2/k_1 + 1)^{\frac{1}{2}} + 1]} \tag{16}$$

Making use of the relation that the equilibrium constant K_h for the hydrolysis reaction is given by the expression

$$K_h = \frac{c_{CH_3COOH}c_{CH_3OH}}{c_{CH_3COOCH_3}c_{H_2O}} = \frac{k_1'}{k_2} = \frac{k_1}{k_2c_{H_2O}}$$

one obtains

$$t = \frac{1}{k_1(4a/K_hc_{H_2O}^\circ + 1)^{\frac{1}{2}}} \ln \frac{2a + x[(4a/K_hc_{H_2O}^\circ + 1)^{\frac{1}{2}} - 1]}{2a - x[(4a/K_hc_{H_2O}^\circ + 1)^{\frac{1}{2}} + 1]} \tag{17}$$

Here $c_{H_2O}^\circ$ represents the concentration of water present, which is treated as a constant in accordance with the assumption made in obtaining Eq. (8) from Eq. (7).

····> *Apparatus.* Thermostats at 25 and 35°; three 250-ml, two 125-ml Erlenmeyer flasks; 5-ml pipette; 100-ml pipette; stopwatch or electric timer; methyl acetate; 2 liters 0.2 N sodium hydroxide; phenolphthalein indicator; 500 ml of standardized 1 N hydrochloric acid; distilled water; ice.

PROCEDURE.   The concentration of methyl acetate at a given time is determined through titration of samples with a standard sodium hydroxide solution; the experimental accuracy depends chiefly on the care used in pipetting and titrating. The sodium hydroxide solution used should be prepared by dilution of a *saturated* stock solution to minimize the amount of carbonate present and hence to reduce the fading of the phenolphthalein end point. It is not necessary, however, to use CO_2-free distilled water, because the amount of carbonate introduced in air-saturated water is negligible when titrating with 0.2 N sodium hydroxide.

A test tube containing about 12 ml methyl acetate is set into a thermostat at 25°C. Approximately 250 ml of standardized 1 N hydrochloric acid is placed in a flask clamped in the thermostat. After thermal equilibrium has been reached (10 or 15 min should suffice), two or three 5-ml aliquots of the acid are titrated with the standard sodium hydroxide solution to determine the exact normality of the sodium hydroxide in terms of the standardized hydrochloric acid. Then 100 ml of acid is transferred to each of two 250-ml flasks clamped in the thermostat and 5 min allowed for the reestablishment of thermal equilibrium. Precisely 5 ml of methyl accetate is next transferred to one of the flasks with a clean, dry pipette; the timing watch is started when the pipette is half emptied. The reaction mixture is shaken to provide thorough mixing.

A 5-ml aliquot is withdrawn from the flask as soon as possible and run into 50 ml of distilled water. This dilution slows down the reaction considerably, but the solution should be titrated at once; the error can be further reduced by chilling the water in an ice bath. The time at which the pipette has been half emptied into the water in the titration flask is recorded, together with the titrant volume. Additional samples are taken at 10-min intervals for an hour; then at 20-min intervals for the next hour and a half. A second determination is started about a quarter hour after the first one to provide a check experiment.

In similar fashion, two runs are made at a temperature of 35°. Because of the higher rate of reaction, three samples are first taken at 5-min intervals, then several at 10-min intervals, and a few at 20-min intervals. It is convenient to start the check determination about a half hour after the first experiment is begun.

CALCULATIONS. The titrant volume at time t, V_t, measures the number of equivalents of hydrochloric acid and acetic acid then present in the 5-ml reaction-mixture aliquot. Let V_∞ represent what the titrant volume per 5-ml aliquot would be *if* the hydrolysis were complete. Then $V_\infty - V_t$ measures the number of equivalents of methyl acetate remaining per 5-ml aliquot at time t, because one molecule of acetic acid is produced for each molecule of methyl acetate hydrolyzed. The corresponding concentration of methyl acetate in moles per liter is $N(V_\infty - V_t)/5$, where N is the normality of the sodium hydroxide solution.

If the reaction actually proceeded to completion, V_∞ could be measured directly by titration of an aliquot from the equilibrium mixture. An appreciable amount of unhydrolyzed methyl acetate is present at equilibrium, however, so V_∞ must be *calculated*.

The volume of the solution initially formed on mixing the 100 ml of 1 N hydrochloric acid with 5 ml of methyl acetate is designated by V_s. At 25°C, V_s is 104.6 ml rather than 105 ml because the solution is not ideal. Let the number of milliliters of sodium hydroxide solution required to neutralize a 5-ml aliquot of the original 1 N hydrochloric acid be V_x. The number of milliliters required to neutralize the hydrochloric acid in 5 ml of the reaction mixture at any time is $V_x \, 100/V_s$, on the assumption that the total volume of the reaction mixture remains constant as the hydrolysis proceeds.

The weight of the 5 ml of methyl acetate is $5\rho_2$, where ρ_2 is the density of methyl acetate (0.9273 g ml^{-1} at 25° and 0.9141 at 35°), and the number of moles in this 5-ml sample is $5\rho_2/M_2$, where M_2 is the molecular weight, 74.08. The number of moles of methyl acetate initially present in any 5-ml aliquot of the reaction mixture is $\dfrac{5\rho_2}{M_2} \dfrac{5}{V_s}$.

Since $1000/N$ ml of sodium hydroxide of normality N is required to titrate the acetic acid produced by the hydrolysis of 1 mole of methyl acetate, $\dfrac{1000}{N} \dfrac{25\rho_2}{M_2 V_s}$ ml will be required for the titration of the acetic acid produced by the complete hydrolysis of the methyl acetate originally contained in any 5-ml sample of the reaction mixture. The total number of milliliters of sodium hydroxide solution V_∞ required to titrate both the hydrochloric acid and the acetic acid produced by the complete hydrolysis of the methyl acetate in a 5-ml sample of the reaction mixture is

$$V_\infty = V_x \frac{100}{V_s} + \frac{1000}{N} \frac{25\rho_2}{M_2 V_s} \tag{18}$$

The value of V_∞ is calculated for each experiment by means of Eq. (18). For each run a tabulation is made of the times of observation and the corresponding values of V_t and $V_\infty - V_t$.

Two graphs are then prepared. For each temperature a plot is made of $\log (V_\infty - V_t)$ versus t; the points obtained in the two runs can be identified by use of circles and squares. The straight line which is considered to best represent the experimental results is drawn for each set of points, and the rate constants for the two temperatures are calculated from the slopes of the two lines, in accordance with Eq. (3). It is not necessary to calculate the actual concentrations of methyl acetate, since a plot of $\log (V_\infty - V_t)$ versus t has the same slope as a plot of $\log [(V_\infty - V_t)(N/5)]$.

Comparison values of k_1 are calculated at each temperature from several sets of points by use of Eq. (3), to illustrate the dependence of the calculated rate constant on the particular pair of points chosen and hence emphasize the advantages of the averaging achieved in the graphical method. It should be noted that it is not significant to substitute an explicit averaging of the values of k obtained from the successive observations by means of Eq. (4).

From the rate constants found for the two temperatures, the heat of activation is calculated by use of Eq. (11).

Practical applications. The rate of a chemical reaction is important in determining the efficiency of many industrial reactions. In organic reactions particularly, where there is the possibility of several reactions going on simultaneously, the kinetic considerations will often be no less important than the equilibrium relationships.

Suggestions for further work. The integration of Eq. (15) to give Eqs. (16) and (17) may be checked to illustrate a typical transformation in chemical kinetics. The integral involved is given in mathematical tables.

The method of least squares may be used (Chapter 20) instead of estimating by eye the "best" straight-line representation of the plot of $\log (V_\infty - V_t)$ versus time.

Different acid concentrations or other acids may be used[7]; the influence of neutral salts may be studied.[8] Nonaqueous solvents may be used,[9] and methyl acetate may be replaced by other esters,[10] higher temperatures being used if necessary.

References

1. S. W. Benson, "Foundations of Chemical Kinetics," McGraw-Hill Book Company, New York, 1960.

2. A. A. Frost and R. G. Pearson, "Kinetics and Mechanisms," John Wiley & Sons, Inc., New York, 1961.
3. E. S. Amis, "Kinetics of Chemical Change in Solution," The Macmillan Company, New York, 1949.
4. Tables of Chemical Kinetics: Homogeneous Reactions, *Natl. Bur. Std. U.S. Circ.* 510, 1951, and *Suppl.* 1, 1956.
5. K. J. Laidler, "Chemical Kinetics," 2d ed., McGraw-Hill Book Company, New York, 1965.
6. E. A. Moelwyn-Hughes, "Kinetics of Reactions in Solutions." Oxford University Press, Fairlawn, N.J., 1947.
7. R. O. Griffith and W. C. M. Lewis, *J. Chem. Soc.*, **109:** 67 (1916).
8. M. Duboux and A. deSousa, *Helv. Chim. Acta*, **23:** 1381 (1940).
9. A. A. Friedman and G. V. Almore, *J. Am. Chem. Soc.*, **63:** 864 (1941).
10. H. S. Harned and R. Pfansteil, *J. Am. Chem. Soc.*, **44:** 2193 (1922).

23   *REACTION OF ETHYL ACETATE WITH HYDROXYL ION FOLLOWED BY ELECTRICAL CONDUCTANCE*

This experiment illustrates the use of a conductimetric method for measuring the progress of a solution reaction and the determination of order and rate constant from such data.

THEORY.[1-6]   The reaction studied in this experiment is

$$CH_3COOC_2H_5 + OH^- \longrightarrow CH_3COO^- + C_2H_5OH \tag{1}$$

Since the actual reaction mechanism may involve several steps, the equation for the overall reaction does not necessarily suggest the correct form for the rate law. However, it has been found that reaction (1) does follow the second-order equation (page 138),

$$\frac{dx}{dt} = k_1(a - x)(b - x) \tag{2}$$

where t = time elapsed from initiation of reaction

x = number of moles per liter reacted at time t

a = initial molar concentration of $CH_3COOC_2H_5$

b = initial molar concentration of OH^-

k_1 = rate constant for reaction (1)

provided the reaction mixture is not so close to its equilibrium state as to require inclusion in Eq. (2) of a term for the reverse reaction.

Integration of Eq. (2) for the case $a \neq b$ leads to the result

$$\ln \frac{b(a - x)}{a(b - x)} = k_1(a - b)t \tag{3}$$

For the case $a = b$ one obtains

$$\frac{x}{a - x} = k_1 a t \tag{4}$$

The temperature dependence of k_1 can be related to $\Delta S^{\ddagger}$ and $\Delta H^{\ddagger}$ for the formation of the activated complex from reactants (page 140). From measurement of k_1 at two or more temperatures, $\Delta H^{\ddagger}$ can be found with the use of Eq. (14) of the preceding experiment.

The experimental problem is to determine x as a function of t for a solution in which reaction proceeds at a constant temperature. The reaction mixture undergoes a marked decrease in conductance with time as hydroxyl ion is replaced by acetate ion. The progress of the reaction can therefore be followed by measurement of conductance.[7]

Apparatus. Two 250-ml glass-stoppered volumetric flasks; modified 250-ml Erlenmeyer flask; conductance bridge; conductance cell; glass-stoppered weighing bottle; ethyl acetate; standardized NaOH; 25-ml burette; 1-ml graduated pipette; 25-ml pipette; 50-ml pipette; thermostats at 25 and 35°; stopwatch or timer.

PROCEDURE.   Chapter 24 may be consulted for information on the theory and practice of conductance measurements. The bridge selected for the present experiment should be of a type which can be balanced and read rapidly; for this reason, a bridge having a continuously adjustable dial of rather wide range is to be preferred to the decade step-switch type, even at some sacrifice in accuracy. The Freas type of conductance cell shown in Fig. 40 (page 172) is suitable.

As some of the experiments are to be based on Eq. (4), solutions of NaOH and of ethyl acetate of equal normality, nominally 0.02 N, are prepared; 250 ml of each is required.

The ethyl acetate solution must be prepared the day it is to be used, because a slow reaction occurs even in the absence of OH^-. A technique which reduces error from volatilization of the ethyl acetate is to weigh the latter in a weighing bottle containing some water (the bottle and water having previously been weighed) and then transfer the sample quantitatively to the volumetric flask for preparation of the final solution. The solution of NaOH of the same normality is prepared by quantitative dilution of standardized stock solution.

Three runs are to be made at 25°C:

(1) $a = 2b$ (2) $2a = b$ (3) $a = b$

and one at 35°C:

(4) $a = b$

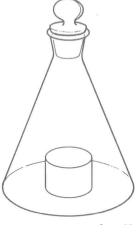

figure 33.
Special mixing flask.

Each is to be preceded by a measurement of resistance of the cell containing NaOH solution diluted with water in such proportion that the concentration of NaOH is equal to that initially present in the *reaction mixture* for that run. The bridge setting so obtained will be very close to that corresponding to zero time in the subsequent run. A prior knowledge of this setting will facilitate finding the balance point after the reaction has started and also provides a valuable check on the extrapolation to zero time required for calculation of the rate constant.

It is essential that the reactants be mixed rapidly and the timer started simultaneously. The use of a special mixing flask† (Fig. 33) will facilitate carrying out this operation. Initially, 25 ml of one reactant is placed in the central compartment, and 25 or 50 ml of the other reactant, as the case may be, is placed in the outer compartment. When the flask, solutions, and empty cell have reached thermal equilibrium with the thermostat, the timer is started and the flask is inverted several times to ensure thorough mixing. A small portion of the reaction mixture is poured into the conductance cell, shaken, and discarded. Then the cell is filled to the mark with the reaction mixture and is returned to the thermostat. The cell resistance R_t is measured at intervals of several minutes, until the rate of change has become relatively slow (about 45 min).

For runs 1 and 2 it is necessary to know R_c, the resistance which the cell would have if the reaction were to reach completion, i.e., until one reactant is used up. It is also helpful to have R_c for runs 3 and 4. While R_c can be found by actual measurement with solutions of NaOH and CH$_3$COONa made up to the appropriate concentrations, it is more easily obtained by calculation as described below. As a matter of principle, it should be noted that R_c is not quite the same as R_∞, the latter being the value of resistance approached asymptotically as the system reacts toward its final equilibrium state. For reaction (1), the equilibrium condition is so far to the right that R_∞ is very close to R_c, but for cases where the distinction is important, R_c rather than R_∞ is the quantity needed for the subsequent analysis of the data.

CALCULATIONS.   For the calculation of x from the conductance data, it is assumed that the conductance of the solution G_t at time t obeys the equation

$$G_t = \frac{1}{1000k} \sum_j c_j l_j$$

$$= \frac{1}{1000k} [(b - x)l_{\text{OH}^-} + xl_{\text{CH}_3\text{COO}^-} + bl_{\text{Na}^+}] \tag{5}$$

† Available commercially from Norman D. Erway, Oregon, Wis.

where l_j = equivalent ionic conductance of species j

c_j = concentration of species j in equivalents per liter

k = cell constant (page 168)

In Eq. (5), we are assuming that the NaOH and CH_3COONa are completely dissociated and that the ethyl acetate, ethanol, and water do not contribute to the conductance. The solutions employed are sufficiently dilute to justify the further assumption that l_{OH^-} and $l_{CH_3COO^-}$ are constant even though the concentrations change during the run.

The value of x ranges from $x = 0$ to $x = c$, where c is the initial concentration of the limiting reactant, that is, a or b, whichever is smaller. For the case of equal concentrations, $c = a = b$. Hence Eq. (5) leads, in all cases, to

$$G_0 - G_t = \frac{x(l_{OH^-} - l_{CH_3COO^-})}{1000k} \tag{6}$$

$$G_0 - G_c = \frac{c(l_{OH^-} - l_{CH_3COO^-})}{1000\ k} \tag{7}$$

Thus x is given by

$$\frac{x}{c} = \frac{G_0 - G_t}{G_0 - G_c} = \frac{(1/R_0) - (1/R_t)}{(1/R_0) - (1/R_c)} \tag{8}$$

By substituting Eq. (8) into (3) and rearranging, one obtains the working equations for the runs with unequal concentrations,

$$\log\frac{AR_t + 1}{BR_t + 1} = \frac{k_1(a - b)t}{2.303} + \log\frac{a}{b} \tag{9}$$

where

$$A = \frac{1}{R_0}\left[\frac{a}{c}\left(1 - \frac{R_0}{R_c}\right) - 1\right]$$

$$B = \frac{1}{R_0}\left[\frac{b}{c}\left(1 - \frac{R_0}{R_c}\right) - 1\right]$$

Similarly, introduction of Eq. (8) into (4) yields an equation for the runs with equal concentrations,

$$\frac{G_0 - G_t}{G_t - G_c} = k_1 at \tag{10}$$

which upon rearrangement becomes

$$G_t = G_c + \frac{1}{k_1 at}(G_0 - G_t) \tag{11}$$

Plots of resistance versus time are prepared for all runs. These are useful for judging the quality of the data and for extrapolation to zero time to obtain R_0. As a check, the values available from the preliminary measurements on the NaOH solutions should be placed on the graphs. If R_c was not measured directly, the ratio R_0/R_c can be calculated for each case from Eq. (5), which reduces to

$$\frac{R_0}{R_c} = \frac{G_c}{G_0} = \frac{(b - c)l_{OH^-} + cl_{CH_3COO^-} + bl_{Na^+}}{b(l_{OH^-} + l_{Na^+})} \tag{12}$$

For computing this ratio, handbook values for ionic conductances at infinite dilution may be used as approximations to l_{Na^+}, l_{OH^-}, and $l_{CH_3COO^-}$. It should not be overlooked that the conductances are dependent on temperature. The measured or calculated values of R_c should be marked on the graphs.

For the cases with $a \neq b$, Eq. (9) is used for a graphical analysis of the data. The quantity on the left-hand side is plotted against time. If the data are consistent with Eq. (2) and the other assumptions made in deriving Eq. (9), these graphs should, within experimental error, fit straight lines with intercepts at log (a/b). Values for the rate constant k_1 may be calculated from the slopes.

For the case $a = b$, the quantity $(G_0 - G_t)/t$ may be plotted against G_t. If the data are consistent with Eq. (11), k_1 may be found from the slope. If an estimate of R_c is available, it can be used to predict the intercept with one axis and thereby improve the accuracy of k_1.

In examining these graphs, it is important to take into account the experimental uncertainty of the points, which is quite different for different regions of the graphs. The computations of this experiment can conveniently be made by an electronic computer.

The enthalpy of activation $\Delta H^{\ddagger}$ is calculated from the values of k_1 at 25 and 35°C.

Suggestions for further work. As an alternative to the conductimetric method, a procedure which may be used to follow the progress of the reaction is to withdraw samples from the reaction mixture at definite intervals, discharge them into excess standard HCl to arrest the reaction, and back-titrate with NaOH. For best results, CO_2-free water should be used to prepare the NaOH solutions. In other respects, the procedure followed may be similar to that described above. A suitable equation for graphical analysis of the data for the case $a \neq b$ is obtained by expressing $a - x$ and $b - x$ as functions of the titration volumes and substituting into Eq. (3); the result of this is

$$\log \left\{ \frac{[V_a(N'/N) + v(a - b)/N] - V_t}{V_a(N'/N) - V_t} \right\} + \log \frac{b}{a} = \frac{k_1(a - b)}{2.303} t \tag{13}$$

where V_t = volume of NaOH solution required to titrate sample in which reaction was stopped at time t

V_a = volume of HCl into which samples were discharged

N = normality of NaOH solution used for titrations
N' = normality of HCl solution
v = volume of each sample taken

It should be noted that, for a given run, the only time-dependent quantity in the first term on the left side of Eq. (13) is V_t.

References

General References 1 to 6 are given on pages 143 and 144.
7. J. Walker, *Proc. Roy. Soc. London, ser. A;* **78:** 157 (1906).

24 INVERSION OF SUCROSE

In this experiment the rate of reaction between sucrose and water catalyzed by hydrogen ion is followed by measuring the angle of rotation of polarized light passing through the solution. The angle of rotation of polarized light passing through the solution is measured with a polarimeter. The reaction is

$$C_{12}H_{22}O_{11} + H_2O + H^+ \longrightarrow C_6H_{12}O_6 + C_6H_{12}O_6 + H^+ \qquad (1)$$

Sucrose $\qquad\qquad\qquad\qquad$ Glucose $\quad$ Fructose

Sucrose is dextrorotatory, but the resulting mixture of glucose and fructose is slightly levorotatory because the levorotatory fructose has a greater molar rotation than the dextrorotatory glucose. As the sucrose is used up and the glucose-fructose mixture is formed, the angle of rotation to the right (as the observer looks into the polarimeter tube) becomes less and less, and finally the light is rotated to the left. The rotation is determined at the beginning (α_0) and at the end of the reaction (α_∞), and the algebraic difference between these two readings is a measure of the original concentration of the sucrose. It is assumed that the reaction goes to completion so that practically no sucrose remains at "infinite" time. At any time t, a number proportional to the concentration c of the remaining sucrose is obtained from the difference between the final polarimeter reading (which has a negative value in this case) and the reading of α_t at time t.

The reaction proceeds too slowly to be measured in pure water, but it is catalyzed by hydrogen ions. The water is in such large excess that its concentration does not change appreciably and the reaction follows the equation for a first-order reaction, even though two different kinds of molecules are involved in the reaction. Thus the reaction-rate constant may be calculated from the equation

$$k = \frac{2.303}{t} \log \frac{\alpha_0 - \alpha_\infty}{\alpha_t - \alpha_\infty} \qquad (2)$$

THEORY.   Guggenheim[1] has described a method for calculating the rate constant of a first-order reaction which does not require waiting for a reading at infinite time. This method is useful if the reaction does not go to completion in one laboratory period and has the added advantage that each plotted point does *not* depend upon a *single* observation of the reading at time infinity.

The Guggenheim method may be applied directly if data are taken at equal time intervals. However, the reaction rate changes so rapidly that the procedure calls for shorter time intervals at the beginning. Accordingly, the values of α_t are plotted as a function of time. The data are then interpolated from the smoothed curves at constant time intervals.

The data are arranged in two sets. For each observation c_1 at time t in the first set, another observation c_2 is taken at time $t + \Delta t$, where Δt is a fixed time interval. If a plot of $\log (c_1 - c_2)$ versus t is prepared, the points will fall on a straight line of slope $-k/2.303$. The constant time interval Δt may be taken as approximately one-half the duration of the experiment. If Δt is too small, there will be a large percentage error in $c_1 - c_2$.

The equation for this method may be derived as follows. From the integrated form of the first-order-reaction differential equation,

$$c = c_0 e^{-kt} \tag{3}$$

The concentrations c_1 and c_2 at two times differing by Δt are

$$c_1 = c_0 e^{-kt} \tag{4}$$
$$c_2 = c_0 e^{-k(t+\Delta t)} \tag{5}$$

Subtracting,

$$c_1 - c_2 = c_0 e^{-kt}(1 - e^{-k\Delta t}) \tag{6}$$

Taking logarithms,

$$\log (c_1 - c_2) = \frac{-kt}{2.303} + \log [c_0(1 - e^{-k\Delta t})] \tag{7}$$

Thus the slope of a plot of $\log (c_1 - c_2)$ versus t is $-k/2.303$.

If instead of measuring concentration directly, some linear function X of the concentration, say, optical rotation, is measured, an equation of the same form as Eq. (7) applies. For example, if

$$X = ac + b \tag{8}$$
$$c_1 - c_2 = \frac{X_1 - X_2}{a} \tag{9}$$

and

$$\log (X_1 - X_2) = \frac{-kt}{2.303} + \log [ac_0(1 - e^{-k\Delta t})] \tag{10}$$

The slope is the same as it would be if actual concentration differences had been plotted.

....→ ***Apparatus.*** Polarimeter (described on page 237); mercury-vapor lamp with filters or sodium-vapor lamp; thermostat and circulating pump; two water-jacketed polarimeter tubes; pure sucrose; 100 ml of 4 N hydrochloric acid; 100 ml of 4 N monochloroacetic acid; three 25-ml pipettes.

PROCEDURE.   Twenty grams of pure cane sugar (sucrose) is dissolved in water (filtered, if necessary, to give a clear solution) and diluted to 100 ml. The sucrose solution, the 4 N hydrochloric acid solution, and the 4 N monochloroacetic acid are placed in the 25° thermostat and allowed to stand a few minutes to come to temperature.

Two jacketed polarimeter tubes, filled with water, are connected in series with the circulating water from a thermostat at 25°. A zero reading is taken with a mercury-vapor lamp and Corning glass filters arranged to transmit only the green light (Chapter 25). A sodium-vapor lamp is equally satisfactory.

Twenty-five milliliters of the sucrose solution is thoroughly mixed with 25 ml of the 4 N monochloroacetic acid solution, and small portions of the solution are used to rinse out one polarimeter tube. The tube is then filled and stoppered. The second polarimeter tube is filled in a similar manner after rinsing, using 25 ml of the sugar solution and 25 ml of the 4 N hydrochloric acid solution. The tubes are filled rapidly, and readings are taken as soon as possible after mixing. The first reading is taken on the hydrochloric acid solution and recorded, and subsequent measurements are recorded about every 10 min for the first hour or so. As the reaction slows down, the observations may be taken less frequently. The observations should extend over a period of 3 hr or more. The reaction goes much more slowly with the monochloroacetic acid, and the readings are taken less frequently. They are taken at convenient intervals of time when the polarimeter is not needed for the hydrochloric acid solution. The polarimeter tube containing hydrochloric acid is replaced by the tube containing monochloroacetic acid.

The final readings (α_∞) are taken after the solutions have stood in a tightly stoppered flask long enough for the reaction to be completed, at least 2 days for the hydrochloric acid and a week for the monochloroacetic acid. A second reading is taken at a later time to be sure that the reaction has been completed. If it is not convenient to obtain the final reading for the monochloroacetic acid, it may be assumed that α_∞ will be the same as for the hydrochloric acid.

CALCULATIONS.   The concentrations of sucrose in terms of $\alpha_t - \alpha_\infty$ are plotted against time in minutes. Then the values of log $(\alpha_t - \alpha_\infty)$ are plotted against time. The best straight lines are drawn through the points for each reaction. The reaction-rate constants k are calculated from the slopes of the lines.

In calculating the results of this experiment, at least one set of data is treated by the Guggenheim method for comparison with the usual method. If the reading at infinite time cannot be obtained conveniently, this method may be used exclusively.

Practical applications. These are discussed under Experiment 22.

Suggestions for further work. Some suggestions for further work are discussed under Experiment 22.

Trichloroacetic acid and sulfuric acid, and other acids, each 4 N, may be used as catalysts. Trichloroacetic acid is about as highly dissociated as hydrochloric acid.

Caution: Trichloroacetic acid is corrosive.

The relative acid strengths of monochloroacetic acid and trichloroacetic acid are to be explained on the basis of molecular structure.

The activation energies may be obtained by running a second set of determinations, using water pumped from a thermostat at 35 or at 15°, and comparing them with those at 25°.

The effect of ionic strength on the rate of this reaction[2] may be investigated. The effect of changing the dielectric constant may be investigated by adding ethanol or dioxane.[3]

Volume changes as measured continuously in a dilatometer may be used to follow the course of a reaction. The hydrolysis of acetal[4,5] is a good example. At 25°, 0.005 M HCl is mixed quickly with enough acetal from a graduated pipette to make the solution 0.15 M with respect to acetal. The solution is transferred immediately to a dilatometer through a tightly fitting stopcock, and the rise of the liquid in the capillary is recorded at frequent intervals. The logarithm (of the final reading minus the reading at time t) is plotted against time, and the rate constant is calculated. A second experiment may be carried out with 0.05 M acetic acid instead of the hydrochloric acid.

The method of least squares (Chapter 20) may be used for calculating the best straight line which represents log $(\alpha_t - \alpha)$ plotted versus time.

Precise measurements of the rate of inversion of sucrose are available.[6,7]

References

1. E. A. Guggenheim, *Phil. Mag.*, (7)**2**: 538 (1926).
2. E. S. Amis and G. Jaffe, *J. Chem. Phys.*, **10**: 598 (1942).
3. E. S. Amis and F. C. Holmes, *J. Am. Chem. Soc.*, **63**: 2231 (1941).
4. J. N. Bronsted and W. F. K. Wynne-Jones, *Trans. Faraday Soc.*, **25**: 59 (1929).
5. F. G. Ciapetta and M. Kilpatrick, *J. Am. Chem. Soc.* **70**: 639 (1948).
6. S. W. Pennycuick, *J. Am. Chem. Soc.*, **48**: 6 (1926).
7. G. Scatchard, *J. Am. Chem. Soc.*, **48**: 2259 (1926).

25   BROMINATION OF ACETONE

The form of the rate law for the bromination of acetone using an acid catalyst is determined. This experiment illustrates the fact that the rate law does not necessarily bear any simple relationship to the stoichiometric equation for the reaction.

THEORY.[1]   The rate law for a reaction cannot be predicted from the balanced equation for the reaction. The rate law can only be determined experimentally. From the form of the rate law certain conclusions can be reached about the mechanism of the reaction.

   The stoichiometric equation for the bromination of acetone is

$$CH_3\overset{\overset{\textstyle O}{\|}}{C}CH_3 + Br_2 = CH_3\overset{\overset{\textstyle O}{\|}}{C}CH_2Br + Br^- + H^+$$

   The reaction rate increases with the concentration of H^+ in acidic solution or with the concentration of OH^- in basic solution. The balanced equation for the reaction occurring in acidic solution involves hydrogen ion as a product, and one would anticipate, therefore, an increasing reaction rate in the course of an experiment carried out in unbuffered solutions.

   The rate of halogenation of acetone is independent of the concentration of halogen, except at very high acidities.[2] The rates of reaction with the different halogens (chlorine, bromine, and iodine) are identical, and the same as the rate of racemization (for the case of optically active ketones), within a few percent.[3-5] These facts can be accounted for in terms of the mechanism

$$CH_3-\overset{\overset{\textstyle O}{\|}}{C}-CH_3 + H^+ \;\overset{K}{\rightleftharpoons}\; \left[CH_3-\overset{\overset{\textstyle O}{\underset{\textstyle}{\|}}}{\underset{}{C}}-CH_3\right]^+ \tag{1}$$
$$\underset{A}{} \qquad\qquad\qquad\qquad \underset{I}{}$$

$$\left[CH_3-\overset{\overset{\textstyle H}{\overset{\textstyle O}{\|}}}{C}-CH_3\right]^+ + H_2O \;\underset{k_{-1}}{\overset{k_1/(H_2O)}{\rightleftharpoons}}\; CH_3-\overset{\overset{\textstyle OH}{|}}{C}=CH_2 + H_3O^+ \tag{2}$$
$$\underset{I}{} \qquad\qquad\qquad\qquad\qquad \underset{E}{}$$

$$CH_3-\overset{\overset{\textstyle OH}{|}}{C}=CH_2 + Br_2 \;\overset{k_2}{\longrightarrow}\; CH_3-\overset{\overset{\textstyle O}{\|}}{C}-CH_2Br + H^+ + Br^- \tag{3}$$
$$\underset{E}{} \qquad\qquad\qquad\qquad \underset{P}{}$$

Since ketones are very weak bases, the equilibrium in the first reaction is unfavorable for the formation of the ion I. Under these circumstances $(I) = K(A)(H^+)$, where K is the equilibrium constant for reaction (1) and the parentheses indicate concentrations.

   The rate equations for enol E and product P are, according to the mechanism, the following:

$$\frac{d(E)}{dt} = k_1(I) - [k_{-1}(H^+) + k_2(Br_2)](E) \tag{4}$$

$$\frac{d(P)}{dt} = k_2(Br_2)(E) \tag{5}$$

These rate equations may be solved for $d(P)/dt$ under steady-state conditions by letting $d(E)/dt = 0$ and substituting $(I) = K(A)(H^+)$. In this way we obtain

$$\frac{d(P)}{dt} = \frac{k_1 k_2 K(A)(H^+)(Br_2)}{k_{-1}(H^+) + k_2(Br_2)} \tag{6}$$

This equation takes on a simpler form if the enol which is formed in the first two steps reacts much more rapidly with halogen than with hydrogen ions; that is, $k_2(Br_2) \gg k_{-1}(H^+)$.

$$\frac{d(P)}{dt} = k_1 K(A)(H^+) \tag{7}$$

This is in accord with the observation that the overall reaction is first order in ketone and acid but independent of the concentration of halogen. It can be seen that in general the apparent second-order rate constant is made up of a combination of k_1 and the equilibrium constant K for the formation of ion I.

A great deal of research has been carried out on this particular reaction, and the above mechanism can be extended to include racemization, deuterium exchange,[6] and catalysis by bases.[7] The overall reaction does not stop with the monobromo-acetone, but it is not necessary to consider the subsequent reactions if only *initial* reaction rates are studied.

····➤ **Apparatus.** Bromine, water, acetone, standard hydrochloric acid (1 *M* or 2 *M*); spectropho-tometer; pipettes; volumetric flasks; ventilated hood.

PROCEDURE.   Since bromine absorbs strongly at the blue end of the visible spectrum, this reaction may be studied conveniently with a spectrophotometer such as that described in Experiment 16.   The absorbancy indices of Br_2 dissolved in distilled water are 160 M^{-1} cm^{-1} at 400 mμ, 100 M^{-1} cm^{-1} at 450 mμ, 30 M^{-1} cm^{-1} at 500 mμ, and 8 M^{-1} cm^{-1} at 550 mμ. The solubility of Br_2 in water is about 0.2 *M* at 25°, and it is more convenient to use water saturated with Br_2 in preparing solutions than to use pure liquid bromine. The concentration of the stock bromine solution is determined, using the absorbancy indices given above. In accurate work it is necessary to determine cell corrections by intercomparing the various cells filled with solvent. In order to obtain the greatest percentage accuracy in the determination of concentration, the absorbancy should be in the range of about 0.2 to 0.7 (percent transmission of about 60 to 20). The spectrophotometer cells and all bromine solutions should be kept covered.

Hydrochloric acid is a convenient catalyst, and a suitable concentration range is 0.05 to 0.5 *M*. It is desired to determine the initial rate law, i.e., the order of the reaction with respect to acetone, hydrogen ion, and bromine, and the rate constant.

In order to avoid complications, only initial velocities are used. A series of experiments is designed to determine the concentration effects. The acetone concentration may be varied in the range 0.1 to 2 M. The bromine concentration may be varied in the range 0.001 to 0.01 M. Some of the higher concentrations should be run first because the initial velocities can be determined in a shorter time. Using a cell holder for four cells, three kinetic experiments can be run simultaneously. In order to obtain accurate results, the spectrophotometer should be equipped with a thermostating arrangement so that the reacting solutions can be held at the desired constant temperature. In the absence of this equipment, the temperatures of the solutions should be measured at the beginning and end of the experiment.

Caution: Experimental operations involving bromine should be carried out in a well-ventilated hood, and solutions containing bromoacetone should be kept stoppered.

This reaction can also be studied titrimetrically. The reaction is stopped by pipetting aliquots into sodium acetate solution to raise the pH. Potassium iodide is added, and the I_2 formed is titrated with sodium thiosulfate.

CALCULATIONS. The reaction rates in moles per liter of Br_2 reacting per second are calculated for each reaction mixture, and the approximate uncertainty is estimated. It is the initial velocity which is desired in each case. Plots of initial velocity versus (acetone), (H^+), and (Br_2) are prepared to determine the order with respect to each of these substances. After the form of the rate law has been determined, the best value of the rate constant is calculated.

Suggestions for further work. The reaction of acetone with iodine may be studied. A 0.01 M KI solution is used as solvent because the solubility of iodine in pure water is so low.

The catalysis of the reaction by acetate–acetic acid buffers may be studied.[8,9] Since the acetate ion is also a catalyst, there will be terms in the rate law for both acetate ion and acetic acid. There is also a term proportional to the product of these concentrations. The reaction is subject to general acid-base catalysis.[10]

The catalysis of the reaction by bases such as pyridine or hydroxyl ion may also be studied.[7]

References

1. L. P. Hammett, "Physical Organic Chemistry," McGraw-Hill Book Company, New York, 1940.
2. L. Zucker and L. P. Hammett, *J. Am. Chem. Soc.*, **61:** 2791 (1939).
3. P. D. Bartlett and C. H. Stauffer, *J. Am. Chem. Soc.*, **57:** 2580 (1935).
4. S. K. Hsi and C. L. Wilson, *J. Chem. Soc.*, **1936:** 623.
5. C. K. Ingold and C. L. Wilson, *J. Chem., Soc.*, **1934:** 773.
6. S. K. Hsi, C. K. Ingold, and C. L. Wilson, *J. Chem. Soc.*, **1938:** 78.
7. P. D. Bartlett, *J. Am. Chem. Soc.*, **56:** 967 (1934).
8. R. P. Bell and P. Jones, *J. Chem. Soc.*, **1953:** 88.
9. H. M. Dawson and J. C. Spivey, *J. Chem. Soc.*, **1930:** 2180.
10. H. M. Dawson, C. R. Haskins, and J. E. Smith, *J. Chem. Soc.*, **1929:** 1884.

Chapter
8

Irreversible
Processes
in
Solution

Liquid viscosities are measured in this experiment by the capillary-flow method and by the falling-ball method.

THEORY.   The flow of a fluid is said to be *laminar* if points fixed in the fluid move smoothly in layers, one layer (lamina) sliding relative to another. The elementary process involved in laminar flow is pictured in Fig. 34, which represents a small region of the fluid. The x axis is chosen to lie in the local direction of flow, the z axis perpendicular to the laminae; v_x represents the velocity of fluid in the x direction. Viscous effects come into play if an element of the fluid is caused to change its shape as it moves. The coefficient of viscosity η is defined by the equation

$$f_x = \eta \frac{\partial v_x}{\partial z} \tag{1}$$

where f_x is the shearing force per unit area exerted in the direction of flow *on the*

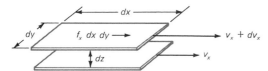

figure 34. The quantities involved in Eq. (1), defining the coefficient of viscosity. Acting on the plane of area dx dy is the shearing force f_x dx dy, which maintains the velocity gradient dv_x/dz. The shearing force per unit area f_x is proportional to dv_x/dz.

element of fluid between two planes at the plane of larger z (an equal and opposite shearing force acts on the opposite face) and $\partial v_x/\partial z$ is the velocity gradient in the z direction. Equation (1) simply expresses the proportionality between the velocity gradient and the shearing force which produces it. The coefficient η depends on temperature, pressure, and composition. A fluid is said to be a *newtonian fluid* if, in laminar flow, η is a constant independent of the velocity gradient. The cgs unit for viscosity (grams per centimeter per second) is called the *poise*.

   Capillary-flow Method.[1-6]   A very satisfactory method for determining η consists of measuring the volume rate of flow of the fluid through a long cylindrical tube of circular cross section. The required equation is obtained by application of Eq. (1). To make the problem tractable, certain simplifying assumptions are introduced:

1. The fluid is assumed to be a newtonian, incompressible fluid.
2. It is assumed that the flow is laminar, with flow lines parallel to the walls. Hence the laminae consist of concentric cylinders coaxial with the tube.

3. The layer next to the wall is assumed to stick to the wall, so that its velocity is zero.
4. The flow is assumed to be steady, in the sense that the velocity of the fluid at any fixed point in the tube is constant in time.

The driving pressure $\mathcal{P}$ between points a and b along a streamline is defined by

$$\mathcal{P} = P_a - P_b + \rho g(h_a - h_b) \tag{2}$$

where P_a, P_b = values of external pressure
$\qquad h_a$, h_b = heights above some reference level
$\qquad \rho$ = density of fluid
$\qquad g$ = acceleration of gravity

$\mathcal{P}$, which is positive when the flow is from a to b, represents the total mechanical force per unit cross section which acts to maintain a steady rate of flow. In the case of steady-state flow through a long cylindrical tube, the fluid is not accelerated and the driving pressure $\mathcal{P}$ is attributable entirely to the shearing work which must be done on the viscous fluid to maintain a steady rate of flow. Thus the shearing force acting on a layer of fluid between two concentric cylinders of radii s and $s + ds$, respectively, is $\pi s^2 \mathcal{P}$. The force in the direction of flow acts on the inner surface of this layer, the area of which is $2\pi s l$, with l being the length of the tube. From Eq. (1), with $dz = -ds$,

$$\frac{\pi s^2 \mathcal{P}}{2\pi s l} = -\eta \frac{\partial v}{\partial s} \tag{3}$$

where v is the velocity of the fluid in the layer of radius s. The velocity is considered to be positive in the direction of flow. Separating variables s and v and integrating from the wall ($s = r$, $v = 0$) to an arbitrary radius, one gets

$$\int_0^v dv = -\frac{\mathcal{P}}{2\eta l} \int_r^s s \, ds \tag{4}$$

$$v = \frac{\mathcal{P}}{4\eta l}(r^2 - s^2) \tag{5}$$

giving v at radius s. Thus the velocity profile is parabolic. The volume of fluid which flows past any cross section per second dV/dt is found by integration over the cross section:

$$\frac{dV}{dt} = \int_0^r 2\pi s v \, ds = \frac{\pi \mathcal{P} r^4}{8\eta l} \tag{6}$$

Provided $\mathscr{P}$ is constant in time, the total volume V which flows past any cross section in time t is

$$V = \frac{\pi \mathscr{P} r^4 t}{8 \eta l} \tag{7}$$

This is known as the *Hagen-Poiseuille equation.*

In the Ostwald viscometer (Fig. 35), the liquid flows through a vertical capillary tube from a reservoir of well-defined, fixed volume at the top into a second reservoir at the lower end. The driving pressure $\mathscr{P}$, defined by Eq. (2) with differences taken between the reservoir surfaces, is $\rho g h$, where h is the difference in height of the surfaces of the two reservoirs, since the external pressure is the same at the surface of both reservoirs. Thus Eq. (7) solved for η becomes

$$\eta = \frac{\pi r^4 g h}{8 V l} \rho t \tag{8}$$

figure 35.

Ostwald viscometer.

To avoid the complications implicit in direct and accurate measurement of the various apparatus dimensions appearing in Eq. (8) and the further problem of a small variation in h during each run, it is common to assume that the equation

$$\eta = A \rho t \tag{9}$$

holds where A is a constant for a given viscometer. Since l is the capillary length, Eq. (8) neglects the driving-pressure drops in the reservoirs. For accurate work, it is necessary to consider the pressure drop expended in accelerating the fluid at the entrance to the capillary tube [see Eq. (18)].

The assumption that the flow is laminar can be checked by calculating the Reynolds number $\mathscr{R}$, which in general is defined by

$$\mathscr{R} = \frac{v \rho a}{\eta} \tag{10}$$

where v = velocity of fluid
$\quad a$ = characteristic dimension of flow system
For the capillary-flow case, v may be taken as the mean fluid velocity $\langle v \rangle$ in the tube,

$$\langle v \rangle = \frac{1}{\pi r^2} \int_0^r (v) 2 \pi s \, ds = \frac{\mathscr{P} r^2}{8 \eta l} = \frac{V}{\pi r^2 t} \tag{11}$$

and a as the tube diameter. The flow may safely be assumed to be laminar for $\mathscr{R} < 2000$; the onset of turbulence usually occurs somewhere within the range 2000 to 10,000.

Provided the flow is laminar, the various assumptions used in obtaining Eq. (8) hold quite well in the interior of the capillary tube but not at the ends. The most important end effect is that which arises from the fact that work must be done on the fluid to accelerate it as it approaches and enters the capillary tube from the entrance reservoir. The driving pressure $\mathscr{P}$ properly includes not only the term $\mathscr{P}'$ for the shearing work done,

$$\mathscr{P}' = \frac{8Vl\eta}{\pi r^4 t} \tag{12}$$

but in addition a term for the work done in accelerating the fluid. If viscous effects are neglected in calculating the latter term, one may use the Bernoulli equation[7]

$$P + \rho g h + \tfrac{1}{2}\rho v^2 = \text{constant along a streamline} \tag{13}$$

which expresses the principle that the work done on an element of volume of an incompressible fluid moving along a streamline equals the total increase in potential and kinetic energy. (The internal energy is assumed to be constant.) Associated with this increase in fluid velocity is a pressure change

$$\Delta P = -\Delta\tfrac{1}{2}\rho\langle v^2\rangle_m \tag{14}$$

where $\langle v^2\rangle_m$ is the mass-averaged value of v^2 (over a cross section), which is zero at the surface of the upper reservoir and has the value

$$\langle v^2\rangle_m = \frac{1}{\rho(dV/dt)} \int_0^r (v^2) 2\pi sv\rho \, ds = \frac{1}{2}\left(\frac{\mathscr{P}'}{4\eta l}\right)^2 r^4 = 2\left(\frac{V}{\pi r^2 t}\right)^2 \tag{15}$$

within the capillary beyond the point at which the velocity profile of Eq. (5) has become established. The increment $\Delta\tfrac{1}{2}\rho\langle v^2\rangle_m$ must be added to the Hagen-Poiseuille pressure drop to give the total driving pressure,

$$\mathscr{P} = \frac{8Vl\eta}{\pi r^4 t} + \frac{1}{2}\rho\langle v^2\rangle_m \tag{16}$$

It may be asked whether a similar application of the Bernoulli theorem should not be made at the exit reservoir; if it were, the deceleration of the fluid would require a pressure increase exactly canceling the decrease of Eq. (14). The answer is that the Bernoulli pressure rise at the exit is neglected, it being assumed that the kinetic energy of the fluid issuing into the exit reservoir is dissipated entirely through viscous effects in the reservoir and passes by heat flow into the surroundings, instead of producing a pressure rise and thereby being transferred to the environment through performance of work. The assumption that the Bernoulli pressure change is the dominant effect at the entrance to the capillary but negligible at the exit is not unreasonable, because the flow scheme is quite different in the two reservoirs, as one may easily verify by adding a dye to the fluid in one reservoir and then observing the flow.[6] The essential correctness of the *kinetic-energy term,* as the last term in Eq. (16) is called, is well established experimentally for the usual conditions of measurement.[2-5]

The equation for η with the kinetic-energy correction included,

$$\eta = \frac{\pi \mathcal{P} r^4 t}{8Vl} - \frac{\rho V}{8\pi l t} \tag{17}$$

is obtained by solving Eq. (16) for η and using Eq. (15) to eliminate $\langle v^2 \rangle_m$. Applied to the Ostwald viscometer, for which $\mathcal{P}$ equals $\rho g h$, Eq. (17) becomes

$$\eta = A\rho t - B\frac{\rho}{t} \tag{18}$$

where

$$A = \frac{\pi r^4 g h}{8Vl} \qquad B = \frac{V}{8\pi l} \tag{19}$$

The apparatus constant B can be estimated reasonably well from the viscometer dimensions, or it can be determined empirically, along with A, from measured flow times and densities for two or more liquids of accurately known viscosity.†

The capillary-flow method has been found useful, with appropriate choice of capillary dimensions, for fluids of viscosity up to 100 poises. For viscosities in a higher range, say, 10 to 10,000 poises, the falling-ball method is generally suitable.

Falling-ball Method.[1,5,8] If a spherical body is caused to move at a sufficiently low velocity v through a fluid, laminar flow takes place. For the case of an incompressible unbounded newtonian fluid with viscosity coefficient η, the retarding force f exerted on the body at low velocities is given by Stokes' equation,

$$f = 6\pi r \eta v \tag{20}$$

where r is the radius of the sphere. If the body falls freely through the fluid, the velocity asymptotically approaches that velocity (the *terminal velocity*) at which the sphere suffers no further acceleration because the retarding force exerted by the

---

† Additional end effects for which corrections are sometimes made arise from the viscous pressure drop in the reservoirs and from the fact that the viscous pressure drop in the capillary tube near the entrance, where the velocity profile is not yet parabolic, is slightly larger than the Hagen-Poiseuille value. Corrections for these effects are made by replacing l in both Eqs. (19) by $l' \approx l + 0.5r$ and inserting a numerical factor $m \approx 1.1$ in the numerator of the expression for B. These corrections are not accurate but will serve as a useful means for gauging the general magnitude of these effects. The numerical constants are often found empirically.[12,13]

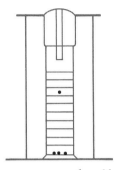

fluid exactly cancels the gravitational force. With allowance for buoyancy, the equation for this balance of forces is

$$\tfrac{4}{3}\pi r^3(\rho - \rho_0)g = 6\pi r\eta v \tag{21}$$

where ρ = density of falling body

ρ_0 = density of fluid

g = acceleration of gravity

The falling-ball viscometer (Fig. 36) consists of a cylindrical tube filled with the fluid of unknown viscosity. A ball of suitable density and radius is allowed to fall along the axis of the tube. The times at which the ball passes regularly spaced horizontal calibration marks are recorded. In the present experiment, the apparatus is constructed from a graduated cylinder, and steel bearing balls are used.

The hydrodynamic equation for this system is complicated by the presence of the walls and by the fact that the velocity may be large enough for Stokes' equation to fail even though the flow remains laminar. A satisfactory solution, in the form of a power-series expansion, has been derived theoretically by Faxen and tested experimentally by Bacon.[8] The equation for η, with only the leading terms included explicitly here, is

$$\eta = \frac{2gr^2(\rho - \rho_0)}{9v}\left[1 - 2.104\left(\frac{r}{R}\right) + 2.09\left(\frac{r}{R}\right)^3 + \cdots\right] \tag{22}$$

where R is the radius of the cylinder. The reason for the existence of the wall effect is that the boundary conditions are different for the cylinder from those for an unbounded medium. A layer of fluid adjacent to the surface of the ball moves with the ball, and fluid farther from the surface moves in such a fashion as to transport fluid upward, from the region ahead of the ball, to its wake. In an unbounded medium, this flow occurs to an appreciable extent throughout a region extending outward from the ball a distance of many times the ball diameter. The presence of the cylinder walls modifies the flow, because the fluid at the walls is at rest.

The most important terms in the Faxen equation after those given in Eq. (22) involve the Reynolds number. For Eq. (22) to be accurate, with these terms neglected, the condition

$$\mathscr{R}^2 \ll \frac{r}{R} \tag{23}$$

must be met, where now

$$\mathscr{R} = \frac{2v\rho r}{\eta} \tag{24}$$

Molecular Theory.[9-11] The molecular theory of viscosity of liquids has not been developed to the point where accurate predictions can be made for liquids of polyatomic molecules. However, the Eyring theory of absolute reaction rates[9] provides a useful basis for correlation of viscosity data and certainly represents a valuable first step toward understanding the molecular processes involved. Applied to the case of viscous flow, the Eyring equation is

$$\eta = \left(\frac{hN_0}{V}\right)e^{\Delta G^{\ddagger}/RT} \tag{25}$$

where h = Planck's constant
$\quad N_0$ = Avogadro's number
$\quad V$ = molar volume = M/ρ
$\quad M$ = molecular weight
$\quad \Delta G^{\ddagger}$ = standard free energy of activation for viscous flow

The exponential dependence on T is typical for processes which require activation. The activation process to which $\Delta G^{\ddagger}$ refers cannot be precisely described, but in general terms it corresponds to the passage of the system into some relatively unfavorable configuration, from which it can then easily go on to the final state of the molecular-flow process. For example, in normal liquids, the activation step may be the creation within the body of the liquid of a vacancy or hole into which an adjacent molecule can move; for associated liquids, it might be the breaking of enough intermolecular bonds to permit a molecule to move into an available vacancy.[10]

The temperature dependence of η implied by Eq. (25) is found by making use of the exact equation

$$\left[\frac{\partial(\Delta G^{\ddagger}/T)}{\partial T}\right]_P = -\frac{\Delta H^{\ddagger}}{T^2} \tag{26}$$

where $\Delta H^{\ddagger}$, the enthalpy change for the activation process, is related to $\Delta G^{\ddagger}$ through

$$\Delta G^{\ddagger} = \Delta H^{\ddagger} - T\,\Delta S^{\ddagger} \tag{27}$$

$\Delta S^{\ddagger}$ being the entropy of activation for viscous flow. The result is

$$\left[\frac{\partial \ln(\eta/\rho)}{\partial(1/T)}\right]_P = \frac{\Delta H^{\ddagger}}{R} \tag{28}$$

where the temperature dependence of V has been acknowledged by replacing V by M/ρ and transferring ρ to the left side before differentiating. Thus, from values of η and ρ at several temperatures, one may calculate $\Delta G^{\ddagger}$, $\Delta H^{\ddagger}$, and $\Delta S^{\ddagger}$.

Equation (25) has been generalized for solutions. Of particular interest are those cases in which mixtures of a pair of liquids show a maximum or minimum in viscosity as a function of mole fraction. Vapor-pressure data for such pairs often show very marked departures from Raoult's law. An equation which ties together these two types of anomalous behavior has been given by Kincaid, Eyring, and Stearn.[9,11]

Apparatus. Ostwald viscometer; stopwatch or timer; thermostat with glass window; 10-ml graduated pipette; density balance or other means for determining densities; acetone, chloroform, or other liquids.

Falling-ball viscometer filled with fluid[†] of known density; precision chrome-steel bearing balls of several diameters, for example, $\frac{1}{16}$, $\frac{5}{64}$, and $\frac{3}{32}$ in.; two timers; switching circuit; micrometer; scale for measurement of fluid height and calibration spacing; forceps; magnet for recovery of balls; hypodermic syringe.

PROCEDURE. The Ostwald viscometer is shown in Fig. 35. Since the glassware is very easily broken, care must be exercised in clamping.

The procedure is based on the use of Eq. (18). The value of the constant B is to be calculated from Eq. (19). A value for V sufficiently accurate for the calculation of B may be obtained by filling the reservoir with a hypodermic syringe. The length l may be measured directly with a steel scale. Then the constant A is to be determined from measurement of the time of flow with water.

After the viscometer has been thoroughly cleaned with *hot* sulfuric acid–potassium dichromate solution, it is thoroughly rinsed by drawing water through it, followed by acetone, and finally is dried by aspirating *clean* air through it. Compressed air is not used because foreign particles or traces of oil may cause serious errors. The viscometer is clamped vertically in the thermostat, and a specified quantity of sample is added from a pipette. The volume of sample used should be such that the liquid surface in the lower reservoir is in the widest part in order that the change in h during a run will be minimized. The same volume of sample is used for all samples.

A dust-free rubber tube is attached to the smaller tube, and by means of a rubber suction bulb the liquid is drawn up into the enlarged bulb and above the upper mark. The liquid is then allowed to flow down through the capillary. The timer is started when the meniscus passes the upper mark and is stopped when it passes the lower mark. If several check determinations on the time of flow do not agree closely, the tube should be cleaned again.

After the measurements with water have been completed, the flow times for a series of other liquids for which η and ρ are accurately known are measured to provide data for a test of Eq. (18). Acetone, ethyl acetate, carbon tetrachloride, and

---

† Polymer liquids having viscosities in the appropriate range are available from Union Carbide Chemicals Co., New York. Flexol plasticizer R-2-H and Niax Pentol LA-475 are suitable.

benzene are suitable. Accurate viscosity and density data for the liquids can be obtained from the "International Critical Tables."[12]

Next, the viscosity of a pure liquid such as *n*-butyl alcohol is determined at a number of temperatures over the range 0 to 45°C in order to find the activation energy for viscous flow. The required density data may be taken from published tables.[12] For the time of flow measurements, it is important to have a thermostat which maintains the temperature reasonably steady over a sufficiently long period for the sample to reach thermal equilibrium and for several flow determinations to be made. Measurements at 0°C may be made with a large unsilvered Dewar flask containing an ice-water bath. The Dewar should be surrounded with a sheet of heavy celluloid or other suitably strong and transparent material to provide protection against the danger of flying glass from an implosion. It is imperative that glasses or safety goggles, preferably of a type having side shields, be worn while this work is being done.

Finally, the viscosity of a polymer liquid is measured by means of the falling-ball viscometer, shown in Fig. 36.† The outermost cylinder is a section of 60-mm Pyrex tubing which is placed around the viscometer to protect it from air drafts. The inner cylinder, which holds the sample, is a standard 100-ml graduated cylinder, having a diameter of roughly 30 mm. The graduated cylinder is mounted perpendicular to the base plate, and a small bubble level fastened to the latter is used to facilitate getting the cylinder vertical. A thermometer (not shown) is taped to the outside wall of the inner cylinder.

The cylinder must be filled well in advance and allowed to stand for hours (or days!) until the bubbles have all escaped. A satisfactory procedure for determining the density of the fluid is to calibrate the cylinder before filling, weigh the cylinder before and after filling, and note the level of the fluid after it has become clear of bubbles and has reached the temperature at which the viscosity is to be measured.

The apparatus must be vertical, so that the balls travel along the axis of the tube. Adequate time should be allowed for the fluid to come to a uniform temperature if the ambient temperature is different from that at which the unit was stored.

Three steel balls of each size to be used are rinsed in benzene, dried on a towel, and not subsequently handled with the fingers. The average weight per ball is found by weighing a set of several of the same size. The diameters of the balls may be taken from the manufacturer's specification or measured carefully with a micrometer, and the densities calculated later from the measured weights. After a practice run, data are obtained for at least two balls of each size.

Since the 10-ml calibration lines of the graduated cylinder run the full circumference, parallax error may be avoided in determining the times at which the ball

---

† The glass parts for a viscometer of this type are available from Norman D. Erway, Oregon, Wis.

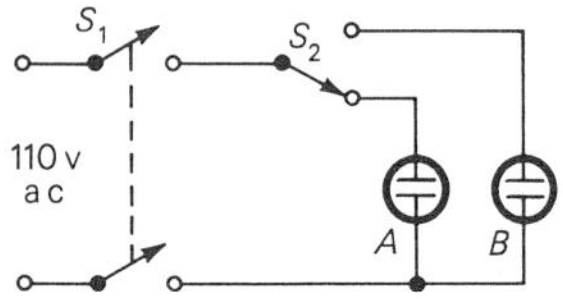

figure 37.

Switching circuit for falling-ball viscometer. The two electric timers are plugged into the outlets A and B.

passes these lines. With the timer switches in the positions shown in Fig. 37, and both timers set to zero, a ball is dropped through the aligning tube at the top. As it crosses the first line, one timer is started by closing switch S_1. As the ball crosses each subsequent line, switch S_2 is thrown to stop one timer and start the other. After each such change, the reading of the stopped timer is recorded, and this timer is then reset to zero. At the end of the run, switch S_1 is opened. The temperature of the fluid is noted. The average distance between calibration lines is measured. The mean cylinder radius may be calculated from the spacing of calibration lines on the assumption that the column has a uniform circular cross section.

After a number of balls have accumulated, they may be drawn up along the wall by the action of a small magnet.

CALCULATIONS.   The constants A and B for the Ostwald viscometer are first calculated from measured values of V, l, and the flow time for water. Then Eq. (18) may be tested and the constants A and B determined more reliably by plotting $\eta/\rho t$ versus $1/t^2$ for the water and the series of other pure liquids for which flow times were measured. The slope should be compared with the value of B obtained from direct measurement of V and l. Although it is difficult to measure r accurately, an estimate of A may be made from Eq. (19) as a rough check.

The flow times for n-butanol are used to calculate η for each temperature. It is essential to use in this calculation density values for the same temperatures as those for which flow times were measured. Values of $\Delta G^{\ddagger}$, $\Delta H^{\ddagger}$, and $\Delta S^{\ddagger}$ for viscous flow at 25°C may then be calculated from Eqs. (25), (28), and (27), respectively. A plot of log (η/ρ) versus $1/T$ is appropriate for determining $\Delta H^{\ddagger}$ from Eq. (28).

The falling-ball data should be examined carefully to determine the average velocity of fall in that region of the tube in which the velocity is sensibly uniform. The viscosity may be calculated from this average. Comparison of results for different ball diameters provides a valuable check on the accuracy of the wall correction.

Practical applications. Studies of the viscosities of high polymers and of solutions of macromolecules have provided interesting information about the structure and motions of these molecules. Nonnewtonian behavior is not uncommon among these systems.

Suggestions for further work. The acetone-chloroform system shows curious viscosity behavior. To observe this, the viscosities of acetone, chloroform, and four binary mixtures (about equally spaced on a mole-fraction basis) should be measured. The solutions may conveniently be made up by volumetric techniques, since density data for the pure liquids may then be used to calculate the mole fractions.

For interpreting the results, the equation of Kincaid, Eyring, and Stearn[9,11] for nonideal mixtures is instructive. In addition to the viscosity and density data for the two pure liquids and their mixtures, the only additional quantities required are the excess free energies of mixing, i.e., above the ideal values. These are easily calculated from vapor-pressure data[13] for the mixtures, provided it is assumed that nonideality in the vapor can be neglected.

References

1. R. B. Bird, W. E. Stewart, and E. N. Lightfoot, "Transport Phenomena," John Wiley & Sons, Inc., New York, 1960.
2. L. Schiller, Strömung in Rohren, in "Handbuch der Experimentalphysik," vol. 4, pt. 4, pp. 39–57, Akademie-Verlag GmbH, Berlin, 1932.
3. S. Erk, Zähigkeitsmessungen, in "Handbuch der Experimentalphysik," vol. 4, pt. 4, pp. 465–468, Akademie-Verlag GmbH, Berlin, 1932.
4. J. F. Swindells, J. R. Coe, Jr., and T. B. Godfrey, *J. Res. Natl. Bur. Std. U.S.,* **48:** 1 (1952).
5. J. F. Swindells, R. Ullman, and H. Mark in A. Weissberger (ed.), "Technique of Organic Chemistry," vol. 1, "Physical Methods of Organic Chemistry," 3d ed., pt. 1, chap. 12, Interscience Publishers, Inc., New York, 1959.
6. N. E. Dorsey, *Phys. Rev.,* **28:** 833 (1926).
7. D. Halliday and R. Resnick, "Physics," combined ed., pp. 376–378, John Wiley & Sons, Inc., New York, 1960.
8. L. R. Bacon, *J. Franklin Inst.,* **221:** 251 (1936).
9. S. Glasstone, K. J. Laidler, and H. Eyring, "The Theory of Rate Processes," pp. 480–516, McGraw-Hill Book Company, New York, 1941.
10. R. E. Powell, W. E. Roseveare, and H. Eyring, *Ind. Eng. Chem.,* **33:** 430 (1941).
11. J. F. Kincaid, H. Eyring, and A. E. Stearn, *Chem. Rev.,* **28:** 301 (1941).
12. "International Critical Tables," McGraw-Hill Book Company, New York, 1928.
13. J. C. Chu, S. L. Wang, S. L. Levy, and R. Paul, "Vapor-Liquid Equilibrium Data," J. W. Edwards, Publisher, Incorporated, Ann Arbor, Mich., 1956.

27 CONDUCTANCE BEHAVIOR OF STRONG ELECTROLYTES

In this experiment practice is obtained in the measurement of the electrical conductance of solutions and in the determination of ionic mobilities by the moving-boundary method.

THEORY.[1-4] When a current flows in an electrolyte solution, charge is carried by the motion of both anions and cations. If an electrolyte solution is placed in a cell containing two parallel electrodes of area A each, separated by a distance l, and the resistance of the cell is found to be R, then the *specific resistance* ρ of the electrolyte solution is

$$\rho = \frac{RA}{l} \tag{1}$$

The specific conductance κ is the reciprocal of the specific resistance. If a cell could be constructed with electrodes of exactly 1 cm² area exactly 1 cm apart, the reciprocal of the cell resistance in reciprocal ohms would be numerically equal to the specific conductance in reciprocal ohms per centimeter. The usual conductance cell does not satisfy these conditions, and so it is convenient to define a constant factor k determined by the cell geometry and called the *cell constant*, such that

$$k = \kappa R \tag{2}$$

where R is the resistance of the actual cell. The numerical value of k for a particular cell is determined experimentally by use of a standard solution of known specific conductance.[5]

The specific conductance will clearly depend on the concentration of the electrolyte. The measurement of conductance is put on an equivalent basis by defining the *equivalent conductance* Λ as

$$\Lambda = \kappa V \tag{3}$$

where V is the volume of solution containing 1 g equiv of solute. In the cgs system, V is in cubic centimeters and Λ is in square centimeters per ohm per equivalent.

The increase of the equivalent conductance of solutions of strong electrolytes with dilution in the low-concentration range is not due to an increase in dissociation, because the dissociation is already complete, but to an increased mobility of the ions. In a concentrated solution of a highly ionized strong electrolyte, the ions are close enough to one another so that any one of them in moving is influenced not only by the electrical field impressed across the electrodes but also by the field of the surrounding ions. The ionic velocities are, then, dependent upon both forces. Arrhenius attempted to treat the electrolytic-conductance behavior of the strong electrolytes in the way in which he had successfully treated the weak electrolytes; such a treatment is, however, inconsistent with the experimental fact, discovered by Kohlrausch, that a plot of the equivalent conductance of a strong electrolyte against the square root of the concentration is very nearly linear. More recently Debye and Hückel and Onsager have been able to calculate the effect of the surrounding ions on the mobility of any given ion and, for dilute solutions, have obtained results entirely consistent with the experimental facts. Complete dissociation is here assumed.

The conductance of these solutions is the result of the movement of ions through the solution to the electrodes. When two electrodes in the solution are made part of a complete electrical circuit, the cations $(+)$ are attracted to the negative pole (cathode) and the anions $(-)$ are attracted to the positive pole (anode). Changes in the equivalent conductance of an electrolyte solution with changes in concentration may result from changes in both the number and the mobility of the ions present. If both anion and cation are monovalent, the overall electrical neutrality of the solution assures that equal numbers of the two ions are present. They will not in general have the same mobility, however, and thus do not share equally in the conduction of current. The *transference number* T_c of the cation is, by definition, the fraction of the current carried by the cation.

The moving-boundary method offers the most accurate method for the determination of transference numbers of both cations and anions, and such a determination, together with a measurement of specific conductivity, may be used to calculate ionic mobilities.

In the moving-boundary method, an initially sharp boundary between two electrolyte solutions having either the same anion or the same cation is subjected to an electric field. If the two solutions have different cations, a boundary will move toward the cathode with the velocity of the cation in the solution into which the boundary moves. In order to obtain a sharp moving boundary, the boundary must be made to move into the solution containing the cation with the higher mobility. The latter solution is called the *leading* solution and the other the *indicator* solution.

In this experiment electrolysis is used to form the indicator solution. The anode in the moving-boundary apparatus is made of metallic cadmium, so that $CdCl_2$ is formed by the passage of current. A boundary between the solutions of cadmium chloride and hydrochloric acid will leave the face of the electrode and move up the tube as illustrated in Fig. 38. Since the cadmium ion has a much lower mobility

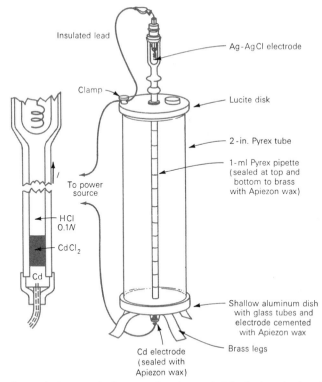

figure 38. Apparatus for the moving-boundary method for the determination of transference numbers.

than the hydrogen ion, this moving boundary is very sharp. As the boundary moves upward, the chloride ion moves downward across the boundary and eventually accumulates around the anode. Electrical neutrality is preserved in the solution near the electrode because only a fraction of the cadmium ions which are formed by the electrolysis of the anode is necessary to maintain the growing column of indicator solution.

If the boundary in a tube of cross-sectional area A moves through a distance Δx in a time Δt, the charge transferred by the cation is equal to $FcA\,\Delta x$, where c is the concentration of the cation in the leading solution† and F is the faraday, 96,490 coulombs equiv^{-1}. The total charge passing through the solution in this time is $I\,\Delta t$, where I is the average current flowing. It then follows that

$$T_c = \frac{FcA\,\Delta x}{I\,\Delta t} \tag{4}$$

If the tube is a 1-ml pipette calibrated in 0.1-ml divisions, then $A\,\Delta x$ is 0.1 ml when the boundary passes from one mark to the next.

The ionic mobility u, which is the ratio of the ion velocity to the electric field strength, may be calculated as follows. The potential difference dE between two points in the tube separated by a distance dx is

$$dE = IdR = \frac{I}{A\kappa}dx \tag{5}$$

where dR is the resistance of the element of solution dx in length. The mobility is obtained by dividing the ion velocity $\Delta x/\Delta t$ by the field strength dE/dx

$$u = \frac{\Delta x/\Delta t}{dE/dx} = \frac{\kappa A\,\Delta x}{I\,\Delta t} \tag{6}$$

In this experiment, the $CdCl_2$ solution below the boundary has a specific conductivity κ, which is lower than that of the hydrochloric acid solution, because it is more dilute, and the Cd^{++} ion has a lower mobility than the H^+ ion. By reference to Eq. (5) it is seen that the electric field strength is greater in the $CdCl_2$ solution below the moving boundary than in the HCl solution above the boundary. Therefore, if H^+ ions diffuse into the $CdCl_2$ solution below the boundary, they will encounter a high field strength and will be rapidly sent up to the boundary. On the other hand, if Cd^{++} ions diffuse ahead of the boundary, they will have a lower velocity than the hydrogen ions because of their lower mobility and will soon be overtaken by the boundary. This so-called *adjusting effect* keeps the boundary sharp.

---

† Clearly the units of c and $A\,\Delta x$ must be consistent. If $A\,\Delta x$ is in cubic centimeters, c must be in equivalents per cubic centimeter.

····→ ***Apparatus.*** Wheatstone bridge assembly; conductance cell; oscillator with a frequency of about 1 kHz; conductance water; 0.1 N solution of hydrochloric acid; thermostat at 25°C; volumetric flasks and pipette; moving-boundary apparatus as shown in Fig. 38; current-regulated dc power supply; methyl violet indicator.

Part A. Conductance Measurements

PROCEDURE. The conductance assembly for this experiment consists of an oscillator to provide an input voltage at about 1 kHz for the bridge, a Wheatstone bridge, an amplifier and earphones or other detector connected to the output terminals of the bridge, and a simplified Wagner grounding circuit, all shown diagrammatically in Fig. 39.

When the bridge has been balanced by adjustment of the resistance R_3 and the ratio R_2/R_1, as indicated by a minimum signal, the cell resistance R_4 is given by $R_4 = (R_2/R_1)R_3$. The ratio R_2/R_1 is adjustable in decimal steps and appears directly on the dial marked with the word "multiply." The resistance R_3 is read in ohms from the four dials.

There is usually some uncertainty in the balance point because of the presence of a background tone, which results from the flow of alternating currents through stray capacitances not shown in Fig. 39. This effect is considerably reduced, and the sharpness of balance correspondingly improved, by the proper adjustment of the Wagner grounding circuit.

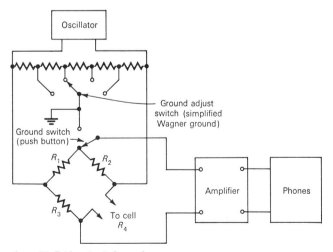

figure 39. Bridge circuit for conductance measurements.

The amplitude of the input voltage may be varied by means of the gain control on the oscillator.

The conductance cell is filled with the solution whose conductance is to be measured. After the elimination of any bubbles in the cell, the latter is placed in the thermostat. While the solution is coming to the desired temperature, a test measurement on a resistor may be made for practice. The resistance to be measured is to be connected to the terminals marked X on the Wheatstone bridge box.

The following directions may be of use in finding the balance point of the bridge; they assume the use of a typical student-type Wheatstone bridge with a Wagner ground auxiliary circuit.

1. The amplifier gain is set to give a tone well below the maximum obtainable.
2. With the X1000 dial not at zero, that position of the multiply switch is selected which gives the minimum intensity, and the X1000 dial is then adjusted for a minimum. If a minimum is not obtained, it may be necessary to try an adjacent position of the multiply switch.
3. The push-button "ground" switch is depressed, and the best "Wagner ground" switch position is selected. The push-button switch is then released.
4. Adjustment of the bridge is continued for minimum intensity, taking the four decade resistors in decreasing order. It is essential that at each step the setting be verified by approach from both the high- and low-resistance sides; proper execution of this step may require thought when the balance point appears to correspond to a dial setting of 0 or 9.
5. An estimate of uncertainty is recorded for each measurement. This should not exceed 0.3 percent, except for pure water.

The Freas-type conductance cell shown in Fig. 40 is particularly suitable for the conductance measurements. The four corners of the thin platinum squares are anchored to a glass frame. The cell is readily filled and emptied, and the volume of solution required is small. For all conductance measurements the cell is immersed in a thermostat, preferably at 25°C regulated to 0.02°.

The electrodes of the conductance cell must have an adherent coating of platinum black and should be immersed in distilled water whenever the cell is not in use. If the electrodes are allowed to dry out, it is difficult to rinse out electrolytes from them, and it is advisable to dissolve off the coating with aqua regia (under the hood) and plate out a fresh deposit. The electrodes and cell are rinsed out thoroughly, first with distilled water and then with conductance water, which is especially pure water prepared by multiple distillations. The conductance cells must be handled with great care; the electrodes must not be touched, and they must not be moved with respect to each other during the course of an experiment.

The cell is filled with conductance water and inspected to make sure there are

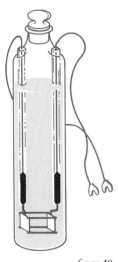

figure 40.

The Freas-type conductance cell.

no air bubbles at the electrodes, and its resistance is measured; it is then rinsed and refilled, and the resistance is measured again. This process is repeated until the resistance has become essentially constant, showing that contaminating electrolytes in the cell have been rinsed out. The cell resistance will not become absolutely constant because the conductance water is very pure and traces of electrolytes insignificant in the later measurements will produce noticeable fluctuations. The specific conductance of the water used should be about 5×10^{-6} ohm^{-1} cm^{-1} or less, corresponding to a resistance of 200,000 ohms or more in a cell of unit cell constant. At these high resistances an accuracy of more than two significant figures is difficult to obtain without special precautions. For the other solutions studied, a precision of the order of a few tenths of a percent should be obtained in the resistance measurements.

When the cell is clean, as shown by a reasonably constant high resistance with conductance water, it is rinsed two or three times with 0.02 N potassium chloride solution and the resistance is then determined with this solution filling the cell. Additional measurements are made on fresh samples of this solution until successive determinations agree closely. The purpose of these measurements is to provide data for calculation of the cell constant.

The cell is emptied and rinsed with the next solution for which the conductance is to be measured. It is advisable to make check determinations on each solution to make sure that the cell was thoroughly rinsed.

Conductance measurements are then made on solutions of hydrochloric acid at concentrations of 0.1, 0.05, 0.025, 0.0125, 0.00625, 0.00312, and 0.00156 N. Enough of the 0.1 and 0.05 N solutions should be prepared for use in the moving-boundary measurements as well. Successive dilutions must be made with great care; otherwise propagation of error will become excessive. They may be made with calibrated volumetric flasks and pipettes.

Part B.   Moving-boundary Measurements

PROCEDURE.   The adaptation of this method for use as a laboratory experiment is described by Longsworth.[6] The glass capillary tube is made of a 1-ml Pyrex pipette graduated every 0.1 ml and having an inside diameter of about 2 mm. The capillary is rinsed several times with 0.1 N hydrochloric acid containing methyl violet indicator. Only enough methyl violet to give a distinguishable color in the capillary tube is required. The indicator should be added to the acid just before the start of the experiment, because the color will fade. The rinsing of the capillary may be aided by the use of a wood or glass rod which just fits in the capillary and is used as a plunger. It is important to dislodge any bubbles at the lower end of the tube. Next the electrode chamber is filled with hydrochloric acid, and the silver–silver chloride electrode is inserted.

It is necessary to immerse the tube in which the boundary moves in a water bath, in order to dissipate the heat which is developed in the tube by the passage of the electric current.

The electrodes are connected to a source of direct current capable of delivering 2 to 4 ma. It is most convenient if the current through the capillary is kept constant during the experiment. If the applied voltage is constant, the current through the capillary will decrease during the experiment, because as the boundary ascends the tube, the length of the column of indicator electrolyte increases correspondingly, and this solution is a poorer conductor than the one it replaces. The potential applied to the cell must therefore be continually increased in order to maintain a constant current. This may be done by one of two methods: (*a*) part of the current supply may be shunted across a rheostat which has a sliding contact, and by manual adjustment of this contact a constant current through the cell may be maintained; (*b*) a power supply designed to produce a constant current with minimum adjustment, such as that of Bender and Lewis,[7] may be used. This method has the advantage that attention may be focused on the determination of the boundary velocity. The current may be measured by means of a low-range milliammeter or, if greater accuracy is required, by measuring the potential drop across an accurately known series resistance with a potentiometer.

Supplies with precise current regulation may also be constructed from readily available operational amplifier modules.[8]

The time, to the nearest second, at which the boundary crosses successive graduations is obtained with the aid of a stopwatch. Four to six values are normally sufficient.

A second experiment is performed with the 0.05 *N* HCl solution. If it is desired that the time intervals be about the same, the current should be reduced proportionately.

CALCULATIONS.   The cell constant for the conductance cell is determined by means of Eq. (2) and the known specific conductance of the potassium chloride solution.[5] The specific conductance for 0.0200 *N* KCl, for example, is 0.002768 ohm^{-1} cm^{-1} at 25°C. The specific conductances of the conductance water and the various HCl solutions are then calculated. The specific conductance of the solute in each case is the difference between the specific conductance of the particular solution and that of the solvent used. The equivalent conductance of the HCl is then calculated for each solution.

A plot is made of equivalent conductance versus the square root of concentration and the equivalent conductance at infinite dilution obtained by extrapolation of the experimental data.

Values of the transference number of the hydrogen ion are calculated for each of the two concentrations on which measurements were made by the use of Eq. (4) and a plot of Ax versus t. In the regions of the tube near the cadmium electrode, the movement of the hydrogen ion may be retarded because of the diffusion of the $CdCl_2$ solution. The hydrogen-ion mobility is calculated from Eq. (6) and the experimentally obtained specific conductance. The chloride transference number and mobility are calculated by using the relation $T_{H^+} + T_{Cl^-} = 1$.

Suggestions for further work. If the current is regulated very closely and the moving boundary tube is calibrated, measurements at successively lower concentrations can be used to study the concentration dependence of the transference number and mobility of the hydrogen ion. The temperature dependence of the mobility may also be investigated, if a modified apparatus is used in connection with thermostats at various temperatures.

Practical applications. Conductance measurements have been used not only for laboratory experiments but also for the determination of the concentration of salts in water and for the flow of water in streams.

The moving-boundary method for ionic mobilities is useful in the study of proteins.

References

1. R. A. Robinson and R. H. Stokes, "Electrolyte Solutions," 2d ed. (revised), Academic Press, Inc., New York, 1965.
2. T. Shedlovsky, Conductometry, in A. Weissberger (ed.), "Technique of Organic Chemistry," vol. 1, "Physical Methods of Organic Chemistry," 3d ed., pt. 4, chap. 45, Interscience Publishers, Inc., New York, 1960.
3. M. Spiro, Determination of Transference Numbers, in A. Weissberger (ed.), "Technique of Organic Chemistry," vol. 1, "Physical Methods of Organic Chemistry," 3d ed., pt. 4, chap. 46, Interscience Publishers, Inc., New York, 1960.
4. D. A. MacInnes, "The Principles of Electrochemistry," Reinhold Publishing Corporation, New York, 1939.
5. G. Jones and B. C. Bradshaw, *J. Am. Chem. Soc.*, **55:** 1780 (1933).
6. L. G. Longsworth, *J. Chem. Educ.*, **11:** 420 (1934).
7. P. Bender and D. R. Lewis, *J. Chem. Educ.*, **24:** 454 (1947).
8. George A. Philbrick Researches, Inc., "Applications Manual for Computing Amplifiers," 2d ed., Nimrod Press, Inc., Boston, 1966.

28   *WEAK ELECTROLYTES*

In this experiment the effects on electrical conductance of the partial dissociation of weak electrolytes is investigated.

THEORY.[1,2]   The principles of electrolytic conduction are discussed in Experiment 27. A factor which affects the equivalent conductance of a solution that was not considered there is the possible limited dissociation of the electrolyte.

Some electrolytes, known as *weak electrolytes*, do not dissociate completely in solution. Instead, there is an equilibrium between ions and associated electrolyte. Acetic acid is a typical weak electrolyte. The apparent equilibrium constant for dissociation may be calculated as

$$K_a = \frac{c_{H^+} c_{A^-}}{c_{HA}} = \frac{\alpha^2 c}{1 - \alpha} \tag{1}$$

where α = degree of dissociation
 c = concentration of the solute
According to the Arrhenius theory, the equivalent conductance at any concentration is related to the degree of dissociation by

$$\alpha = \frac{\Lambda}{\Lambda_0} \tag{2}$$

where Λ = equivalent conductance at concentration c
 Λ_0 = equivalent conductance at infinite dilution
In the case of a weak electrolyte the value of Λ_0 cannot be obtained by the extrapolation to infinite dilution of results obtained at finite concentration, because Λ is a rapidly varying and nonlinear function of $\sqrt{c}$. Instead, Λ_0 is obtained by the use of the law of Kohlrausch:

$$\Lambda_{0,HR} = \Lambda_{0,HCl} + \Lambda_{0,NaR} - \Lambda_{0,NaCl} \tag{3}$$

This is equivalent to saying that for a simple, binary electrolyte like HR

$$\Lambda_{0,HR} = l_{0,H^+} + l_{0,R^-}$$

The equivalent conductance at infinite dilution is a sum of ionic contributions, but l_{0,H^+}, for example, is independent of the electrolyte from which the hydrogen ions are obtained. Since HCl and the salts NaR and NaCl are all strong electrolytes, the values on the right-hand side of Eq. (3) can all be obtained by extrapolation, as discussed in Experiment 27.

The apparent equilibrium constant K_a, expressed in Eq. (1), is equal to the true equilibrium constant, which is expressed in terms of activities, only for ideal solutes. For real systems at finite concentration

$$K = K_a \frac{\gamma_{H^+} \gamma_{A^-}}{\gamma_{HA}} \tag{4}$$

where γ_i is the activity coefficient of species i. Since $\gamma_i \rightarrow 1$ for infinitely dilute solutions,

$$\lim_{c \to 0} K_a = K \tag{5}$$

Measurements of K_a over a range of concentrations thus provide a way of estimating K and the activity coefficient factor $\gamma_{H^+}\gamma_{A^-}/\gamma_{HA}$ at the various concentrations. To a good approximation γ_{HA} may be considered to be unity. The value of the product $\gamma_{H^+}\gamma_{A^-}$ is, by definition, $\gamma_{\pm}^2$, the square of the mean ionic activity coefficient for the weak electrolyte HA.

Apparatus. Conductance cell; Wheatstone bridge assembly and oscillator; conductance water; $0.02000\ N$ potassium chloride solution (or other solution of accurately known specific conductance); $0.05\ N$ solutions of acetic acid and monochloroacetic acid.

PROCEDURE.   The use of the conductance cell and bridge is described in Experiment 27. The cell constant is determined from measurements on a solution of known specific conductivity, as described there.

If the normalities of the acids are not known accurately, they are determined by titration with standard base. The conductance of a $0.05\ N$ solution of acetic acid is determined, and a $0.025\ N$ solution is then prepared by quantitative dilution with conductance water. In this fashion, conductance measurements are made on 0.05, 0.025, 0.0125, 0.00625, 0.00312, and 0.00156 N acetic acid solutions. A similar sequence of measurements is made for the monochloroacetic acid solutions.

CALCULATIONS.   For each concentration of each system measured, the equivalent conductance of the solute is calculated as described in Experiment 27. For each sequence of measurements, a plot is made of equivalent conductance versus concentration.

For each of the acids the apparent equilibrium constant for dissociation is calculated for each concentration and the values of the true equilibrium constants estimated. The equivalent conductances at infinite dilution are obtained from values of $\Lambda_{0,HCl}$, $\Lambda_{0,NaCl}$ and $\Lambda_{0,NaR}$ taken from the literature[3] or from measurements made on the salts and extrapolated as in Experiment 27.

Practical applications. Dissociation constants and other properties can be determined by measurements of conductance.

Suggestions for further work. The influence of substitution and structure on the apparent dissociation constants of organic acids may be studied. For example, the dissociation constants of mono- and dichloroacetic acids and propionic acid may be determined and compared with the dissociation constant of acetic acid. In the same way the influence of substituting amino or nitro groups into benzoic acid may be studied.

References

1. General References are given in Experiment 27.
2. D. A. MacInnes and T. Shedlovsky, *J. Am. Chem. Soc.*, **54:** 1429 (1932).
3. "International Critical Tables," vol. VI, McGraw-Hill Book Company, New York, 1928.

Chapter 9

Electromotive Force

In this experiment the standard electrode potential of the silver–silver chloride electrode is determined from electromotive-force measurements on a cell consisting of it and a hydrogen electrode. Measurements are made on other cells also. Experience is gained in the use of a potentiometer.

THEORY.[1-5]   An electrochemical cell is a device whereby the decrease in free energy of a system in a spontaneous process may be made a source of electrical work. The process involved may be an ordinary chemical reaction, the transfer of a constituent from one concentration level to another, etc.; the essential requirement is that it must be possible to accomplish it as the resultant of an oxidation step and a reduction step, each of which takes place separately at an appropriate electrode. The electromotive force of the cell depends upon the change of state of the cell system and also upon the closeness of approach to reversibility permitted by the intrinsic characteristics of the electrode processes themselves and by the degree of irreversibility introduced by the manner in which the cell is used. In the present case we shall be concerned with cells which are considered capable of reversible performance and shall assume that the electromotive-force measurements are made by the potentiometric method, which permits the current drawn from the cell in the voltage measurement to be reduced to a level for which an adequate approach to reversible cell performance is obtained. It also permits a true test of reversibility, since the direction of the cell reaction and current flow can be changed at will.

In order to correlate a cell voltage with a cell reaction as written on paper, it is necessary to follow certain conventions. For a given cell reaction, the corresponding cell is designated by specifying, at the left, the electrode at which oxidation takes place and, at the right, the electrode at which reduction takes place when the reaction takes place in the direction specified. (For a given cell expression, the conventional cell reaction can be written down directly in terms of these same requirements.) If the spontaneous cell reaction takes place in the direction specified, ΔG is negative and the right-hand electrode will be the positive electrode. Correspondingly, in this case, a positive sign is required for the voltage of the cell, since the electrical work, which is proportional to the cell voltage, is done at the expense of the free energy of the cell system. Conversely, if the right-hand electrode is found to be negative, the cell electromotive force should be given a negative sign, and the cell reaction postulated does not take place spontaneously.

The voltage of the cell may be considered to be the resultant of two single-electrode potentials. By convention, the voltage is then calculated as the single-electrode potential for the electrode written at the right minus the single-electrode potential for the electrode written at the left in the cell expression. These conventional electrode potentials are then reduction potentials. The absolute values of such electrode potentials cannot be determined, and so the hydrogen electrode with hydrogen gas at unit fugacity and hydrogen ion at unit activity is taken as the reference electrode with a defined potential of zero. A comprehensive summary of theoretical and experimental information on this and other reference electrodes has been provided by Ives and Janz.[4]

For unit activity of all species taking part in the cell reaction, the electromotive force is designated as $E°$ and is called the *standard electromotive force*. This standard electromotive force then can be expressed in terms of two standard electrode potentials. The determination of the standard electrode potential for a given electrode then requires a knowledge of the value of $E°$ for a cell containing this electrode and a second electrode for which the standard potential is known. Thus, for the cell

Pt, H_2; HCl, molality m; AgCl, Ag   (1)

the standard electromotive force $E°$ is equal to the standard electrode potential of the silver–silver chloride electrode, since the standard electrode potential of the hydrogen electrode is zero. Extensive tabulations of standard electrode potentials are available.[3-5]

The calomel electrode, often used as a reference electrode, does not conform completely to the conventional standard-state specifications. The mercury and mercurous chloride are present as the pure liquid and solid, respectively, and thus are at unit activity relative to the ordinary standard-state choice. The activity of the chloride ion is not unity, however, but varies with the concentration of potassium chloride used. This variation is then accounted for in the different reference electrode potentials given for the calomel electrode for different potassium chloride concentrations.

The electrode potential changes with the activities of the ions. The fundamental equation governing the effect of activity of ions on the voltage is

$$E = E° - \frac{RT}{nF} \ln Q \tag{2}$$

where R = gas constant, 8.314 joules deg^{-1} mole^{-1}
    F = faraday, 96,490 coulombs equiv^{-1}
    n = number of faradays transferred for reaction as written
    Q = activity quotient

$$Q = \frac{a_G^g a_H^h}{a_A^a a_B^b} \tag{3}$$

for the generalized reaction $aA + bB = gG + hH$.

The activities can be expanded in terms of activity coefficients and molalities as

$$a_i = \gamma_i m_i \tag{4}$$

where a_i = activity of species i

γ_i = activity coefficient of species i

m_i = molality of species i

For cell (1), the cell reaction is

$$\tfrac{1}{2}H_2 + AgCl = Ag + H^+ + Cl^-$$

The activities of Ag and AgCl are unity because of the presence of the solid phases, and $a_{H^+}a_{Cl^-}$ may be written as $\gamma_\pm^2 m^2$ where $\gamma_\pm$ is the mean ionic activity coefficient and m is the molality. At atmospheric pressures it is satisfactory to assume that the fugacity of hydrogen gas is given by its pressure in standard atmospheres, thus

$$E = E° - \frac{RT}{F} \ln \frac{\gamma_\pm^2 m^2}{p_{H_2}^{\frac{1}{2}}} \tag{5}$$

The Debye-Hückel theory predicts that $\ln \gamma_\pm$ is proportional to the square root of ionic strength, for low concentrations. For a solution of a uni-univalent electrolyte like HCl, the ionic strength is equal to the molar concentration, so that

$$E = E° - \frac{RT}{F} \ln \frac{m^2}{p_{H_2}^{\frac{1}{2}}} - BT\sqrt{c} \tag{6}$$

where B = a constant

c = molar concentration of HCl

The electrolytes involved in the two electrode reactions may not be the same. In that case a salt bridge, usually of KCl solution, is introduced. For a cell consisting of a metal electrode and its monovalent ion in solution and a saturated calomel electrode,

$$M; M^+(a) \| KCl\ (sat);\ Hg_2Cl_2, Hg \tag{7}$$

the electromotive force is given by

$$E = -E°_{M^+;M} - \frac{RT}{F} \ln a_{M^+} + E°_{KCl(sat);\,Hg_2Cl_2,\,Hg} + E_j \tag{8}$$

The liquid-junction potential E_j is made small by the action of the KCl salt bridge which couples the calomel electrode to the solution in which it is immersed.

The salt bridge does not completely eliminate the liquid-junction potential, however, and thus an exact value for an electrode potential cannot be calculated from measurements on a cell of type (7).

For cells involving divalent metal ions,

$$M; M^{++}(a) \| KCl(sat);\ Hg_2Cl_2, Hg \tag{9}$$

the electromotive force is given by

$$E = -E°_{M^{++};M} - \frac{RT}{2F} \ln a_{M^{++}} + E°_{KCl(sat);\,Hg_2Cl_2,\,Hg} + E_j \tag{10}$$

The activities in Eqs. (8) and (10) may be related to molalities by Eq. (4). The individual activity coefficients of the ions cannot be determined separately from thermodynamic measurements. The mean ionic activity coefficient $\gamma_\pm$ is measurable. For a typical strong electrolyte $A_{\nu_+}B_{\nu_-}$, which dissociates into ν_+ cations A and ν_- anions B,

$$\gamma_\pm = (\gamma_A^{\nu_+}\gamma_B^{\nu_-})^{\frac{1}{\nu}} \tag{11}$$

where $\nu = \nu_+ + \nu_-$.

It is usual to substitute $\gamma_\pm$ for γ_+ in determining the metal ion activities; this is only an approximation.

Apparatus. Potentiometer and accessories; hydrogen electrode and tank of hydrogen; silver–silver chloride electrode; approximately 1 N HCl of accurately known molality; calomel electrode (saturated potassium chloride); various metal electrodes and 0.100 M solution of corresponding salts, e.g., Cu, $CuSO_4$; Pb, $Pb(NO_3)_2$; Cd, $CdCl_2$.

PROCEDURE.   The electromotive forces of a cell with a hydrogen electrode and a silver–silver chloride electrode are measured for HCl solutions of 1.0, 0.1, 0.05, 0.025, 0.0125, and 0.00625 N. The construction of the cell is illustrated in Fig. 41. If silver–silver chloride electrodes are not available, they may be prepared from a piece of platinum wire (No. 26) about 7 mm in length coiled into a helix and sealed into a glass tube as illustrated in Fig. 41. The electrode is cleaned in warm 6 N nitric acid, and silver is electrodeposited from a silver nitrate solution. The surface of the deposit is then converted to silver chloride by electrolysis as the anode in 1 N

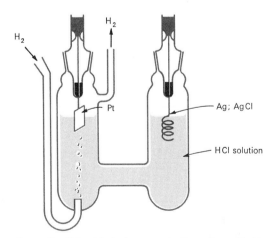

figure 41. Cell with hydrogen and silver–silver chloride electrodes.

HCl solution. Too thick a coat of silver chloride will make the electrode response sluggish. The electrodes are immersed in distilled water for storage.

These electrodes are subject to an aging effect during the first 20 to 30 hr after preparation and are sensitive to traces of bromide in the solutions. In the presence of air the potential is slightly more positive than for an air-free solution, probably because of a slight decrease in the concentration of chloride within the interstices of the electrode by the reaction

$$2Ag + 2HCl + \tfrac{1}{2}O_2 = 2AgCl + H_2O$$

All these effects have to be taken into account in work of the highest precision.

The electromotive force of this cell is determined by means of the potentiometer. In Fig. 42 is shown the principle of the potentiometer in which the electromotive forces or potentials of two cells A and B may be compared. A wire RS, of uniform and high resistance, is stretched along a linear scale. The current is supplied by cell B, whose electromotive force is larger than that of cell A. Since the wire is of uniform resistance and the same current passes through each section of it, there will be a uniform fall of potential per unit length in the direction R to S. To measure an unknown electromotive force (cell A), a second circuit containing a key, galvanometer, and sliding contact is necessary. The negative terminal of cell A is connected opposite the negative terminal of cell B, at R. The sliding contact T is moved along the wire until there is no deflection of the galvanometer G when the key K is pressed. If the sliding contact is moved too far to the right, the galvanometer will deflect in one direction; if too far to the left, the galvanometer will deflect in the other direction. If the potential drop per unit length of the slide-wire is known from the potential difference between R and S and the distance between R and S, the electromotive force of cell A may be determined directly from the length RT. Then, when the setting at T gives no galvanometer deflection, the ratio RT/RS gives the ratio of the voltage of cell A to the potential difference between R and S.

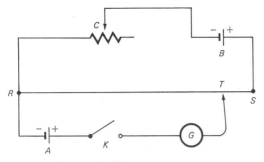

figure 42. The principle of the potentiometer.

Potentiometers are designed so that the fall of potential per unit length of wire is adjusted to some decimal fraction of a volt, and the unknown voltage is then read directly from the scale. This calibration is accomplished with the rheostat *C*, by using a Weston cell (page 510) in place of cell *A*. The Weston standard cell has a voltage of 1.0186 volts at 20°. The point *T* is moved to a position such that there are 1018.6 divisions of the wire between *R* and *S*. The current from the cell *B* through the wire is then changed by adjustment of resistance *C* until the galvanometer shows no deflection, signifying that the fall of potential along *RT* is 1.0186 volts and the difference of potential per unit length is 1 mv. The potentiometer with the rheostat *C* having been adjusted against the standard cell, the readings thereafter are given directly in voltages.

It is important to record which is the positive electrode, i.e., the electrode connected to the positive terminal of the potentiometer when the circuit is balanced. If there is any doubt as to which is positive, the circuit may be compared with one in which an ordinary dry cell is used. In this cell the zinc is negative and the carbon electrode is positive. The negative terminal is the one that gives a blue color when the wires are both touched to a piece of moist litmus paper. If the galvanometer always deflects in the same direction no matter how the potentiometer is set, the terminals of the unknown cell must be reversed to obtain a point of balance.

All measurements should be made at a constant temperature, preferably 25°C. The barometric pressure is recorded.

The potential of a cell consisting of the silver–silver chloride electrode and a calomel electrode is then determined. In the commercially available type of calomel electrode, a thread of glass fibers wet with KCl solution performs the function of a salt bridge. The two electrodes are immersed in a 0.1 *N* HCl solution.

Other metallic electrodes are prepared from strips or rods of pure metals. For each electrode, a 0.1 *M* solution of a soluble salt of the same metal is prepared by carefully weighing out the salts with due allowance for water of crystallization, or better, by determining the concentration by analytical methods. The potential between each metal electrode and the calomel electrode is determined in 0.1 *M* and in 0.01 *M* solution. The 0.01 *M* solutions may be prepared by quantitative dilution of the 0.1 *M* solutions.

For best results the aqueous solutions should be bubbled out with a stream of purified nitrogen to remove dissolved oxygen.

CALCULATIONS.   The molalities of the hydrochloric acid solutions are calculated from known normalities by use of density data, which may be obtained from a handbook.

The partial pressure of hydrogen above the solution is taken to be equal to the barometric pressure less the vapor pressure of water. In more accurate work the vapor pressure of the solution would have to be known. The partial pressure should be expressed in standard atmospheres. A plot of

$$E + \frac{RT}{F} \ln \frac{m^2}{p_{H_2}^{\frac{1}{2}}} \qquad \text{versus} \qquad \sqrt{c}$$

is made and extrapolated to zero concentration to obtain $E°$ for the silver–silver chloride electrode.

The mean ionic activity coefficient of HCl at each concentration is determined.

The standard electrode potentials of the other electrodes are determined. Values of mean ionic activity coefficients for several salts are given in Table 1. Liquid-junction potentials must necessarily be neglected. All values should be compared with accepted values.

Table 1. Mean Activity Coefficients of Electrolytes at 25°C

Electrolyte	Concentration		
	0.001 *M*	0.01 *M*	0.1 *M*
Cadmium chloride	0.819	0.524	0.228
Copper sulfate	0.69	0.40	0.16
Lead nitrate	0.89	0.69	0.37
Silver nitrate	0.95	0.90	0.731
Zinc sulfate	0.700	0.387	0.150
Sodium chloride	—	0.9032	0.7784

Practical applications. The values of the standard potentials of electrodes and the activities of electrolytes are obtained in the manner described here. A knowledge of $E°$ for different electrodes is essential for developing new electrochemical batteries and fuel cells.

Suggestions for further work. The values of $E°$ for other electrodes may be determined by making measurements at several concentrations and extrapolating to infinite dilution. The effect on the electrode potential of hard metals by bending and straining may be studied.

References

1. H. S. Harned and B. B. Owen, "The Physical Chemistry of Electrolyte Solutions," 3d ed., Reinhold Publishing Corporation, New York, 1958.
2. J. O'M. Bockris, "Electrochemistry," Academic Press Inc., New York, 1954.
3. F. Daniels and R. A. Alberty, "Physical Chemistry," 3d ed., Chap. 7, John Wiley & Sons, Inc., New York, 1966.
4. D. J. G. Ives and G. J. Janz "Reference Electrodes," Academic Press Inc., New York, 1961.
5. W. M. Latimer, "The Oxidation States of the Elements and Their Potentials in Aqueous Solutions," Prentice-Hall, Inc., Englewood Cliffs, N.J., 1952.

30   THE GLASS ELECTRODE

The use of the glass electrode for the determination of pH is illustrated by the titrations of phosphoric acid and glycine. The preparation of buffers of given pH and the determination of buffer capacity are introduced.

THEORY.   The glass electrode has several advantages over the hydrogen electrode for the measurement of pH. The glass electrode functions in both oxidizing and reducing media and in the presence of proteins and sulfur compounds, all of which interfere with the use of platinized platinum.

The glass electrode consists of a thin membrane of soft glass enclosing a dilute solution of potassium chloride and acetic acid in which is immersed a platinum wire coated with Ag-AgCl. The variation of the potential of a glass electrode with varying hydrogen-ion concentration is the same as that of a hydrogen electrode.† A number of theories for the action of a glass electrode have been proposed.[1] A saturated calomel electrode is used in conjunction with the glass electrode, so that the cell may be represented diagrammatically as follows:

Glass membrane   Salt bridge

$$\text{Ag, AgCl; KCl, } CH_3CO_2H \vdots \text{unknown solution} \| \text{KCl } (sat); Hg_2Cl_2, Hg$$

The construction of commercially available electrodes is illustrated in Fig. 43. The electromotive force of this cell cannot be measured with a potentiometer because of the high resistance of the glass membrane. For this reason an electronic voltmeter is used. The glass-electrode potential changes 0.0591 volt per pH unit at 25°C, and pH meters are graduated directly in terms of pH. According to the simplified theory, two solutions of the same hydrogen-ion activity, with the glass membrane interposed, should show no potential difference. However, glass electrodes usually do show a small potential (asymmetry potential) under these conditions, and for this reason it is necessary to set the pH meter periodically by using a buffer of known pH. Some useful buffers for this purpose studied by MacInnes[2] are given in Table 1.

Table 1. *pH Values of Standard Buffers*

Buffer	12°C	25°C	38°C
CH_3CO_2H (0.1 N), CH_3CO_2Na (0.1 N)	4.650	4.640	4.635
Potassium acid phthalate (0.05 M)	4.000	4.000	4.015

† At pH's above 10 it is necessary to correct for response to Na^+ ions unless special glasses are used.

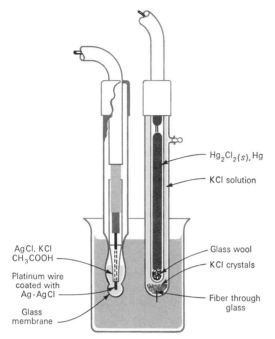

figure 43. Glass-electrode–calomel-electrode assembly.

The pH of a mixture of a weak acid or base and its salt may be calculated with a reasonable degree of precision from the ordinary mass-action equilibrium formulation

$$K_c = \frac{[H^+][A^-]}{[HA]} \tag{1}$$

where HA represents a weak acid and brackets denote molar concentrations. Taking the logarithm of this equation and rearranging, we obtain

$$pH = pK + \log \frac{[A^-]}{[HA]} \tag{2}$$

where pK is equal to $-\log K_c$. Note that if the concentrations of the acidic and basic forms of the buffer are equal, $[A^-] = [HA]$, and

$$pH = pK = -\log K_c$$

This fact may be used to determine ionization constants of rather weak acids and bases. The pH of the buffer depends upon the ratio of the concentrations of these two forms, and not on the total amounts. However, the capacity of a buffer to resist changes of pH produced by the addition of acid or alkali depends upon the concentrations of the two forms present. The slope of a plot of equivalents of acid or base added per liter of buffer versus pH is sometimes called the *buffer capacity*. The buffer capacity[3] may be calculated from the concentrations of the salt and undissociated acid, using

$$\frac{dB}{d(\text{pH})} = \frac{2.3[\text{A}^-][\text{HA}]}{[\text{A}^-] + [\text{HA}]} \tag{3}$$

where B represents the number of equivalents of acid or base added per liter of buffer. The buffer capacity is a maximum at pH = pK. At 1 pH unit away from the pK, a buffer is about 33 percent as effective.

The pK value of a weak acid determined with Eq. (2) depends upon the salt concentration, since the equilibrium expression (1) has been written in terms of concentrations rather than activities. The value of the ionization constant of a weak acid or base determined by titration is referred to as the apparent ionization constant, to distinguish it from the thermodynamic ionization constant obtained by extrapolation to infinite dilution.

The titration of a polybasic acid such as phosphoric acid, using a pH meter, may be used to evaluate the ionization constants. The successive ionizations of phosphoric acid may be represented as follows:

$$\text{H}_3\text{PO}_4 \rightleftharpoons \text{H}_2\text{PO}_4^- + \text{H}^+$$

$$K_1 = \frac{[\text{H}_2\text{PO}_4^-][\text{H}^+]}{[\text{H}_3\text{PO}_4]} = 7.52 \times 10^{-3} \qquad \text{p}K_1 = 2.124 \qquad \text{at } 25° \tag{4}$$

$$\text{H}_2\text{PO}_4^- \rightleftharpoons \text{HPO}_4^{--} + \text{H}^+$$

$$K_2 = \frac{[\text{HPO}_4^{--}][\text{H}^+]}{[\text{H}_2\text{PO}_4^-]} = 6.22 \times 10^{-8} \qquad \text{p}K_2 = 7.206 \qquad \text{at } 25° \tag{5}$$

$$\text{HPO}_4^{--} \rightleftharpoons \text{PO}_4^{3-} + \text{H}^+$$

$$K_3 = \frac{[\text{PO}_4^{3-}][\text{H}^+]}{[\text{HPO}_4^{--}]} = 4.79 \times 10^{-13} \qquad \text{p}K_3 = 12.32 \qquad \text{at } 25° \tag{6}$$

The third dissociation takes place at such a high pH that it cannot be studied in dilute aqueous solutions. The first two end points can be recognized by the large change in pH for a small addition of base.

At the pH at which the second acid group is half neutralized, the hydrogen-ion activity is equal to the equilibrium constant K_2.

The pK values for ionizations (4) to (6) are the thermodynamic values.[4] At 0.1 ionic strength the value of pK_2 is 6.80. The ionic strength is half the sum of the concentrations of the ions, each multiplied by the ion valence squared.

In aqueous solution pure glycine exists as a dipolar ion (zwitterion), as evidenced by the high dielectric constant which is measured for such a solution. When glycine is titrated with acid, the hydrogen ions react with the carboxyl group as illustrated by the reverse of Eq. (7). When half the carboxyl groups have reacted, the hydrogen-ion activity is equal to the equilibrium constant K_1. The equivalent pH is $-\log K_1 = 2.35$. When neutral glycine is titrated with sodium hydroxide, the hydrogen on the amino group is titrated as indicated by Eq. (8). The equilibrium expressions for these reactions are written as follows:

$$^+H_3NCH_2CO_2H \rightleftharpoons {}^+H_3NCH_2CO_2^- + H^+ \tag{7}$$

Form present in   Dipolar ion
acid solution

$$K_1 = \frac{[^+H_3NCH_2CO_2^-][H^+]}{[^+H_3NCH_2CO_2H]} = 4.47 \times 10^{-3} \qquad pK_1 = 2.35 \text{ at } 25°$$

$$^+H_3NCH_2CO_2^- \rightleftharpoons H_2NCH_2CO_2^- + H^+ \tag{8}$$

Dipolar   Form present in
alkaline solution

$$K_2 = \frac{[H_2NCH_2CO_2^-][H^+]}{[^+H_3NCH_2CO_2^-]} = 1.66 \times 10^{-10} \qquad pK_2 = 9.78 \text{ at } 25°$$

The three dissociable hydrogens of phosphoric acid are equivalent in H_3PO_4, and so the question as to which H^+ dissociates first does not arise. In the case of glycine, there are two possibilities, and various types of evidence, including the high dielectric constant of neutral solutions of glycine, indicate that the carboxyl group is the stronger acid group.

····> **Apparatus.** pH meter with glass electrode and calomel electrode; bottle of standard buffer; 0.1 N acetic acid; 0.1 N sodium acetate; 0.1 M phosphoric acid; glycine.

Part A. Titration of Phosphoric Acid

PROCEDURE AND CALCULATIONS. Many satisfactory pH meters are commercially available. The procedure to be followed depends on the particular model used.

Twenty-five milliliters of 0.1 M H_3PO_4 is titrated with 0.1 N NaOH. The pH is measured after each addition of about 5 ml of base, except near the end points, where more readings are taken. In order to determine the pH range for the color change of a typical indicator, a few drops of phenolphthalein solution are added before the titration. The color change of the indicator is noted during the titration. The pH of the solution is plotted versus volume of sodium hydroxide added, and the ionization constant K_2 is calculated.

Part B. Buffers

PROCEDURE AND CALCULATIONS.   The following buffer solutions are carefully prepared, using volumetric equipment:

Solution	ml 0.10 N HAc	ml 0.10 N NaAc
1	95	5
2	50	50
3	5	95

The pH of each solution is determined with the pH meter and compared with
the values calculated from Eq. (2). Since acetic acid is a weak acid, it does not contribute appreciably to the acetate concentration of these buffers, so that the acetate-
ion concentration is determined only by the amount of sodium acetate added. Four
milliliters of 0.1 N sodium hydroxide is added to each of the above buffers, and the
pH again measured. Four milliliters of 0.1 N sodium hydroxide is added to 100 ml
of distilled water, and the pH change noted. The ratios $\Delta B/\Delta(\text{pH})$ of the numbers
of equivalents of base added per liter of buffer to the changes in pH are calculated
and compared with the buffer capacities calculated by using Eq. (3).

Part C. Titration of Glycine

PROCEDURE AND CALCULATIONS.   Two approximately 250-mg portions of
glycine are weighed and dissolved in 30 ml of distilled water. One portion is titrated
with 0.1 N NaOH, and the other with 0.1 N HCl, the pH being recorded at six to
eight intervals. No definite end points are obtained, however, since they occur in
such strongly acidic or basic solutions that a large amount of the added acid or base
is required to change the pH.

The titration curves are plotted by graphing the volume of acid added to the
left of the origin on the horizontal axis and the volume of base added to the right
with the pH plotted as ordinate. In a second figure the pH is plotted versus the number
of equivalents of acid or base which have reacted per mole of glycine. The number
of equivalents of hydrogen ion or hydroxyl ion which have reacted with glycine is
the difference between the number of equivalents of acid or base added and the number of equivalents remaining free in solution (calculated from the pH and the volume
of solution being titrated). The calculations are most conveniently arranged in tabular form. The corrected titration curve is plotted, and the pK values of glycine calculated from it.

Practical applications. The rates of many reactions depend markedly upon the pH, and
therefore solutions in which such reactions are studied must be buffered. Industrially, pH is

frequently controlled by automatic devices which add acid or base, depending upon the potential of a glass electrode. The pH meter is particularly important in biological research.

Suggestions for further work. The glass electrode may be used for the measurement of the pH of a wide variety of miscellaneous substances, e.g., milk, sour milk, blood, orange juice, lemon juice, water extract of soil, tap water, tap water from which carbon dioxide has been expelled. If a mixture of acids which have pK's differing by about 2, e.g., hydrochloric acid, acetic acid, and lactic acid is titrated with the aid of a pH meter, it is possible to determine the amount of each acid present.

Measurements of pH may be used to determine the degree of hydrolysis of salts, provided the salts have been carefully purified and pure water is used. While aqueous solutions of salts of strong acids and strong bases are neutral, solutions of salts of strong acids and weak bases are acidic, and solutions of salts of weak acids and strong bases are basic. To illustrate this, the pH values of solutions of sodium chloride, sodium acetate, ammonium acetate, and aniline hydrochloride may be measured. The degree of hydrolysis x is calculated from $x = [\text{OH}^-]/c$ or $x = [\text{H}^+]/c$, depending upon whether the solution is basic or acidic. The concentration of the salt in equivalents per liter is represented by c. The experimentally determined degrees of hydrolysis are compared with the values calculated from the ionization constants in the literature.

References

1. R. G. Bates, "Electrometric pH Determinations," John Wiley & Sons, Inc., New York, 1954.
2. D. A. MacInnes, "The Principles of Electrochemistry," chap. 15, Reinhold Publishing Corporation, New York, 1939.
3. J. T. Edsall and J. Wyman, "Biophysical Chemistry," vol. I, chap. 8, Academic Press Inc., New York, 1958.
4. L. P. Hammett, "Introduction to the Study of Physical Chemistry," pp. 335ff., McGraw-Hill Book Company, New York, 1952.

31 FREE ENERGY AND THE EQUILIBRIUM CONSTANT

This experiment illustrates the important relation between free energy and the equilibrium constant and interrelates chemical and electrical measurements through thermodynamics.

THEORY.[1,2] One of the most important equations of physical chemistry is that which interrelates the standard free-energy change $\Delta G°$ and the thermodynamic equilibrium constant K for a chemical reaction:

$$\Delta G° = -RT \ln K \tag{1}$$

This equation is exact only if the equilibrium constant K is expressed in terms of activities; it is often useful as an approximation when activities are replaced by concentrations or partial pressures.

The standard state for a gas is the pure gas in the hypothetical ideal-gas state at 1 atm pressure at the given temperature; for a pure liquid or solid, it is usually the pure condensed phase at the given temperature and a pressure of 1 atm. For a solution constituent considered to act as a solvent, the standard state is taken as the pure liquid at the temperature and pressure of the solution; for a constituent treated as a solute, it is so selected that the activity coefficient, defined as the ratio of activity a_2 to the concentration x_2 on the scale chosen, will approach unity as x_2 approaches zero.

The standard free-energy change for a reaction may be calculated indirectly by addition of the standard free-energy changes for any sequence of reactions which together are equivalent to the total reaction. This procedure may be facilitated by the tabulation of the standard free energy of formation, which for a particular compound is the free-energy change for the given temperature and for standard-state conditions for all species involved, for the reaction in which 1 mole of the compound in question is formed from its elements. It is rather more common, however, to calculate the standard free-energy change for a given temperature from the relation

$$\Delta G^\circ = \Delta H^\circ - T \Delta S^\circ \tag{2}$$

The standard heat of reaction ΔH° is calculated from tabulated standard heats of formation, and the standard entropy change ΔS° from standard entropies obtained by use of the third law of thermodynamics. An important modern application of such indirect calculations of equilibrium constants is in the prediction of the performance to be expected from proposed rocket propellants, in which the calculation of the equilibrium composition for complex reaction systems plays an important part.[3] It should be understood, however, that kinetic as well as thermodynamic considerations may be important in determining the composition of a system.

The free-energy change for a chemical reaction may be determined most directly, and frequently with high accuracy, if the reaction can be made to take place reversibly in an electrochemical cell for which the electromotive force can be measured. For a reversible process at constant temperature and pressure, the decrease in free energy of the system, $-\Delta G$, is equal to w_{net}, the work other than PV work done on the surroundings. For the cell, w_{net} is the electrical work and is equal to nFE, when E is the electromotive force of the cell, F is the faraday, and n is the number of faradays of electricity which flow through the cell circuit for completion of the cell reaction as written. Letting a_j, ν_j represent the activity and number of moles of typical product j, with a_i, ν_i playing parallel roles for typical reactant i, for the cell reaction

$$\Delta G = \Delta G^\circ + RT \ln \frac{\prod_j a_j^{\nu_j}}{\prod_i a_i^{\nu_i}} \tag{3}$$

The symbol Π denotes a product of factors. Since $\Delta G = -nFE$, and defining $\Delta G° = -nFE°$,

$$E = E° - \frac{RT}{nF} \ln \frac{\prod\limits_j a_j^{\nu_j}}{\prod\limits_i a_i^{\nu_i}} \tag{4}$$

The calculation of the standard electromotive force $E°$, which corresponds to standard-state conditions for all species involved, requires a knowledge of the thermodynamic activities of all species taking part in the reaction as well as the actual electromotive force E.

The electromotive force of a cell may be resolved into the algebraic difference of two electrode potentials, each of which is set equal to a standard electrode potential (evaluated on a scale established by setting the standard potential of the hydrogen electrode equal to zero) plus the appropriate term involving the activities of the species taking part in the pertinent electrode reaction. If one of the electrodes is the calomel electrode, the activity of its chloride ion does not appear in the expression for the electromotive force; while it is not unity, it is accounted for in the numerical value of the electromotive force recorded as "standard" for this electrode.

Cells with liquid junctions between the different electrolytes cannot be given exact thermodynamic treatment, but they can give a useful result if the liquid junctions have been practically eliminated by a salt bridge. Such cells are used widely in pH measurements; they are to serve in another way in this experiment.

The reaction

$$Fe^{++} + Ag^+ = Ag + Fe^{3+} \tag{5}$$

is particularly suitable for testing Eq. (1), because the equilibrium is quickly reached, the equilibrium constant is easily obtained by volumetric analysis, and the free-energy change may be calculated from the voltage of the cell

$$Pt; Fe^{++}, Fe^{3+} \| Ag^+; Ag \tag{6}$$

····➤ *Apparatus.* Two 150-ml glass-stoppered bottles; platinum electrode; silver electrode; two half-cells (Fig. 44); crystallizing dish; calomel cell; potentiometer assembly; 0.1 M ferric nitrate in 0.5 M nitric acid; 0.1 M silver nitrate; 0.1 M potassium thiocyanate, ferrous sulfate, barium nitrate; purified nitrogen or carbon dioxide.

Part A. Titration Method

PROCEDURE. Precipitated silver is prepared by dissolving about 7 g of silver nitrate in water and adding an excess of copper wire. The precipitate of silver is filtered and rinsed with distilled water until the rinsings give no test for copper ion with ammonia solution. The yield is split between the two glass-stoppered bottles, to each of which is added 100 ml of a solution which is 0.100 M in ferric nitrate and 0.05 M in nitric acid. The nitric acid reduces hydrolysis of the ferric salt. Purified nitrogen or carbon dioxide from a tank is bubbled slowly through the solution for a few minutes to sweep out dissolved oxygen; the glass joint is greased, and the bottle tightly stoppered. Oxidation of ferrous ions by dissolved air constitutes one of the greatest difficulties in this experiment.

The two bottles are heated to about 50°; they are removed and shaken at frequent intervals and then set aside to stand for at least 24 hr. Part B may be performed while equilibrium is being attained.

When equilibrium has been reached, the solutions are analyzed for silver ions. A 25-ml sample of the solution is titrated with 0.1 M potassium thiocyanate, the ferric nitrate already in solution serving as an indicator. The potassium thiocyanate solution is standardized with the 0.1 M silver nitrate solution, an equal volume of the ferric nitrate solution being added as indicator.

The titrations should be made as soon as the stoppers are removed in order to avoid air oxidation of ferrous ion.

Part B. Potentiometric Method

PROCEDURE. Twenty milliliters of 0.1 M ferric nitrate in 0.05 M nitric acid is mixed with 20 ml of freshly prepared ferrous nitrate solution. The latter is prepared by mixing equal portions of 0.2 M ferrous sulfate and 0.2 M barium nitrate. The barium sulfate is allowed to settle for a few minutes in a stoppered vessel; the resulting solution is decanted into one of the half-cells, and the rubber stopper which holds the platinum electrode is sealed tightly in place. It is just as important to prevent air oxidation here as in the procedure for Part A, and to this end the use of purified nitrogen or carbon dioxide over the solutions may be advisable. The presence of a small amount of suspended barium sulfate in the solution should not affect the results.

Into the other half-cell are placed 0.1 M silver nitrate and the silver electrode, with the rubber stopper seating tightly. The sidearms in both half-cells must be completely filled.

The two half-cells and the calomel cell are now set up as shown in Fig. 44. The crystallizing dish contains saturated ammonium nitrate solution, which acts as a salt bridge.

The principle of the potentiometer should be fully understood (Experiment 29 and Chapter 24).

Three potentials are determined between the following pairs of electrodes: silver against calomel, ferrous-ferric against calomel, and silver against ferrous-ferric.

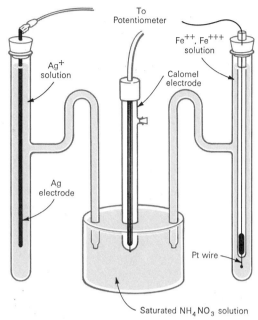

figure 44. Measurement of electrode potentials for calculating the free-energy change in chemical reactions.

In each case, time should be allowed for a steady potential to be reached. The sign of the cell voltage should be recorded as the sign of the right-hand electrode for the cell as written (page 179).

The silver residues from both procedures are to be placed in a special recovery bottle.

CALCULATIONS.   The concentration of silver ion in the equilibrium solution is calculated from the potassium thiocyanate titration. The concentration of the ferrous ion is the same as that of the silver ion, and the concentration of ferric ion is calculated by subtracting the concentration of ferrous ion from the concentration of ferric ion originally present. The *apparent equilibrium constant* K_{app} is then calculated as follows:

$$K_{app} = \frac{c_{Fe^{3+}}}{c_{Fe^{++}}c_{Ag^+}}$$

The term apparent equilibrium constant is used because concentrations rather than activities are used except for the metallic silver, whose activity does not appear explicitly since it is unity for this pure solid phase.

The electrochemical reactions at the two electrodes corresponding to the cell given by Eq. (6) are

$$Fe^{++} = Fe^{3+} + e$$

and

$$Ag^+ + e = Ag$$

The reduction potential $E^\circ_{Fe^{3+},Fe^{++};Pt}$ may be calculated from the electromotive force E_I of the cell

$$\text{Hg, Hg}_2\text{Cl}_2\text{; KCl}(sat)\,\|\,\text{Fe}^{3+}(a_{Fe^{3+}}),\ \text{Fe}^{++}(a_{Fe^{++}});\ \text{Pt} \qquad\qquad \text{(I)}$$

anode   *cathode*

Assuming that the junction potential is negligible,

$$E_I = -E^\circ_{KCl(sat);Hg_2Cl_2,Hg} + E^\circ_{Fe^{3+},Fe^{++};Pt} - 0.0591 \log \frac{a_{Fe^{++}}}{a_{Fe^{3+}}}$$

where $E^\circ_{KCl(sat);Hg_2Cl_2,Hg}$ is the reduction potential for the saturated calomel electrode.

If $a_{Fe^{++}}$ is set equal to $a_{Fe^{3+}}$ when $c_{Fe^{++}} = c_{Fe^{3+}}$,

$$E^\circ_{Fe^{3+},Fe^{++};Pt} = E_I + E^\circ_{KCl(sat);Hg_2Cl_2,Hg}$$

The value of $E^\circ_{Fe^{3+},Fe^{++};Pt}$ found is compared with that given in tables.

The reduction potential $E^\circ_{Ag^+;Ag}$ may be calculated from the electromotive force E_{II} of the cell

$$\text{Hg, Hg}_2\text{Cl}_2\text{; KCl}(sat)\,\|\,\text{Ag}^+(a_{Ag^+});\ \text{Ag} \qquad\qquad \text{(II)}$$

anode   *cathode*

Assuming that the junction potential is negligible,

$$E_{II} = -E^\circ_{KCl(sat);Hg_2Cl_2,Hg} + E^\circ_{Ag^+;Ag} - 0.0591 \log \frac{1}{a_{Ag^+}}$$

Thus

$$E^\circ_{Ag^+;Ag} = E_{II} + E^\circ_{KCl(sat);Hg_2Cl_2,Hg} + 0.0591 \log \frac{1}{a_{Ag^+}}$$

Then for the cell

$$\text{Pt; Fe}^{++}(a_{Fe^{++}}),\ \text{Fe}^{3+}(a_{Fe^{3+}})\,\|\,\text{Ag}^+(a_{Ag^+});\ \text{Ag}$$

$$E^\circ = -E^\circ_{Fe^{3+},Fe^{++};Pt} + E^\circ_{Ag^+;Ag}$$

$$= 0.0591 \log K \qquad\qquad \text{(III)}$$

The value of K is calculated from this equation and compared with the value of K obtained by direct analysis. The agreement can be only approximate, because analytically determined concentrations, rather than activities, are used for calculating K from the equilibrium mixture and for determining $E^{\circ}_{\mathrm{Fe^{3+},Fe^{++};Pt}}$, but since the solutions are fairly dilute, the error is not great. The contact potential between the unlike solutions is another source of considerable error.

Practical applications. An equilibrium constant for a reaction can be calculated when the standard free-energy change is known. In accumulating tables of free energies for this purpose, the direct electromotive-force measurement of reversible cells constitutes one of the most valuable methods. The equilibrium constants for various reactions may be calculated from the standard oxidation-reduction potentials.

Suggestions for further work. The results may be made considerably more accurate by carrying out the measurements with a series of more dilute solutions and evaluating log K and E° by extrapolation to infinite dilution, where the concentrations and activities become identical. More accurate determination of ferric iron is advisable, using reduction with zinc and titration with potassium permanganate.

The oxidation of hydroquinone by silver ion is an excellent reaction[2] to study because the equilibrium constant can be determined accurately by iodiometric titration and because dissolved oxygen from the air does not affect the results. The reaction is

and the cell by which the equilibrium constant can be calculated is

Pt, quinhydrone; $H^+ \| Ag^+$; Ag

The dissociation pressure of copper oxide or mercuric oxide may be calculated from electromotive-force measurements of suitable cells.

References

1. G. N. Lewis and M. Randall, "Thermodynamics," 2d ed., rev. by K. S. Pitzer and L. Brewer, McGraw-Hill Book Company, New York, 1961.
2. R. Livingston and J. J. Lingane, *J. Chem. Educ.*, **15:** 320 (1938).
3. R. L. Wilkins, "Theoretical Evaluation of Chemical Propellants," Prentice-Hall, Inc., Englewood Cliffs, N.J., 1963.

32   THERMODYNAMICS OF ELECTROCHEMICAL CELLS

The electromotive force of a cell is measured at different temperatures, and the heat of the reaction is calculated by means of the Gibbs-Helmholtz equation.

THEORY.   The following reaction[1] is studied:†

$$Zn(Hg) + PbSO_4(s) = ZnSO_4 \ (0.02 \ m) + Pb(Hg) \tag{1}$$

The enthalpy change for this reaction may be measured by combining the reactants in a calorimeter under conditions where only work of expansion is performed, or it may be calculated indirectly from electrical measurements. Since the measurement of electrical quantities is very precise, the latter method is a useful supplement to the direct calorimetric method.

Instead of placing zinc amalgam directly in contact with lead sulfate and carrying out the reaction irreversibly, the reaction may be brought about reversibly in the electrochemical cell

$$Zn(Hg); \ ZnSO_4 \ (0.02 \ m); \ PbSO_4(s), \ Pb(Hg) \tag{2}$$

The cell represented by (2) is particularly suited for experimental study. It is a cell *without transference;* i.e., it has no liquid junction and therefore no uncertain junction potential, provided the effect of the slight solubility of lead sulfate is considered negligible.

Electrical work is equal to the product of the potential and the charge carried through the circuit. When the cell operates reversibly, the electrical work done is determined by the Gibbs free-energy change accompanying the cell reaction:

$$\Delta G = -nFE \tag{3}$$

where E = cell electromotive force for reversible operation

n = number of faradays of electricity passed through the external circuit for occurrence of the cell reaction as written

F = faraday, 96,490 coulombs equiv^{-1}

Thus, for the above reaction ΔG is the Gibbs free-energy change per mole. It can be determined by measuring the cell potential for reversible operation, such as is adequately approached in voltage measurements with a potentiometer. The use of the potentiometer is described in Experiment 29.

At constant temperature and pressure, the heat effect accompanying the direct irreversible reaction, in which only PV work is done, is equal to ΔH, the change in enthalpy for the process; while for the reversible execution of the reaction it is equal to $T \ \Delta S$, where ΔS is the corresponding change in entropy. The difference in these two quantities determines the electrical work done in the reversible process, since for the specified conditions of constant temperature and pressure $\Delta G = \Delta H - T \ \Delta S$.

---

† The symbol (*s*) refers to the solid state. Zn(Hg) and Pb(Hg) are two-phase systems consisting of a solid intermetallic compound in equilibrium with a saturated liquid amalgam.

If $\Delta S > 0$, the electrical work done is greater than that equivalent to ΔH; the energy balance is achieved through heat absorbed by the cell from its surroundings in constant-temperature operation. If $\Delta S < 0$, heat is given up to the surroundings in the reversible operation of the cell at constant temperature and the electrical work becomes less than that equivalent to ΔH.

Values of ΔG, ΔS, and ΔH for the cell reaction are calculated from measured values of E and $(\partial E/\partial T)_P$ using Eqs. (3) to (5):

$$\Delta S = -\left(\frac{\partial \Delta G}{\partial T}\right)_P = nF\left(\frac{\partial E}{\partial T}\right)_P \tag{4}$$

$$\Delta H = \Delta G + T\,\Delta S = -nFE + nFT\left(\frac{\partial E}{\partial T}\right)_P \tag{5}$$

Apparatus. H-type cell, preferably with sintered-glass disk in connecting arm; mercury; granular zinc; granular lead; zinc sulfate; lead sulfate; mortar and pestle; thermostats at several temperatures from 0 to 40°, or rapidly adjustable thermostat in this range; potentiometer assembly.

PROCEDURE. The H-type cell is shown in Fig. 45. For preparing these cells, sintered-glass disks sealed into straight tubes are available from supply houses. The coarse grade of sintered glass is preferred. Alternatively, an open tube plugged with clean glass wool may be used. The purpose of this disk or plug is to prevent solid lead sulfate from contaminating the zinc half-cell. Contact with the amalgam electrode is obtained by platinum wires sealed into the end of a glass tube. Mercury is placed in this tube, and the leads from the potentiometer dip into the mercury.

About 500 ml of 0.02 m ZnSO$_4$ is prepared. To 100 ml of this solution is added about 2 g of lead sulfate, and the mixture is shaken vigorously.

In all precise work with electrochemical cells, oxygen must be carefully excluded from the cell. If purified nitrogen or carbon dioxide is available, the solutions of ZnSO$_4$ and of ZnSO$_4$ saturated with PbSO$_4$ are swept out while the amalgams are being prepared.

The amalgams are prepared by grinding the granular metal with mercury under a little dilute (0.5 N) H$_2$SO$_4$ in a mortar. The amalgams should contain about 6 percent of Zn or Pb by weight. The sulfuric acid prevents an oxide scum from forming on the surface and hastens the amalgamation. Some grinding of the zinc with mercury should be done before adding the acid; otherwise the granules will tend to float on the acid. The amalgams are carefully rinsed with distilled water and with three or four portions of ZnSO$_4$ solution; they are then transferred to their respective arms of the cell. If the zinc amalgam has thickened to form a sludge, moderate warming will render it mobile.

The ZnSO$_4$ solution is placed over the zinc amalgam, and the ZnSO$_4$ solution saturated with PbSO$_4$ is placed over the lead amalgam, care being taken not to allow

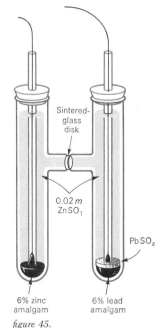

Sintered-glass disk

0.02 m ZnSO$_4$

PbSO$_4$

6% zinc amalgam

6% lead amalgam

figure 45.

Electrochemical cell for determination of thermodynamic functions.

excessive mixing of the two solutions. The platinum leads are then introduced, with the platinum completely immersed in the amalgam.

The potential is to be determined at several temperatures in the range 0 to 40°. Temperature intervals of 10 or 15° are convenient. For careful work the cell should be placed in a thermostat. The cell may take so long to reach equilibrium that a manually controlled bath cannot be recommended. A well-stirred ice bath in an insulated jacket is used for the 0° measurement.

Readings of the potential at each temperature are taken at intervals until a value constant within a few tenths of a millivolt is obtained; this value is taken to be the cell potential for reversible operation at the particular temperature. If erratic operation or continual drift of the cell voltage occurs, a new cell should be set up.

The first set of readings may be recorded starting at 0° and increasing the temperature. A check run is then made with descending temperature, starting at the highest point in the previous set.

CALCULATIONS.   The potential is plotted against the absolute temperature, and $(\partial E/\partial T)_P$ is obtained by drawing a tangent to the curve. The estimated reliability of the voltage measurements should be considered in drawing the curve.

Values of ΔG, ΔS, and ΔH are calculated for 25°C by use of Eqs. (3) to (5), for both joules and calories as the energy unit. The heat for the direct irreversible reaction is compared with that for the reversible case.

The value obtained for ΔH is compared with that obtained from the heats of formation.[2] The heats of formation of the saturated zinc and lead amalgams and the heat of dilution of 0.02 m ZnSO$_4$ may be neglected[1,3,4] in this calculation.

Practical applications. The relation between ΔG and ΔH discussed in this experiment was studied down to low temperatures by Richards and led to the first expression of what is now known as the third law of thermodynamics.

Suggestions for further work. Various other cells may be studied, including a copper-zinc cell, in which ΔH and ΔG are nearly equal, a copper-lead cell, where ΔH is less than ΔG, and a silver-zinc cell, where ΔH is greater than ΔG.

Cadmium and cadmium sulfate may be used in place of zinc and zinc sulfate in the apparatus of this experiment.[3]

By extending these measurements to several different concentrations, it is possible to obtain the $E°$ of the cell, the activity coefficients, heats of transfer, and the partial and integral heats of dilution of ZnSO$_4$ (Refs. 1, 5) or of CdSO$_4$ (Ref. 6).

In the case of the cadmium-lead system, thermodynamic data may be obtained for reactions involving the pure metals, using measurements of the electromotive force of the Cd-Cd amalgam couple by LaMer and Parks[7,8] and of the Pb-Pb amalgam couple by Gerke[3]. The latter reference contains a wealth of practical information on the techniques of precise potential measurements.

References

1. W. J. Clayton and W. C. Vosburgh, *J. Am. Chem. Soc.*, **58:** 2093 (1936).
2. I. A. Cowperthwaite and V. K. LaMer, *J. Am. Chem. Soc.*, **53:** 4333 (1931).
3. R. H. Gerke, *J. Am. Chem. Soc.*, **44:** 1684 (1922).
4. Selected Values of Chemical Thermodynamic Properties, *Natl. Bur. Std. U.S. Circ.* 500, 1952.
5. V. K. LaMer and I. A. Cowperthwaite, *J. Am. Chem. Soc.*, **55:** 1004 (1933).
6. V. K. LaMer and W. G. Parks, *J. Am. Chem. Soc.*, **53:** 2040 (1931).
7. V. K. La Mer and W. G. Parks, *J. Am. Chem. Soc.*, **55:** 4343 (1933).
8. V. K. LaMer and W. G. Parks, *J. Am. Chem. Soc.*, **56:** 90 (1934).

Chapter 10

Dielectric
and
Optical
Properties
of
Matter

The next four experiments are concerned with the measurement of dielectric properties of polar substances. Some common elements of the basic theory relating dielectric effects to molecular properties will be outlined here.

It was pointed out by Debye that molecules of the so-called *polar* substances, though electrically neutral, possess a nonvanishing electric dipole moment even in the absence of applied fields and that the magnitude of the molecular dipole moment could be found from dielectric-constant data. The measurement and study of molecular dipole moments continues to be an important step in characterizing the properties of molecules.

Let us begin by defining the term dipole moment. To do so, we consider a cluster of charges for which the net charge is zero. The dipole moment $\mathbf{p}$ of such a group of charges is a vector quantity,† the components of which are defined by

$$p_x = \sum_i q_i x_i \qquad p_y = \sum_i q_i y_i \qquad p_z = \sum_i q_i z_i$$

where x_i, y_i, z_i are the coordinates of charge q_i. It is easily shown that $\mathbf{p}$ is independent of the location of the origin of the coordinate system if the net charge $\Sigma_i q_i$ is zero.

It is usual to recognize two distinct processes which may occur when a substance is subjected to the influence of an electric field: (*a*) conduction and (*b*) polarization. Conduction is the transport of charged particles over distances large compared with molecular dimensions. The term polarization refers to the displacement over relatively short distances (of the order of a molecular diameter or less) of charges which are bound within some more or less permanent, though not rigid, aggregate of charged particles such as a neutral molecule. The charge displacements involved in the polarization process never carry a charge very far from the position it would have in the absence of the field.

The state of polarization of a substance is characterized by a vector quantity $\mathbf{P}$, called the polarization. The vector $\mathbf{P}$ can be defined as the average resultant dipole moment per unit volume due to bound charge, this average being taken over an element of volume which, though small by ordinary standards, nevertheless contains a relatively large number of molecules. The value of $\mathbf{P}$ in a given small region of the substance depends on the electric field $\mathbf{E}$ in the same region.

The concept of the field $\mathbf{E}$ within matter requires clarification. The field $\mathbf{E}$ at any point *in a vacuum* is defined by the statement that $\mathbf{E}\,dq$ is the force which would be exerted on a "test" charge dq of infinitesimal magnitude placed at the given point. The field defined in exactly this same way within matter fluctuates wildly over atomic-scale dimensions. The macroscopic field $\mathbf{E}$ at a point *within matter* is therefore defined as the average of this atomic-scale field over a small region in the neighborhood of the given point, this region being chosen small by ordinary standards yet large enough to contain many molecules.

† The notation used for vectors is illustrated in the Appendix.

For *isotropic* substances, **P** lies in the same direction as **E**, and for this case the susceptibility χ is defined by

$$P = \chi E \tag{1}$$

where P and E are the magnitudes of **P** and **E**, respectively. The dielectric constant ϵ is†

$$\epsilon = 1 + 4\pi\chi \tag{2}$$

Subsequent discussion will be limited to the isotropic case.

Two equations for P are fundamental for the interpretation of dielectric constants:

$$P = N\langle p_E \rangle \tag{4}$$

$$= \frac{\epsilon - 1}{4\pi} E \tag{5}$$

where P = polarization (in direction of the field E)

N = number of molecules per unit volume

p_E = component in direction of E of total dipole moment of a molecule

The angle brackets mean that the quantity enclosed is to be averaged over the thermal motions of the molecules and also over the different species of molecules present in the case of a mixture. Equation (4), it should be noted, relates P to molecular properties, while Eq. (5), which follows from Eqs. (1) and (2), relates P to the directly measurable macroscopic quantity ϵ.

When the field E is applied suddenly, a certain amount of time is required before the equilibrium degree of polarization becomes established. The static dielectric constant ϵ_s is the value of ϵ measured in either a constant field E or in a field changing slowly enough that the substance is essentially at equilibrium at every instant.§ We shall consider this case first.

Three contributions to $\langle p_E \rangle$ are to be considered. The first arises if the molecule has a dipole moment even in the absence of an applied field. This moment is

---

† Maxwell's displacement field vector **D** at any point is related to **E** and **P** by

$$\mathbf{D} = \mathbf{E} + 4\pi\mathbf{P} \tag{3a}$$

For isotropic substances, **D**, **E**, and **P** all have the same direction at any point and Eq. (3a) may be replaced by an equation relating the magnitudes,

$$D = E + 4\pi P \tag{3b}$$

The dielectric constant ϵ is usually defined for the isotropic case by $D = \epsilon E$, which then leads directly to Eq. (2). These equations are in the cgs-esu system of units.

§ For most liquids or gases at ordinary temperatures, this criterion still permits ϵ_s to be measured at rather high frequencies. In the present experiment, for example, measurement of ϵ at roughly 1 MHz yields ϵ_s to all intents and purposes.

called the *permanent* dipole moment, and its magnitude is represented by the symbol μ. Let this dipole moment lie in a direction which makes an angle ϑ with the field direction. Then the contribution of the permanent dipole term to $\langle p_E \rangle$ is

$$\mu \langle \cos \vartheta \rangle \tag{6}$$

Because of thermal agitation, a single molecule will in a period of time assume all possible orientations with reference to a fixed direction. In the absence of an applied field, these orientations have equal probabilities and $\langle \cos \vartheta \rangle$ vanishes. In the presence of a field, however, orientations in which the dipole is close to alignment with the field will be favored over those in which the dipole is opposed to the field. The average value of $\cos \vartheta$ in a field E_i is

$$\langle \cos \vartheta \rangle = \frac{\mu E_i}{3kT} \tag{7}$$

where k = Boltzmann constant
$\quad\quad T$ = absolute temperature

Here the average field acting on the given molecule has been designated E_i. It is related to, but is not equal to, the field E defined above. The field E_i, called the *internal* or *local* field, may be visualized as the average field at the location of a molecule, produced by all charges except those of the given molecule.

For typical values of the quantities in Eq. (7), this average is of the order of 1×10^{-5}. It is clear, therefore, that the alignment is far from being complete; rather there is only a very slight degree of preference shown for orientation with the field. Nevertheless, for polar molecules, this mechanism gives the dominant contribution to $\langle p_E \rangle$, of magnitude

$$\frac{\mu^2}{3kT} E_i \tag{8}$$

In addition to this term, associated with orientation of the permanent molecular dipole moments, additional polarization results from distortion of the molecule produced by the externally applied field. Specifically, this distortion includes (*a*) deformation of the same sort as that involved in vibrational motions, namely, changes in bond lengths and angles, and (*b*) displacement of the average positions of the electrons relative to the nuclear framework. The field-dependent part of the molecular dipole moment produced in this way is called the *induced* moment. When averaged over all orientations of the molecule, it lies in the same direction as, and is closely proportional to, E_i. These terms in $\langle p_E \rangle$ are therefore written as

$$(\alpha_v + \alpha_e) E_i \tag{9}$$

where α_v = mean molecular vibrational or atomic polarizability
$\quad\quad \alpha_e$ = mean molecular electronic polarizability

The complete expression for $\langle p_E \rangle$ is therefore

$$\langle p_E \rangle = \left(\frac{\mu^2}{3kT} + \alpha_v + \alpha_e \right) E_i \qquad (10)$$

Before Eq. (10) can be used, it is necessary to have a relation between E_i and E. Debye used an expression, derived earlier by Lorentz,

$$E_i = \frac{\epsilon + 2}{3} E \qquad (11)$$

which is satisfactory only if the concentration of polar molecules is very low. Therefore the following equations are accurate for gases and are reasonably satisfactory for very dilute solutions of polar molecules in nonpolar solvents but are not useful for polar liquids. Substitution of (11) in (10) and of the resulting expression for $\langle p_E \rangle$ into (4) yields

$$P = N \left(\frac{\mu^2}{3kT} + \alpha_v + \alpha_e \right) \frac{\epsilon_s + 2}{3} E \qquad (12)$$

Then equating the right sides of (12) and of (5), eliminating N through the substitution

$$N\overline{V} = N_0 = \text{Avogadro's number}$$

where $\overline{V} = M/\rho = $ molar volume

$\rho = $ density

$M = $ molecular weight

and collecting the measurable quantities on the left side leads to

$$\frac{\epsilon_s - 1}{\epsilon_s + 2} \overline{V} = \frac{4\pi N_0}{3} \left(\frac{\mu^2}{3kT} + \alpha_v + \alpha_e \right) \qquad (13)$$

A quantity called the molar polarization $\mathscr{P}$ is defined as

$$\mathscr{P} = \frac{\epsilon - 1}{\epsilon + 2} \frac{M}{\rho} = \frac{\epsilon - 1}{\epsilon + 2} \overline{V} \qquad (14)$$

The left side of Eq. (13) is the static value of $\mathscr{P}$.

Equation (13) has been found to be quite accurate for most gases. Thus measurements of dielectric constant and density of a polar gas as a function of temperature may be combined to permit a determination of both the permanent dipole moment μ and the mean total polarizability $\alpha_v + \alpha_e$. From Eqs. (13) and (14), a graph of $\mathscr{P}_s$ versus $1/T$ will give a straight line with slope $4\pi N_0 \mu^2 / 9k$ and intercept $(4\pi N_0/3)(\alpha_v + \alpha_e)$.

33 DIPOLE MOMENT FROM GAS DIELECTRIC-CONSTANT MEASUREMENTS

In this experiment, dielectric-constant measurements are made on gases to permit the determination of the dipole moment of a molecule by a method free from the effects of solute-solvent interactions, which are present in a determination by the solution method described in Experiment 34.

THEORY.[1]   The molar polarization for a gas may be calculated (page 206) by use of the Clausius-Mosotti equation:

$$\mathscr{P}(T,P) = \frac{\epsilon(T,P) - 1}{\epsilon(T,P) + 2} \overline{V}(T,P) \tag{1}$$

where $\epsilon(T,P)$ = dielectric constant of gas at T, P
$\overline{V}(T,P)$ = molar volume of gas at T, P

For a real gas at moderate to low pressures, the molar volume can be approximated by use of the ideal-gas law, and since the value of the dielectric constant for a gas at low pressures differs but slightly from unity, the molar polarization can be represented as

$$\mathscr{P}(T,P) \approx \frac{\epsilon(T,P) - 1}{3} \frac{RT}{P} \tag{2}$$

Measurement of the dielectric constant at a series of pressures can then permit an extrapolation to zero pressure to obtain a result $\mathscr{P}°(T)$ which reflects the properties of the isolated molecules, uninfluenced by any effects of intermolecular forces. Since according to theory (page 206)

$$\mathscr{P}°(T) = \frac{4\pi}{3} N_0 \frac{\mu^2}{3kT} + \frac{4\pi}{3} N_0 \alpha \qquad \alpha = \alpha_v + \alpha_e \tag{3}$$

a plot of $\mathscr{P}°(T)$ versus $1/T$ will permit calculation of the polarizability α from the intercept and of the dipole moment μ from the slope.

If measurements are made at only one pressure P, the value of $\mathscr{P}(T,P)$ used as an approximation to $\mathscr{P}°(T)$ should be calculated by use of Eq. (1) and the appropriate equation-of-state data for the gas concerned.

····▸ ***Apparatus.*** Heterodyne-beat circuit with precision capacitor; gas dielectric-constant cell; tank of compressed gas (sulfur dioxide, hydrogen chloride, etc.) with reducing valve; cell-filling manifold with manometer and vacuum pump; Dry Ice, ice, steam generator or other temperature-control.

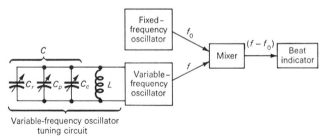

figure 46. Principle of the heterodyne-beat method. C_c, dielectric-constant cell; C_p, precision capacitor; C_r, rough-tuning capacitor.

PROCEDURE.   The principle on which the heterodyne-beat method is based is schematically represented in Fig. 46. A radio-frequency signal of constant frequency f_0, generated by a fixed-frequency oscillator, and a second signal of frequency f, generated by a variable-frequency oscillator, are fed into a mixer (page 596), whose function is to produce in its output voltage a component of frequency $|f - f_0|$. This difference frequency, or beat frequency, will be in the audio range when f and f_0 are nearly equal and can then be detected by earphones or other suitable means.

The frequency of the variable oscillator will be given very closely by the relation

$$f = \frac{1}{2\pi\sqrt{LC}}$$

where L and C represent the inductance and total capacitance,

$$C = C_c + C_p + C_r$$

of the variable-frequency-oscillator tank circuit (Fig. 46). Since L is fixed, the frequency is determined by C. For an arbitrary setting of C_p, the beat frequency $|f - f_0|$ will usually be above the audio range. As C_p is then varied, a beat note will first be heard as a high-pitched tone when the difference $|f - f_0|$ enters the audio range. As C_p is further changed, the beat frequency continuously decreases through the audio range until it becomes too low to be heard, passes through zero, and then again increases on the other side. For precise location of the setting of C_p corresponding to zero beat, it is necessary also to have a meter, tuning eye, or other device capable of indicating the beats when the beat frequency is too low to be audible.

Provided f_0 and L are constant and other factors† having incidental effects on

---

† The factors most likely to lead to frequency instability are variation in the plate or filament supply voltages and changes in the temperature of critical circuit components. In addition, poor mechanical construction or inadequate shielding may lead to erratic behavior of the oscillator.

f are sufficiently stable, the condition for zero beat is equivalent to

$$C_c + C_p + C_r = \text{constant}$$

Thus changes in C_c are accurately measured by finding the change in C_p required to restore a condition of zero beat. The capacitor C_r is adjusted as may be necessary to bring the zero-beat settings within the calibrated range of C_p, but C_r must obviously not be changed when an increment in C_c is being measured.

The potential accuracy of the heterodyne-beat method is very high. Differentiation of the frequency-determining relation gives, for constant L,

$$\frac{\Delta f}{f} = -\frac{1}{2}\frac{\Delta C}{C}$$

If f_0 is 1 MHz and a beat frequency of 1 Hz can be distinguished from zero beat, the detectable change of capacitance in the circuit is 1 part in 500,000.

For work with gases it is most practical to use a fixed-plate cell for which the capacitance is to the necessary degree of approximation directly proportional to the dielectric constant ϵ of the filling gas. The required minimization of the amount of other dielectric employed can conveniently be achieved in the concentric-cylinder capacitor designs shown in Fig. 47, for which the capacitance, ignoring end effects and contribution of any plate separators, is given in picofarads, for length l (cm),

$$C = 0.2416l\left(\frac{1}{\log r_2/r_1} + \frac{1}{\log r_4/r_3}\right)\epsilon \tag{4}$$

For metal plates, the separators may be of quartz, mica, or synthetic sapphire spheres,† with assembly through force fit obtained either directly or through temperature equalization of sections appropriately cooled and heated before assembly; for most work stainless steel can be recommended as a construction material. An adequately rigid unit is required, but excessive mass is to be avoided because it will unnecessarily slow the rate of attainment of thermal equilibrium. A suitable capacitor can also be assembled as shown from sections of ordinary Pyrex tubing onto which conductive metal coatings have been fired by use of one of the proprietary media now available.§ While the required cutting and slotting of the cylinders can most easily be done on a standard glass cutoff machine, satisfactory results can be obtained by use of the diamond saws common in amateur lapidary workshops. The metal coatings must be applied before assembly of the capacitor. If the fired coating is too thin, it may be used as a base for electroplated reinforcement.

It is preferable to mount the cell as shown in a metal container which can be

---

† Available from the M. A. Miller Mfg. Co., Libertyville, Ill. 60048.

§ Bright Platinum 05, Hanovia Liquid Gold Division, Engelhard Industries, Inc., East Newark, N.J., is recommended.

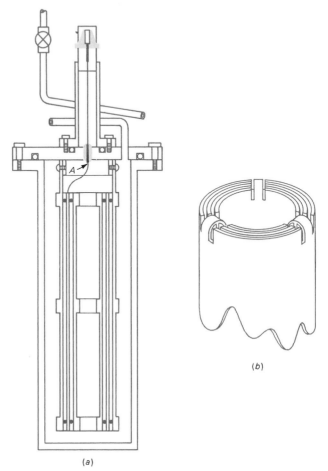

figure 47. (a) Metal coaxial cylinder cell. (b) End assembly, glass coaxial cylinder cell.

grounded; if a glass container is used, appropriate external shielding must be provided to establish definitively the capacitance to ground of all cell elements. Provided solid dielectric between the plates makes a negligible contribution to the cell capacitance, for a given temperature and pressure the dielectric constant of a filling gas can be calculated as

$$\epsilon(T,P) = 1 + \frac{\Delta C}{C_0} \qquad (5)$$

where C_0 = capacitance of cell with vacuum dielectric

$\quad\Delta C$ = increase of capacitance of cell on admission of gas

The value of ΔC may be determined most easily by use of a precision compensating capacitor in parallel with the cell. Because $\epsilon - 1$ is small here, a special capacitor such as the General Radio type 1422ME is then essential. In an alternative approach the necessary sensitivity in adjustment may be obtained by appropriate use of a series combination of a fixed capacitor and a variable capacitor, as suggested, for example, by Hartshorn,[2] provided that in the calculations due account is taken of stray capacitance to ground (designated here by Greek subscripts) in addition to the discrete circuit components. In Fig. 48 are shown several possible circuit arrangements, together with the relations appropriate for the calculation of ΔC from the experimentally measured quantities; the proper circuit configuration to use will be determined by the particular components available.

The value of C_0 is established by measurement of the change in capacitance resulting from opening the lead connection at point A in Fig. 47. This measurement may be made by the heterodyne-beat method using a balancing capacitor of adequate range or by means of an impedance bridge. For the concentric-cylinder configuration recommended, the value of C_0 should change only slightly with temperature; the validity of this assumption must be checked, however, if a wide range of temperature is involved. Procedures for evaluation of the additional capacitances involved in the use of the compensating capacitor arrangements of Fig. 48b and c are discussed by Zahn[3] and by Greene and Williams.[4]

The heterodyne-beat circuit of Chien,[5] which has a record of successful use in dipole-moment measurements by the solution method, is also suitable for the present application. A compact and convenient alternative based on the schematic of Fig. 49 can provide the advantages of modular construction, etched circuit-board assembly

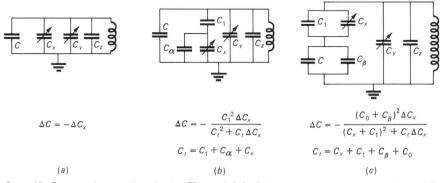

$$\Delta C = -\Delta C_x$$

$$\Delta C = -\frac{C_1^2\,\Delta C_x}{C_t^2 + C_t\,\Delta C_x}$$

$$C_t = C_1 + C_\alpha + C_x$$

$$\Delta C = -\frac{(C_0 + C_\beta)^2\,\Delta C_x}{(C_x + C_1)^2 + C_t\,\Delta C_x}$$

$$C_t = C_x + C_1 + C_\beta + C_0$$

$\quad\quad(a)\quad\quad\quad\quad\quad\quad\quad\quad(b)\quad\quad\quad\quad\quad\quad\quad\quad(c)$

figure 48. Compensating capacitor circuits. The symbol C_y designates a coarse tuning capacitor, and C_z represents lead capacitance; these elements do not enter into the calculation of ΔC.

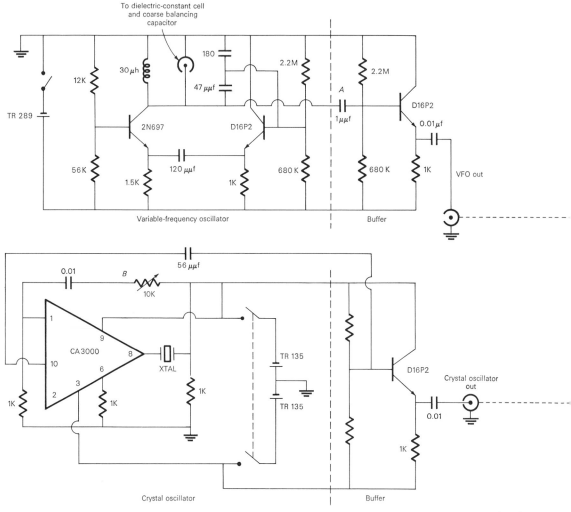

figure 49. Heterodyne-beat circuitry for dielectric-constant measurements. A, twisted wire gimmick or discrete capacitor; B, adjust trimmer for symmetry of oscillator waveform; C, switch closure removes 1-kHz reference

to permit rigid mounting of components and fixed lead positions, low power dissipation, a phase detector (page 634) for frequency comparison instead of the conventional mixer stage, and an audio-frequency reference oscillator for improved precision in making null adjustments.

With this, as with any heterodyne-beat circuit, obtaining adequate frequency

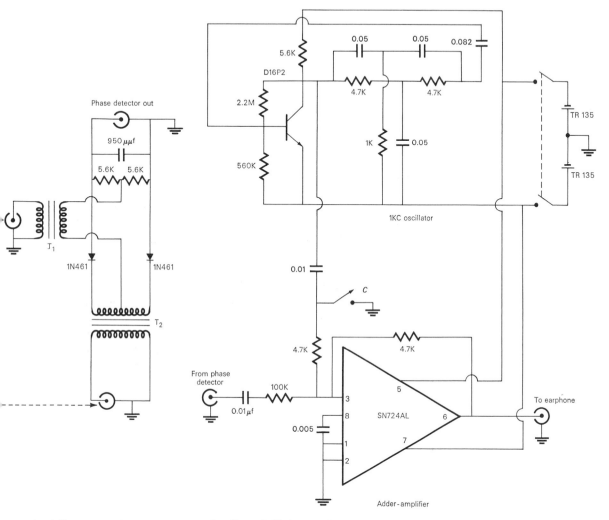

signal. T_1, 30 turns primary, 45 turns secondary No. 30 Beldsol wire on Indiana General CF102-Q-1 core.
T_2, 30 turns primary, 60 turns secondary No. 30 Beldsol wire on Indiana General CF102-Q-1 core.

stability for the variable-frequency oscillator is the major problem encountered. Variation of this frequency due to causes other than change of medium in the dielectric-constant cell will introduce an error, the magnitude of which can be estimated from the following relations corresponding to the several circuit arrangements of Fig. 48:

$$\frac{\delta C}{C} = \begin{cases} 2\left(1 + \dfrac{C_x + C_y + C_z}{C}\right)\dfrac{\delta f}{f} & \text{Fig. 48}a \\[2em] 2\left(1 + \dfrac{C_y + C_z}{C} + \dfrac{C_1}{C}\dfrac{C_x + C_\alpha}{C_x + C_\alpha + C_1}\right)\dfrac{\delta f}{f} & \text{Fig. 48}b \\[2em] 2\left(1 + \dfrac{C_y + C_z}{C}\right)\left(1 + \dfrac{C_\beta}{C}\right)\left(1 + \dfrac{C + C_\beta}{C_1 + C_x}\right)\dfrac{\delta f}{f} & \text{Fig. 48}c \end{cases} \tag{6}$$

For gas pressures near 1 atm, the value of $\epsilon - 1$ will commonly be less than 0.01, so that for a precision of 1 percent in the evaluation of the molar polarization $\delta C/C$ must be reduced at least to near 1 part in 10^5, including the effects of any uncertainty in null setting and any frequency-drift contribution. The primary causes of instability of the variable-frequency oscillator include variations in supply voltage and temperature. Use of the very stable mercury cells specified should eliminate the first cause, while proper thermal insulation and allowance of adequate warm-up time can take care of the second; rigid leads in the tank circuit and adequate shielding are also essential.

If the null setting is made at zero beat, there can be difficulty due to the tendency of the less stable oscillator (the variable-frequency oscillator) to become locked in synchronism with the more stable, with a resultant dead band over a range of setting of the compensating capacitor. The undesired coupling responsible for this effect can be adequately minimized by careful shielding and isolation of the variable-frequency oscillator by use of a small capacitor to couple it to the buffer stage employed. The problem can be circumvented completely by using as a reference a stable fixed audio frequency (commonly about 1000 Hz) with which the beat frequency is compared. If these two signals are combined, as shown in Fig. 49 (the principle of the adder stage is explained in any reference on operational amplifiers), a near match in their frequencies will result in a periodic variation in sound intensity which can permit a null setting to be made to within a fraction of a hertz. The lock-in problem is eliminated because the two oscillator frequencies are then separated by the audio reference frequency. Obviously, in the use of this method the null setting must be made consistently on the same side of the zero beat setting. If an oscilloscope is available, the null setting may be determined visually rather than aurally. In this case the reference-oscillator output is applied to the external input terminals of the horizontal-deflection section of the oscilloscope, and the beat-frequency signal to the vertical input terminals, with the corresponding beam-deflection amplitudes set equal. A stationary Lissajous pattern will be generated when the ratio of the two frequencies is a rational number; when the frequencies are matched, the pattern will be an ellipse, which for an appropriate phase angle δ between the voltages reduces to a straight line ($\delta = 0$) or a circle ($\delta = 90°$).

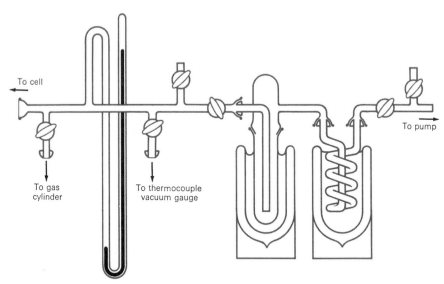

To cell

To gas
cylinder

To thermocouple
vacuum gauge

To pump

figure 50. Gas manifold for filling dielectric-constant cell.

The arrangement of the gas-handling apparatus is indicated in Fig. 50. Ample clearance should be provided beneath the dielectric-constant cell to permit easy interchange of thermostatic media. The gas manifold may be made of glass, in which case O-ring joints are suggested, or it may be assembled by use of Swagelock or Gyrolock fittings or their equivalent; a suitable protective trap should be placed ahead of the vacuum pump. The lead from the cell to the precision capacitor and from the latter to the variable-frequency oscillator circuitry must be rigid and adequately shielded. Double-shielded coaxial cable (RG 71) will normally be adequate for this purpose; the equivalent may be made up from RG 58 coaxial cable set inside a section of $\frac{1}{4}$-in. copper tubing, the inner diameter of which has been increased slightly by etching with nitric acid.[6]

To initiate a series of measurements, a room-temperature water bath is placed around the dielectric-constant cell, which is then evacuated. A set of three or four determinations of the corresponding null setting of the precision capacitor is then made and the results recorded. The cell is next filled to a pressure of nominally 1 atm with the gas to be studied; the approach to thermal equilibrium will be reflected in the progressive shift in the new null-point setting of the precision capacitor. When thermal equilibrium has been attained, the capacitor setting is again determined as above, and these readings are recorded, together with the temperature and pressure. The cell is then reevacuated to permit a check on the capacitance-increment measurement. These steps are repeated as necessary to obtain consistent results.

Further determinations are made for other temperatures in the range suitable for the gas under study. Adequate care must be taken at each such level to ensure that thermal equilibrium is reached. The rate-determining step here is heat transfer to and from the capacitor plates; this process can be expedited by the presence of gas at a pressure of several centimeters of mercury.

CALCULATIONS.   For a given temperature and pressure, the dielectric constant of the gas is calculated by use of Eq. (5) and the molar polarization in accordance with Eq. (1). Critical-constant data[7] for the gas may be used in the calculation of the molar volume by means of the Berthelot, van der Waals, or Redlich-Kwong[8] equation of state.

A plot is made of $\mathscr{P}(T,P)$ versus $1/T$, and the corresponding values of the dipole moment and polarizability are calculated from the slope and intercept, respectively, of the straight line considered to give the best representation of the data. The intercept is compared with the molar refraction as calculated from refractive-index data for the gas, and the dipole moment with the accepted value. The molar refraction is defined in Eq. (12), Experiment 34, page 219.

Practical applications. Gas-phase dielectric-constant measurements permit accurate characterization of isolated polar molecules and provide one approach to the study of intramolecular rotation.

Suggestions for further work. Measurements at several pressures may be made to permit evaluation of $\mathscr{P}^{\circ}(T)$ by an extrapolation based on Eq. (2). The frequency stability of the variable-frequency oscillator may be checked by use of a frequency counter. Measurements may be extended to a gas for which the molecular dipole moment is relatively small, or the absence of a permanent dipole moment for a molecule like carbon dioxide may be demonstrated. Comparison may be made of different compensating capacitor arrangements.

References

1. M. Davies, "Some Electrical and Optical Aspects of Molecular Behavior," Pergamon Press, New York, 1965.
2. L. Hartshorn, *Proc. Phys. Soc.*, **36:** 399 (1924).
3. C. T. Zahn, *Phys. Rev.*, **24:** 400 (1924).
4. E. W. Greene and J. W. Williams, *Phys. Rev.*, **42:** 119 (1932).
5. J.-Y. Chien, *J. Chem. Educ.*, **24:** 494 (1947).
6. P. Bender, D. L. Flowers, and H. L. Goering, *J. Amer. Chem. Soc.*, **77:** 3463 (1955).
7. K. A. Kobe and R. E. Lynn, Jr., *Chem. Rev.*, **52:** 117 (1953).
8. O. Redlich and J. N. S. Kwong, *Chem. Rev.*, **44:** 233 (1949).

34   DIPOLE MOMENT FROM DIELECTRIC-CONSTANT MEASUREMENTS ON SOLUTIONS

The heterodyne-beat method is used for measurement of the dielectric constant of a series of dilute solutions in a nonpolar solvent for the purpose of determining the solute dipole moment by the Debye solution method.

THEORY.[1-4]   The Debye theory relating the static dielectric constant to molecular properties is outlined in an earlier section (pages 203 to 206). For determining the molecular dipole moment, it is best, when feasible, to make dielectric-constant measurements for the gas over a range of temperatures and use the method of Experiment 33. However, it is not always practicable to make such measurements, especially if the substance of interest lacks sufficient volatility in a conveniently accessible temperature range or if it is chemically unstable as a gas. In such cases, a reasonably satisfactory method is the solution method, also originally developed by Debye.[1] If reasoning similar to that leading to Eq. (13) (page 206) is repeated, there is obtained for the solution case the equation

$$\mathcal{P} = X_1 \mathcal{P}_1 + X_2 \mathcal{P}_2 \tag{1}$$

with

$$\mathcal{P} = \frac{\epsilon_s - 1}{\epsilon_s + 2} \, \overline{V} = \frac{\epsilon_s - 1}{\epsilon_s + 2} \left(\frac{X_1 M_1 + X_2 M_2}{\rho} \right) \tag{2}$$

$$\mathcal{P}_1 = \frac{4\pi N_0}{3} \left(\alpha_{v_1} + \alpha_{e_1} \right) \tag{3}$$

$$\mathcal{P}_2 = \frac{4\pi N_0}{3} \left(\frac{\mu_2^2}{3kT} + \alpha_{v_2} + \alpha_{e_2} \right) \tag{4}$$

where X_1, X_2 = mole fractions
$\quad\quad M_1, M_2$ = molecular weights
Subscript 1 refers to the solvent, and 2 to the solute, while ϵ_s, $\overline{V}$, and ρ are the static dielectric constant, average molar volume, and density of the solution, respectively. The quantity $\mathcal{P}$ is expressed in Eq. (2) in terms of directly measurable quantities, and for dilute solutions $\mathcal{P}_1$ is approximated by $\mathcal{P}_1^\circ$, the molar polarization of the pure solvent,

$$\mathcal{P}_1^\circ = \frac{\epsilon_1 - 1}{\epsilon_1 + 2} \frac{M_1}{\rho_1} \tag{5}$$

so that $\mathcal{P}_2$ may be calculated from the expression

$$\mathcal{P}_2 = \frac{\mathcal{P} - X_1 \mathcal{P}_1^\circ}{X_2} \tag{6}$$

However, as a result mainly of interactions among polar solute molecules, $\mathcal{P}_2$ calculated from Eq. (6) is found to vary with concentration. Since the magnitude of this interaction depends on the distance between the polar solute molecules, extrapolation of $\mathcal{P}_2$ to zero solute mole fraction yields a value $\mathcal{P}_2^\circ$ which is free from the effect of such interactions. (This procedure is comparable to the use of the method of limiting densities for the accurate determination of the molecular weight of a real gas.) Formally, $\mathscr{P}_2^0$ may be written as

$$\mathscr{P}_2^0 = \lim_{X_2 \to 0} \frac{\mathscr{P} - X_1 \mathscr{P}_1^0}{X_2} \tag{7}$$

Then Eq. (4) is replaced by

$$\mathscr{P}_2^0 = \frac{4\pi N_0}{3}\left(\frac{\mu^2}{3kT} + \alpha_v + \alpha_e\right) \tag{8}$$

$$= \mathscr{P}_{2\mu}^0 + \mathscr{P}_{2v}^0 + \mathscr{P}_{2e}^0$$

where μ, α_v, and α_e pertain to the solute molecule at infinite dilution in the given solvent.

Before μ can be calculated from $\mathscr{P}_2^0$, it is necessary to separate the dipole term in Eq. (8) from the others. In the solution case, the temperature method of determining μ is found to be unreliable, presumably because of temperature-dependent interactions between solute and solvent. Another method is therefore used, which depends on the behavior of ϵ as a function of frequency.

Figure 51 shows qualitatively how ϵ varies with frequency for a typical polar substance. The frequency dependence of $\mathscr{P}$ is similar. Equation (13) on page 206,

$$\frac{\epsilon_s - 1}{\epsilon_s + 2}V = \frac{4\pi N_0}{3}\left(\frac{\mu^2}{3kT} + \alpha_v + \alpha_e\right) \tag{9}$$

holds in region A. Beyond the upper end of this region, the field is alternating so rapidly that there is insufficient time in the period of a half-cycle for molecular-orientation polarization to develop. In the relatively flat region B, where the value

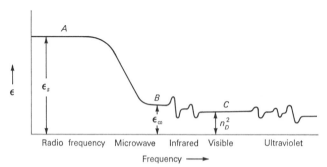

figure 51. Frequency dependence of ϵ for a typical polar molecule. The frequency scale is only qualitative.

of ϵ is usually denoted by ϵ_∞, there is no appreciable contribution to $\langle p_E \rangle$ from dipolar orientation and Eq. (9) is replaced by

$$\frac{\epsilon_\infty - 1}{\epsilon_\infty + 2} \overline{V} = \frac{4\pi N_0}{3} (\alpha_v + \alpha_e) = \mathscr{P}_v + \mathscr{P}_e \tag{10}$$

In region C, only the electronic contribution remains.

In the infrared and visible regions, ϵ can be determined by optical methods through application of the equation[5]

$$\epsilon = n^2 \tag{11}$$

where n is the index of refraction and ϵ and n are to be measured at one and the same frequency. To find $\mathscr{P}_v + \mathscr{P}_e$, it would therefore be appropriate to measure ϵ_∞ or n at far-infrared frequencies, but this is at present a rather difficult type of measurement. On the other hand, n is easily measured at the frequency of the sodium D line to obtain

$$\left(\frac{n_D^2 - 1}{n_D^2 + 2} \right) \overline{V} = \frac{4\pi N_0}{3} \alpha_e = \mathscr{P}_e \tag{12}$$

where n_D is the index of refraction measured with light of the sodium D line.

The quantity on the left-hand side of Eq. (12) is called the *molar refractivity;* though it is best to measure it, this quantity can be estimated from tables of atomic refractivities. For the solution quantity $\mathscr{P}_{2e}$, it is satisfactory to use the value of $\mathscr{P}_e$ calculated from n_D for the pure liquid, so long as $\mathscr{P}_e$ is reasonably small compared with the dipole term, though in principle one should determine refractive indices for the solutions and use an equation similar to Eq. (7).

As $\mathscr{P}_v$ is at present so difficult to measure and is often quite small, the common practice is to neglect it. For small rigid molecules this procedure probably does not introduce much error. However, for large or rather flexible molecules, or those having relatively small μ, the neglect of $\mathscr{P}_v$ is not justifiable.

In summary, then, the Debye equation for the solution case, as usually applied, is

$$\frac{4\pi N_0}{3} \frac{\mu^2}{3kT} = \mathscr{P}_2^\circ - \mathscr{P}_{2v}^\circ - \mathscr{P}_{2e}^\circ = \mathscr{P}_{2\mu}^\circ \tag{13}$$

or

$$\mu = 0.01273 \times 10^{-18} \sqrt{\mathscr{P}_{2\mu}^\circ T}$$

where $\mathscr{P}_2^\circ$ is evaluated from Eq. (7) and $\mathscr{P}_{2e}^\circ$ from (12) and $\mathscr{P}_{2v}^\circ$ is neglected. With the quantities of Eq. (13) expressed in cgs units, the value of μ is obtained in electrostatic units. The Debye unit often used for μ is equivalent to 10^{-18} esu.

····→ ***Apparatus.*** The heterodyne-beat apparatus described in the preceding experiment (page 208) is excellent for the present purpose as well. The circuit of Chien[6] is also very satisfactory. Several other suitable circuits are available.[3] Additional requirements are a special dielectric cell for solutions, with thermostat, an analytical balance, a refractometer, a sample of the substance (such as nitrobenzene or acetonitrile) whose dipole moment is to be determined, and about 500 ml of a suitable nonpolar solvent (such as benzene).

PROCEDURE.   Dielectric-constant measurements are made on benzene and on dilute solutions of acetonitrile in benzene. Concentrations suggested for the work are 1, 2, 3, and 4 mole percent of solute.

To minimize the error due to stray capacitance, it is best to use a cell consisting of a variable capacitor, designed to give a convenient difference in capacitance between maximum and minimum settings. The cell described in Experiment 35 (page 229) is satisfactory for this purpose. Two sturdy pins in the top plate serve as positive stops to guarantee that the maximum and minimum settings will be completely reproducible. To eliminate disturbances from stray fields and from variations in stray capacitance, the cell is enclosed within a heavy metal shield, and coaxial cable is used for the connection to the oscillator. The entire assembly should be well constructed mechanically to avoid any undesired relative motion of parts of the cell.

The dielectric constant is calculated as the ratio of two capacitance increments,

$$\epsilon = \frac{C_{b,\text{liq}} - C_{a,\text{liq}}}{C_{b,\text{air}} - C_{a,\text{air}}} \tag{14}$$

where $C_{b,\text{liq}} - C_{a,\text{liq}}$ is the difference in cell capacitance for the two settings with the sample present and $C_{b,\text{air}} - C_{a,\text{air}}$ is the corresponding difference with air as dielectric. The dielectric constant of air is here being considered to be unity, as it virtually is in comparison with values for liquids.

For the accurate measurement of capacitance required for determination of dipole moments, the heterodyne-beat method is probably the best available. The samples encountered in using the Debye method will usually meet the requirement of having a very low conductance. (At 1 MHz, the precision is impaired if the specific conductance exceeds about 10^{-8} ohm^{-1} cm^{-1}.) The principle of the heterodyne-beat method and details of procedure are given in the preceding experiment.

Each solution is made up by weighing the solute in a 100-ml volumetric flask, adding solvent to the calibration mark, and then weighing the solution. The weight of the flask filled with pure solvent should also be obtained. These weighings should be made with an analytical balance. Both mole fractions and densities are to be found from these data. It is best to perform the weighings for all samples in the same volumetric flask, since in this way the normal calibration error in the flask will cause a practically constant error in the density which will not appreciably affect the final results with the method of calculation outlined below.

The sample temperature should be controlled to within a few tenths of a degree, as it is important that all measurements of ϵ_s and of ρ be made at the same temperature.

Finally, the refractive index of the pure solute liquid is measured, and its density is either measured (with a pycnometer or density balance) or obtained from the literature.

CALCULATIONS.   For the computation of the molecular dipole moment, it is necessary first to evaluate the molar polarization $\mathscr{P}_2^\circ$ of the solute at infinite dilution. One could in principle calculate $\mathscr{P}_2$ by Eqs. (2), (5), and (6), plot these values against X_2, and extrapolate to $X_2 = 0$. This procedure is not satisfactory, however, because the curvature of the graphs, together with the increasing effect of experimental error on the points as X_2 becomes small, makes the extrapolation unreliable.

The evaluation of $\mathscr{P}_2^\circ$ may be accomplished analytically with good accuracy by a procedure due to Hedestrand.[7] Assume that, for the dilute solutions involved, ϵ_s and ρ can be accurately expressed as linear functions of X_2

$$\epsilon_s = \epsilon_1 + aX_2 \tag{15}$$
$$\rho = \rho_1 + bX_2 \tag{16}$$

where ϵ_1 and ρ_1 are the static dielectric constant and the density of the actual solvent sample used, at the temperature at which the measurements have been made. The values of the coefficients a and b may be obtained from plots of ϵ_s and ρ against X_2.

Expressions (15) and (16) may be substituted into the equation

$$\mathscr{P}_2 = \frac{1}{X_2}\left(\frac{\epsilon_s - 1}{\epsilon_s + 2}\frac{X_1 M_1 + X_2 M_2}{\rho} - X_1 \frac{\epsilon_1 - 1}{\epsilon_1 + 2}\frac{M_1}{\rho_1}\right) \tag{17}$$

which is an explicit form of Eq. (6). In this way there is obtained an equation for $\mathscr{P}_2$ as a function of X_2 which is valid in the dilute-solution range. The limiting value of $\mathscr{P}_2$, that is, $\mathscr{P}_2^\circ$, may then be evaluated by letting X_2 approach zero; the result is

$$\mathscr{P}_2^\circ = A(M_2 - Bb) + Ca \tag{18}$$

where

$$A = \frac{\epsilon_1 - 1}{\epsilon_1 + 2}\frac{1}{\rho_1}$$

$$B = \frac{M_1}{\rho_1}$$

$$C = \frac{3M_1}{(\epsilon_1 + 2)^2 \rho_1}$$

A discussion of this and of several other analytical extrapolation procedures is given by Böttcher.[2]

The orientation polarization $\mathscr{P}^0_{2\mu}$ is then calculated, and from it the value of the dipole moment of the molecule is obtained from Eq. (13).

The experimental result for μ should be compared with tabulated values[8,9] for both the gas and for the solute molecule in the particular solvent used. There usually is an appreciable difference between the gas and solution values, resulting at least in part from a real change in the molecular charge distribution due to the effect of the surrounding solvent molecules.[2,4]

Suggestions for further work. The dipole moment of a less polar molecule such as chloroform may be determined. Careful attention to detail is required for accurate results in such work. Different extrapolation procedures may be tried with one set of data.[2,4,10-12].

Solvent effects on the dipole moment can be studied by measuring the dipole moment for the same molecular species in several solvents. The theory of Onsager accounts quite well for the differences between gas and solution values for a number of cases where it has been tested.[1,4]

References

1. P. Debye, "Polar Molecules," Dover Publications, Inc., New York, 1929.
2. C. J. F. Böttcher, "Theory of Electric Polarisation," Elsevier Press, Inc., Houston, 1952.
3. C. P. Smyth in A. Weissberger (ed.), "Technique of Organic Chemistry," vol. 1, "Physical Methods of Organic Chemistry," 3d ed., pt. 3, chap. 29, Interscience Publishers, Inc., New York, 1960.
4. C. P. Smyth, "Dielectric Behavior and Structure," McGraw-Hill Book Company, New York, 1955.
5. F. A. Jenkins and H. E. White, "Fundamentals of Optics," 3d ed., eq. (23p), p. 481, McGraw-Hill Book Company, New York, 1957.
6. J.-Y. Chien, *J. Chem. Educ.*, **24:** 494 (1947).
7. G. Hedestrand, *Z. Physik. Chem.*, **B2:** 428 (1929).
8. L. G. Wesson, "Tables of Electric Dipole Moments," The Technology Press of The Massachusetts Institute of Technology, Cambridge, Mass., 1948.
9. A. L. McClellan, "Tables of Experimental Dipole Moments," W. H. Freeman and Company, San Francisco, 1963.
10. I. F. Halverstadt and W. D. Kumler, *J. Am. Chem. Soc.*, **64:** 2988 (1942).
11. W. Kwestroo, F. A. Meijer, and E. Havinga, *Rec. Trav. Chim.*, **73:** 717 (1954).
12. E. A. Guggenheim, *Trans. Faraday Soc.*, **45:** 417 (1949); **47:** 573 (1951).

35 *DIELECTRIC CONSTANTS OF POLAR LIQUIDS; RESONANCE METHOD*

The dielectric constants and refractive indices of a number of pure liquids are measured. The results are interpreted with the Onsager equation. The resonance method is used for the dielectric-constant measurements.

THEORY.†   The Debye theory relating the dielectric properties of a substance to the molecular dipole moment is outlined in the preceding experiment. A major limitation of the Debye theory is that it is applicable only when the concentration of polar molecules is low—thus, only to polar gases and to dilute solutions of polar solutes in nonpolar solvents.

A significant advance in dielectric theory was made by Onsager,[1] who obtained an equation which is found to be remarkably good even for pure polar liquids, with the exception of associated liquids such as alcohols or water.[2,3]

It is useful for purposes of comparison to write several equations which would follow from Debye's theory if it *were* to be applied to the case of a pure polar liquid:

$$\frac{\epsilon_s - 1}{\epsilon_s + 2} \overline{V} = \frac{4\pi N_0}{3} \left(\frac{\mu^2}{3kT} + \alpha_v + \alpha_e \right) \tag{1}$$

$$= \frac{4\pi N_0}{3} \frac{\mu^2}{3kT} + \frac{\epsilon_\infty - 1}{\epsilon_\infty + 2} \overline{V} \tag{2}$$

or

$$\frac{4\pi N_0}{3} \frac{\mu^2}{3kT} = \left(\frac{\epsilon_s - 1}{\epsilon_s + 2} - \frac{\epsilon_\infty - 1}{\epsilon_\infty + 2} \right) \overline{V} \tag{3}$$

where N_0 = Avogadro's number
       μ = molecular permanent dipole moment
       α_v = molecular vibrational polarizability
       α_e = molecular electronic polarizability
       ϵ_s = static (equilibrium) dielectric constant
       ϵ_∞ = "high-frequency" dielectric constant (page 218)
       $\overline{V}$ = molar volume = M/ρ
       M = molecular weight
       ρ = density
       k = Boltzmann constant

The chief limitation of the Debye equation arises from the use of the Lorentz equation for the internal field (page 206). It is assumed in Debye's derivation that a given polar molecule is acted on by a constant field which is the time average of the field at a molecule due to all charges outside this molecule.

Onsager improved the calculation of the field acting on a molecule by taking into account the fact that the local field at a given polar molecule is a function of the orientation angle θ (page 205) of the given molecule. The reason for this is that the given molecule influences the polarization of its neighbors, which in turn make important contributions to the field at the given molecule. To see this more

---

† The present discussion necessarily assumes that the reader is familiar with the basic principles outlined in Experiment 34, especially pp. 217–219.

clearly, one may visualize calculating the local field at a given polar molecule by holding this molecule at a particular value of θ and averaging over the thermal motions of all the other molecules, and then repeating this process for various different values of θ; if account is taken of the effect of the given molecule on its neighbors, the resulting local field will be found to depend on θ. Thus the Onsager theory leads to a different expression for $\langle \cos \theta \rangle$ from that of Debye. Also, explicit allowance is made in Onsager's derivation for the fact that the apparent permanent dipole moment of a molecule immersed in the liquid is slightly different from that of an isolated molecule. Onsager's equation replacing Eq. (3) may be written as

$$\frac{4\pi N_0}{3} \frac{\mu^2}{3kT} = \frac{(2\epsilon_s + \epsilon_\infty)(\epsilon_s + 2)}{3\epsilon_s(\epsilon_\infty + 2)} \left(\frac{\epsilon_s - 1}{\epsilon_s + 2} - \frac{\epsilon_\infty - 1}{\epsilon_\infty + 2} \right) \bar{V}$$

$$= \frac{(\epsilon_s - \epsilon_\infty)(2\epsilon_s + \epsilon_\infty)}{\epsilon_s(\epsilon_\infty + 2)^2} \bar{V} \tag{4}$$

where μ is the dipole moment of the isolated molecule. Since in most cases ϵ_∞ can be replaced by n_D^2 (page 219) without serious error, the quantities on the right side are all readily measurable. Equation (4) has been tested by calculating values of μ from it for a large number of polar liquids and comparing these with values found by the gas or the solution method of Debye, or by spectroscopic methods; very good agreement has regularly been found for "normal" liquids, but not for those liquids which from other properties had already been identified as "associated" liquids.[4]

Thus for normal liquids, Eq. (4) offers an attractive means of finding μ, since it requires only a single measurement of ϵ_s of moderate accuracy, whereas either the gas or the solution method of Debye calls for a series of measurements of high accuracy. The resulting values of μ, while not as reliable as those found by measurements on the gas, are probably not significantly inferior to solution values provided the given substance is not associated.

The chief importance, however, of Onsager's equation is not so much in providing a short cut to finding molecular dipole moments as in offering an insight into the nature of molecular interactions in liquids. Whereas the Debye equation neglects the correlation between the orientation of a molecule and that of its neighbors, Onsager's equation takes account of this correlation to the extent that it is due to dipole-dipole interactions. The wide success of Onsager's equation for normal liquids is evidence that in such liquids the dipole-dipole interaction is the dominant one producing correlation in orientation of neighboring molecules. If for a given liquid Onsager's equation does not give the correct value for μ, one has strong evidence that there exists in this liquid some specific type of interaction, such as hydrogen bonding or complex formation, which has an important influence on the relative orientation of neighboring molecules.

····> **Apparatus.** Resonance-type apparatus for measurement of dielectric constant; pycnometer or density balance; refractometer; series of pure polar-liquid samples, about 100 ml of each.

PROCEDURE.   The dielectric constants, refractive indices, and densities are to be found for a number of pure liquids at 25°C. A suitable list is chloroform, methyl ethyl ketone, ethanol, acetone, acetonitrile, trichloroethylene, *n*-propyl alcohol, toluene.

The refractive indices may be measured with an Abbe refractometer (page 65). The densities may be found from the literature or measured by means of a pycnometer or density balance.

The dielectric-constant measurements are based on the equation $\epsilon = C/C_0$ (page 230). In order to circumvent difficulties due to stray capacitance, it is convenient to use a dielectric cell which is arranged so that measurements are made of capacitance differences between two fixed positions (*a* and *b*) of the rotor plates of a variable capacitor. The dielectric constant of the liquid sample is then obtained from the equation

$$\frac{\epsilon_{\text{liq}}}{\epsilon_{\text{ref}}} = \frac{C_{b,\text{liq}} - C_{a,\text{liq}}}{C_{b,\text{ref}} - C_{a,\text{ref}}} \tag{5}$$

where $C_{b,\text{liq}} - C_{a,\text{liq}}$ = capacitance increment between positions *a* and *b* measured with plates immersed in liquid sample

$C_{b,\text{ref}} - C_{a,\text{ref}}$ = corresponding increment measured with plates immersed in a reference substance of known dielectric constant

The reference measurements are usually made either with air, for which the value of ϵ at 25°C, 760 mm, is 1.00053, which may be considered unity for the present purpose, or with benzene, for which the value of ϵ at temperature $t°$C is $\epsilon = 2.274 + 0.0020(25 - t)$ (Ref. 5). The sample of benzene must be of high purity if a really accurate measurement is involved. The advantage of using benzene rather than air is that the former gives a much larger difference ($C_{b,\text{ref}} - C_{a,\text{ref}}$), which is therefore measurable with greater fractional accuracy.

In this experiment, the capacitance increments are to be measured by a resonance method.† Its chief advantage lies in the fact that the apparatus is relatively simple and can easily be used over a wide frequency range.[5,6] On the other hand, it is less accurate than the heterodyne-beat method and, at a given frequency, cannot be used with solutions having as high a conductance as can the bridge method.

The standard apparatus assembly for the resonance method is schematically

---

† The heterodyne-beat method of Experiment 33 is also very satisfactory for the present experiment; in general, the sample cell should have a somewhat lower capacitance increment ($C_{b,\text{air}} - C_{a,\text{air}}$) than that used for Experiment 33 because much larger values of ϵ are encountered.

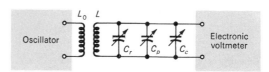

figure 52. *Schematic representation of standard resonance apparatus. The electronic voltmeter must have a high input impedance at the frequency of the oscillator so that it will not load the circuit.*

represented in Fig. 52. The dielectric-constant cell forms a part of a resonant circuit comprising a coil L and a parallel combination of capacitors: the dielectric-constant cell C_c, a calibrated precision condenser C_p, and a coarse-tuning-range condenser C_r. This circuit is coupled inductively (page 593) to the coil L_0 of an oscillator which generates an ac voltage at a frequency f. (The frequency is commonly in the radio-frequency region, say 1 MHz, though a wide range of frequencies can be used.) The voltage e_s induced in the secondary coil L depends on the current i_0 through the oscillator coil L_0 and on the coefficient of mutual inductance M (page 593). The equation in terms of a complex e_s and i_0 is $e_s = -iM\omega i_0$, where $\omega = 2\pi f$. For *loose* coupling, the oscillator frequency f and current i_0 remain constant when the capacitors in the secondary circuit are tuned. Furthermore, M depends only on the geometry of the coils. Therefore e_s is practically unaffected by tuning of the capacitors.

The equivalent of the secondary circuit is shown in Fig. 53 for the case of loose coupling. We now examine the behavior of this circuit as the capacitors are tuned. The following relations involving the secondary circuit current i_s and series impedance Z (page 586) hold:

$$i_s = \frac{e_s}{Z} \tag{6}$$

$$Z = i\omega L + R + \frac{1}{i\omega C} \tag{7}$$

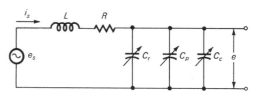

figure 53. *Equivalent of the resonance circuit, for the case of loose coupling, with induced voltage e_s and measured voltage e indicated.*

where C = parallel combination = $C_c + C_p + C_r$

$\omega = 2\pi f$

R = effective series resistance, mainly that in the coil

Hence

$$i_s = \frac{e_s}{R + i(\omega L - 1/\omega C)} \tag{8}$$

The voltage across the capacitor is

$$e = i_s \frac{1}{i\omega C} \tag{9}$$

$$= \frac{e_s/i\omega C}{R + i(\omega L - 1/\omega C)} \tag{10}$$

A vacuum-tube voltmeter placed across the capacitors measures $|e|$, the magnitude of e, given by

$$|e|^2 = ee^* = \left(\frac{e_s}{\omega C}\right)^2 \left[R^2 + \left(\omega L - \frac{1}{\omega C}\right)^2\right]^{-1} \tag{11}$$

where e^* is the complex conjugate of e. Hence

$$|e| = \frac{e_s/\omega C}{[R^2 + (\omega L - 1/\omega C)^2]^{\frac{1}{2}}} \tag{12}$$

As C is varied, $|e|$ goes through a maximum when the condition $\omega L = 1/\omega C$ is met, provided R is small ($R \ll \omega L$).

So long as e_s, ω, and L are held constant, the resonance condition then corresponds to a constant value for $C = C_c + C_p + C_r$. If the system is initially at resonance, and then C_c is changed, the change in C_c can be measured by finding the change in C_p required to restore the condition of resonance.

An alternative circuit[7] is shown in Fig. 54. The 6E5 tuning-eye tube acts both as an ocillator tube and as a resonance indicator. The piezoelectric quartz crystal in the grid circuit acts like a very sharply resonant circuit.

As the capacitance in the plate tuning circuit is increased from below the resonance value, oscillations begin when the natural frequency of the plate circuit approaches that of the quartz crystal. The dc plate current then decreases, the dc plate voltage rises, and the tuning eye of the tube begins to close. As the capacitance is further increased, a point is reached at which oscillation ceases abruptly, with a corresponding sudden increase in the shadow angle. The critical capacitance setting at which the shadow angle abruptly widens is quite reproducible and is taken as the reference point for capacitance measurements. As the crystal resonance frequency is

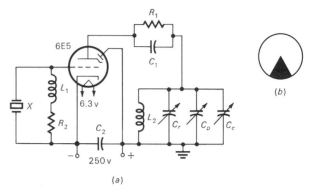

(a)

figure 54. (a) Resonance circuit for measurement of dielectric constants: $R_1 = 150,000$ ohms; $R_2 = 40,000$ ohms; $L_1 = 2.5$-mh radio-frequency choke; $X = 1$-MHz quartz crystal; $C_p = 150$-pf variable air capacitor, calibrated precision type; $L_2 = $ oscillator tank inductance: 60 turns No. 25 enameled wire, close wound on $1\frac{1}{2}$-in.-diameter coil form; $C_1 = 0.001$-μf 450 WVDC (working voltage dc); $C_2 = 0.01$-μf 450 WVDC; $C_c = $ dielectric-constant cell; $C_r = 150$-pf variable air capacitor. (b) Face of 6E5 tuning-eye tube. The shadow angle (blackened area) widens as the dc plate potential increases. As an alternative to the tuning-eye indicator, a milliammeter may be inserted in the plate circuit.

very stable and L is practically constant, this critical condition of resonance again corresponds to $C = C_c + C_p + C_r = $ constant.

Instead of the tuning-eye tube, a milliammeter which indicates the plate current of the oscillator tube may be used to find the critical capacitance setting corresponding to resonance. As the resonance condition is approached with a gradually increasing value of capacitance, the tube current gradually decreases to a minimum value and then very suddenly jumps to a large value as the oscillation stops.

Whichever method is used to indicate the resonance condition, the measuring procedure is essentially the same. The rotor of the dielectric cell containing the reference substance is set in position *a* (minimum capacitance). The precision condenser is set near the upper limit of its range, and the range condenser is tuned to bring the circuit close to resonance, as shown by the response of the electronic voltmeter, tuning eye, or plate current meter, as the case may be. The precision condenser is then adjusted carefully to locate the critical capacitance setting. The dielectric-cell rotor is then set to position *b* (maximum capacitance), and the condition of resonance is reestablished by setting the precision condenser, *the range condenser remaining unaltered.* The capacitance increment of the cell, $\Delta C_c = C_{b,\mathrm{ref}} - C_{a,\mathrm{ref}}$, is equal to the decrease in the capacitance of the precision condenser required to compensate for it. These same operations are carried out with the liquid sample as dielectric, and the dielectric constant calculated by means of Eq. (5). For accurate work the precision capacitor must be carefully calibrated.

A suitable dielectric-constant cell may be made from a small variable condenser, the number of rotor and stator plates of which is chosen to give a convenient capacitance increment. The guiding consideration here is that the capacitance increment with a sample having the highest dielectric constant to be measured must not exceed the range of the precision condenser. The condenser, which should have two rotor bearings, is mounted below the metal plate which forms the top of the cell; the stator plates are insulated, and the rotor plates are grounded. An insulated arm is fastened to the rotor shaft, and two brass pins are driven into the supporting plate to provide reproducible minimum and maximum capacitance settings. To minimize pickup of stray voltages and to guard against disturbing effects of variations in stray capacitance between the capacitor and exterior objects, the cell should be surrounded by a grounded metal shield, fastened to the top plate for support, and the connection to the oscillator should be made with coaxial cable. The liquid sample is contained in a lipless beaker or a truncated polyethylene bottle which fits closely inside the shield can.

CALCULATIONS.   Molecular dipole moments are calculated from Eq. (4) and, for comparison, from Eq. (3). The Onsager values are compared with reported[3,8-11] values of μ, based on measurements by the Debye gas or solution methods or by spectroscopic methods, and any discrepancies are discussed with due regard for the estimated experimental uncertainty.

Practical applications. The principle of the resonance method is employed in a number of common electronic instruments such as the grid-dip meter and the Q meter. These can be used for rapid measurement of dielectric properties over a wide range of frequencies when high accuracy is not required.

Suggestions for further work. Onsager[1] gives an equation for a solution of a polar solute in a nonpolar solvent. This equation may be tested by making appropriate measurements on solutions. It may be applied to the case of a solute which is associated as a pure liquid to determine the degree to which such association is important in the solutions.

For accurate measurements, the precision capacitor must be calibrated. It is an instructive exercise to perform this calibration after devising a suitable procedure. The method used will depend on the equipment available. If a precision capacitor and a sensitive bridge are available, the calibration may be made at audio frequencies with satisfactory results. Smyth[3,5] describes a procedure for use with the heterodyne-beat method which may be adapted for use with resonance equipment.

References

1. L. Onsager, *J. Am. Chem. Soc.*, **58:** 1486 (1936).
2. C. J. F. Böttcher, "Theory of Electric Polarisation," Elsevier Press, Inc., Houston, Texas, 1952.

3. C. P. Smyth, "Dielectric Behavior and Structure," McGraw-Hill Book Company, New York, 1955.

4. L. C. Pauling, "Nature of the Chemical Bond," 3d ed., chap. 12, Cornell University Press, Ithaca, N.Y., 1960.

5. C. P. Smyth in A. Weissberger (ed.), "Technique of Organic Chemistry," vol. 1, "Physical Methods of Organic Chemistry," 3d ed., pt. 3, chap. 29, Interscience Publishers, Inc., New York, 1960.

6. L. Hartshorn, "Radio-frequency Measurements by Bridge and Resonance Methods," John Wiley & Sons, Inc., New York, 1941.

7. P. Bender, *J. Chem. Educ.*, **23:** 179 (1946).

8. C. H. Townes and A. L. Schawlow, "Microwave Spectroscopy," app. VI, McGraw-Hill Book Company, New York, 1955.

9. L. G. Wesson, "Tables of Electric Dipole Moments," The Technology Press of the Massachusetts Institute of Technology, Cambridge, Mass., 1948.

10. A. L. McClellan, "Tables of Experimental Dipole Moments," W. H. Freeman and Company, San Francisco, 1963.

11. A. A. Maryott and E. R. Smith, Table of Dielectric Constants of Pure Liquids, *Natl. Bur. Std. U.S. Circ.* 514, 1951.

36   DIELECTRIC CONSTANT OF A SOLID AS A FUNCTION OF TEMPERATURE

THEORY.[1,2]   The dielectric constant of a substance may be defined as

$$\epsilon = \frac{C}{C_0} \tag{1}$$

where C = capacitance between a pair of conductors immersed in the substance

C_0 = capacitance for same arrangement of conductors *in vacuo*

The change in capacitance from C_0 to C is due to the polarization of the substance under the influence of the field present when the capacitor plates are charged. Polarization is brought about by shifts, over distances of the order of molecular dimensions, in the average positions of the charged particles of which the substance is constituted. Thus $\epsilon - 1$ is a measure of the extent of such polarization per unit field.

For polar liquids, the dominant contribution to the polarization is generally that due to a slight alignment in the field direction, on the average, of molecules possessing permanent dipole moments. Smaller contributions arise from field-induced distortion of bond lengths and bond angles and from shifts in the electron charge distribution relative to the nuclear framework. These three contributions, it may be noted, are associated, respectively, with rotational, vibrational, and electronic coordinates of the molecule.

As the temperature is lowered, most polar liquids show a gradual increase in dielectric constant, due partly to an increased degree of alignment in the field and partly to an increase in density. Upon freezing, there is observed in most cases an abrupt decrease in dielectric constant because freedom of the molecules to rotate in the field is lost. The small remaining dielectric effect is due to vibrational and electronic terms, which are often not much changed by a change in state, and to lattice polarizability.

It is not uncommon, however, for a polar liquid to exhibit only a small change in dielectric constant upon freezing, and then to show a marked drop at some lower temperature. This phenomenon is attributed to the retention in the solid state of sufficient rotational freedom to allow reorientation of molecular dipoles in a field. This freedom is lost only at a lower temperature at which a crystalline phase transition takes place. To affect the dielectric constant appreciably, the rotation, whether it involves all or only a part of the molecule, must be such as to cause a change in the orientation or magnitude of the molecular dipole moment. The general correctness of this intepretation has been verified by the study of other properties of the same solids.[1-4] For example, an increase in rotational freedom is often accompanied by a marked increase in entropy and in heat capacity, an increase in symmetry of the crystal structure, and a decrease in the width of lines in the nuclear-magnetic-resonance spectrum of the solid.

Such rotational effects are observed most often for substances the molecules of which are approximately spherical or cylindrical in shape. Thus, if there are attached, to a single carbon atom, four atoms or groups not all the same but of approximately the same size, the molecule will be polar and yet may exhibit rotation in the solid phase. For example, as a substituent in an organic molecule, a methyl group has about the same effective size as a chlorine or bromine atom. Thus rotation is observed in solid phases of $C(CH_3)_3Cl$ and $C(CH_3)_3Br$.

In the present experiment, the dielectric constant of a polar substance is measured over a range of temperatures above and below the freezing point. The experimental arrangement is shown in Fig. 55. At a series of temperatures, estab-

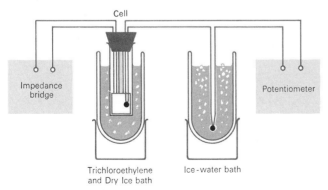

figure 55. Apparatus for measurement of dielectric constant as a function of temperature. The thermocouple junctions are shown as solid black circles.

lished by a bath of Dry Ice in trichloroethylene, the capacitance C of the cell and connecting leads is measured by means of an impedance bridge. Then ϵ is calculated from

$$C = C_1 + \epsilon C_2 \qquad (2)$$

where C_1 = capacitance of connecting leads
$\qquad C_2$ = capacitance of empty cell alone

The quantities C_1 and C_2 are determined by two measurements of C, one with the cell empty and one with it filled with a liquid of known dielectric constant. Equation (2) applies to both these measurements. The temperature is measured by means of a thermocouple immersed in the sample.

The sample under investigation, particularly while in the liquid state, may have an appreciable conductance. The cell will be equivalent electrically to a network (Fig. 56) consisting of a resistor in *parallel* with a capacitor. The equivalent resistance R and capacitance C depend only on properties of the medium (specific conductance and dielectric constant) and the geometry of the cell. The resistance is given by Eq. (2) of Experiment 27,

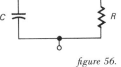

figure 56.

Network equivalent to cell filled with sample.

$$R = \frac{k}{\kappa} \qquad (3)$$

where k = cell constant
$\qquad \kappa$ = specific conductance of medium

while the capacitance, including that of the connecting leads, is given by Eq. (2). The cell impedance Z for a given frequency f is determined by R and C through the relation (page 588)

$$\frac{1}{Z} = \frac{1}{R} + i\omega C \qquad (4)$$

where $\omega = 2\pi f$.

The measurement of R and C may be accomplished by means of an impedance bridge, which is a generalized Wheatstone bridge having impedances (page 586) rather than resistances as branch elements. A typical bridge for measurement of resistance and capacitance is shown in Fig. 57. The device whose impedance is to be measured constitutes branch 4; the unknown impedance is Z_4. The input to the bridge is a sinusoidal (sine-wave) voltage, typically at a frequency of 1000 Hz.

The operation of balancing the bridge consists of adjusting R_1, R_2, and R_3 for zero output voltage (or, practically, for minimum output) as indicated by the detector. The condition for null output is

$$\frac{Z_1}{Z_2} = \frac{Z_3}{Z_4} \qquad (5)$$

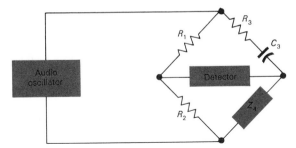

figure 57. *Schematic diagram of impedance bridge used for dielectric-constant measurements. The cell is branch 4.*

where Z_1, Z_2, Z_3, Z_4 are the branch impedances. For the type of bridge shown, the known impedances are

$$Z_1 = R_1 \qquad Z_2 = R_2 \qquad Z_3 = R_3 + \frac{1}{i\omega C_3}$$

It is convenient to write Z_4 formally as

$$Z_4 = R_4 + \frac{1}{i\omega C_4} \tag{6}$$

$$= \frac{1}{i\omega C_4}(1 + iD) \tag{7}$$

where

$$D = \omega R_4 C_4 \tag{8}$$

One may visualize R_4 and C_4 as the parameters of a fictitious resistor and capacitor which, connected in *series*, would give the total impedance Z_4. The dissipation factor D is seen to be the ratio of R_4 to $1/\omega C_4$, the latter being the magnitude of the reactance of capacitor C_4 at the frequency of measurement.

In principle, from the values of R_1, R_2, R_3, and C_3 at balance, one could calculate R_4 and C_4, or alternatively, C_4 and D. The bridge dials are usually calibrated in terms of the latter pair of variables. The equations required for calculation of R and C from the measured quantities C_4 and D may be obtained from the equation

$$Z = Z_4$$

where Z is given by Eq. (4) and Z_4 by Eq. (7). The result is

$$R = R_4 \frac{1 + D^2}{D^2} \qquad C = \frac{C_4}{1 + D^2} \tag{9}$$

The value of R is often not of interest because it is sensitive to the presence of impurities in the sample.

····➤ **Apparatus.**† Dielectric-constant cell; impedance bridge; thermocouple; potentiometer; samples, e.g., 1,1,1-trichloroethane; Dewar flask, trichloroethylene; Dry Ice; Variac-controlled electric heater (200-watt radiant immersion type recommended).

PROCEDURE.   The dielectric constant of 1,1,1-trichloroethane, CCl_3CH_3, is to be measured over the temperature range -70 to $0°C$. It is recommended that a graph of electromotive force versus temperature for the thermocouple be prepared in advance from handbook data.

The apparatus shown in Fig. 55 is assembled with the trichloroethylene§ bath initially at room temperature. The ground terminal of the bridge should be connected to the outer electrode if the cell electrodes have a coaxial cylindrical arrangement. For the determination of C_1 and C_2, the capacitance C is measured with the cell empty and with the cell filled with benzene at room temperature. The temperature of the bath is noted.

The bath is removed, and finely powdered Dry Ice is added, a few grams at a time, with stirring, until there is an excess. (Rather violent ebullition of bath liquid is apt to occur.)

The sample is placed in the cell, and the cold bath slowly raised into place. The object is to freeze the sample from the bottom upward in such a way as to avoid the formation of air bubbles, cracks, or voids within the solid sample. For highest accuracy, the cell should be fitted with a stopcock and standard-taper connection so that it may be attached to a vacuum manifold, and the sample distilled into it with air completely excluded.

The heater is turned on to produce a slow rate of rise of temperature. The capacitance C is measured at approximately $5°$ intervals. For each measurement, the electric heating is stopped and the temperature monitored until both temperature and capacitance are reasonably steady. Near phase changes, additional points should be recorded; to hold the temperature steady for this purpose it may be necessary to add a few grams of Dry Ice. It is desirable to obtain values of

---

† The design of the cell must be such as to prevent entrance of moist air from the room. A suitable cell available commercially is Type 2TN50 conductance cell manufactured by J. C. Balsbaugh, Marshfield Hills, Mass. This cell requires about 40 ml of sample and has an air capacity of 50 pf. It may be modified by a glass blower to permit insertion of the thermocouple leads through the top. The thermocouple may be soldered to the ground electrode of the cell for mechanical support and improved thermal contact.

Coaxial thermocouple wire, well suited to this application, is manufactured by the Precision Tube Co., North Wales, Pa.

The impedance bridge chosen for this experiment should operate at a frequency above 10 kHz for best results.

§ Trichloroethylene is mildly toxic; inhalation of appreciable amounts of the vapor is to be avoided.

C intermediate between the values for the pure phases on either side of the transition.

For best results, the final experiment should be preceded by a preliminary trial, the object of which is to acquire facility with the apparatus and to locate approximately the temperatures at which phase changes occur.

If the conductance of the sample exceeds a certain level, it becomes difficult or even impossible to balance the bridge precisely.† The best procedure in such cases is to employ a higher frequency, say 20 kHz. This can be done by using an external oscillator as generator for the bridge. As a rule, the conductance of the sample decreases rapidly as the temperature is lowered.

After the conclusion of the work, the cell should be rinsed with a solvent and dried with a current of air.

CALCULATIONS.   The dielectric-constant data are tabulated, and ϵ plotted as a function of temperature. The results are interpreted qualitatively and correlated with other available data, such as heat capacities,[3] crystal-structure data,[5] or nuclear-magnetic-resonance-line widths,[4] which have a bearing on the question of rotation in the solid state.

Practical applications. The study of dielectric properties has contributed much to our understanding of the structure and behavior of solids.

Suggestions for further work. Data may be taken with a slowly falling temperature. Since supercooling is common, it is often possible to obtain dielectric data for the supercooled liquid for a considerable range below the melting point.

It is instructive to calculate the concentration of an ionized impurity which would be needed to give a value of 0.1 for D for the case of the liquid phase of the sample studied. The correct order of magnitude will be obtained if the equivalent conductance of the impurity is taken as 50 cm² ohm⁻¹ equiv⁻¹.

The relationship $4\pi k C_0 = 1$ (esu units) or $k C_0 = \epsilon_0 = 8.854 \times 10^{-12}$ farad m⁻¹ (mks-coulomb units), easily derived for the parallel-plate case but actually valid for any cell geometry, is useful for estimating effects of sample conductance.

References

1. C. J. F. Böttcher, "Theory of Electric Polarisation," pp. 399–409, Elsevier Press, Inc., Houston, Texas, 1952.
2. C. P. Smyth, "Dielectric Behavior and Structure," McGraw-Hill Book Company, New York, 1955.
3. L. M. Kushner, R. W. Crowe, and C. P. Smyth, *J. Am. Chem. Soc.*, **72:** 1091, 4009 (1950).
4. J. G. Powles and H. S. Gutowsky, *J. Chem. Phys.*, **21:** 1695 (1953).
5. R. W. G. Wyckoff, "Crystal Structures," 2d ed., Interscience Publishers, Inc., New York, 1963.

† The null becomes broad unless $D \ll 1$. For small D, Eq. (9) leads to $D \simeq 1/\omega RC$. Thus, for example, to obtain a sharp null with $C = 100$ pf and $f = 1$ kHz requires $R \gg 1.6$ megohms.

37   *OPTICAL ROTATORY DISPERSION*

The rotation of plane-polarized light may be determined by the use of a polarimeter. The variation of rotation with wavelength is called *optical rotatory dispersion,* and the study of this effect may be used to obtain information on molecular structure.

THEORY.[1]   Electromagnetic radiation consists of sinusoidally varying electric and magnetic fields, the directions of which lie in mutually perpendicular planes. The oscillations are transverse; those in the electric and magnetic fields are perpendicular to the direction of propagation. If the electric component, for example, is restricted to a single plane (as illustrated in Fig. 58*a*), the light is said to be plane-polarized. In ordinary light the electric component has all possible orientations, and none is preferred. This is because the individual atoms and molecules which are radiating act independently. Such unpolarized light may also be considered to consist of two plane-polarized waves which are at right angles to each other and have a completely random phase relationship; i.e., at any instant the phase difference between the two waves is equally likely to have any value between 0 and 2π.

   In optically *isotropic* crystals the index of refraction is independent of the direction of propagation of the light through the crystal. In *anisotropic* crystals, such as calcite ($CaCO_3$), an incident beam of unpolarized light is split into two beams in the crystal. It is found that these two beams are plane-polarized, their planes of vibration being at right angles. The velocities of propagation of these two components through the crystalline medium are different because of the difference in the index of refraction of the medium for the two differently polarized rays. This makes possible the elimination of one component, so that plane-polarized light is obtained. A Nicol prism is constructed by cutting a calcite prism in half along a suitable diagonal plane and cementing the sections together with Canada balsam. At the

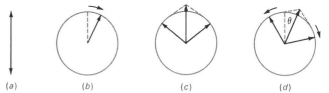

figure 58. (a) Electric vector of plane-polarized light moving in a direction perpendicular to the page; (b) circularly polarized light in which the electric vector rotates around the direction of propagation; (c) plane-polarized light considered to be the resultant of two vectors representing circularly polarized light with opposite senses of rotation; (d) rotation of the plane of polarization of plane-polarized light due to the fact that one type of circularly polarized light is propagated through the medium at a higher velocity than the other.

calcite–Canada balsam interface, one component is totally reflected to the side, where it is absorbed by a black coating applied to the prism. The other component, for which the refractive index of the balsam and the calcite are almost equal, is freely transmitted.

In order to get a better understanding of the rotation of the plane of polarization of light by certain solutions and crystals, it is useful to consider circularly polarized light. If plane-polarized light is allowed to pass into a suitably cut slab of calcite, the light will be separated into two waves which are plane-polarized at right angles to each other and have equal amplitudes but are propagated in the same direction through the crystal. The waves travel through the crystal at different speeds, and so there will be a phase difference between them when they emerge from the crystal. If the thickness of the crystal is chosen so that this angle is 90°, the emerging light is circularly polarized and the ground crystal is referred to as a *quarter-wave plate*. Such a combination of two plane-polarized waves can be represented by a vector rotating about the direction of propagation as shown in Fig. 58*b*. Plane-polarized light may be considered to be made up of two oppositely rotating circularly polarized beams which are in phase. As shown in Fig. 58*c*, the resultant of the addition of these two vectors remains in a single plane.

The optical rotation by a gas, liquid, or solid may be considered to arise from a difference in the velocity of clockwise and counterclockwise circularly polarized light. As shown in Fig. 58*d*, the resultant of the addition of the two vectors representing circularly polarized light is rotated through an angle θ if one component travels more rapidly than the other.

Molecules which can be distinguished from their mirror images are able to rotate the plane of polarized light when it passes through them. The presence in the molecule of an asymmetric carbon atom (one for which all four attached groups are different) leads to such a structure and is the most common, but not the only, cause of optical activity.

Thus, in the study of the physical chemistry of the proteins it is found that there is present an intrinsic optical activity, resulting from the contributions of the individual amino acids, whatever the overall conformation of the protein; it derives from the asymmetric carbon atoms. In the cases of those proteins with α-helical content, however, there is a second and additional contribution to the optical rotation which originates in the asymmetry of the α helix itself. From observations of the magnitude of this second contribution it has become possible to make an estimate of the fraction of the polypeptide which is in the helical conformation.

The magnitude of the optical rotation is measured with a source of monochromatic light and a polarimeter, which consists primarily of two Nicol prisms, between which the optically active substance is placed. When the second Nicol, known as the *analyzer*, is placed at right angles to the first, no light can pass

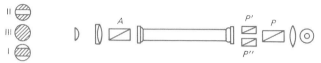

figure 59. Polarimeter.

through if the cell contains an optically inactive substance. When a substance that is capable of rotating the plane of polarized light is inserted between the Nicols, light can again be seen through the analyzer. The angle through which the analyzer must be turned to darken the field again is represented by α. If the analyzer is turned clockwise (as seen by the observer) to restore darkness, the substance is said to be dextrorotatory. If darkness is restored when the analyzer is turned counterclockwise, it is levorotatory.

Because of the error inherent in locating the point of minimum intensity, it is better to employ a scheme in which the eye is required only to compare one field with another field of nearly the same intensity, as is done in half-shadow and triple-shadow polarimeters.

The principle of the Landolt-Lippich triple-shadow polarimeter is illustrated in Fig. 59. Behind the large polarizing Nicol prism P are placed two auxiliary Nicols, P' and P'', whose planes of polarization have been adjusted to make an angle θ with that of the principal polarizing prism P. The angle θ may be adjusted to optimum conditions, which will depend on the intensity of the light and the transparency of the liquid. When the analyzer A, whose orientation is indicated by the instrument scale, is turned so that it is at right angles with the main polarizing Nicol, the central strip of the field, as viewed through the magnifying eyepiece, is dark and the sides are lighter, as shown at I. When the analyzer is turned through the small angle θ to cross with the smaller Nicols, the sides are dark and the central strip is lighter, as shown at II. When the analyzing Nicol is turned back through half of this small angle, it gives a uniform field as shown at III. This proper setting is readily found, and the corresponding reading of the scale is recorded. The double-field polarimeter (Laurent type) employs only one auxiliary Nicol prism, which covers half the field of the polarizing prism. The field of view is thus divided into two parts, and the reference analyzer setting is again that which gives a uniformly illuminated field. The scale is usually graduated directly into quarters of degrees, and with the aid of verniers and a magnifying lens, the angles may be read to $0.01°$.

The magnitude of the optical rotation is affected by the concentration of the solution, the length of the path of the light in the solution, the wavelength of the light, the temperature, and the nature of the solvent. The specific rotation $[\alpha]_\lambda^t$ for a given wavelength λ at a given temperature t is defined by the relation

$$[\alpha]_\lambda^t = \frac{\alpha}{lc} = \frac{\alpha}{lp\rho}$$

where α = observed angle of rotation

l = length of light path, decimeters

c = concentration of solute, grams solute per milliliter solution

p = concentration of solute, grams solute per gram solution

ρ = density of solution, g ml^{-1}

The specific rotation depends upon the wavelength of the light, and this dependence is called *optical rotatory dispersion*. The dispersion of optical rotation is closely related to light absorption; in the vicinity of absorption bands, the optical rotation usually increases rapidly and then decreases through zero to give an opposite rotation as the wavelength is changed. This is referred to as the *Cotton effect*. These changes are more sensitive than most properties to changes in molecular conformation. This tool has recently been developed into a very powerful one for studying molecular structures of complicated organic molecules.[2]

Apparatus. Polarimeter; sodium-vapor lamp and mercury-vapor lamp with filters, or other sources of monochromatic light; sugar solutions.

PROCEDURE. Solutions containing approximately 5, 10, and 15 g of sugar per 100 ml are prepared, using volumetric flasks. The crystallized sucrose should be heated to 105°, cooled in a desiccator, and weighed out accurately.

Monochromatic light sources must be used with the polarimeter because of optical rotatory dispersion. To demonstrate this phenomenon for sucrose, the sodium-vapor lamp is supplemented as a light source by a low-intensity mercury-vapor lamp, with which filters† are used to isolate the strong lines in the visible region at 5780 Å (yellow), 5460 Å (green), and 4358 Å (blue). For this purpose it would be advantageous to use a monochromator so that the wavelength could be varied continuously. The light source should be placed at the proper focal distance from the end of the polarimeter (about 20 cm) and should not be close enough to heat the instrument. It must be carefully positioned on the optical axis of the instrument to ensure uniform illumination of the polarizer.

The polarimeter tube is rinsed and filled with distilled water, as full as possible and the cap is screwed on, not tightly enough to cause strain, as this would produce an additional optical rotation. Any small air bubble remaining is driven up into an enlargement, above the line of vision. The glass plates at the ends must be clean, and the exposed surface must be dry. The analyzer is rotated until the field is uniformly illuminated, and several readings are taken. The average gives the zero point. The setting of the analyzer should always be approached from the same direc-

† Wratten filters may be obtained from the Eastman Kodak Co., for the yellow line No. 22, for the green line No. 77, for the blue line No. 40. The glass-filter combinations supplied by the Corning Glass Works may also be used, as described in Chapter 25. An excellent liquid filter solution for the 4358 Å line is described in Experiment 41. Some polarimeters are equipped with a removable glass filter, for use with the sodium D line source, which must be removed when other wavelengths are employed.

tion in order to avoid backlash. The zero reading is subtracted from the readings on the optically active material. It should be taken at the beginning and end of each set of determinations.

The tube is next rinsed two or three times with a sugar solution and filled as before; three or more readings are taken. For each solution the rotation is measured for each wavelength of light available.

CALCULATIONS.   The specific rotation of sucrose is calculated from the observed optical rotations for each of the wavelengths employed. The results are compared with the values given in tables.[3] The change with temperature of the specific rotation of sucrose in water solution is approximately -0.02 percent $°C^{-1}$ in the neighborhood of room temperature and is essentially independent of wavelength.

Plots are made of optical rotation versus concentration for each wavelength and of specific rotation versus wavelength.

Practical applications. Optical rotation is used in identifying materials and in determining the structure of organic compounds.[2] It finds important applications in quantitative analysis, as, for example, in the determination of the concentration of sugar in solutions.[4] Certain chemical changes may be followed without disturbing the system, as, for example, in the rate of inversion of cane sugar by catalysts (described in Experiment 24). The helix-coil transformation in polypeptides may be studied by measurements of optical rotation.[5]

Suggestions for further work. Other substances which are optically active, such as tartaric acid, may be studied in the same manner as sugar. Nonaqueous solutions may be used, e.g., camphor in benzene, carbon tetrachloride, and acetone.

References

1. W. Kauzmann, "Quantum Chemistry," Academic Press Inc., New York, 1957.
2. C. Djerassi, "Optical Rotatory Dispersion," McGraw-Hill Book Company, New York, 1960.
3. "International Critical Tables," vol. II, McGraw-Hill Book Company, New York, 1928.
4. F. J. Bates et al., "Polarimetry, Saccharimetry and Sugars," *Natl. Bur. Std. U.S. Circ.* C440, 1942.
5. J. T. Yang and P. M. Doty, *J. Am. Chem. Soc.*, **79:** 761 (1957).

Chapter 11

Spectroscopy

38 SPECTROMETRY AND SPECTROGRAPHY

The calibration and use of a spectrometer or spectrograph are illustrated in this experiment. The study of typical emission spectra is used to emphasize theoretical and practical applications.

THEORY.[1-3] The passage of polychromatic light through a prism or its reflection from a ruled grating results in the dispersion of the light into its various wavelengths. The visible range of the spectrum so produced extends from the violet at about 4000 Å to the deep red at about 7500 Å. The angstrom unit Å, named after the Swedish physicist Ångström, is defined as 10^{-8} cm (see Appendix, page 566).

For many purposes it is convenient to characterize spectral lines in terms of *wave number* $\bar{\nu}$, which is the reciprocal of the wavelength in centimeters. For example, the wave number of the green line in the mercury arc spectrum is $1/(5460.73 \times 10^{-8}$ cm), or 18,312.6 cm^{-1}.

Specially sensitized photographic plates, made by the Eastman Kodak Co., are available for overlapping wavelength ranges from the far ultraviolet to the near infrared; for the visible spectrum panchromatic film is also useful. Glass prisms and lenses restrict observations to wavelengths from about 3600 to 10,000 Å. For the ultraviolet region quartz or fluorite optics are used, while infrared prism spectrometers employ NaCl, KBr, and CsBr prisms with special detectors. Reflection gratings avoid the problem of light absorption by prisms and can provide high dispersion and resolution of the spectrum.

The conditions required for the production of emission spectra (high-temperature–low-voltage arc or high-voltage discharge) are such that only for atoms and very simple molecules can the emission spectrum be studied; for complex molecules the methods of absorption spectroscopy must be used. For atoms the emission lines originate in a change in the electronic energy of the atom. According to quantum mechanics, the electronic energy of a particular kind of atom can have only certain discrete and characteristic values. For a gas such as helium, under ordinary conditions all but a negligible fraction of the atoms will be in the lowest electronic-energy state. In the high-voltage gas discharge tube, however, atoms are raised to various high-energy states. When such an excited atom drops back to a lower energy level, the energy balance is maintained by the emission of radiation, which is observed as one of the characteristic emission lines of the atom. The frequency of the emitted radiation is given by the quantum condition

$$h\nu = hc\bar{\nu} = E_2 - E_1 \tag{1}$$

where h = Planck's constant
ν = frequency

E_2 = energy of higher energy state

E_1 = energy of lower energy state

c = velocity of light

$\tilde{\nu}$ = wave number

The emission spectrum of a diatomic molecule, as obtained from a discharge tube, is a superposition of the spectra of the molecule and of the atoms produced by its dissociation. Because changes in quantized energies of vibration and rotation are possible for molecules as well as changes in electronic energy, the emission spectrum of a diatomic molecule is quite complex. It consists of a series of bands of lines, each band corresponding to a particular change in electronic energy combined with various smaller changes in rotational and vibrational energies.

The theoretical calculation of the electronic energy levels of complex atoms is quite difficult, but for atomic hydrogen, the simplest atom, the following result has been obtained for the wave numbers $\tilde{\nu}$ of the emission lines:

$$\tilde{\nu} = \Re\left(\frac{1}{n_1^2} - \frac{1}{n_2^2}\right) \tag{2}$$

where $\Re$ = Rydberg constant, 109,677.76 cm^{-1}

n_1, n_2 = integral quantum numbers characterizing initial and final energy states

For a given value of n_1, successive higher values of n_2 produce a series of lines. For $n_1 = 2$, this series of lines lies in the visible region of the spectrum. The Lyman series, for which $n_1 = 1$, is found in the ultraviolet range, and other series corresponding to $n_1 = 3$, 4, etc., lie in the infrared region. It is interesting to note that the relation expressed by Eq. (2) was found empirically by Balmer to represent the visible emission lines of atomic hydrogen long before the first theoretical derivation of the formula was achieved by Bohr.

Apparatus. Spectrograph; panchromatic film; plateholder; mercury-vapor lamp; argon, helium, hydrogen, nitrogen, and mercury-argon discharge tubes; discharge-tube transformer; photographic developer and fixer solutions; sample undeveloped film; microscope comparator.

PROCEDURE. The spectrograph focusing is checked by examination of the spectrum from a mercury-vapor lamp† placed in front of the slit. The slit width should be set to give a narrow line (about 0.1 mm) but should not be so small as to require inconveniently long exposures; a slit height of about 5 mm is recommended. The lines should be in good focus in all parts of the spectrum; if this is not the case, further adjustment of the instrument should be made in accordance with the instructions furnished by the manufacturer, or with *expert* assistance.

The panchromatic film, which is employed because of its sensitivity to the

† A convenient lamp and transformer are available from Oriel Optical Corporation, Stamford, Conn.

entire visible range of the spectrum, must be handled in complete darkness. The film is placed in the holder with the emulsion side out; this operation is facilitated by preliminary examination of a sample undeveloped film. Some holders are designed for use with cut film only, but others accommodate film or plates. In the latter case the film is placed in a metal film sheath for support before being put into the plateholder.

The spectrograph shutter is closed, the holder is attached to the camera, and the black slide covering the film is withdrawn. By means of the rack-and-pinion control provided, the plateholder position is adjusted so that the top edge of the film is in position for the first exposure. The shutter is then opened, and the spectrum of the mercury-vapor lamp is recorded. The proper exposure times for this and the other spectra studied depend on the characteristics of the particular spectrograph and light sources used. Approximate exposure times for the various spectra should be specified by the instructor as reference data for this experiment. For a new source, a set of trial exposures varying between wide limits may first be taken, and the optimum exposure time selected on the basis of these results.

The plateholder is moved up 1 cm, as indicated on the adjacent scale, and the next spectrum recorded. Exposures are thus taken of the argon, helium, hydrogen, and nitrogen discharge-tube spectra. The discharge tubes should be placed in position immediately in front of the slit.

Caution: The operating voltage for these tubes is several thousand volts.

A switch in the transformer primary circuit is used to control the discharge tube; the intensity can be varied by a resistance connected in the primary circuit. Alternatively, an autotransformer may be used to supply the primary voltage.

A second mercury spectrum is recorded as the last exposure on the film. The order in which the spectra are taken is recorded.

The plateholder is taken to the darkroom. Three trays containing, respectively, D-19 developer solution, distilled water, and F-5 acid fixing solution are placed in a sequential order, and their positions memorized. In total darkness, the film is developed, with sensitized side facing upward, for 5 min, carefully rinsed in the distilled water tray, and then fixed for 15 min with frequent agitation. At this time the light may be turned on. Care should be taken to avoid scratching the sensitized side. The film is then washed in running water for 30 min and air-dried. Further details of the practice and theory of the photographic process are given on page 566.

A straight line is marked on the film with a needle, connecting a sharp line in the upper mercury spectrum with the corresponding line in the lower. The distance of each of the mercury lines from this reference line is obtained using a comparator; the film is mounted between two pieces of *plate* glass, to keep it flat, with the emulsion side up. A preliminary dispersion curve is drawn through a plot of wavelength versus displacement in millimeters from the reference line; the wavelengths of the mercury lines are obtained from Fig. 155. With this curve, lines in the helium spectrum listed in handbooks are identified; their positions relative to the reference line provide additional points to define the dispersion curve more accurately.

The displacements of several lines in the argon spectrum from the reference line are determined, together with those of the several lines of the Balmer series identified in the hydrogen spectrum. The comparator settings should all be approached from the same direction to eliminate difficulties from backlash and looseness in adjustment.

If a comparator is not available, an enlargement of the film may be made and the line positions measured with an accurate steel rule.

CALCULATIONS.   The wavelengths found for the argon lines by means of the dispersion curve are compared with literature values. The wavelengths for the Balmer lines are calculated by use of the theoretical formula of Eq. (2) and compared with those found experimentally. No measurements on the nitrogen spectrum are made, but the features of the spectrum are carefully noted.

Practical applications. The spectrograph has been one of the most useful tools in the advancement of science, particularly in the fields of chemistry, physics, and astronomy. With it, most of the elements and many compounds may be identified and a quantitative analysis obtained, even with minute quantities. It has aided in establishing the structure of organic compounds. It has been responsible for the discovery of many of our elements. It has made possible a determination of the composition and temperature of the sun and stars. Even the velocities of some of the stars have been calculated with its help. Intelligent advances in photochemistry demand a thorough knowledge of absorption spectra, and a spectrometer furnishes the best source of monochromatic illumination for controlled experiments in that branch of physical chemistry.

The nature of the absorption spectrum, whether continuous or discontinuous, is of value in interpreting the mechanism of the molecular absorption and the nature of certain photochemical reactions.

Suggestions for further work. The absorption spectrum of potassium permanganate or of a dye solution may be obtained by placing an absorption cell in front of the slit and illuminating it with a small frosted electric light bulb. Potassium permanganate and especially salts of neodymium and praesodymium give fairly sharp bands in sufficiently dilute solutions.

Several suitable experiments on band spectra of diatomic molecules have been suggested by Davies.[4]

References

1. W. R. Brode, "Chemical Spectroscopy," John Wiley & Sons, Inc., New York, 1947.
2. G. R. Harrison, R. C. Lord, and J. R. Loofbourow, "Practical Spectroscopy," Prentice-Hall, Inc., Englewood Cliffs, N.J., 1948.
3. G. Herzberg, "Atomic Spectra and Atomic Structure," 2d ed., Dover Publications, Inc., New York, 1944.
4. M. Davies, *J. Chem. Educ.*, **28:** 474 (1951).

Chapter
12

Molecular
Spectroscopy

HETERONUCLEAR DIATOMIC MOLECULE

The use of physical methods for the determination of molecular structure continues to grow in importance as improvements in apparatus facilitate more accurate and detailed observations. In this experiment an infrared spectrometer is used in the study of the rotation-vibration energy-level scheme for a diatomic molecule. The use of isotopic substitution is illustrated.

THEORY.[1-3]   The determination of the characteristic energy levels for a diatomic molecule such as hydrogen chloride requires consideration of a complex dynamical system of electrically interacting particles consisting of two nuclei and a large number of electrons. The quantum-mechanical problem is simplified by acceptance of the principle known as the *Born-Oppenheimer approximation.* Because of the marked difference in mass between the electron and even the smallest nucleus, the electrons in a molecule complete many cycles of their motion in any time interval long enough to permit the more slowly moving nuclei to achieve a finite change in their positions. It is thus possible to consider the electron motion as establishing a potential function which determines the motion of the nuclei. This electronic potential function $U(r)$ may be constructed by solving the Schrödinger wave equation for the electron motion for each of a continuous succession of fixed configurations of the nuclei, corresponding in turn to different values of r, the internuclear distance. A typical potential-energy curve for the ground electronic-energy state of a stable diatomic molecule is shown in Fig. 60. For this curve the value $U(r)$ corresponding to a particular internuclear distance r is found as the lowest energy value for the system consistent with the Schrödinger wave equation for the motion of the electrons for fixed distance r between the nuclei. In principle, then, the potential-energy curve $U(r)$ can be calculated theoretically for a molecule, but in practice its form must be established experimentally except for the simplest cases, such as H_2. It should be noted, however, that the function $U(r)$ will be the same for different isotopic species of the same compound, such as $H^{35}Cl$ and $H^{37}Cl$.

The determination of the permitted energy levels ultimately reduces to finding the values of W for which the radial equation

$$\frac{1}{r^2}\frac{d}{dr}\left[r^2\frac{dR(r)}{dr}\right] + \frac{8\pi^2\mu}{h^2}\left[W - \frac{J(J+1)h^2}{8\pi^2\mu r^2} - U(r)\right]R(r) \equiv 0 \qquad (1)$$

admits of solutions for $R(r)$, the radial factor in the wave function for the system, which are single-valued, continuous, quadratically integrable, etc., as required for the interpretation of wave functions as probability distribution functions. In this equation μ, the *reduced mass* of the two-particle system, is calculated from the

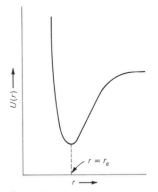

figure 60.

Potential-energy curve for ground electronic state of a stable diatomic molecule. r_e is the equilibrium internuclear distance in the molecule.

masses M_1, M_2 of the two atoms as

$$\frac{1}{\mu} = \frac{1}{M_1} + \frac{1}{M_2} \tag{2}$$

The quantity $h^2 J(J+1)/8\pi^2\mu r^2$, in which J is a quantum number which takes on the values 0, 1, 2, 3, etc., may be regarded as a centrifugal-energy term arising from the rotational motion of the system; the product μr^2 is the moment of inertia of the system for internuclear distance r. Since the precise form of the function $U(r)$ is not known, it is convenient to introduce a Taylor series expansion about the equilibrium internuclear distance $r = r_e$, corresponding to the minimum in $U(r)$, for both it and the centrifugal-energy term. For a function $f(x)$ the expansion about the point $x = x_0$ is given by

$$f(x) = f(x_0) + f'(x_0)(x - x_0) + \frac{1}{2!}f''(x_0)(x - x_0)^2$$

$$+ \frac{1}{3!}f'''(x_0)(x - x_0)^3 + \frac{1}{4!}f''''(x_0)(x - x_0)^4 + \cdots \tag{3}$$

It follows that for $r - r_e = \xi$,

$$U_J \equiv U(r) + \frac{h^2 J(J+1)}{8\pi^2\mu}\frac{1}{r^2} = U(r_e) + \frac{h^2 J(J+1)}{8\pi^2\mu r_e^2}$$

$$+ 2\pi^2\mu\nu_e^2\xi^2 + a\xi^3 + b\xi^4 - \frac{h^2 J(J+1)}{4\pi^2\mu_e^3}\xi + \frac{3h^2 J(J+1)}{8\pi^2\mu r_e^4}\xi^2 + \cdots \tag{4}$$

where the frequency ν_e is defined by the relation

$$2\pi^2\mu\nu_e^2 = \frac{1}{2}\left[\frac{d^2 U(r)}{dr^2}\right]_{r=r_e}$$

and the coefficients a, b by

$$a = \frac{1}{3!}\left[\frac{d^3 U(r)}{dr^3}\right]_{r=r_e} \qquad b = \frac{1}{4!}\left[\frac{d^4 U(r)}{dr^4}\right]_{r=r_e}$$

Since the minimum in $U(r)$ occurs at $r = r_e$,

$$\left[\frac{dU(r)}{dr}\right]_{r=r_e} \equiv 0$$

In terms of the variable ξ, and the substitution $rR(r) = S(\xi)$, Eqs. (1) and (4) lead to

$$\frac{d^2 S(\xi)}{d\xi^2} + \frac{8\pi^2\mu}{h^2}(W - U_J)S(\xi) \equiv 0 \tag{5}$$

If the series expansion (4) for U_J is limited to the first three terms, the wave equation (5) can be solved exactly to give for the energy levels the expression

$$W \text{ (ergs)} = U(r_e) + \frac{h^2 J(J+1)}{8\pi^2 \mu r_e^2} + h\nu_e(v + \tfrac{1}{2}) \tag{6}$$

The first term simply fixes a reference point for measurement of energy for the molecule. The second is the quantum-mechanical expression appropriate for the rotational energy levels of a *rigid* two-particle system for fixed internuclear distance r_e. The third term, in which the vibrational quantum number v can have the values 0, 1, 2, 3, . . . , represents the energy levels of a one-dimensional harmonic oscillator. Corresponding to this energy-level formula, the functions $S(\xi)$ are identical in form with those for the one-dimensional harmonic oscillator.

The relation (6) provides a first approximation to the desired result, but to obtain an expression which can accurately characterize the energy-level system for a real molecule, the effect of the other terms in series (4) must be determined. On the assumption that these terms are reasonably small, so that the energy levels for the system will lie close to those given by the rigid-rotator harmonic-oscillator approximation, standard methods of perturbation theory may be employed to give the following result:

$$W_{v,J,}(\text{cm}^{-1}) = \frac{1}{hc} W_{v,J}(\text{ergs}) = E(r_e) + \tilde{\nu}_e(v + \tfrac{1}{2}) + B_e J(J+1)$$

$$- x_e \tilde{\nu}_e(v + \tfrac{1}{2})^2 - \alpha_e(v + \tfrac{1}{2})J(J+1) - D_e J^2(J+1)^2 \tag{7}$$

where c is the velocity of light and where

$$\tilde{\nu}_e = \frac{\nu_e}{c}$$

$$E(r_e) = \frac{U(r_e)}{hc} + \frac{3bh}{128\pi^4\mu^2 c^3 \tilde{\nu}_e^2} - \frac{7a^2 h}{1024\pi^6\mu^3 c^5 \tilde{\nu}_e^4}$$

$$B_e = \frac{h}{8\pi^2 c\mu r_e^2} \qquad D_e = \frac{h^3}{128\pi^6\mu^3 r_e^6 c^3 \tilde{\nu}_e^2} = \frac{4B_e^3}{\tilde{\nu}_e^2}$$

$$x_e = \frac{3}{2}\frac{h}{16\pi^4\mu^2 c^3 \tilde{\nu}_e^3}\left(\frac{5}{2}\frac{a^2}{4\pi^2\mu c^2 \tilde{\nu}_e^2} - b\right)$$

$$\alpha_e = \frac{3h^2}{32\pi^4\mu^2 r_e^4 c^2 \tilde{\nu}_e}\left(\frac{-2ar_e}{4\pi^2\mu c^2 \tilde{\nu}_e^2} - 1\right)$$

The term in x_e reflects the anharmonicity of the vibrational motion; that in α_e is a vibration-rotation interaction term which may be associated with the change in the effective value of the moment of inertia of the molecule as the vibrational energy is increased. The quantity D_e is called the *centrifugal-distortion coefficient*, since it appears because of the tendency of the internuclear distance for the nonrigid molecule to increase as the rotational energy increases.

The expression (7) for the energy levels is not exact. In a higher approximation there would appear terms proportional to $(v + \frac{1}{2})^3$, etc., and further vibration-rotation interaction effects such as a dependency of the centrifugal-distortion coefficient on the vibrational quantum number. Such refinements are rarely justified by the accuracy of the spectroscopic data available.

Defining the rotational coefficient B_v corresponding to value v for the vibrational quantum number as

$$B_v = B_e - \alpha_e(v + \tfrac{1}{2}) \tag{8}$$

one obtains

$$W_{v,J}(\text{cm}^{-1}) = E(r_e) + \tilde{\nu}_e(v + \tfrac{1}{2}) - x_e\tilde{\nu}_e(v + \tfrac{1}{2})^2 + B_v J(J + 1) - D_e J^2(J + 1)^2 \tag{9}$$

Consider the difference in energy of the states characterized by $v = v''$, $J = J''$ and $v = v'$, $J = J'$, respectively,

$$W' = E(r_e) + (v' + \tfrac{1}{2})\tilde{\nu}_e - x_e\tilde{\nu}_e(v' + \tfrac{1}{2})^2 + B_{v'}J'(J' + 1) - D_e J'^2(J' + 1)^2$$

$$W'' = E(r_e) + (v'' + \tfrac{1}{2})\tilde{\nu}_e - x_e\tilde{\nu}_e(v'' + \tfrac{1}{2})^2 + B_{v''}J''(J'' + 1) - D_e J''^2(J'' + 1)^2$$

$$\Delta W = W' - W'' = (v' - v'')\tilde{\nu}_e - x_e\tilde{\nu}_e(v' - v'')(v' + v'' + 1)$$

$$+ B_{v'}J'(J' + 1) - B_{v''}J''(J'' + 1)$$

$$- D_e[J'^2(J' + 1)^2 - J''^2(J'' + 1)^2] \tag{10}$$

A transition between these two states may be accomplished by absorption of a photon of wave number $\tilde{\nu}$ such that $\tilde{\nu} = \Delta W$, *provided* that:

1. The molecule is heteronuclear, so that vibrational or rotational motion of the molecule creates an oscillating electric moment.
2. The rotational quantum number J changes so that $J' = J'' \pm 1$.

There is no restriction on the change in the vibrational quantum number v. The principal near-infrared rotation-vibration absorption band occurs for the so-called *fundamental transition*, for which $v'' = 0$ (ground vibrational state) and $v' = 1$ (first excited vibrational state). The first overtone band ($v'' = 0$, $v' = 2$) and the second overtone band ($v'' = 0$, $v' = 3$), etc., are found to be progressively weaker.

For a given change in vibrational quantum number, the set of absorption frequencies corresponding to transitions in which the rotational quantum number J changes by $+1$ constitutes the R branch of the rotation-vibration absorption band;

the set for which $\Delta J = -1$ constitutes the P branch. The absorption frequencies for the two branches may be summarized as follows:

R branch:

$$J' = J'' + 1 \qquad J'' = 0, 1, 2, 3, \ldots$$

$$\tilde{\nu}_R = (v' - v'')\tilde{\nu}_e - x_e\tilde{\nu}_e[(v' - v'')(v' + v'' + 1)] + (B_{v'} + B_{v''})(J'' + 1)$$

$$+ (B_{v'} - B_{v''})(J'' + 1)^2 - 4D_e(J'' + 1)^3 \quad (11)$$

P branch:

$$J' = J'' - 1 \qquad J'' = 1, 2, 3, \ldots$$

$$\tilde{\nu}_P = (v' - v'')\tilde{\nu}_e - x_e\tilde{\nu}_e[(v' - v'')(v' + v'' + 1)] - J''(B_{v'} + B_{v''})$$

$$+ J''^2(B_{v'} - B_{v''}) + 4D_eJ''^3 \quad (12)$$

Equations (11) and (12) can be combined in appropriate ways to facilitate the calculation of the parameters characterizing the energy-level scheme for the molecule.

Now let $R(J)$, $P(J)$ represent, respectively, the wave numbers of the lines in the R and P branches which *originate* in the state characterized by $J'' = J$. Then from Eqs. (11) and (12) it follows that

$$\tfrac{1}{2}[R(J) + P(J + 1)] = (v' - v'')\tilde{\nu}_e[1 - x_e(v' + v'' + 1)]$$

$$+ (B_{v'} - B_{v''})(J + 1)^2 \qquad J = 0, 1, 2, \ldots \quad (13)$$

$$R(J) - P(J) = 4B_{v'}(J + \tfrac{1}{2}) - 4D_e[(J + 1)^3 + J^3] \qquad J = 1, 2, 3, \ldots \quad (14)$$

$$R(J - 1) - P(J + 1) = 4B_{v''}(J + \tfrac{1}{2}) - 4D_e[(J + 1)^3 + J^3]$$

$$J = 1, 2, 3, \ldots \quad (15)$$

Note that the difference $R(J) - P(J)$ in Eq. (14) is selected because both lines then originate in transitions from identically the same lower state, i.e., same values for v'', J'', and hence the properties of the upper state can be determined without a knowledge of the coefficients for the lower state. Conversely, the transitions associated with the lines $R(J - 1)$, $P(J + 1)$ lead to the same *upper* state. For a given band the values for the band origin, $(v' - v'')\tilde{\nu}_e[1 - x_e(v' + v'' + 1)]$, and $B_{v'} - B_{v''}$ may be determined as the intercept and slope, respectively, of a plot of $\tfrac{1}{2}[R(J) + P(J + 1)]$ against $(J + 1)^2$. A graphical method based on Eqs. (14) and (15) may then be used for evaluation of $B_{v'}$, $B_{v''}$, and D_e. A plot of $[R(J) - P(J)]/(J + \tfrac{1}{2})$ versus $J^2 + J + 1$ leads to values for $B_{v'}$ and D_e, and a similar plot based on Eq. (16) yields values for $B_{v''}$ and D_e. The two values thus obtained for

D_e should agree, and the difference $B_{v'} - B_{v''}$ should be consistent with the result given by use of Eq. (13).

Because the rotational quantum number must change for an allowed transition, there is a discontinuity in the line spacing at the center of the band. For the harmonic-oscillator, rigid-rotator model, this so-called *zero gap* would be exactly twice the common line spacing in the P and R branches, which in turn is twice the common value (in this approximation) of the rotational coefficient B for the two vibrational states involved. For an actual molecule, the inevitable difference in the values of the rotational constant for the two vibrational states involved results in different line spacings in the P and R branches. The influence of the centrifugal-distortion term is more difficult to demonstrate, as it will be masked by experimental error unless very accurate measurements have been made.

····→ **Apparatus.** Infrared spectrometer; precision rule; gas tanks (HCl, DCl, etc.) with appropriate valves; spectrometer cells with NaCl windows; cell-filling manifold.

PROCEDURE.[4]   Specialized research spectrometers permit the determination of the wavelengths of emission or absorption lines in atomic and molecular spectra with accuracy measured in parts per million or better, at wavelengths from the ultraviolet to the infrared region of the spectrum. At the present time, however, even standard commercial instruments can provide results whose interpretation can be accomplished quantitatively only in terms of the energy-level scheme defined by Eq. (7).

The basic features of an infrared spectrometer are discussed in Part Two, pages 553 to 558. The operating instructions for the particular instrument to be used in the laboratory should be carefully read and thoroughly understood before any work is attempted.

The spectrometer cell is taken from its storage desiccator, and the windows are checked for transparency and freedom from cracks. If they have become clouded through attack by moisture, they must be repolished before use.† The cell is attached to the filling manifold (Fig. 61), as is the hydrogen chloride tank. Dewar flasks filled with liquid nitrogen are placed about the cold traps; the second trap is recommended simply as an additional insurance element in providing protection for the vacuum pump. The main tank valve is opened, with the metering valve M closed. The pump is then turned on to evacuate the manifold and cell. Stopcocks P and C are now closed; leakage in the manifold proper can be detected by the response of the manometer. If the manifold proves gastight, stopcock C is opened to permit a similar check of the cell. After any leaks detected have been eliminated *with the assistance of the staff,* the cell and manifold are reevacuated, stopcock P is

† This procedure can be simplified by use of a kit available from the Beckman Instrument Co.

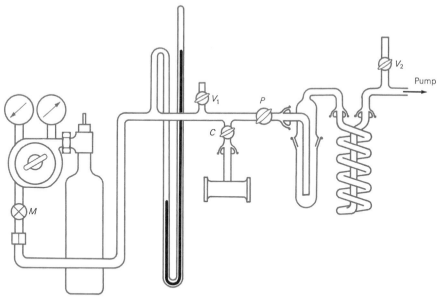

figure 61. Cell-filling manifold.

closed, and valve M is opened *slightly* to admit gas to a pressure of a few centimeters of mercury. Stopcock P is reopened to pump out the flushing gas and then
closed. The cell is now filled to the appropriate pressure (as specified for the particular cell-spectrometer combination in use), after which valve M and stopcock C
are closed. The manifold is evacuated again then, with stopcock P closed, vented
through V_1. The vacuum pump is turned off and vented through stopcock V_2. The
Dewar flasks are removed from around the cold traps, which are then disconnected
and transferred immediately to a hood to warm up.

The cell is now removed from the manifold and put in its holder on the
spectrometer. The wavelength range of interest is located by rapid scan, and the
absorption band of interest is recorded at slow scan rate. After the desired results
have been obtained, the cell is reconnected to the manifold, and the cold traps are
replaced. The cell can then be evacuated, flushed with air, and returned to its
storage desiccator or else refilled after replacement of the hydrogen chloride tank
with one of deuterium chloride.† Proceeding in the above fashion, records such as
those shown in Figs. 62 and 63 are obtained.

The typical infrared spectrometer is designed to give a spectrum presentation
linear either in wavelength or wave number. By linear interpretation the position of

† Available, for example, from the Matheson Co., Joliet, Ill.

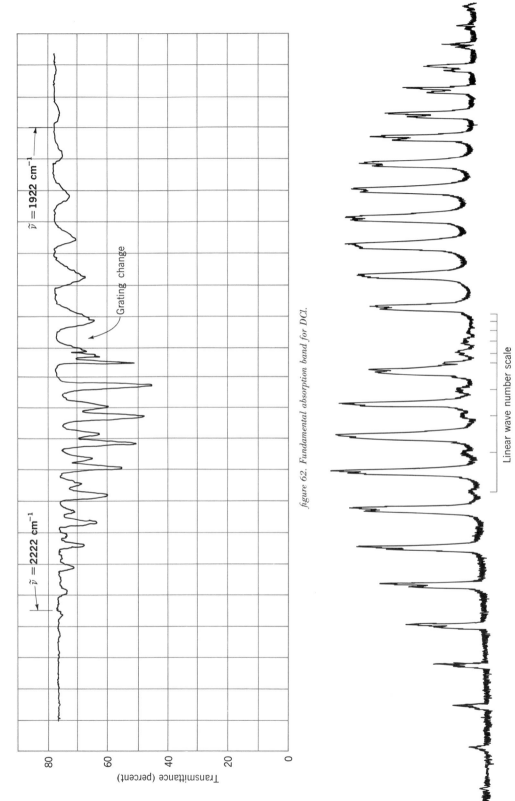

figure 62. Fundamental absorption band for DCl.

figure 63. Fundamental absorption band for HCl.

each of the lines in a given absorption band is determined. For the P and R branches separate tabulations are then made of line *wave number* versus the rotational quantum number J for the state in which the transition originates.

CALCULATIONS. Assuming that the measurements have been made on the fundamental absorption band, for data of moderate accuracy the molecular parameters of interest are calculated as follows. The rotational constant B_1 for the first excited vibrational level is calculated as the average of the set of values obtained for $[R(J) - P(J)]/4(J + \frac{1}{2})$. The value of B_0 is determined in similar fashion as the average of the set of values of $[R(J - 1) - P(J + 1)]/4(J + \frac{1}{2})$. The band origin, here equal to $\bar{\nu}_e(1 - 2x_e)$, is obtained as the intercept of a plot versus $(J + \frac{1}{2})^2$ of $[R(J) + P(J + 1)]/2$; the value of $(B_1 - B_0)$ given by the slope of this plot should be compared with that yielded by the values of B_1 and B_0 obtained previously. The feasibility of a successful estimation of the centrifugal distortion coefficient D_e from the data available is considered, and the calculation made if appropriate. Note that a rough prediction of the value of D_e can be made from the relation given with Eq. (7), using B_0 and the band origin for the fundamental band as approximations to B_e and $\bar{\nu}_e$, respectively.

The values of B_e and α_e can now be calculated; supplementary data for this purpose are given in Table 1. From the value of B_e the equilibrium internuclear distance r_e is obtained; Table 2 provides values of the necessary isotopic masses. For the calculation of $\bar{\nu}_e$ and $x_e\bar{\nu}_e$ the band origin must be known for more than one band. The additional data required may be taken from Table 1.

The results obtained should verify that the value of r_e is the same for the different isotopic species (within limits of experimental uncertainty), and that the zero-order vibrational frequencies satisfy the relation

$$\frac{\bar{\nu}_{e,\text{H}^{35}\text{Cl}}}{\bar{\nu}_{e,\text{D}^{35}\text{Cl}}} = \left(\frac{\mu_{\text{D}^{35}\text{Cl}}}{\mu_{\text{H}^{35}\text{Cl}}}\right)^{\frac{1}{2}} \tag{16}$$

Table 1. Supplementary Spectroscopic Results for $H^{35}Cl$, $D^{35}Cl$[a]

	$H^{35}Cl$	$D^{35}Cl$
Band origins:		
2-0 band	5667.98 cm^{-1}	4128.43 cm^{-1}
3-0 band	8346.78	
Rotational constants:		
B_2	9.835	5.168
B_3	9.535	

[a] Based on data given in Refs. 9 and 10.

Table 2. Isotopic Masses[a]

Isotope	Mass	Isotope	Mass
^{1}H	1.007825	^{35}Cl	34.9689
^{2}H	2.014102	^{37}Cl	36.9659
^{12}C	12.000000	^{79}Br	78.9183
^{16}O	15.9949	^{81}Br	80.9163
^{19}F	18.9984	^{127}I	126.9045
^{32}S	31.9721		

[a] Based on summary given in Ref. 8.

Equation (16) is a necessary consequence of the fact that the electronic potential function $U(r)$, and hence $[d^2U(r)/dr^2]_{r=r_e}$, is the same for the two isotopic species.

Practical applications. An accurate knowledge of the energy-level system for a molecule is essential in the calculation of thermodynamic properties by the methods of statistical thermodynamics.[5,6] The velocity of light may be calculated from B_0 values as given in terms of frequency by microwave spectroscopy and in wave number by infrared studies.[7]

Suggestions for further work. A general check of the spectrometer calibration may be made by use of accurate spectral data.[11,12] Measurements may be extended to HBr, DBr or HI, DI, or other diatomic molecules.

References

1. L. D. Landau and E. M. Lifschitz, "Quantum Mechanics," Addison-Wesley Publishing Company, Inc., Reading, Mass., 1958.
2. G. W. King, "Spectroscopy and Molecular Structure," Holt, Rinehart and Winston, Inc., New York, 1964.
3. G. Herzberg, "Molecular Spectra and Molecular Structure," vol. I, "Spectra of Diatomic Molecules," 2d ed., D. Van Nostrand Company., Inc., Princeton, N.J., 1950.
4. A. H. Nielsen in D. Williams (ed.), "Methods of Experimental Physics," vol. 3, Academic Press Inc., New York, 1964.
5. J. G. Aston and J. J. Fritz, "Thermodynamics and Statistical Mechanics," John Wiley & Sons, Inc., New York, 1959.
6. R. E. Pennington and K. A. Kobe, *J. Chem. Phys.,* **22:** 1442 (1954).
7. E. K. Plyler, L. R. Blaine, and W. S. Connor, *J. Opt. Soc. Am.,* **45:** 102 (1955).
8. D. H. Rank, W. B. Birtley, D. P. Eastman, B. S. Rao, and T. A. Wiggins, *J. Opt. Soc. Am.,* **50:** 1275 (1960).
9. D. H. Rank, D. P. Eastman, B. S. Rao, and T. A. Wiggins, *J. Opt. Soc. Am.,* **52:** 1 (1962).
10. R. W. Kiser, "Introduction to Mass Spectrometry and Its Applications," Prentice-Hall, Inc., Englewood Cliffs, N.J., 1965.
11. K. N. Rao, C. J. Humphreys, and D. H. Rank, "Wavelength Standards in the Infra-red," Academic Press Inc., New York, 1966.
12. "Tables of Wave Numbers for Calibration of Infra-red Spectrometers," *Pure Appl. Chem.,* **1:** 537 (1961).

40 INFRARED AND RAMAN SPECTRA OF TRIATOMIC MOLECULES

The complementary nature of infrared and Raman spectra and the influence of molecular symmetry on selection rules in rotation-vibration spectra are illustrated through studies on nonlinear and linear triatomic molecules.

THEORY.[1-4] In many cases it is possible to interpret usefully the infrared and Raman spectra of a polyatomic molecule in terms of a model which resolves the molecular energy into additive electronic, translational, and rotation-vibration contributions. Since at ordinary temperatures only a negligible fraction of the molecules will be in excited electronic states, ordinary observations reflect the properties of molecules in the ground electronic-energy state. It is changes in rotational-vibrational energy for these molecules which are detected in the molecular spectra considered here.

A further approximation step introduces the assumption that the rotation-vibration contribution to the energy can itself be written as the sum of a rotational energy characteristic of the molecule as a rigid body and a separate vibrational energy, with no rotation-vibration interaction. It should be recognized that even exact calculations for such a model yield at most a first approximation to the properties of a real molecule. A more complete discussion is given in Experiment 39 (page 247).

For a rigid linear triatomic molecule the expression for the permitted rotational energy values is identical in form to that for the rigid diatomic rotator,

$$W_{\text{rot}}(\text{cm}^{-1}) = \frac{h}{8\pi^2 cI} J(J+1) = BJ(J+1) \tag{1}$$

where I = moment of inertia of molecule
c = velocity of light
B = rotational constant

A nonlinear triatomic molecule is in general an asymmetric top for which no such simple formula can be derived, and the determination of the permitted energy values is a difficult process.

In considering the vibrational-energy problem, note first that for a set of three points the configuration in space at any time can be defined exactly by giving the three cartesian coordinates for each of the points or by fixing values of nine independent combinations of these coordinates from which the latter can be determined uniquely. Three such combinations are required to fix the position of the center of mass in space. To fix the orientation of the equilibrium configuration of the molecule two more are needed for the linear case and three more for the nonlinear case.

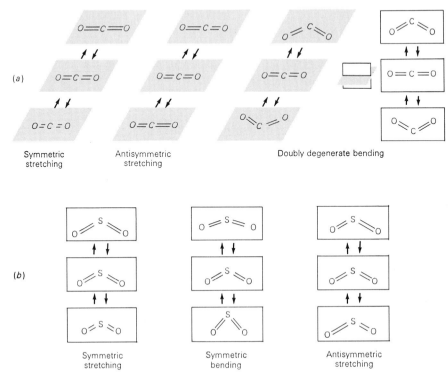

figure 64. *Normal modes of vibration for* (a) *carbon dioxide and* (b) *sulfur dioxide.*

It follows that any possible vibrational motion of the molecule can be described in terms of four independent combinations of atom displacements for the linear case and three for the nonlinear case. Each such combination defines a *normal mode of vibration* for the molecule, and accounts for one *degree of freedom* for vibration. Any possible vibrational motion corresponds to a superposition, with appropriate amplitudes and phases, of the set of normal modes for the molecule. Sets of normal modes for CO_2 and SO_2 are shown in Fig. 64.

Prediction of the vibrational-energy scheme for a molecule requires specification of the potential-energy function governing the vibrational motion. The forces in a molecule act primarily to maintain bond lengths and interbond angles at characteristic values, and for small-amplitude vibrations the vibrational potential energy can be expressed to a very useful degree of approximation as a quadratic function of the so-called *internal displacement coordinates,* which are simply changes in bond lengths and angles in the molecule. In addition, for small-amplitude vibrations the internal-displacement coordinates are linearly related to the changes in the cartesian coordinates of the atoms, which simplifies accounting for the kinetic energy associated with the vibrational motion. The Schrödinger wave equation can then be solved through a change of variables from the internal-displacement coordinates to a set of appropriate linear combinations, called *normal coordinates*, which characterize the normal modes of vibration. The corresponding expression for the vibrational energy is

$$W_{vib}(cm^{-1}) = \sum_{i=1}^{\phi} (v_i + \tfrac{1}{2})h\bar{\nu}_i \qquad v_i = 0, 1, 2, 3, \ldots$$

where ϕ is the number of degrees of freedom for vibration. For a nonlinear molecule $\phi = 3n - 6$, n being the number of atoms present; for a linear molecule $\phi = 3n - 5$.

Here v_i is the vibrational quantum number for the ith degree of freedom (or ith normal mode). The quantity $\bar{\nu}_i$ represents the characteristic vibrational frequency (expressed in wave numbers) for the normal mode. The determination of the set of such frequencies for the molecule, and the assignment of each frequency to the appropriate normal mode, is one of the goals of molecular spectroscopy. Commonly there is less resistance to change of interbond angle than to change of bond length, so that bending modes typically are characterized by lower frequencies than are stretching modes.

Inspection of Fig. 64 will show that an oscillating electric moment is conferred on the CO_2 molecule by the asymmetric stretching mode ν_3 and the doubly degenerate bending mode ν_2 but not by the symmetric stretching vibration ν_1. In accordance with the basic selection rule for the electric-dipole spectrum (page 267) the fundamental vibrational transitions (page 250) will be permitted in the infrared absorption spectrum for the two former modes but *not* for the symmetric stretching mode. While it is not so obvious, it is further true that the polarizability of the CO_2 molecule is, to the small-vibration approximation, unaltered by the asymmetric stretching and bending modes but is changed by the symmetric stretching mode. Only for the symmetric stretching mode, then, is the fundamental transition Raman active. Here again is an example of the mutual-exclusion rule, for a molecule with a center of symmetry, discussed in Experiment 41, page 273.

The rotational fine structure of the permitted infrared bands is determined by selections rules governed by the direction of the oscillating electric moment due to the vibrational motion. For the asymmetric stretching mode the transition moment (oscillating electric moment) is parallel to the symmetry axis of the molecule. In this case the rotational selection rule is $\Delta J = \pm 1$, and the absorption band, which is termed a *parallel band*, has a P branch ($\Delta J = -1$) and an R branch ($\Delta J = +1$) with a pronounced central minimum. For the degenerate bending mode the transition moment is perpendicular to the symmetry axis of the equilibrium configuration of the molecule. In this case the selection rule gives $\Delta J = 0$ or ± 1. The resulting absorption band, termed a *perpendicular band*, then has a strong central maximum, a Q branch, resulting from superposition of transitions for which $\Delta J = 0$ for various values of J; P and R branches are present as well. Even where the fine structure is not resolved, the band contours can permit differentiation between perpendicular and parallel bands of linear molecules and thus facilitate assignment of frequencies to normal modes of vibration.

It should be remarked that the perpendicular band for the fundamental transition for the bending mode of CO_2 has superimposed on it a set of other (weaker) perpendicular bands due to transitions originating in excited vibrational states rather than the ground state. For the interpretation of the observed results[5] it is necessary to take into account the Fermi resonance effect, described below.

In the case of SO_2, it will be noted that all three normal modes will produce the periodic variation in the dipole moment of the molecule required for infrared activity, and it is also true that all three will be Raman active. It will be noted that the transition moment due to the vibrational motion is parallel to the C_2 symmetry axis for the bending mode and the symmetric stretching mode. Since this is the direction of the principal axis of inertia (page 277) of intermediate moment of inertia, the rotational selection rules result in what is termed a type B band for this asymmetric-top molecule; the type B band has a complex fine structure, but it does not have a strong central branch. The antisymmetric stretching mode produces an oscillating electric moment which lies in the plane of the molecule and is perpendicular to the C_2 axis. Its direction therefore is that of the principal axis of inertia of least moment. The type A band which results is again of complicated fine structure but, in contrast to the type B band, will often show a moderately strong central maximum with adjacent maxima on either side. The proper classification of an unresolved band for an asymmetric-top molecule will often be difficult to accomplish, however.

The only Raman active fundamental transition for CO_2 is that for the totally symmetric stretching mode. Actually two closely spaced lines are observed instead of one, because it just happens that for this molecule ν_1 is almost precisely equal to $2\nu_2$. There is then an accidental degeneracy between the first excited level for the symmetric stretching mode and the first overtone level for the bending mode. A strong second-order perturbation interaction occurs which accounts for the observed results. This phenomenon is often termed *Fermi resonance*, since it was first explained by Fermi.

In the Raman spectrum of SO_2, all three normal modes are active. Accurate depolarization factor measurements (page 266) can identify the lines associated with the totally symmetric stretching and bending modes, since for them the depolarization factor ρ will be less than $\frac{6}{7}$, while for the asymmetric stretching mode it will be equal to $\frac{6}{7}$.

····→ ***Apparatus.*** Infrared spectrometer; spectrometer gas cell with KBr windows; cell-filling manifold; Raman spectrometer; Raman sample tube; photographic plates and supplies; CO_2 and SO_2 cylinders with reducing valves.

PROCEDURE.   The infrared absorption spectra are obtained for CO_2 and SO_2 in accordance with the procedure described previously (page 252). The appropriate filling pressures for the cell must be specified for the particular cell-spectrometer combination used. While NaCl windows are satisfactory for work with CO_2, KBr windows must be used to permit detection of the lowest-frequency fundamental absorption band for SO_2. Results such as shown in Figs. 65 and 66 should be obtained.

It is, of course, preferable to record the Raman spectra for the compounds in the gas phase; this requires, however, a multitraversal gas Raman tube[6] and a spectrograph of high light-gathering power if unreasonable exposure times are to be avoided. A simpler procedure is to obtain the spectra from the liquid phase, if feasible, as described in Experiment 41. For this purpose an all-glass Raman tube, as shown in Fig. 67a, is suitable for SO_2. Since the vapor pressure of SO_2 is about 5 atm at the temperature (about 30°C) which might be reached in the excitation unit, the tube must be of adequate strength, properly annealed, and carefully handled to avoid scratches. The tube can be filled by use of the gas manifold employed with the infrared spectrometer cells. The SO_2 is frozen out with a Dry Ice cooling mixture; the gas-metering valve is then closed and the Raman tube sealed off by someone *with adequate experience in the kind of operation involved.* In filling

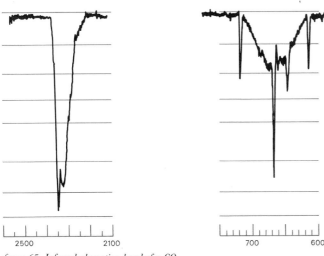

2500   2100   700   600

figure 65. Infrared absorption bands for CO_2.

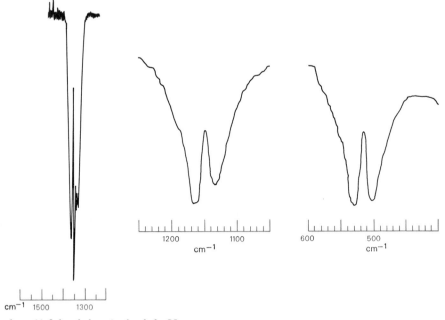

figure 66. Infrared absorption bands for SO₂.

the tube, account must be taken of the thermal expansion of the medium as it warms up to the temperature it will reach in the excitation unit. An alternative Raman tube design which offers some advantages for use with media such as liquid SO_2 is shown in Fig. 67*b*.

Caution: Extreme care should be taken in handling Raman tubes filled with liquefied gases such as SO_2. Safety glasses or a protective face shield should be worn at all times.

For liquid CO_2 the vapor pressure at room temperature is nearly 60 atm, which makes its confinement in glass particularly hazardous. Unless facilities for obtaining the Raman spectrum of the gas are available, Raman data for this compound should be taken from the literature.

The carbon disulfide molecule also is a linear symmetric molecule and can provide results comparable with those for CO_2, including the complication of Fermi resonance. This phenomenon is clearly reflected in the Raman spectrum, which can easily be obtained from liquid CS_2. It is necessary to have an infrared spec-

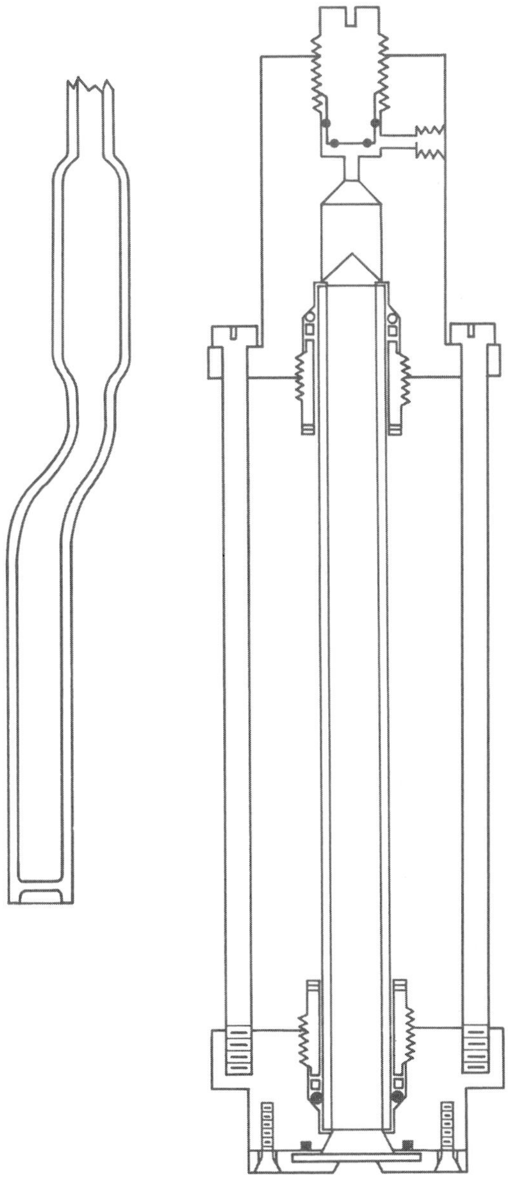

(a)   (b)

figure 67. (a) Glass Raman tube for liquid SO₂; (b) Raman tube assembly for high-pressure samples.

trometer whose range extends below 400 cm^{-1} to reach the lowest-frequency funda-mental absorption band of CS_2.

Caution: Carbon disulfide is a very volatile liquid with an extremely low flash point. It must be handled with extreme care, to avoid fire and explosion.

The exposure times appropriate for the particular spectrograph–excitation unit combination employed should be specified by the laboratory staff. The procedure in obtaining the Raman spectra otherwise corresponds to that described for Experi-ment 41. The crystal-violet–*p*-nitrotoluene filter solution recommended there will serve also to protect the SO_2 and CS_2 from photochemical degradation by ultraviolet light.

CALCULATIONS.   The wave numbers at the various band centers are tabulated separately for CO_2 and SO_2, together with a description of the band contour. The Raman lines are identified and their frequency shifts (in wave numbers) are determined as described on page 273. Assignments are then made of frequencies to normal modes of vibration from the two molecules and of the origin of any combi-nation or overtone bands detected (page 266). The experimental results are com-pared with data from the literature.[5,7,8]

Practical applications. Infrared and Raman spectra are basic tools in determining the fundamental vibrational frequencies of a molecule, the nature of the vibrational potential func-tion, and structural parameters such as bond lengths, etc.

Suggestions for further work. The search for combination and overtone bands may be expedited by recording spectra at higher gas pressures than appropriate for observation of the fundamental absorption bands. The infrared absorption spectrum of acetonitrile vapor may be studied; the structure of the band at 1059 cm^{-1} is particularly interesting.[9] A depolarization factor study may be made for SO_2, as described in Experiment 41.

References

1. N. B. Colthup, L. H. Daly, and S. E. Wiberly, "Introduction to Infra-red and Raman Spectros-copy," Academic Press Inc., New York, 1964.
2. J. C. D. Brand and J. C. Speakman, "Molecular Structure; The Physical Approach," Edward Arnold (Publishers) Ltd., London, 1960.
3. A. H. Nielsen in D. Williams (ed.), "Methods of Experimental Physics," vol. 3, Academic Press Inc., New York, 1964.
4. B. P. Stoicheff in D. Williams (ed.), "Methods of Experimental Physics," vol. 3, Academic Press Inc., New York, 1964.
5. P. E. Martin and E. F. Barker, *Phys. Rev.*, **41:** 291 (1932).
6. H. L. Welsh, C. Cumming, and E. J. Stansbury, *J. Opt. Soc. Am.*, **41:** 712 (1951).
7. J. M. Taylor, W. S. Benedict, and J. Strong, *J. Chem. Phys.*, **20:** 1884 (1952).
8. R. D. Shelton, A. H. Nielsen, and W. H. Fletcher, *J. Chem. Phys.*, **21:** 2178 (1953).
9. P. Venkateswarlu, *J. Chem. Phys.*, **19:** 293 (1951).

41 RAMAN SPECTRA OF POLYATOMIC MOLECULES

The Raman spectra of several compounds are obtained. The Raman frequency shifts are measured and compared with the infrared absorption frequencies of the compounds. There is shown the marked difference between the value of the depolarization factor for the Raman line associated with the totally symmetric mode of vibration for carbon tetrachloride and those for the other lines.

THEORY.[1-3]   When a transparent and homogeneous medium is traversed by a beam of light, laterally diffused radiation may be observed. This phenomenon, termed the *scattering of light by the medium,* is a universal property of matter and had been under experimental and theoretical investigation for a number of years when in 1928 Raman discovered, in the scattered light, weak radiation of discrete frequencies not present in the monochromatic incident light and characteristic of the material under investigation. The term *Raman effect* refers to the production of these altered frequencies, whose complement constitutes the Raman spectrum.

The Raman effect arises from an exchange of energy between the scattering molecule and a photon of the incident radiation, which results in a transition of the molecule from one of its discrete energy states to another and a compensating change in the energy, and hence in the frequency, of the photon. The fundamental equation is

$$h\nu + E_1 = h\nu' + E_2 \tag{1}$$

where h = Planck's constant
  ν = frequency of incident photon
  ν' = frequency of scattered photon
E_1, E_2 = initial and final energies of molecule

The Raman line of frequency ν' is called a *Stokes line* if $\nu > \nu'$ and an *anti-Stokes line* if $\nu' > \nu$. The Stokes lines correspond to transitions in which the molecule is raised from a lower to a higher energy state at the expense of the photon, the anti-Stokes lines to transitions in which the molecule drops from an excited state to a lower energy level and gives up energy to the photon. Hence any permitted transition can give rise to both a Stokes and an anti-Stokes line, of which the former will be stronger because of the relatively small number of molecules in the higher energy states. Theory and experiment are in good agreement on the ratio of the intensities of the Stokes and anti-Stokes lines corresponding to a given transition.[1]

The difference in frequency between the Raman line and the exciting line is independent of the frequency of the incident light and is a measure of the separation of two energy states of the molecule. It is called the *Raman frequency shift,* or Raman frequency, and is ordinarily expressed in wave numbers $\tilde{\nu}$, or cm^{-1} (page 249).

Thus

$$\Delta \tilde{\nu} = \frac{\nu - \nu'}{c} = \frac{E_2 - E_1}{hc} \tag{2}$$

where c is the velocity of light. For polyatomic molecules, only changes in the vibrational contributions to the energy are ordinarily observed in the Raman effect. The total vibrational contribution is the sum of the contributions of all the vibrational degrees of freedom of the molecule; for a particular vibrational degree of freedom, this contribution can have only values given by

$$E_i = (n + \tfrac{1}{2})hc\tilde{\nu}_i \qquad n = 0, 1, 2, 3, \ldots \tag{3}$$

where $\tilde{\nu}_i =$ corresponding fundamental vibrational frequency, cm^{-1}
$n =$ vibrational quantum number

It follows from Eqs. (2) and (3) that in the Raman spectrum of a polyatomic molecule there will be found:

1. Frequency shifts equal to fundamental vibrational frequencies of the molecule, corresponding to transitions between adjacent energy levels associated with a single vibrational frequency. These lines are ordinarily the strongest Raman lines.
2. Frequency shifts equal to linear combinations (sums and differences) of several fundamental frequencies, due to simultaneous changes in the energy associated with the several modes of vibration concerned.
3. Frequency shifts equal to integral multiples of the fundamental vibrational frequencies, due to the less common transitions between non-adjacent levels associated with a single frequency. These lines are usually very weak.

Corresponding to each fundamental vibrational frequency there is a *normal mode of vibration*, the complete description of which involves the specification of the motion undergone by each atom in the molecule. Any vibrational motion of the molecule can be represented as a superposition of the different normal modes with appropriate amplitudes. In general, *all* atoms in the molecule may be involved in each normal mode of vibration, but it has been found experimentally and explained theoretically that the presence of various groups in the molecule can give rise to characteristic vibrational frequencies irrespective of the nature of the rest of the molecule. Thus all aliphatic nitriles have a characteristic frequency of approximately 2100 cm^{-1} which is associated with the stretching of the carbon-nitrogen triple bond. These group frequencies are often useful in the identification of structural features through the Raman spectrum.

Significant results can be obtained from a study of the Raman lines obtained with appropriate excitation conditions. Let the incident light be nonpolarized and confined to a beam traveling in the Y direction perpendicular to the axis of the

Raman tube, which is considered to coincide with the X axis. Let the scattered light be resolved into components $I_\perp$ polarized perpendicular to the XZ plane and $I_\parallel$ polarized parallel to the XZ plane. Then the *depolarization factor* ρ for the Raman line is the ratio $I_\perp / I_\parallel$. Considering lines associated with the fundamental transition from the ground state to the first excited vibrational level for a single mode of vibration, the depolarization factor will be less than $\frac{6}{7}$ if the symmetry of the equilibrium configuration of the molecule is preserved throughout the whole cycle of the vibrational motion involved; otherwise ρ will be equal to $\frac{6}{7}$. For the totally symmetric modes of molecules such as carbon tetrachloride and sulfur hexafluoride which belong to cubic point groups, the depolarization factor is zero; for the totally symmetric modes of other molecules, it will be greater than zero but less than $\frac{6}{7}$. The determination of the polarization states of Raman lines is thus of great help in the assignment of frequencies to particular modes of vibration.

Details of the single-exposure method, in which the two components of the scattered light are recorded simultaneously, have been given by Cleveland.[4] The use of a convenient two-exposure method introduced by Edsall and Wilson[5] has been described also by Crawford and Horwitz.[6] It should be emphasized that accurate depolarization-factor measurements are difficult to make. The recent improvements in photoelectric detection systems make it possible to avoid the problems of quantitative photographic photometry.

Intramolecular vibrations also give rise to absorption bands in the infrared region of the spectrum at frequencies equal to fundamental vibrational frequencies and their harmonics and combinations. The quantum theory permits a prediction from the structure of a molecule of the number of fundamental vibrational frequencies, etc., that will be observed in the Raman spectrum and in the infrared absorption spectrum. Different rules are found to apply to the two different types of spectra, which thus yield complementary information in the study of molecular vibrations. A given vibrational frequency may be detected only in the Raman effect, only in the infrared spectrum, or in both. Conversely, from a comparison of the infrared and Raman spectra of a compound, important information concerning the structure of the molecule may be obtained.

For a permitted transition in either spectrum there is required the presence of an oscillating electric moment in the molecule. In the case of the infrared absorption spectrum, this oscillating moment results directly from motion of the molecule. In the Raman case, it is the result of a periodic change in the polarizability of the molecule, due to molecular motion, and a consequent periodic change in the electric moment induced in the molecule by the electric field associated with the incident radiation. For example, consider first the symmetrical stretching mode of the CCl_4 molecule, in which the four C—Cl bonds symmetrically contract and expand in length. This mode is not infrared active, because the full symmetry of the equilibrium configuration is maintained throughout the whole vibrational cycle and the electric moment of the molecule does not change, but remains zero throughout the cycle. On the other hand, as the size of the molecule changes, the ease with which an applied electric field can perturb the structure changes, and this change in polarizability leads to Raman activity.

For a heteronuclear diatomic molecule it is clear that vibrational motion will change the electric dipole moment and also the polarizability. For the HCl molecule, both a near-infrared absorption spectrum and a vibrational Raman spectrum are observed. For a homonuclear molecule such as N_2 the electric moment will obviously stay zero through the whole vibrational cycle, but the polarizability must again be expected to change. Hence, N_2 will have a vibrational Raman spectrum, but no infrared absorption spectrum.

Most studies of Raman spectra have been made on materials in the liquid state, where the higher concentration of molecules than in the gas leads to more intense scattering. Interactions between molecules in the condensed phase commonly have surprisingly little effect on the vibrational spectrum, except where highly specific interactions such as hydrogen bonding or complex formation are involved. The introduction of special techniques for the Raman spectroscopy of gases has been an important recent development in this field.[7,8]

Apparatus. Spectrograph; photographic plates or film; mercury-vapor lamps and transformers; Raman tubes; filter jacket; filter solution; chloroform, deuterochloroform, benzene, carbon tetrachloride or other liquids; argon calibration lamp and transformer.

PROCEDURE.   A typical apparatus assembly is shown in Fig. 68a. A spectrograph of fairly large aperture is preferable, but the common wavelength spectrometer of aperture about $f/16$ gives satisfactory results with a slit width of 0.1 to 0.2 mm. A grating spectrograph can provide the advantage of a dispersion essentially linear in wavelength, with resultant convenience in wavelength determination. Since the Raman spectra will be recorded at slit openings larger than appropriate for the comparison spectra, the spectrograph slit should be of the bilateral type, in which the two jaws move symmetrically with respect to center. Since any motion of the plate holder may cause the plate to shift position inside as well, a Hartmann diaphragm (Fig. 68b) should be provided to permit a given set of Raman and comparison spectra to be recorded with the plate holder fixed.

The intense source of light necessary is conveniently provided by a bank of commercial mercury-vapor lamps A,† which yield sharp lines and comparatively little continuous background in the visible region. The special transformers used

† The A-H2 lamps previously recommended have been discontinued. Suitable substitutes may be found in the Uviarc series (GE); the UA-3 lamp has been found quite satisfactory.

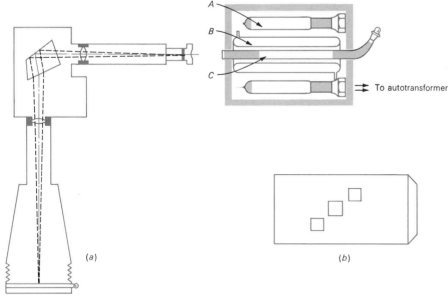

figure 68. (a) Apparatus for Raman spectroscopy; (b) Hartmann diaphragm.

with these lamps are designed to have a "drooping" voltage-current characteristic; i.e., they provide a high open-circuit voltage which automatically drops to a safe working level when the lamp fires and current is drawn. The lamps contain some argon to aid in starting, and coated electrodes. The radiation emitted is almost exclusively from the mercury present, but faint lines may be noted originating from the inert gas and from the electrodes. From a cold start it takes the lamps about 5 min to reach full brilliance. After being turned off, they will not restart immediately; a moderate cooling period must be allowed first.

Less convenient but of higher radiation quality are dc mercury arc lamps with water-cooled electrodes;[9] sharp lines of high intensity with very favorable line-to-continuous-background intensity ratio can be obtained. Microwave excitation of alkali-metal vapors has been recommended where wavelengths other than those of the mercury spectrum are needed. The laser promises to become of increasing importance as a Raman excitation source, with the advent of the argon laser providing excitation frequencies in the blue-green region of the spectrum to supplement the helium-neon laser with its 6328-Å emission.

The power dissipation within the excitation unit will typically run from 500 watts up. To maintain satisfactory conditions within the unit, it is necessary to circulate air through the enclosure and to provide an adequate flow of water through a water jacket surrounding the sample tube and filter. A pressure reducer should be placed between the water jacket and the city water line, with the water valve located just ahead of the reducer. A moderate but *steady* flow rate is required; checks should be made at intervals to see that this condition is maintained. The proper setting of the pressure reducer may correspond to no noticeable deflection of the associated pressure gauge.

The liquid to be studied is contained in the Raman tube *C* (Fig. 68), constructed from a section of Pyrex tubing with a plane window sealed on one end. Such tubes are readily available at moderate cost from manufacturers of scientific glassware. When sample size is not a limitation, it will be found convenient to use a tube of moderately large diameter, 15 mm for example. The curved section at the rear of the tube is intended to minimize *reflection* of light into the spectrograph; the liquid level must be raised into the offset section. Since the temperature of the sample in the excitation unit will inevitably rise above room temperature, in filling the Raman tube allowance must be made for the subsequent thermal expansion of the liquid. Except for the window and the section directly opposite the lamps, the tube is painted black,† again to aid in reducing reflection rather than scattering of light into the spectrograph. The outer surfaces of the tube should be kept clean, and the liquids used should be free from fluorescent impurities.

The mounting for the Raman tube should make possible accurate and reproducible positioning of the tube so that its axis coincides with the axis of the spectrograph collimator. Improper alignment of the Raman tube is the primary source of unsatisfactory results in this experiment. Efficient use of the scattered light, and consequent reduction of required exposure times, can be achieved through use of a suitable condensing lens selected in accordance with the recommendations of Nielsen.[10] A stable but adjustable mounting for the condensing lens is necessary, with a reference scale to permit accurate repositioning.

In the visible region of the spectrum, the radiation from a typical mercury arc is concentrated primarily in the lines at 4047, 4358, 5460, 5769, and 5790 Å, with weaker lines appearing, for example, at 4078, 4916, 4960 Å, etc. An effectively monochromatic light source at 4358 Å may be obtained by use, in jacket *B* (Fig. 68) surrounding the Raman tube but inside the water jacket, of a filter solution containing 0.01 percent crystal violet and 4 percent *p*-nitrotoluene in ethyl alcohol.

Caution: Fire hazard.

Alternatively, a filter jacket may be ring-sealed to the Raman tube, or an appropriate coating may be applied to the outside of the tube.[11] For the green line at 5460 Å

---

† Velvet Coating 101-C10 Black, Reflective Products Division, 3M Company, St. Paul, Minn., has given excellent results.

as exciting line, the filter problem becomes more difficult. To minimize interference from the strong yellow lines (5769, 5790 Å) a saturated solution of neodymium chloride in slightly acidified ethanol can be used.

Caution: Fire hazard.

More elaborate filters are described in the literature.[12,13]

Eastman spectrographic plates are recommended: type 103a-J is suggested for use when Hg λ 4358 Å is the exciting line and type 103a-F for spectra excited by Hg λ 5460 Å radiation. These plates should be ordered with antihalation backing to minimize the consequences of the inevitable overexposure of the exciting line. Tri-X panchromatic film can also be used. These plates (or film) should be handled in *total darkness.* In loading the plate holder, the emulsion side of the plate must face forward. This side can be distinguished by touch, since the back of the plate is smoother. (Alternatively, the tip of the tongue applied to the *corner* of the plate will stick to the emulsion side but not to the other.) The emulsion side of the Tri-X pan film will face forward when the film is held in the right hand with the edge serrations at the top right-hand side. A metal film sheath should be used with the cut film. During any spectrum recording the room lights should be *off*, to avoid fogging the plate through stray light entering the spectrograph slit.

To start the experimental work, first the air- and water-circulation systems and then the mercury-vapor lamps are turned on. While the lamps are warming up, the plate holder is loaded and put into place on the spectrograph. The filter solution appropriate to the excitation frequency to be used is placed inside the water jacket, and the Raman tube is filled with the desired sample and put into position. The spectrograph slit is set at 0.1 mm opening and the Hartmann diaphragm adjusted for illumination of the central region of the slit. The spectrograph shutter is checked to see that it is set for time exposure, and the condensing lens position is compared with that required. The room lights are then turned off, the plate-holder slide is raised, and the spectrograph shutter is opened to start an exposure which is continued for 30 min (or other time as directed). During this period, instructions should be obtained on the use of the comparator, which will be used later in the wavelength determinations.

At the end of this time, the shutter is closed and the slit setting changed to 0.03 mm. Here again, as in all such settings, the slit width should be decreased below the desired value and then increased as appropriate. The argon discharge lamp† is moved onto the spectrograph optic axis and switched on. The Hartmann diaphragm is repositioned for use of its top opening, and the lamp image is centered

---

† A convenient lamp and transformer combination is available from Oriel Optics Corporation, Stamford, Conn., 06902.

on the slit by adjustment of the lamp position. The condensing lens is moved as necessary to improve the slit illumination. The shutter is opened for a 1-min exposure time, then closed and the diaphragm reset for use of its lower window. A second calibration spectrum is now recorded at a somewhat longer exposure time, about 4 min, to accommodate the weaker reference lines.

The plate-holder slide is then pushed down and the plate-holder position shifted 15 mm. The argon lamp is removed, the slit again set at 0.1 mm, and the Hartmann diaphragm and condensing lens repositioned. The sample tube is removed and replaced with another containing a second liquid. In this fashion Raman and comparison spectra are recorded for chloroform, deuterochloroform, and benzene, or other compounds as desired.

After the last comparison spectrum has been recorded, the plate-holder slide is pushed down and the mercury lamps are turned off. The plate holder is removed from the camera. In *total darkness* the plate is taken out, developed, fixed, and dried. Care should be taken to place the plate in the developing tray with the emulsion side *up*. A 2-min development time in D-19 developer is recommended; the tray should be rocked gently during this time. The plate is then rinsed in distilled water and transferred to a tray containing F-5 fixer, where it should remain until the unexposed emulsion has cleared completely; the room lights may be turned on in the later stage of the fixing process. It is then washed in tap water for a time at least equal to the time of fixing, after which it can be rinsed with distilled water and dried.

While comparator measurements are being made on this first plate, two Raman spectra of carbon tetrachloride are recorded on a second plate. In the first there is placed around the Raman tube a cylinder of Polaroid sheet with optic axis parallel to the tube axis; in the second the Polaroid sheet axis is perpendicular to the Raman tube axis. Each 1-hr exposure is recorded through the central window of the Hartmann diaphragm, with the plate holder shifted appropriately between exposures. A comparison spectrum may be added after the second Raman spectrum, with the exposure time selected on the basis of earlier experience. The plate is then processed as described above.

The "American Institute of Physics Handbook"[14] provides an excellent tabulation of the wavelengths of the lines of the argon and mercury spectra. Identification of lines in the calibration spectra can be greatly facilitated by wavelength estimates based on the known wavelengths and positions of the mercury blue, green, and yellow lines. An argon line α, slightly lower in wavelength than any Raman line, can then be selected as a reference. Measurements are then made of the distances from this line to the various Raman lines and to each of a series of argon lines spaced at moderate intervals along the wavelength range of interest to a final argon line β. If the Raman lines are of low intensity, it may be helpful to make a photographic enlargement on which the measurements are made.

CALCULATIONS. If a grating spectrograph is used, the differences between the actual wavelengths of the intermediate argon lines and those calculated on the assumption of a linear wavelength-displacement relation between lines α and β can then be plotted to give a correction curve for use in the determination of the wavelengths of the Raman lines.

For spectra recorded with a prism instrument, the dispersion curve is markedly nonlinear; this nonlinearity is somewhat less pronounced when the plot is based on the wave numbers rather than wavelengths of the lines. The wave numbers of the Raman lines can thus be determined with moderate accuracy by interpolation on a plot of line wave number versus displacement from the reference argon line. Better results can be obtained by use of the Hartmann interpolation formula

$$\lambda = \lambda_0 + \frac{C}{d - d_0}$$

where λ = wavelength of unknown line
λ_0, d_0, C = constants calculated from the comparator readings and known wavelengths of three calibration lines
d = comparator reading for unknown line

The Raman frequency shifts for the compounds are then determined and compared with the principal infrared absorption frequencies given in Table 1.

For chloroform and deuterochloroform agreement should be obtained between the Raman shifts and the infrared frequencies listed. For benzene, no such coincidences actually occur; this is an example of the so-called *mutual-exclusion rule* for molecules which, like benzene, possess a center of symmetry. In such a case the Raman active fundamentals are not infrared active and vice versa. It is readily seen that application of this rule can be complicated by the uncertainties in the Raman and infrared data.

The experimental results will show that four of the fundamental frequencies for deuterochloroform are nearly matched in chloroform. The two cases in which a marked difference is seen arise from normal modes of vibration in which, because of the small mass of the hydrogen or deuterium atom compared to the rest of the molecule, the motion is primarily associated with the hydrogen (or deuterium) atom, a bending motion for the lower of the two frequencies and a stretching of the C—H or C—D bond for the higher. If for this stretching motion the molecules are considered to act as pseudo-diatomic molecules with one heavy and one light atom, theory (page 255) would predict that the ratio of the two frequencies would be approximately equal to $\sqrt{2}$. There again the differences in the vibrational frequencies for the isotopic species reflect the effects of the differences in the isotopic masses, with the potential-energy function the same in the two cases.

The depolarization factors for the Raman lines of carbon tetrachloride are given by the ratios of the intensities of the components of the perpendicular and parallel spectra recorded. The marked differences between the depolarization factor for the $\Delta\tilde{\nu} = 459$ cm^{-1} line ($\rho \sim 0$) and those for the other lines ($\rho \sim \frac{6}{7}$) should be noted. The $\Delta\tilde{\nu} = 459$ cm^{-1} is associated with the totally symmetric breathing vibration of the molecule, in which the four C—Cl bonds symmetrically expand and contract in length; at any arbitrary stage of this motion the full symmetry of the equilibrium configuration of the molecule is preserved.

Table 1. *Principal Infrared Absorption Frequencies of Chloroform and Benzene*[18]
$\tilde{\nu}_{\text{vac}}$, cm^{-1}

CHCl$_3$	C$_6$H$_6$
260	671
364	1037
667	1485
760	1807
1205	1964
3033	3045
	3099

Practical applications. Many uses have been discovered for the Raman spectra, and these are described in the voluminous literature published since the effect was discovered in 1928. A knowledge of the fundamental vibrational frequencies of the molecules is required for the theoretical calculation of the thermodynamic properties of gases by statistical methods, and the structures of molecules can be deduced through a study of the Raman and infrared spectra. Raman spectra have also found application in the qualitative[15] and quantitative[16] analysis of multicomponent systems and in the determination of the degree of dissociation of strong electrolytes in aqueous solution.[17]

Suggestions for further work. The Raman spectra of other liquids may be determined. A comparison of the Raman spectra of the two geometric isomers *cis-* and *trans*-dichloroethylene provides an interesting study. Infrared data are available for these compounds.[18]

References

1. H. A. Szymanski (ed.), "Raman Spectroscopy: Theory and Practice," Plenum Press, New York, 1967.
2. B. P. Stoicheff in D. Williams (ed.), "Methods of Experimental Physics," vol. 3, Academic Press Inc., New York, 1964.
3. G. W. King, "Spectroscopy and Molecular Structure," Holt, Rinehart and Winston, New York, 1964.
4. F. F. Cleveland, *J. Chem. Phys.*, **13:** 101 (1945); see also P. Bender and P. A. Lyons, *ibid.*, **18:** 438 (1950).
5. J. T. Edsall and E. B. Wilson, Jr., *J. Chem. Phys.*, **6:** 124 (1938).
6. B. L. Crawford, Jr., and W. Horwitz, *J. Chem Phys.*, **15:** 268 (1947).
7. B. P. Stoicheff, *Advan. Spectroscopy*, **1:** 91–174 (1959).

8. H. L. Welsh, E. J. Stansbury, J. Romanko, and T. Feldman, *J. Opt. Soc. Am.*, **45:** 388 (1955).

9. H. L. Welsh, M. F. Crawford, J. R. Thomas, and G. R. Love, *Can. J. Phys.*, **30:** 577 (1952).

10. J. R. Nielsen, *J. Opt. Soc. Am.*, **20:** 701 (1930); **37:** 494 (1947).

11. G. Glockler and J. F. Haskin, *J. Chem. Phys.*, **15:** 759 (1947).

12. R. F. Stamm, *Ind. Eng. Chem. Anal. Ed.*, **17:** 3181 (1945).

13. J. Brandmuller and H. Moser, "Einführung in die Raman Spektroscopie," Dr. Dietrich Steinkopff Verlag, Darmstadt, 1962.

14. D. E. Gray (ed.), "American Institute of Physics Handbook," sec. 7, McGraw-Hill Book Company, New York, 1963.

15. F. F. Cleveland, *J. Am. Chem. Soc.*, **63:** 622 (1941).

16. E. J. Rosenbaum, C. C. Martin, and J. L. Lauer, *Ind. Eng. Chem. Anal. Ed.*, **18:** 731 (1946).

17. O. Redlich and J. Biegeleisen, *J. Am. Chem. Soc.*, **65:** 1883 (1943).

18. G. Herzberg, "Infra-red and Raman Spectra of Polyatomic Molecules," D. Van Nostrand Company, Inc., Princeton, N.J., 1950.

42 STARK EFFECT IN THE ROTATIONAL SPECTRUM OF A MOLECULE

THEORY.†   The energy-level diagram for a polyatomic molecule is represented rather schematically in Fig. 69. We shall be concerned here with pure rotational

† A brief commentary on references suggested for various aspects of the theory outlined here appears at the end of this section, preceding the list of references.

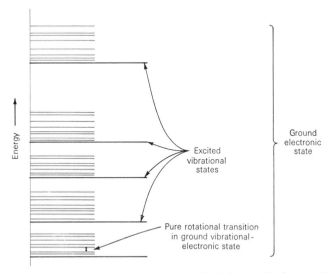

figure 69. A portion of the schematic energy-level diagram for the ground electronic state of a polyatomic molecule. A similar but not identical pattern of levels exists for the various stable excited electronic states.

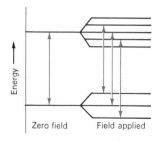

Zero field   Field applied

figure 70.

Energy-level scheme for two rotational levels showing Stark splittings (exaggerated) produced by application of an electrostatic field.

transitions, i.e., transitions between different rotational levels of the same vibrational and electronic state. The application of an electrostatic field causes each rotational level of a molecule to split up into sublevels, each with slightly different energy (Fig. 70). As a result, a rotational absorption line is split into a group of closely spaced lines. This phenomenon is known as the *molecular Stark effect*. From an analysis of Stark effect splittings, the permanent dipole moment (page 205) of a molecule can be found.

Although a molecule is a complex system of electrons and nuclei, we shall represent it by a simple model: a set of point masses, corresponding to the nuclei, held in a rigid arrangement by bonds of negligible mass. The fact that the actual molecule vibrates is taken into account only to the extent of recognizing that quantities which are functions of internuclear distances are, in effect, mean values for the given vibrational state and will therefore vary slightly from one vibrational state to another.† Furthermore, the translation of the center of mass can be neglected in considering the rotational motion.

It is convenient to classify molecules into four categories, each of which exhibits certain characteristic rotational properties. Molecules in which all atoms lie on a single straight line are called *linear*. Nonlinear molecules are further classified according to the number n of axes of threefold or higher symmetry, as shown in Table 1. (The last column is given for later reference.) For a symmetric top, the unique axis of threefold or higher symmetry is called the *figure axis*.

Table 1. Classification of Molecules According to the Number n of Axes of Threefold or Higher Symmetry

Type	n	Examples	Relations among moments of inertia
Asymmetric top	None	H_2O, CH_2Cl_2, NOCl (bent)	$I_x \neq I_y \neq I_z$
Symmetric top	One	CH_3Cl, PF_3, BrF_5, BF_3	$I_x = I_y \neq I_z$
Spherical top	More than one	CH_4	$I_x = I_y = I_z$

Let the cartesian coordinate system x, y, z be fixed to the molecule with origin at the center of mass, considered to be at rest. The kinetic energy of rotation T_r of the molecule is

$$T_r = \frac{1}{2} \sum_i m_i v_i^2 \tag{1}$$

† The justification for this procedure and a discussion of higher-order terms which must be considered in a more detailed analysis of vibration-rotation spectra have been given by Wilson; the results are summarized in Ref. 12.

where v_i is the magnitude of the instantaneous velocity of the ith nucleus and m_i is its mass. The vector representing the velocity of the ith nucleus is $\mathbf{v}_i = \boldsymbol{\omega} \times \mathbf{r}_i$, where $\boldsymbol{\omega}$ is the angular velocity of rotation of the molecule, the same for all nuclei, and $\mathbf{r}_i$ is the vector from the origin to the ith nucleus.[1] Through the substitution

$$v_i^2 = \mathbf{v}_i \cdot \mathbf{v}_i = (\omega_y z_i - \omega_z y_i)^2 + (\omega_z x_i - \omega_x z_i)^2 + (\omega_x y_i - \omega_y x_i)^2 \tag{2}$$

the kinetic energy of rotation may be expressed in terms of the components of $\boldsymbol{\omega}$ along the axes x, y, and z:

$$T_r = \tfrac{1}{2}I_{xx}\omega_x^2 + \tfrac{1}{2}I_{yy}\omega_y^2 + \tfrac{1}{2}I_{zz}\omega_z^2 - I_{xy}\omega_x\omega_y - I_{yz}\omega_y\omega_z - I_{zx}\omega_z\omega_x \tag{3}$$

where

$$I_{xx} = \sum_i m_i(y_i^2 + z_i^2) \qquad\qquad I_{xy} = \sum_i m_i x_i y_i$$

$$I_{yy} = \sum_i m_i(x_i^2 + z_i^2) \qquad\qquad I_{yz} = \sum_i m_i y_i z_i \tag{4}$$

$$I_{zz} = \sum_i m_i(x_i^2 + y_i^2) \qquad\qquad I_{zx} = \sum_i m_i z_i x_i$$

The coefficients I_{xx}, I_{yy}, I_{zz}, called *moments of inertia*, and I_{xy}, I_{yz}, I_{zx}, called *products of inertia*, depend on the structure of the molecule and on the choice of coordinate system x, y, z. It is always possible to choose the orientation of the coordinate axes relative to the molecule in such a way that the products of inertia all vanish.[2] Axes chosen in this way are called *principal axes,* and the moments about these axes are referred to as *principal moments of inertia.* It is customary to use only a single subscript to denote a principal moment; thus the moment about x is written I_x if x is a principal axis. For a symmetric top, the identification of the principal-axis system is trivial because the figure axis and any axis perpendicular to it are principal axes.

The relations given in the last column of Table 1, though not immediately obvious, are necessary consequences of the symmetry properties specified in the second column.

The equations which govern the rotational motion take on their simplest form when expressed in terms of angular-momentum variables.[1] The angular momentum **P** of a rigid body is defined by

$$\mathbf{P} = \sum_i \mathbf{r}_i \times \mathbf{p}_i = \sum_i m_i \mathbf{r}_i \times \mathbf{v}_i \tag{5}$$

where $\mathbf{p}_i = m_i\mathbf{v}_i$ is the linear momentum of the ith nucleus. In terms of components in the x, y, z system, e.g.,

$$P_x = \sum_i m_i(y_i v_{zi} - z_i v_{yi}) \tag{6}$$

With the use in Eq. (6) of expansions such as

$$v_{zi} = (\boldsymbol{\omega} \times \mathbf{r}_i)_z = \omega_x y_i - \omega_y x_i \tag{7}$$

it may be seen that *in the principal-axis system* the components of $\mathbf{P}$ are simply related to the angular-velocity components:

$$\begin{aligned} P_x &= I_x \omega_x \\ P_y &= I_y \omega_y \\ P_z &= I_z \omega_z \end{aligned} \tag{8}$$

The kinetic energy, expressed in terms of the angular-momentum components *in this system*, becomes

$$T_r = \frac{P_x^2}{2I_x} + \frac{P_y^2}{2I_y} + \frac{P_z^2}{2I_z} \tag{9}$$

Consider now the case of a symmetric top with z the figure axis. Since $I_x = I_y$,

$$\begin{aligned} T_r &= \frac{1}{2I_x}(P_x^2 + P_y^2) + \frac{1}{2I_z}P_z^2 \\ &= \frac{1}{2I_x}\mathbf{P}^2 + \frac{1}{2}\left(\frac{1}{I_z} - \frac{1}{I_x}\right)P_z^2 \end{aligned} \tag{10}$$

where

$$\mathbf{P}^2 = P_x^2 + P_y^2 + P_z^2 \tag{11}$$

A symmetric top is called *prolate* if $I_z < I_x = I_y$ and *oblate* if $I_z > I_x = I_y$. The sign of the second term in T_r differs for these two cases.

The classical motion, i.e., the motion prescribed by Newton's laws, can be rather easily visualized for a symmetric top subject to no externally applied torques. The molecule rotates with a constant angular velocity ω_z (and hence with a constant angular momentum P_z) about the figure axis; at the same time, the figure axis precesses, i.e., describes a circular cone, with uniform angular velocity about the direction of the total angular-momentum vector $\mathbf{P}$, which is fixed in direction as well as magnitude.[2,3] This motion is depicted in Fig. 71, where X, Y, Z is a space-fixed coordinate system. In this motion, the quantities

$$\mathbf{P}, P_z, P_X, P_Y, P_Z \tag{12}$$

remain constant. From Eq. (10), T_r also is constant. The values of these constants for any given case may be thought of as being determined by the initial conditions.

Since the spectroscopic transitions to be considered later consist of changes in energy of the molecule produced by the action of an electromagnetic field, it is

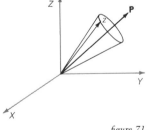

figure 71.

Classical motion of a symmetric top. The figure axis z precesses about $\mathbf{P}$, which is constant in magnitude and in direction.

instructive to examine briefly the classical theory of this effect. If an oscillating electric field $E_Z(t)$, polarized along the space-fixed Z axis, acts on the molecule, the latter is subjected to a torque[1] $\boldsymbol{\tau}$,

$$\boldsymbol{\tau} = \boldsymbol{\mu} \times \mathbf{E}(t) \tag{13}$$

where $\boldsymbol{\mu}$ is the permanent dipole moment of the molecule (page 205). For a symmetric-top molecule, $\boldsymbol{\mu}$ must lie in the direction of the z axis. Newton's equation of motion for the case of rotational motion of a rigid body is[1]

$$\frac{d\mathbf{P}}{dt} = \boldsymbol{\tau} \tag{14}$$

where $\boldsymbol{\tau}$ is the torque due to external forces acting on the body. The torque therefore causes T_r and several of the quantities of (12) to change. It is to be noted, however, that as $\boldsymbol{\tau}$ has no component in either the z or the Z direction, the components P_z and P_Z are not altered even with the field applied. The changes in T_r will be important only if the frequency of alternation of the field is close to that of some component of the rotational motion (classical resonance effect).

The quantum-mechanical treatment of the top begins with a recognition of the fact that the variables

$$\mathbf{P}^2, P_z, P_Z \tag{15}$$

can simultaneously have precisely specified values. As a manifestation of the uncertainty principle, the components P_x, P_y, P_X, and P_Y cannot also at the same time have precisely specified values. The eigenvalues of $\mathbf{P}^2$, P_z, and P_Z are, respectively,[4-6]

$$\mathbf{P}^2: \quad J(J+1)\hbar^2 \qquad J = 0, 1, 2, 3, \ldots \tag{16}$$

$$P_z: \quad K\hbar \qquad K = 0, \pm 1, \pm 2, \ldots, \pm J \tag{17}$$

$$P_Z: \quad M\hbar \qquad M = 0, \pm 1, \pm 2, \ldots, \pm J \tag{18}$$

where $\hbar = h/2\pi$

$h =$ Planck's constant

States with precisely specified values of $\mathbf{P}^2$, P_z, and P_Z are therefore labeled by J, K, and M and written symbolically as Ψ_{JKM}.

The *stationary states* of the symmetric top are states which satisfy the eigenvalue equation

$$\mathcal{H}\Psi = W\Psi \tag{19}$$

where $\mathcal{H}$ is the hamiltonian operator, which is derived from the classical expression for the energy [Eq. (10)] by replacing each quantity with the corresponding quantum-mechanical operator. The values of W satisfying (19) are the energy levels. It is clear that the states Ψ_{JKM} are eigenstates of $\mathcal{K}$, and hence these are the stationary states for a symmetric top. The energy eigenvalues are immediately obtained from Eqs. (10), (16), (17), and (19) as

$$W = \frac{1}{2I_x} J(J + 1)\hbar^2 + \frac{1}{2}\left(\frac{1}{I_z} - \frac{1}{I_x}\right) K^2 \hbar^2 \tag{20}$$

It should be noted that W does not depend on M or on the sign of K. The rotational energy is often written in the form

$$W = hBJ(J + 1) + h(A - B)K^2$$
$$A = \frac{h}{8\pi^2 I_z} \qquad B = \frac{h}{8\pi^2 I_x} \tag{21}$$

where A and B are called *rotational constants*, or, loosely, *reciprocal moments*. As defined in Eq. (21), they have dimensions of frequency.

If nonrigidity of the molecule is taken into account, the rotational-energy expression (9) must be augmented by the addition of terms involving higher powers of the angular momentum. The quantum-mechanical result for a symmetric top is[3,7-10]

$$W = hBJ(J + 1) + h(A - B)K^2 - hD_K K^4 - hD_{JK}J(J + 1)K^2$$
$$- hD_J J^2(J + 1)^2 \tag{22}$$

where the coefficients D_K, D_{JK}, and D_J are called *centrifugal-distortion constants*. The effects of these terms are roughly equivalent to a dependence of the effective rotational constants on J and K. It is not necessary to consider centrifugal distortion further in the present experiment, but Eq. (22) has been introduced here because it is often encountered in the literature.

If the molecule is subjected to an electromagnetic field of frequency ν, transitions from one stationary state to another can occur with the absorption or emission of energy. The probability of occurrence of such a transition per unit time is a maximum if the frequency satisfies the Bohr condition,

$$h\nu = W' - W'' \tag{23}$$

where W' = energy of upper state
$\quad\quad\quad W''$ = energy of lower state

The probability of transition is proportional to μ^2. Transitions which can be produced by an oscillating electric field along the Z direction acting on the dipole moment along the z direction are summarized by the selection rules

$$\Delta J = J' - J'' = 1 \tag{24}$$
$$\Delta K = K' - K'' = 0 \tag{25}$$
$$\Delta M = M' - M'' = 0 \tag{26}$$

Use of Eqs. (21) and (23) to (25) then leads to

$$\nu = B[J'(J' + 1) - J''(J'' + 1)] = 2BJ' \qquad (27)$$

Since typical values of B are in the range 10^3 to 10^5 MHz the frequencies are in the microwave and far-infrared regions.

If a constant electrostatic field in the space-fixed Z direction is applied, the classical energy equation becomes

$$W_{\text{class}} = \frac{1}{2I_x} \mathbf{P}^2 + \frac{1}{2}\left(\frac{1}{I_z} - \frac{1}{I_x}\right) P_z^2 - \mu E_Z \cos\theta \qquad (28)$$

where θ is the angle between the z and Z axes. The energy levels in the field are the eigenvalues of the quantum-mechanical hamiltonian operator corresponding to Eq. (28); these eigenvalues may in principle be found as the roots of a determinant called the *secular determinant* for the given system. As the roots of the determinant derived from Eq. (28) cannot be found in closed form, advantage is taken of the fact that the last term is ordinarily small enough for an approximation method (perturbation theory) to be used with excellent results. The energy levels in the field are obtained in this way as a power series in the variable $\mu E_Z/hB$; since typical values for this ratio are much less than unity, the convergence is extremely good, and in fact it is rarely necessary to go beyond the second-order term. The result for the energy levels in the field, to the second order, is[7-10]

$$W = hBJ(J + 1) + h(A - B)K^2 - \frac{KM}{J(J + 1)}\mu E_Z + W^{(2)} \qquad (29)$$

with

$$W^{(2)} = \begin{cases} -\dfrac{\mu^2 E_Z^2}{6hB} & \text{for } J = 0 \\[2mm] \dfrac{\mu^2 E_Z^2}{2hB}\left\{\dfrac{[3K^2 - J(J + 1)][3M^2 - J(J + 1)]}{J^2(J + 1)^2(2J - 1)(2J + 3)} - \dfrac{M^2 K^2}{J^3(J + 1)^3}\right\} & \text{for } J > 0 \end{cases}$$

The third and fourth terms on the right side of Eq. (29) are, respectively, the linear and quadratic Stark effect terms. The superscript on $W^{(2)}$ denotes that this term is of second order in E_Z.

Some discussion of this result may be helpful. To the extent that the second-order term $W^{(2)}$ can be neglected, it is accurate to say that the stationary states are the same as in the absence of the field and that $\mu \cos\theta$ in Eq. (28) can be replaced by its average value calculated for the unperturbed (zero-field) states Ψ_{JKM}. Classically, the time average of the precessing vector $\boldsymbol{\mu}$ is equal to its component along

P; the projection of this component in turn onto Z then gives the time average of $\mu_Z = \mu \cos \theta$ as

$$\langle \mu \cos \theta \rangle_{av} = \mu \left[\frac{K}{\sqrt{J(J+1)}} \right] \left[\frac{M}{\sqrt{J(J+1)}} \right] = \mu \frac{KM}{J(J+1)} \tag{30}$$

since the cosines of the two angles of projection are P_z/P and P_Z/P, respectively. This result offers some insight into the first-order term in Eq. (29). The second-order term is to be understood as follows. Application of the field results in a slight change in the stationary states, such that the average value of $\mu \cos \theta$ is changed by a small increment proportional to E_Z; there results from this a change in energy proportional to E_Z^2.

So long as the field is weak, the intensities of transitions produced by an oscillating field are the same as in the absence of the electrostatic field. With the electrostatic and oscillating fields both in the Z direction, the selection rules (24) to (26) still apply and the transition frequencies become

$$\nu = 2BJ' - \left[\frac{K'M'}{J'(J'+1)} - \frac{K''M''}{J''(J''+1)} \right] \frac{\mu E_Z}{h} + \cdots \tag{31}$$

$$= 2BJ' + \left[\frac{K'M'}{J'(J'^2 - 1)} \right] \frac{2\mu E_Z}{h} + \cdots \tag{32}$$

where only the first-order Stark effect term has been given explicitly. In Eq. (32), μ and E_Z are in electrostatic units and ν is in hertz. If, instead, μ is expressed in Debye units and E_Z in volts per centimeter, Eq. (32) is replaced by

$$\nu = 2BJ' + \frac{K'M'}{J'(J'^2 - 1)} (1.0070 \times 10^6)\mu E_Z + \cdots \tag{33}$$

with ν still in hertz. The relative intensities of the different M components for given values of J and K are proportional to $(J')^2 - M'^2$.

It may be mentioned that for linear molecules, Eq. (29) may be used with K set equal to zero. For linear molecules, and for the $K = 0$ states of symmetric tops, the Stark effect is therefore of second order in the field.

CALCULATIONS.   The spectrum of a line of the symmetric-top molecule CH_3CF_3 at several values of applied electrostatic field appears in Fig. 72. The value of J is confirmed from the qualitative appearance of the Stark pattern. The frequencies are determined and used for the calculation of μ. Values of rotational constant and moment of inertia are calculated from the frequency of the zero-field line.

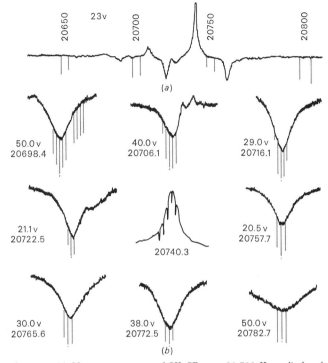

figure 72. (a) *Microwave spectrum of CH_3CF_3 near 20,700 Hz as displayed by a Stark modulation-type spectrograph (page 546), at a modulation amplitude of 23 volts. Zero-field absorption lines produce upward deflections; Stark components produce downward deflections. The $J = 1 \rightarrow 2$ transition for the ground vibrational state is near 20,740 Hz, and the corresponding line from molecules in an excited torsional-vibrational state appears near 20,710 Hz. Frequency marks (sharp spikes) are at 2.50 Hz above and below the frequency values shown.* (b) *Expanded patterns of main line and its two Stark components at several voltages. The frequency values shown refer to the marks designated by dots. The interval between adjacent frequency marks is 0.50 Hz for the Stark components and 0.20 Hz for the main line. Frequency increases to the right, as in part* (a), *in all cases. The frequency calibration marks are accurate to 0.02 Hz. The average spacing between the electrode and the parallel walls of the waveguide (Fig. 167) is 0.464 ± 0.005 cm. The voltages are accurate to ±0.1 volt. (Sample and waveguide calibration data courtesy of K. R. Lindfors.)*

Suggestions for further work. If the oscillating electric field is in the X direction, and hence perpendicular to the electrostatic field E_Z, the selection rule on M becomes[3,7-10]

$$\Delta M = M' - M'' = \pm 1 \tag{34}$$

The student may calculate and plot the Stark pattern for the 20,740 MHz line of CH_3CF_3 for this case. The relative intensities for this case are proportional to[3,7-10]

$$(J' \pm M' - 1)(J' \pm M') \tag{35}$$

where the upper signs are for $\Delta M = +1$ and the lower for $\Delta M = -1$.

References

Certain of the concepts of classical physics employed above are introduced by Halliday and Resnick[1] at an elementary level and with vector notation. Symon[2] gives an excellent classical treatment of the rotating top. The properties of angular momentum in quantum mechanics are developed by Eyring, Walter, and Kimball,[4] by Landau and Lifschitz,[5] and, from a somewhat different viewpoint, by Pauling and Wilson.[6] Spectroscopic applications are discussed at an elementary level by Brand and Speakman[11] and at a more advanced level by Herzberg;[3] Gordy, Smith, and Trambarulo;[7] Townes and Schawlow;[8] Sugden and Kenney;[9] and Wollrab.[10] An excellent general reference on the separation of translational, rotational, and vibrational motions is Wilson, Decius, and Cross.[12] Comprehensive tabulations of molecular constants measured by microwave spectroscopy are given by Townes and Schawlow[8] (appendix VI) and by Starck.[13]

1. D. Halliday and R. Resnick, "Physics," combined ed., pp. 208, 213, 245–249, John Wiley & Sons, Inc., New York, 1960.
2. K. R. Symon, "Mechanics," 2d ed., Addison-Wesley Publishing Company, Inc., Reading, Mass., 1960.
3. G. Herzberg, "Infra-red and Raman Spectra of Polyatomic Molecules," D. Van Nostrand Company, Inc., Princeton, N.J., 1945.
4. H. Eyring, J. Walter, and G. E. Kimball, "Quantum Chemistry," pp. 39–47, John Wiley & Sons, Inc., New York, 1944.
5. L. D. Landau and E. M. Lifschitz, "Quantum Mechanics," Addison-Wesley Publishing Company, Inc., Reading, Mass., 1965.
6. L. Pauling and E. B. Wilson, Jr., "Introduction to Quantum Mechanics," McGraw-Hill Book Company, New York, 1935.
7. W. Gordy, W. V. Smith, and R. F. Trambarulo, "Microwave Spectroscopy," pp. 89–103, 154–159, John Wiley & Sons, Inc., New York, 1953.
8. C. H. Townes and A. L. Schawlow, "Microwave Spectroscopy," pp. 48–82, 248–268, McGraw-Hill Book Company, New York, 1955.
9. T. M. Sugden and C. N. Kenney, "Microwave Spectroscopy of Gases," D. Van Nostrand Company, Inc., Princeton, N.J., 1965.
10. J. E. Wollrab, "Rotational Spectra and Molecular Structure," Academic Press Inc., New York, 1967.
11. J. C. D. Brand and J. C. Speakman, "Molecular Structure," pp. 74–83, Edward Arnold (Publishers) Ltd., London, 1960.
12. E. B. Wilson, Jr., J. C. Decius, and P. C. Cross, "Molecular Vibrations," pp. 11–14, 273–275, McGraw-Hill Book Company, New York, 1955.
13. B. Starck in K.-H. Hellwege (ed.), "Molecular Constants from Microwave Spectroscopy," Landolt-Börnstein, "Zahlenwerke und Funktionen aus Naturwissenschaften und Technik," group II, vol 4, Springer-Verlag New York Inc., New York, 1967.

Chapter 13

Magnetic Resonance

In classical mechanics, angular momentum is associated with the motion of a body about a center, like the earth around the sun, and with the spinning of a body on an axis, like the daily rotation of the earth. Angular momentum is a vector quantity, with both magnitude and direction. The direction is that of the axis about which rotation occurs. In quantum mechanics angular momentum is also important, and it was recognized that only certain values of angular momentum are possible in a consistent theory which agrees with experimental observations. If the square of the angular momentum of a quantum-mechanical object is measured, it will always be found to have the value

$$J^2 = j(j + 1)\hbar^2 \tag{1}$$

where $\hbar = h/2\pi$ (h = Planck's constant)

$$j = 0, \tfrac{1}{2}, 1, \tfrac{3}{2}, 2, \ldots$$

The component of angular momentum in a particular direction, say the z direction, will be observed to be

$$J_z = m\hbar \qquad m = -j, -j + 1, \ldots, j - 1, j \tag{2}$$

If one component of the angular momentum has a definite value, then the other components cannot simultaneously have well-defined values. The angular momentum due to the orbital motion of an electron about a nucleus is associated with integral values of j: 0, 1, 2, ... Associated with an elementary particle such as an electron, proton, or neutron is an intrinsic angular momentum corresponding to $j = \tfrac{1}{2}$, which is present even when the particle is at rest. It is called the *spin angular momentum* and corresponds closely, but not exactly, to the angular momentum associated with the spinning of a classical particle about its axis.

An atom or a nucleus which is built up from many elementary particles may have either an integral, 0, 1, 2, . . . , or a half-integral, $\tfrac{1}{2}, \tfrac{3}{2}, \ldots$, angular-momentum quantum number. There is a tendency for angular momenta to combine in such a way that they cancel each other out. Typical nuclear-spin quantum number values, denoted by I, are 0 for ^{12}C and ^{16}O, $\tfrac{1}{2}$ for ^1H and ^{19}F, 1 for ^2H (D) and ^{14}N, $\tfrac{3}{2}$ for ^{11}B and ^{39}K; etc. In molecules, the electronic spin angular momenta nearly always cancel each other. In a molecule with an odd number of electrons this is not completely possible, and at least one electron must remain unpaired. Many transition-metal complexes have more than one unpaired electron.

A charged particle with nonzero angular momentum will also have a magnetic moment associated with it. The magnetic moment is proportional to (and in a direction the same as or exactly opposite to) the angular momentum

$$\boldsymbol{\mu} = \gamma \mathbf{j} \tag{3}$$

The proportionality constant γ is called the *magnetogyric* or *gyromagnetic ratio*. For the electron γ is negative; for the proton it is positive.

If a system with a nonzero magnetic moment is placed in a magnetic field, which can be taken to define the z direction, then states with different values of the quantum number m will have different energies. For a particle with spin $\frac{1}{2}$, like a proton or an electron, in a field **H** of magnitude H

$$E_{\pm} = -\boldsymbol{\mu} \cdot \mathbf{H} = \pm \tfrac{1}{2}|\gamma|H \qquad (4)$$

If a sample containing unpaired spin-$\frac{1}{2}$ particles is placed in a uniform magnetic field and allowed to come to thermal equilibrium, the populations of the two magnetic energy levels will be related by the Boltzmann factor

$$\frac{n_+}{n_-} = e^{-(E_+ - E_-)/kT} = e^{-|\gamma|H/kT} \qquad (5)$$

Since for an ordinary magnetic field strength and temperature $|\gamma|H \ll kT$, this can be approximated by expanding the exponential

$$\frac{n_+}{n_-} \approx 1 - \frac{|\gamma|H}{kT} \qquad (6)$$

We see that there is a slightly greater population in the lower energy level.

If the sample is now exposed to radiation with a frequency ν_0 such that

$$h\nu_0 = \Delta E = |\gamma|H$$

then transitions between the two levels will be induced. The *probability* per unit time W of an induced transition upward is the same as that of an induced transition downward, but since $n_- > n_+$, the number of transitions upward per unit time $n_- W$ will be slightly greater than the number $n_+ W$ of downward transitions. This means that the sample will absorb energy from the radiation field at a rate

$$P = h\nu_0 n_- W - h\nu_0 n_+ W = h\nu_0 W \, \Delta n$$

where $\Delta n = n_- - n_+$.

If the spin system had no way of losing energy to its surroundings, the excess of upward transitions would quickly equalize the populations and steady state would be attained with $n_+ = n_-$. However, there are other interactions which tend to return the spin system to equilibrium, a process usually assumed to be of first order:

$$\frac{d\Delta n}{dt} = -k(\Delta n - \Delta n_{\text{eq}}) = \frac{\Delta n_{\text{eq}} - \Delta n}{T_1} \qquad (7)$$

where Δn_{eq} is the equilibrium value of Δn in the absence of radiation. $T_1 = k^{-1}$, known as the *spin-lattice relaxation time*, is the time characterizing the return of the spin system to equilibrium.

The two effects combine to give a net rate of change for the population difference

$$\frac{d\Delta n}{dt} = -2W\,\Delta n + \frac{\Delta n_{eq} - \Delta n}{T_1} \tag{8}$$

For the steady state, designated by subscript ss, $d\Delta n/dt = 0$ so that

$$\Delta n_{ss} = \frac{\Delta n_{eq}}{1 + 2WT_1} \tag{9}$$

and the steady-state absorption of power by the system is

$$P_{ss} = \Delta n_{eq} \frac{W}{1 + 2WT_1} \, h\nu_0 \tag{10}$$

It can be shown that W is proportional to the intensity of radiation at frequency ν_0, provided the intensity is not too great.

References

1. M. W. Hanna, "Quantum Mechanics in Chemistry," chap. 9, W. A. Benjamin, Inc., New York, 1965.
2. A. Carrington and A. D. McLachlan, "Magnetic Resonance," Academic Press Inc., New York, 1966.
3. J. D. Pople, W. G. Schneider, and H. J. Bernstein, "High-resolution Nuclear Magnetic Resonance," McGraw-Hill Book Company, New York, 1959.

43   ELECTRON SPIN RESONANCE

Electron spin resonance (ESR) spectra are obtained for several free radicals. Experience is gained in the use of a phase-sensitive detector and in magnetic field and frequency measurements.

THEORY.[1,2]   The general theory of magnetic resonance, which has just been discussed, will now be specialized to the case of a free radical, i.e., a molecule with a net electronic spin angular momentum of $\frac{1}{2}\hbar$. In this case

$$E_\pm = \pm\tfrac{1}{2}g\beta H = m_s g\beta H \qquad \Delta E = h\nu = g\beta H \tag{1}$$

where m_s is the quantum number for the z component of the electron spin angular momentum. The constant β is the Bohr magneton, $e\hbar/2mc$, and is equal to $(9.2731 \pm 0.0002) \times 10^{-21}$ erg gauss^{-1} in cgs units. The dimensionless constant g is very close to 2, but its exact value varies from one radical to another. For a free electron $g = 2.0023$.

In addition to its interaction with the external magnetic field, the electron magnetic moment will interact with other nearby magnetic moments. ESR experiments are usually carried out in dilute systems, where different free radicals are too far apart for their electronic magnetic moments to affect each other. The electron and nuclear moments in the same radical will interact.

One interaction between the electronic and nuclear magnetic moments is the same as the classical interaction between two bar magnets. It depends upon the orientation of the radical, however, and as the radical tumbles rapidly and randomly in solution, it averages to zero. Another interaction, which is explained by relativistic quantum mechanics but has no classical analog, is isotropic and thus remains nonzero even for radicals tumbling in solution. It is of the form

$$A' \mu_e \cdot \mu_N \tag{2}$$

where μ_e is the electron magnetic moment, μ_N the nuclear magnetic moment, and A' is a constant proportional to the probability of finding the unpaired electron at the position of the nucleus.

We consider as an example a nucleus of spin 1 such as ^{14}N. The component of nuclear spin angular momentum in a specified direction for such a nucleus can be only ± 1 or 0 times $\hbar$ ($m_I = \pm 1$ or 0). The nuclear magnetic moment is also proportional to the nuclear spin angular momentum, and thus the possible interaction energies of the electron and nuclear magnetic moments are $\pm \bar{A}/2$ and 0, where $\bar{A}$ is proportional to A'. For usual magnetic field strengths $\bar{A}$ is much less than $g\beta H$.

The approximate energy-level scheme for such a system is shown in Fig. 73. If H were held constant and the frequency ν varied, the absorption spectrum shown in Fig. 74 would result. In practice, for reasons of experimental convenience, the frequency is kept fixed at a value ν_0 and the strength of the external field H varies. This is illustrated in Fig. 75. The different nuclear spin levels vary with H in slightly different ways because of the direct interaction of the nuclear magnetic moments with the external field. This is of no consequence in most ESR experiments,

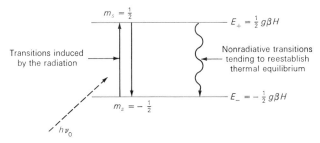

figure 73. Fundamental ESR transitions.

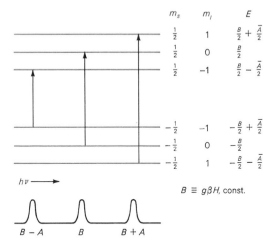

figure 74. Energy levels with a spin-1 nucleus and transitions for fixed field and variable frequency (not to scale).

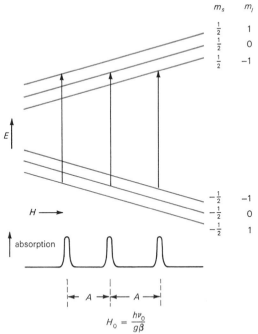

figure 75. Energy levels and transitions for fixed frequency and variable field (not to scale).

however, because nuclear spin orientations do not change in an ESR experiment and thus this interaction energy does not appear in the observed transition energies. The spectrum, which now appears as a function of the field strength H, is the same as that which would be observed if H were kept fixed and ν varied. Note, however, that the transitions as labeled by the m_I value occur in reversed order with increasing independent variable. The hyperfine interaction energy is customarily reported in units of field strength (gauss) rather than in units of energy ($A = \bar{A}/g\beta$).

As noted above, the hyperfine interaction energy is proportional to the probability of finding the unpaired electron at the position of the nucleus. (Strictly speaking, it is proportional to the probability of finding an electron there with spin up minus that of finding one there with spin down.) If there are several magnetic nuclei in the radical, the hyperfine interactions add, and each is proportional to the probability of finding an unpaired electron at that nucleus. The ESR spectrum of a free radical thus gives a map of the probability distribution for finding unpaired electron spin. The exact value of g also varies from one radical to another in a way which can often be approximately calculated. Electron spin resonance thus provides a way of identifying free radicals and of testing proposed electronic structures.

The energy expressions given above are really valid only when $g\beta H \gg \bar{A}$ or, equivalently, $H \gg A$. In most research applications H is approximately 3500 gauss, while for free radicals in solution A is seldom greater than 20 gauss. The spectrometer used in the laboratory experiment, however, uses a much lower field, $H \sim 120$ gauss. The approximation then breaks down, and the two splittings in the three-line spectrum of Figs. 74 and 75 are not equal. This happens because the hyperfine interaction has the effect of mixing different states. The state with $m_s = \frac{1}{2}$, $m_I = 0$, for example, is replaced by a state that is primarily $m_s = \frac{1}{2}$, $m_I = 0$, but also containing a small amount of $m_s = -\frac{1}{2}$, $m_I = 1$. When this is properly taken into account, it can be shown that the three transition field strengths at fixed frequency ν_0 are

$$H_{-1} = \frac{h\nu_0}{g\beta} + A - \frac{g\beta}{2h\nu_0} A^2 + \cdots$$

$$H_0 = \frac{h\nu_0}{g\beta} - \frac{g\beta}{h\nu_0} A^2 + \cdots \tag{3}$$

$$H_{+1} = \frac{h\nu_0}{g\beta} - A - \frac{g\beta}{2h\nu_0} A^2 + \cdots$$

The quadratic terms are essentially $-A^2/2H$ and $-A^2/H$, and thus become negligible when $H \gg A$.

Although diphenylpicrylhydrazyl (DPPH) has many nuclei with magnetic

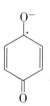

moments, its ESR spectrum does not show hyperfine structure when a solid sample is used. This is because interactions between adjacent radicals in the solid have the effect of averaging over all hyperfine levels to produce a single line. Hyperfine structure would be observed if the sample were a dilute solution of DPPH, in which the radicals are far apart. The structural formula of DPPH is shown in the margin.

Fremy's salt is potassium peroxylamine disulfonate, $K_2ON(SO_3)_2$. In dilute solution the spectrum of the $ON(SO_3)_2{}^{--}$ ion shows hyperfine structure due to the ^{14}N nucleus with spin 1. There are no other magnetic nuclei in the ion, assuming only the common isotopes ^{16}O and ^{32}S are present. The stable organic radical di(t-butyl) nitroxide, $(C_4H_9)_2NO$, behaves almost as if the nitrogen were the only magnetic nucleus present. This is because the unpaired spin is almost entirely in the NO region, and very little of it reaches the protons.

The p-benzosemiquinone radical ion shown at the left is stable for a short time in basic solution. In this radical there are four spin-$\frac{1}{2}$ hydrogen nuclei, and the total number of nuclear spin states is $2^4 = 16$.

The energy levels in this case will be given by

$$E = g\beta Hm_s + \bar{A}m_s M_I + \cdots \tag{4}$$

where $M_I = \sum_{n=1}^{4} (m_I)_n \tag{5}$

is the quantum number for the z component of the total nuclear spin angular momentum.

In an ESR experiment we observe the transitions for which $\Delta m_s = \pm 1$, $\Delta M_I = 0$, so that

$$h\nu = \Delta E = g\beta H + \bar{A}M_I + \cdots \tag{6}$$

The value of $\bar{A}$ for p-benzosemiquinone is small enough for higher-order terms corresponding to the quadratic parts of (3) to be neglected. Equation (6) can be solved for the resonance field values $H(M_I)$ at a fixed frequency ν_0.

There is one nuclear spin state with $M_I = 2$ and one with $M_I = -2$, having all $m_I = \frac{1}{2}$ and all $m_I = -\frac{1}{2}$, respectively. There are four states each with $M_I = \pm 1$, having three of the $m_I = \pm\frac{1}{2}$ and one $m_I = \mp\frac{1}{2}$, and there are six states having $M_I = 0$, corresponding to the various choices of two $m_I = \frac{1}{2}$ and two $m_I = -\frac{1}{2}$. Since all nuclear spin states are essentially equally populated, the spectrum will consist of five lines in the intensity ratio $1:4:6:4:1$.

The central three lines are thus actually superpositions of lines due to several transitions which occur at the same field for a fixed frequency. When higher-order terms are taken into account, it is found that these different lines do not actually come at exactly the same field, but the displacement is much less than the line width in the present case (and in most cases), so that in actual practice it can be ignored.

(It should be noted that the indication of an unpaired electron at one position in a molecular structure does not mean that it is confined to that point. Many other structures could be drawn, and the unpaired electron spin is actually distributed throughout the molecule.)

Most ESR work on free radicals is done at magnetic fields of about 3500 gauss and frequencies of 9.5×10^9 Hz in the microwave region.[3] These high frequencies lead to greater sensitivity in the spectrometer and to spectra which are more readily interpreted. However, the experimental procedure is simpler and the equipment less expensive if lower frequencies are used. In the Alpha ESR-10 a frequency of approximately 340 MHz is employed, and the magnetic field strength is reduced accordingly.

The apparatus is indicated schematically in Figure 76. The heart of the system is the 340-MHz oscillator and the magnet. A direct current flows through the Helmholtz coils producing a steady magnetic field. In this field is a coil which is a part of the tuned circuit in the 340-MHz oscillator. If there is a paramagnetic sample in this coil and the magnetic field is of the right magnitude, energy will be absorbed by the sample from the oscillator circuit. Suppose now that in addition to the steady field there is a field oscillating at 60 Hz. Then whenever the field passes through the resonance value, absorption will occur and a signal at 60 Hz will be generated. This signal can be displayed on an oscilloscope.

Unfortunately, however, the signal will be very small for most samples. It will often be of a magnitude less than that of the random noise in the circuit and will thus be invisible. Another detection scheme is used to alleviate the situation. The magnetic field is modulated at 100 kHz over a very small region. If this region lies in the range where absorption occurs, a signal at 100 kHz will be produced with amplitude proportional to the first derivative of the absorption signal. This is illustrated in Figure 77. This signal then passes through a phase-sensitive detector (page 634) which gives a dc output. Such a detector is sensitive only to an input at or near the modulation frequency. Since only a small fraction of the random noise will have this frequency, the signal-to-noise ratio is greatly improved. It can be further improved by a filter on the output which responds to the slowly varying signal as the dc field is swept but not to more rapidly varying noise.

If 60-Hz modulation of a large amplitude is applied at the same time as the 100-kHz modulation, the output from the phase detector can be observed on the oscilloscope. If the 60-Hz modulation is turned off and the dc field is slowly swept through the resonance region, the signal can be recorded. If an *XY* recorder is used,

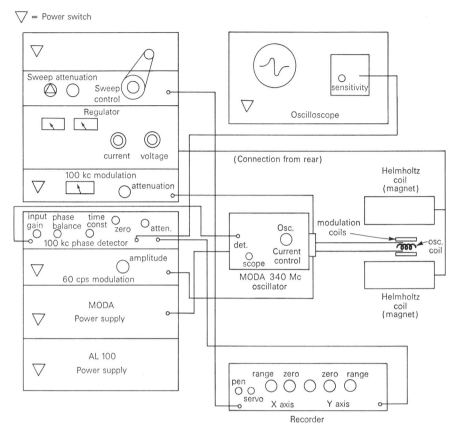

figure 76. The ESR spectrometer system.

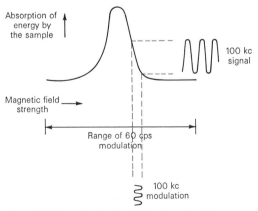

figure 77. Absorption signal and the effect of modulation.

a voltage from the field sweep unit is fed into the recorder X axis to provide a sweep which is very closely proportional to the magnetic field strength.

The frequency of the main spectrometer oscillator (nominally at 340 MHz) can be determined to about five figures with the aid of a heterodyne frequency meter (page 623). The principles involved in this type of measurement are mixing (page 596) and harmonic generation (page 597).

The Gertsch model FM-1B frequency meter contains an oscillator which is tunable from 20 to 40 MHz. The oscillator output is passed through a diode which generates harmonics of appreciable intensity to well beyond the twelfth. A sample voltage picked up at the spectrometer oscillator coil is coupled to the same diode and mixes with these harmonics. The frequency-meter oscillator frequency is adjusted until some harmonic is close enough to the spectrometer frequency for the beat frequency to be in the audio range. The voltage at the beat frequency is amplified and applied to an earphone, where it produces audible sound. When the frequency meter has been adjusted carefully for zero beat, its frequency is exactly a submultiple of the spectrometer frequency. The frequency of the frequency-meter oscillator can then be determined in a second zero-beating step, in which this oscillator is compared with a crystal-controlled oscillator (which provides a very stable reference frequency at 1 MHz) and a second variable oscillator which is accurately calibrated.

When the harmonic number is not known, the procedure used is to determine the zero-beat frequencies for two adjacent harmonic numbers. Thus, if the two frequency meter settings for zero beat are f_1 and f_2, corresponding to harmonic numbers n_1 and n_2, respectively, then

$$n_1 f_1 = n_2 f_2 = \nu_0 \tag{7}$$

where ν_0 is the spectrometer frequency, assumed to have remained constant. If there are adjacent zero-beat points, with $f_2 > f_1$, then $n_1 = n_2 + 1$ and n_1 can be determined from

$$n_1 = \frac{f_2}{f_2 - f_1} \tag{8}$$

As a check, a third zero-beat reading may be made.

Additional special equipment would be required to make an accurate absolute measurement of the magnetic field strength at any point in the sweep. Accurate *relative* measurements can be made quite easily by measuring the current flowing through the field coils. This is done by using a potentiometer to measure the potential drop across a standard resistor in series with the coils. The magnet used in the Alpha ESR-10 does not use ferromagnetic pole pieces. This means that more current is required to produce a given field strength at the sample than would be required if pole pieces were used. It has the advantage that the field strength is accurately proportional to the current I

$$H = KI \qquad (9)$$

where the factor K depends only on the geometry of the coils, including the number of turns, and on the position of the sample in the field.

⸱⸱⸱➤ ***Apparatus.*** ESR spectrometer, Alpha ESR-10† or equivalent; oscilloscope; XY or strip chart recorder; frequency meter; potentiometer; 1-ohm standard resistor; diphenylpicrylhydrazyl; Fremy's salt or di(t-butyl) nitroxide; p-benzoquinone; 0.1 N NaOH; hexane, isopropyl alcohol.

PROCEDURE.   All power switches are turned on, including those of the oscilloscope and the recorder (to power-on chart-off position; servo switch should be in the off position). Several minutes are allowed for warm-up while the various controls are located and identified.

The DPPH sample is placed in the coil at the end of the master-oscillator detector-amplifier (MODA) arm. The sample itself must actually be in the coil. The selector switch on the MODA is set to the "oscilloscope" position and the oscillator current set to a minimum (far counterclockwise). The MODA output is connected to the input of the oscilloscope and the trace located. The oscillator current is then increased until noise appears. The best operating point is about one-eighth turn beyond the minimum level of oscillation indicated by the appearance of noise on the oscilloscope. This should be about three-fourths turn clockwise. The 60-Hz modulation is increased and the magnet current adjusted until the signal appears on the oscilloscope. The magnet current should be noted so that this field strength can be easily reobtained when desired.

The MODA is switched to the "phase detector" position and its output connected to the input of the 100-kHz phase detector. The oscilloscope output of the phase detector is connected to the input of the oscilloscope. The integration time should be set to its minimum value. When the 100-kHz modulation is turned up, a derivative signal should appear on the oscilloscope. The phase balance control is adjusted to maximize this signal. The 100-kHz modulation amplitude and the input gain of the phase detector are adjusted to obtain a symmetric signal with a reasonable signal-to-noise ratio. The signal which appears is accurately proportional to the derivative of the absorption only when the modulation is much less than the line width. The modulation amplitude should be temporarily turned up to a high level and the resulting distortion of the waveform observed. The effect of increasing the time constant should also be observed.

Recording with an XY Recorder. When the sweep control is turned, the signal

† Alpha Scientific, Oakland, Calif.

should move across the face of the oscilloscope. The 60-Hz modulation is now reduced to zero, and this unit is turned off. A piece of graph paper is positioned on the recorder and the power switch moved to the chart-hold position. The sensitivity controls should be set to 0.1 volt in.$^{-1}$ on the X axis and 5 mv in.$^{-1}$ on the Y axis. The servo switch can then be turned on and the X and Y zero controls used to position the pen over the paper (with the pen raised). The ESR sweep-unit controls (sweep attenuation) are adjusted so that one turn of the dial sweeps an appropriate field range (several times the line width for DPPH). The recorder should just cover the full X-axis range on one full sweep. The input gain, output attenuation, and modulation amplitude are adjusted to give a signal of reasonable size with a good signal-to-noise ratio. When a slow sweep rate is used, the time constant can be increased to reduce the noise. With the sweep control set at the beginning of a sweep the drive belt is connected (slow or medium speed). The recorder pen is released and the sweep drive motor turned on to record a spectrum. At the end of the sweep the pen should be raised before repositioning.

A 0.005 M solution of Fremy's salt in 0.1 N NaOH solution is prepared in sufficient quantity to fill one of the small sample tubes.

Note: At the end of the period this solution should be kept in the sample tube tightly stoppered for use in the next period.

The DPPH sample is removed from the spectrometer and the Fremy's salt solution inserted. The sweep range is adjusted so that all three lines on the spectrum can be obtained on about half the width of the chart, and a spectrum is recorded.

A solution of di(t-butyl) nitroxide, 0.001 to 0.005 M in hexane, may be used in place of the Fremy's salt solution. This solution is stable and can be kept indefinitely in a sealed glass tube.

The remainder of the first period (if any) should be used in gaining familiarity with the function and location of the various controls so that the spectrometer can be used rapidly on the second day. It is also advisable to practice using the Gertsch frequency meter and the current-measuring potentiometer.

On the second day, a saturated solution of p-benzoquinone in isopropyl alcohol is prepared in a 50-ml flask. After the instrument has warmed up and the magnet current is set in the resonance region, the Fremy's salt solution is inserted. A fresh sheet of chart paper is placed on the recorder. The magnet current and sweep are adjusted (with the pen raised) so that the three-line spectrum is centered and covers about half the width of the chart. *After this adjustment is made, the settings on the magnet current, sweep attenuation and X-axis zero and X-axis gain must not be changed; on a strip chart recorder, the chart speed should not be changed.*

The Y-axis zero is set to put the pen in the upper third of the chart. The sample is then removed and the 100-kHz modulation set to two divisions on the 0.01 scale or a little less and the time constant set to 0.1. These settings may be modified to produce the best signal.

The sweep is positioned ready to start a sweep with the belt in position. The input gain is increased until a small amount of noise (pen tremor) is observed. A sample tube is filled three-fourths full of the quinone solution and three drops of 0.1 N NaOH solution is added. The tube is stoppered and inverted to mix the solution. (In basic solution in the presence of air the benzoquinone is quickly converted to the p-benzosemiquinone radical ion.) The tube is then inserted in the spectrometer, the pen is lowered and a sweep started immediately. If a satisfactory spectrum is not obtained, the process is repeated. When a good spectrum of this radical is obtained on the upper third of the chart, the Fremy's salt solution is reinserted and a spectrum recorded on the central third of the chart. A DPPH spectrum is recorded on the lower third and the DPPH sample left in place.

With the recorder pen raised and the sweep motor off, the belt is removed so that the sweep control can be moved easily by hand. The sweep control is adjusted to place the pen over the center of the DPPH resonance. The current is measured at this point and the frequency of the spectrometer oscillator determined. The current is also measured with the pen positioned over the center of each of the three Fremy's salt lines and over the center of the middle line of the semiquinone. It may be necessary to prepare a fresh quinone-NaOH mixture.

Strip Chart Recorder. It is not possible with a strip chart recorder to position the spectra one above the other. In this case they are recorded successively. The current measurements are made after all the spectra have been recorded. Each sample is inserted in turn and the appropriate lines located.

The frequency of the MODA oscillator is determined. An instructor should be consulted regarding the details of this measurement.

CALCULATIONS.   The spectrometer oscillator frequency ν_0 is determined by rounding off the value of n_1 from Eq. (8) to the nearest integer and substituting in Eq. (7).

The g value for DPPH has been determined to be 2.0037 ± 0.0001. Equation (1) can be used to determine the value of H_0, the field at the center of the DPPH resonance, from the measured value of ν_0. The proportionality constant K in Eq. (9) is evaluated from H_0 and the current I_0 measured at the center of the DPPH line. A scale calibration factor r for the chart, in gauss per scale division or gauss per inch, can be determined from the currents $I_{\pm 1}$ and the chart positions $x_{\pm 1}$ for the end lines of the Fremy's salt resonance

$$r = K\frac{I_1 - I_{-1}}{x_1 - x_{-1}} \tag{10}$$

Values of A for Fremy's salt and the semiquinone are calculated from measured currents or chart spacings. For Fremy's salt, because of second-order effects, the two spacings are unequal. However, from Eqs. (3) and (9)

$$A = \tfrac{1}{2}(H_1 - H_{-1}) = \frac{K}{2}(I_1 - I_{-1}) \tag{11}$$

For the semiquinone second-order shifts are negligible, and A is given simply as the product of the observed spacing between adjacent lines and r.

Because second-order effects are not negligible in the Fremy's salt spectrum, the current which would correspond to resonance in the absence of hyperfine splitting is not I_0 but

$$I_0' = I_1 + I_{-1} - I_0 \tag{12}$$

This and the current at the center of the semiquinone spectrum can be used to obtain g values. They are most accurately obtained in terms of the difference from the DPPH value

$$g_i - g_{\text{DPPH}} = \Delta g = g_{\text{DPPH}} \cdot \frac{I_{\text{DPPH}} - I_i}{I_i} \tag{13}$$

where g_i = gyromagnetic ratio for substance i

I_i = current value for center of resonance for substance i

The accepted values of g and A for Fremy's salt are 2.00550 ± 0.00005 and 13.0 ± 0.1 gauss. For di(t-butyl) nitroxide, $g = 2.0063$ and $A = 15.18$ gauss.

References

1. M. Bersohn and J. C. Baird, "An Introduction to Electron Paramagnetic Resonance," W. A. Benjamin, Inc., New York, 1956.
2. A. Carrington and A. D. McLachlan, "Magnetic Resonance," Academic Press Inc., New York, 1966.
3. C. S. Poole, Jr., "Electron Spin Resonance: A Comprehensive Treatise on Experimental Techniques," Interscience Publishers, Inc., New York, 1967.

44   *NUCLEAR MAGNETIC RESONANCE*

THEORY.[1-4]   Of primary interest in high-resolution nuclear magnetic resonance studies are spectra of nuclei of spin $\frac{1}{2}$, in particular ^{1}H, ^{19}F, and ^{13}C. For such a nucleus in a magnetic field of intensity H_0 the spatial quantization of the nuclear spin angular momentum leads, as indicated above, to a set of two energy levels of spacing $\gamma h H_0/2\pi$, where γ is the gyromagnetic ratio of the nucleus. For an ensemble of such nuclei the rate of adjustment of the populations of the two levels to consistency with the Boltzmann distribution for thermal equilibrium is governed by the spin-lattice relaxation time T_1 introduced above (page 287). Transitions between these levels can be induced by application of a magnetic field $\mathbf{H}_1$, of circular frequency $\omega = 2\pi\nu = \gamma H_0$, circularly polarized in a plane perpendicular to the main magnetic field direction. The resultant NMR signal, as obtained, for example, by use of a crossed-coil spectrometer (page 540), is found to be shifted in phase relative to the $\mathbf{H}_1$ field; the magnitudes u of the signal component in phase with $\mathbf{H}_1$ and v of the signal component $90°$ out of phase with $\mathbf{H}_1$ can be usefully characterized for appropriate conditions through the relations

$$u = \gamma H_1 M_0 T_2 \frac{(\gamma H_0 - \omega)T_2}{1 + T_2^2(\gamma H_0 - \omega)^2 + \gamma^2 H_1^2 T_1 T_2} \tag{1}$$

$$v = \gamma H_1 M_0 T_2 \frac{1}{1 + T_2^2(\gamma H_0 - \omega)^2 + \gamma^2 H_1^2 T_1 T_2} \tag{2}$$

where $H_1 = $ intensity of $\mathbf{H}_1$ field

$$M_0 = \Delta n_{\text{eq}} \frac{\gamma h}{4\pi}$$

These relations represent predictions based on the Bloch[11] equations for steady-state conditions and presume that the system has had time to adjust to the conditions $(T; H_0, H_1$ fields) to which it has been subjected. The relaxation times T_2 and T_1 here account for the effects of interactions within the ensemble of spins and between the spin system and its surroundings. The magnetic field in the vicinity of a particular nucleus fluctuates continually as the result of the motion of other nuclei, etc., in its neighborhood. The components of this fluctuation spectrum at frequencies near the transition frequency are effective in inducing transitions between the energy levels involved, and are reflected in the value of T_1. The second relaxation time, the transverse relaxation time T_2, accounts for the effects of the low-frequency components of the fluctuation spectrum, including the effect of inhomogeneity of the magnetic field with its resultant change in energy-level spacing as the position of the nucleus in the field is changed, by diffusion, for example. For typical samples and conditions in high-resolution proton NMR spectroscopy, T_1 and T_2 will have roughly the same value, of the order of a few seconds.

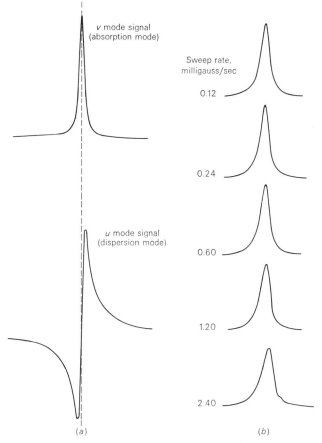

figure 78. (a) *Slow-passage signals from protons in a dilute aqueous solution of ferric chloride;* (b) *effect of sweep rate on NMR signal.*

By use of a phase detector (page 634) referenced to the source of the H_1 field, the signal components u and v can be measured separately. If, at constant ω and H_1, the value of H_0 is considered to be changed slowly enough for effectively steady-state conditions to be always maintained, Eqs. (1) and (2) can be used to predict the shapes of the resulting spectrum lines. Such "slow-passage" results are shown in Fig. 78a. If the sweep rate, i.e., the rate of change of the H_0 field, is progressively increased, the line shape will be changed, as shown in Fig. 78b, where for the highest sweep rate there can be seen an indication of the onset of the transient effect known as "wiggles" or "ringing" that is so marked in Fig. 84 (page 312). The character of this ringing, i.e., its amplitude and persistence, depends on the value of T_2 and the sweep rate. As indicated above, the effective T_2 depends not only on the sample but also on the homogeneity of the H_0 field. With a sample of appropriately long intrinsic T_2, experimental adjustments of field homogeneity can thus be expedited by use of the ringing pattern as a quality index.

The v-mode, or absorption, signal is used in high-resolution spectroscopy (note the breadth of the u-mode, or dispersion, signal in Fig. 78a). From Eq. (2) it is seen that as the H_1 field intensity is increased from zero, the peak v-mode signal first grows in proportion but subsequently goes through a maximum value followed by a continuous decrease. This behavior is associated with the perturbation of the energy-level populations through transitions induced by the H_1 field. At sufficiently high H_1 values the populations become effectively equalized, and no net energy absorption is detected.

At low H_1, that is, when $\gamma^2 H_1^2 T_1 T_2 \ll 1$, the full width of the absorption line at half height is predicted by Eq. (2) to be $2/\gamma T_2$, or $(0.075/T_2)$ milligauss for protons. This provides one approach to the measurement of T_2; a result which characterizes the sample, however, is obtained only if the magnetic field is sufficiently homogeneous. A determination of T_1 (Refs. 5 and 6) can be made on the premise that the peak amplitude P of the absorption signal will be proportional to Δn, the difference in population of the two levels. In Fig. 79 are shown the v-mode signals, at 1-sec intervals, from a sample of degassed benzene in the period just after its introduction into the H_0 field. The slope of a plot of t versus log $[P(\infty) - P(t)]$ equals $-1/2.303T_1$, since the process is assumed to be of first order. Alternatively the sample can be subjected to a strong H_1 field, of intensity high enough to equalize the level populations, which is then removed and the approach to equilibrium monitored by observation of the recovering signal with a suitably weak H_1 field. The purpose of degassing the benzene sample is to remove dissolved oxygen, whose paramagnetic character leads to a marked shortening of the value of T_1.

For experimentally realizable magnetic field strengths the absorption frequency v lies in the radio-frequency range. The gyromagnetic ratio for the proton, for example, equals 2.675×10^4 sec^{-1} gauss^{-1}; this means an absorption frequency of 60 Mc (60 $\times$ 10^6 sec^{-1}) for a 14,100-gauss field, the field strength employed in

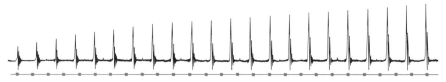

figure 79. Growth of NMR absorption signal on introduction of sample into the field.

most high-resolution proton NMR work. Modern superconducting magnet systems have made it possible to use fields as high as 50 kilogauss, and 23-kilogauss field strengths are routinely available with electromagnets as well.

In ordinary optical-absorption spectroscopy, the energy levels are fixed, and the frequency of the radiation field must be adjusted to fit the energy per quantum to the energy-level spacing. It will be clear from Eq. (2) that an NMR spectrum can be obtained by such a frequency sweep for a specific constant value of the magnetic field intensity H_0, as well as by the field-sweep method considered above, in which the energy-level spacing is adjusted to match the energy per quantum of the radiation field.

The important chemical applications arise from effects encountered when the nuclei studied are present in molecules. Electron circulation in the neighborhood of nucleus A in a molecule modifies the field that otherwise would be sensed by the nucleus. This effect is commonly a diamagnetic shielding effect and is quantitatively described by the *shielding constant* σ for the nucleus

$$H_A = H_0(1 - \sigma_A) \tag{3}$$

where H_A is the field acting at nucleus A.

The value of the shielding constant is sensitive to the average electron density in the neighborhood of the nucleus. For protons the greatest difference in shielding constant is about 1×10^{-5}, so that all information of chemical interest in high-resolution proton NMR spectra is compressed into a frequency range of 1 part in 100,000 of the nominal absorption frequency for constant H_0 (or of the nominal H_0 for constant frequency). It is thus necessary to achieve a stability of the order of parts per billion in the control of the magnetic field strength and radiation field frequency to obtain the results now available in high-resolution NMR work.

Nuclei which have the same shielding constant are called *chemically equivalent*. Chemical equivalence necessarily characterizes nuclei in a symmetrically equivalent set, i.e., a set which is permuted by the symmetry operations for the molecule. It can also be produced by the averaging effect of sufficiently rapid intramolecular motion, as, for example, the equivalence of the protons in a methyl group due to the internal rotation about the carbon-carbon bond. In addition, sufficiently rapid exchange of a nucleus between two nonequivalent sites can result in a single resonance peak at an intermediate position, instead of the two separate peaks which would be obtained if the residence time were long enough in each of the two environments. In this connection, "long" means large compared with the reciprocal of the chemical shift (in cycles per second) between the two environments.

The chemical shift is a measure of the difference of line positions in a spectrum due to a difference in shielding constants for nuclei in different environments.[2,7,8]

For fixed frequency of the radiation field, the relative chemical shift may be defined as

$$\delta = \frac{H_r - H}{H_r} \times 10^6 \tag{4}$$

where H = resonance field for signal specified

H_r = resonance field for reference signal

For proton spectra, the reference signal commonly used is the single sharp peak from tetramethylsilane, the protons of which are very highly shielded. The definition given above then makes the chemical shift positive for protons in almost any other environment and of progressively greater magnitude as the shielding constant becomes smaller. The chemical shift may also be given in field units as

$$H_r - H = H_r \delta \times 10^{-6} \tag{5}$$

For practical purposes the nominal field strength H_0 can be substituted for H_r on the right-hand side, since $\delta \times 10^{-6}$ is so small. Typical chemical-shift results for protons are shown in Fig. 80.

The spacing of lines in an NMR spectrum is measured by the so-called sideband method. The magnetic field to which the sample is subjected is varied by passing an audio-frequency current through coils whose axis coincides with the direction of the main magnetic field. The modulation field, whose intensity can be expressed as $H_m \cos 2\pi\nu_m t$ (ν_m = modulation frequency), can be introduced into the Bloch equations[9] or treated as a time-dependent perturbation in a quantum-mechanical treatment.[10] In either case the prediction is the appearance of additional lines in the spectrum, symmetrically disposed about each line otherwise present. The intensities of these sidebands in relation to the corresponding parent line, now termed the *centerband*, is determined by the modulation index $\beta = \gamma H_m/2\pi\nu_m$. The spacing within each set of sidebands is uniformly, for normal modulation amplitude H_m, equal to the modulation frequency when expressed in frequency units through the relation $\delta\nu = (\gamma/2\pi)\delta H$. If β is small, only the first sidebands will be of appreciable intensity. For constant radio frequency ν_0, the first lower sideband will appear for a field strength H_- such that $\nu_0 - \nu_m = (\gamma/2\pi)H_-$, the center band for H_0 such that $\nu_0 = (\gamma/2\pi)H_0$, and the first upper sideband for H_+ such that $\nu_0 + \nu_m = (\gamma/2\pi)H_+$. Since the modulation frequency can readily be measured, a simple method for spectrum calibration is available, as illustrated in Fig. 81a. The sample involved contained chloroform and tetramethyl silane (TMS). An audio-frequency modulation at 450 cps was applied, and a field-sweep spectrum was recorded starting at an approximately low magnetic field strength. The first signal encountered is the first lower sideband of the TMS line. After it was recorded, the modulation was removed as the sweep continued through the chloroform line, after

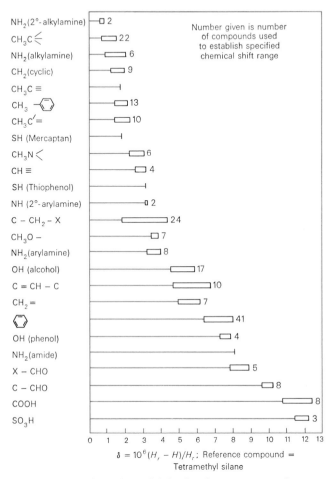

figure 80. Proton relative chemical-shift values for various structural environments. Results given are based on data of Meyer, Gutowsky, and Saika from
measurements made primarily on pure-liquid samples. (L. H. Meyer, A. Saika,
and H. S. Gutowsky, J. Am. Chem. Soc., **75:** 4567 (1953.)

which a modulation field at 420 cps was applied to obtain the corresponding sideband
from the TMS. Assuming a linear sweep, the chloroform line can be determined by
interpolation to be 436 cps below the TMS line at 60 Mc, corresponding to
a 102.5 milligauss difference in their resonance fields at this frequency.

To this point the relations written assume a common value for the magnetic
field intensity for all nuclei which contribute to the signal. This condition is quite

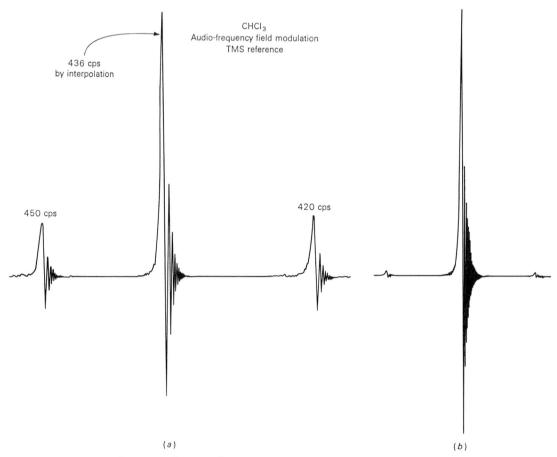

436 cps
by interpolation

CHCl₃
Audio-frequency field modulation
TMS reference

450 cps

420 cps

(a)   (b)

figure 81. (a) *Spectrum calibration by use of sidebands generated by field modulation;* (b) *spinning sidebands due to rotation of sample in inhomogeneous field.*

difficult to satisfy to the required degree of approximation. Noting that the relaxation times are of the order of seconds in high-resolution proton NMR spectroscopy, the nuclei actually respond to the field they experience averaged over an appreciable period of time. The effect of field gradients can then be reduced by mechanical circulation of the nuclei through an inhomogeneous field[11,12] by spinning the sample tube about its axis, so that to the greatest extent possible all the nuclei spend equal amounts of time in regions of field strength above and below average by the same amount, and thus all sense the same average field strength. The effective field homogeneity can be strikingly improved in this way. Success in this approach requires that the gradients whose effects are being minimized be small; in particular, it should be noted that there will be no effect on gradients in the direction of the spinning axis.

Because of this sample spinning, the nuclei are subjected to a magnetic field which varies periodically at the spinning rate, normally about 30 cps. Sideband generation must then be expected, with the intensity of the spinning sidebands determined by the field gradients and their position relative to the center band determined by the spinning rate. In Fig. 81*b* are shown such sidebands, exaggerated in intensity above an acceptable level by deliberate distortion of the field.

The peak *v*-mode signal, according to Eq. (2), depends not only on the number of nuclei contributing (through M_0 and Δn_{eq}) but also on the relaxation times. Since the latter are not the same in different molecules, or even for different nuclei in the same molecule, relative peak heights do not properly reflect relative numbers of nuclei. For slow-passage conditions and sufficiently low values of H_1, it is readily shown that the integrated area of the absorption line is independent of the relaxation times, and hence the relative numbers of nuclei contributing to different signal lines can be determined by integration of the spectrum. This is done by use of the signal to generate a charging current for a capacitor; the change in voltage across the capacitor due to line traverse is fixed by the integral of this current with respect to time, and thus provides a result proportional to the line area. While practical work is not done under the conditions specified above, the equivalent result can nevertheless be obtained.[13] Successful integration of a spectrum is, however, one of the most difficult operations in NMR spectroscopy.

NMR spectroscopy does not have the sensitivity available in ultraviolet and infrared spectroscopy. The normal limitation in this sense can be substantially removed by use of the technique called *time averaging*.[14] Because of the stability of modern NMR spectrometers and the consequent reproducibility of spectrum scans, it is possible to use a digital computer to add algebraically the signals generated in replicate scans of a given spectrum. The noise contributions, which vary randomly from one spectrum to another, tend to cancel out while the actual spectrum line contributions add coherently. The result is an improvement in signal-to-noise ratio that is roughly equal to the square root of the number of scans. The advantage gained is illustrated in Fig. 82.

The difference in shielding constants for nuclei in different electronic environments divides the high-resolution NMR spectrum into a sequence of groups of lines. It is advantageous to work at as high a magnetic field and corresponding radio-frequency as possible to minimize insofar as possible the overlapping of different groups in order to simplify the interpretation of the spectrum. The fine structure in a particular group arises from a phenomenon termed *spin-spin coupling*; the energy of a system of nuclei in a magnetic field depends not only on the magnetic field

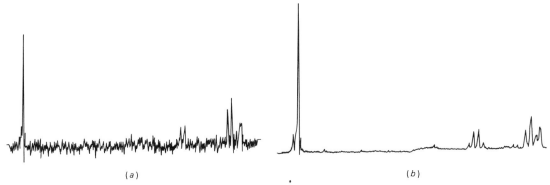

figure 82. Improvement of signal-to-noise ratio by time averaging. (a) *Single scan, 0.1 percent solution of ethyl benzene;* (b) *resultant spectrum from 100 scans.* (Varian Associates.)

strength but also on the relative orientations of the individual nuclear magnets. It has been found possible to account for the latter effect by the inclusion in the hamiltonian function, representing the total energy of the system of nuclei in the field, of a remarkably simple field-independent term to obtain

$$\mathcal{K} = -\left(\sum_i \boldsymbol{\mu}_i \cdot \mathbf{H}_i + \sum_{i<j} \xi_{ij}\, \boldsymbol{\mu}_i \cdot \boldsymbol{\mu}_j \right) \tag{6}$$

The first sum accounts for the interactions of the individual nuclear magnets with the field; the second, whose special form ensures that the interaction of each pair of nuclei is counted only once, accounts for the interactions between the nuclei. The constant factor ξ_{ij} measures the strength of the interaction of nucleus i with nucleus j. Explicitly recognizing the existence of the chemical-shift effect and the proportionality between nuclear magnetic moment and nuclear spin angular momentum, expressing I_z and $\mathbf{I}_i$ in units of $\hbar$, and $\mathcal{K}$ in units of cycles per second, Eq. (6) is converted to

$$\mathcal{K} = -\left[\sum_i \frac{\gamma_i}{2\pi}\, (1 - \sigma_i) H_0\, (I_z)_i + \sum_{i<j} J_{ij}\mathbf{I}_i \cdot \mathbf{I}_j \right] \tag{7}$$

The quantity J_{ij} is the *spin-spin-coupling constant* for nuclei i and j. For protons, it ranges in magnitude from zero to about 30 cps. These spin interactions do not take place directly between the nuclei but rather constitute an electron-coupled effect which is transmitted through the chemical-bond chain from one nucleus to the other. It is found that spin couplings are rarely transmitted effectively through more than a few chemical bonds. The magnitude of the coupling constant characterizing the interaction of two nuclei depends of course on the gyromagnetic ratios of the nuclei and also on structural considerations such as the H—C—H angle for two protons bonded to the same carbon atom or the dihedral angle between the two H—C—C planes for the interaction of protons bonded to adjacent carbon atoms. The magnitudes of spin-coupling constants have been predicted with some success by quantum-mechanical methods supplementing the information obtained empirically through studies of molecules of known structure. In Table 1 are summarized typical values[7,15,16] of coupling constants for proton interactions in organic molecules. The complication of the spectrum resulting from the spin interactions will commonly be helpful in the elucidation of the structure of a molecule from the observed NMR spectrum.

Table 1. Coupling Constants for Proton-Proton Interaction

Structural grouping	Configuration		J, cps
Ethyl	CH_3CH_2R		6–8
Olefinic	C=C with H, H	gem	1–2
	H, H C=C	cis	8–12
	C=C H, H	trans	17–18
Benzene derivatives	H, H (ortho)	ortho	5–8
	H, H (meta)	meta	1–3
	H, H (para)	para	1
Cyclohexane derivatives	CHR—CHR′	axial-axial	5–8
		equatorial-equatorial or axial-equatorial	2–3

For two chemically nonequivalent nuclei A, B of a given kind, Eq. (7) becomes

$$\mathcal{H} = -\left\{ \left(\frac{\gamma}{2\pi} H_0 \right) \left[(1 - \sigma_A)(I_z)_A + (1 - \sigma_B)(I_z)_B \right] + J_{AB} \mathbf{I}_A \cdot \mathbf{I}_B \right\} \tag{8}$$

Since there are two orientations relative to the field possible for each nucleus, there are four discrete energy levels for the two-proton system. The energies of these levels are again sought as the eigenvalues of the quantum-mechanical operator corresponding to the specified hamiltonian function. Because all three components of the spin angular momentum for each nucleus are involved in this hamiltonian, the result is not as simple as that presented earlier (page 290). Standard quantum-mechanical procedures lead to the following results for the case of spin $\frac{1}{2}$:

$$
\begin{aligned}
W_1 &= -\tfrac{1}{2}(\nu_A + \nu_B + \tfrac{1}{2}J_{AB}) \\
W_2 &= -\tfrac{1}{2}\{[(\nu_A - \nu_B)^2 + J_{AB}^2]^{\frac{1}{2}} - \tfrac{1}{2}J_{AB}\} \\
W_3 &= \tfrac{1}{2}\{[(\nu_A - \nu_B)^2 + J_{AB}^2]^{\frac{1}{2}} + \tfrac{1}{2}J_{AB}\} \\
W_4 &= \tfrac{1}{2}(\nu_A + \nu_B - \tfrac{1}{2}J_{AB})
\end{aligned}
\tag{9}
$$

Here $\nu_A = (\gamma/2\pi)(1 - \sigma_A)H_0$; it represents the resonance frequency which would be found for nucleus A if there were no spin coupling to nucleus B. It is found that there are four allowed transitions between these levels:

$$
\begin{aligned}
\Delta W_{1 \rightarrow 3} &= W_3 - W_1 = \tfrac{1}{2}\{\nu_A + \nu_B + [(\nu_A - \nu_B)^2 + J_{AB}^2]^{\frac{1}{2}} + J_{AB}\} \\
\Delta W_{2 \rightarrow 4} &= W_4 - W_2 = \tfrac{1}{2}\{\nu_A + \nu_B + [(\nu_A - \nu_B)^2 + J_{AB}^2]^{\frac{1}{2}} - J_{AB}\} \\
\Delta W_{1 \rightarrow 2} &= W_2 - W_1 = \tfrac{1}{2}\{\nu_A + \nu_B - [(\nu_A - \nu_B)^2 + J_{AB}^2]^{\frac{1}{2}} + J_{AB}\} \\
\Delta W_{3 \rightarrow 4} &= W_4 - W_3 = \tfrac{1}{2}\{\nu_A + \nu_B - [(\nu_A - \nu_B)^2 + J_{AB}^2]^{\frac{1}{2}} - J_{AB}\}
\end{aligned}
\tag{10}
$$

The relative intensities of the corresponding absorption lines are given by

$$
I_{1 \rightarrow 3} = I_{3 \rightarrow 4} = \frac{(1 - R)^2}{1 + R^2}
$$

$$
I_{1 \rightarrow 2} = I_{2 \rightarrow 4} = \frac{(1 + R)^2}{1 + R^2}
\tag{11}
$$

where $R = \dfrac{J_{AB}}{\nu_A - \nu_B + [(\nu_A - \nu_B)^2 + J_{AB}^2]^{\frac{1}{2}}}$

Since the energies above are expressed in cycles per second, the energy differences give directly the corresponding absorption frequencies for the system for the constant applied field H_0. It can be seen that for very weak coupling, the first two transitions can be associated with nucleus A, the other two with nucleus B, but for finite J_{AB}, all four characterize the two-nucleus system as a whole. The appearance

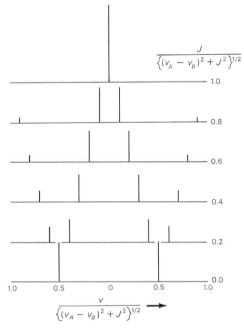

figure 83. Calculated spectra for two nonequivalent nuclei of spin $\frac{1}{2}$. The zero of the horizontal scale is located at the frequency $\nu = \frac{1}{2}(\nu_A + \nu_B)$.

of the spectrum depends only on the absolute value of the ratio $J_{AB}/(\nu_A - \nu_B)$, and it is not possible from the spectrum to determine either the sign of the coupling constant or which nucleus is more shielded, although chemical intuition can often answer the latter question. Typical results calculated for the two-spin system are shown in Fig. 83.

The appearance of such a multiplet pattern will depend not only on the identity of the molecule concerned but also on the medium in which it is present. This is ordinarily the result of a medium effect on the chemical shift, as reflected in the spectra shown in Fig. 84a. A comparison of spectra obtained at two different field strengths is made possible through Fig. 84b.

When the coupling is small compared with the chemical shift, the spectrum reduces to two doublets of individual spacing J_{AB}, with centers separated by the chemical shift $\nu_A - \nu_B$ (Fig. 85). This result corresponds to the first-order perturbation theory in the calculation of the energy levels, for which the interaction energy effectively reduces to $J_{AB} (I_z)_A (I_z)_B$. To this degree of approximation, the coupling of nucleus B to nucleus A is equivalent to adding to the z component of the mag-

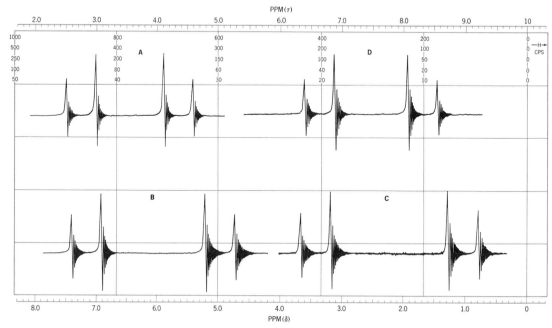

figure 84a. NMR spectrum of 2,3-dibromothiophene A. Pure liquid. B. 50% solution in acetonitrile. C. 10% solution in acetonitrile. D. 20% solution in CCl₄.

netic field acting on nucleus A the quantity $m_B J_{AB}$, where m_B is the quantum number describing the orientation of nucleus B relative to the field. For spin $\frac{1}{2}$ there are two orientations possible, and because they differ in energy by so small an amount, their populations are practically equal. In the experimental spectrum which constitutes a study of a very large number of the two-spin systems considered, in half the cases nucleus A sees one orientation for B, in the remainder the other orientation for B, accounting for the splitting of the resonance of A into two lines of equal intensity and separation equal to the coupling constant. The extension of this first-order prediction to more complex systems is considered below.

In the other extreme, where the chemical shift becomes small compared with the spin-coupling constant, the spectrum gives no indication of the existence of the coupling. In general, there will be no evidence in the spectrum of spin couplings between nuclei in a chemically equivalent set provided that the coupling to any nucleus outside the set is exactly the same for all nuclei in the set. If this special condition is satisfied, the set of nuclei is said to be magnetically equivalent as well as chemically equivalent. Such magnetic equivalence may be inherent in the symmetry of the molecule containing the nuclei but can also result from the averaging

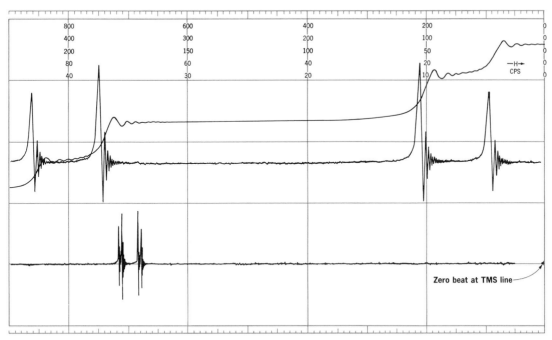

—H→
CPS

Zero beat at TMS line

figure 84b. NMR spectrum of 2,3-dibromothiophene at 100 Mc sec⁻¹ (nominal 23-kgauss field).

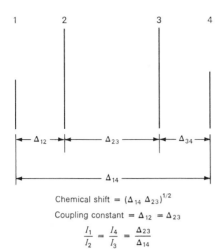

$$\text{Chemical shift} = (\Delta_{14}\,\Delta_{23})^{1/2}$$

$$\text{Coupling constant} = \Delta_{12} = \Delta_{23}$$

$$\frac{I_1}{I_2} = \frac{I_4}{I_3} = \frac{\Delta_{23}}{\Delta_{14}}$$

figure 85. Calculation of chemical shift and coupling constant from AB pattern.

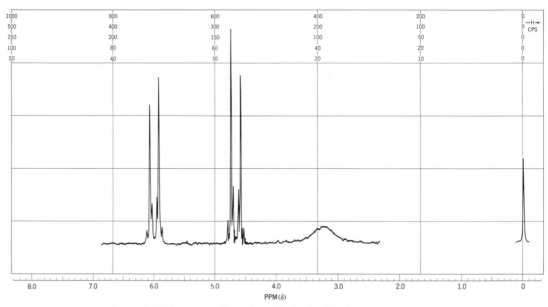

figure 86. NMR spectrum of paranitroaniline in CH₃CN solution.

effect of intramolecular motion. Thus the rotation of a methyl group about the carbon-carbon bond makes the three methyl protons magnetically as well as chemically equivalent.

The effect of the coupling of a nucleus B to a nucleus A can be observed only if nucleus B maintains a particular orientation relative to the field for a sufficiently long time. Exchange of nuclei between molecules or reorientation due to electric field gradients of a nucleus having an electric quadrupole moment can reduce the mean lifetime of a spin state to the point where an otherwise effective spin coupling disappears. For an appropriate intermediate range in lifetime, a marked broadening of the spin multiplet will occur, as is often observed in the spectra of molecules having spin coupling of protons to deuterium or nitrogen as shown in Fig. 86.

The foregoing principles provide a basis for the explanation of the structure of the spectrum of ethanol as shown in Fig. 87a. The sample studied contained a slight amount of acid to catalyze exchange of hydroxyl protons between molecules. The integrated intensities of the three groups of lines stand in the ratio of 1:2:3, confirming their assignment to the hydroxyl, methylene, and methyl protons, respectively. The appearance of the hydroxyl peak at the lowest applied field can be predicted on the basis of the relatively high electronegativity of oxygen, which produces a reduction of electron density in the neighborhood of the hydroxyl proton,

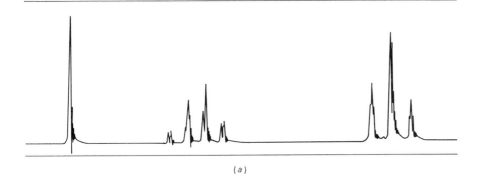

(a)

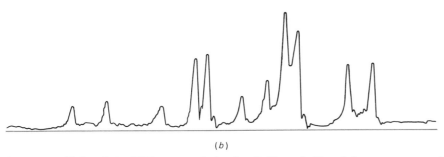

(b)

figure 87. (a) High-resolution NMR spectrum of ethanol at 60 Mc sec^{-1}; (b) methylene-group spectrum of ethanol, 12 Mc sec^{-1} spectrometer frequency. (Varian Associates, Instrument Division.)

and hence reduces its screening constant. The single hydroxyl peak results from the elimination of spin-coupling effects by the rapid exchange of hydroxyl protons between molecules.

The methylene and methyl line groups have approximately the quartet and triplet structures, respectively, that the first-order-perturbation prediction would suggest. The basic line spacing is seen to be essentially constant through the two multiplet groups, which are well separated in this spectrum, obtained with a spectrometer frequency of 60 Mc sec^{-1}. The two proton sets considered individually are chemically and magnetically equivalent because of the internal rotation about the carbon-carbon bonds in the molecule. The multiplet structure then results only from interactions between the two sets. The first-order effect of the methylene protons on the methyl protons, as noted above, depends on the z component of the resultant spin angular momentum of the two methylene protons, each of which can have, with equal probability, $m = +\frac{1}{2}$ or $m = -\frac{1}{2}$. The resultant value of I_z (in units of $h/2\pi$, of course) can then be $+1$, 0, or -1. The value 0 is twice as probable as either of the other two, since the resultant $+1$ can arise only from $m_1 = +\frac{1}{2}$, $m_2 = +\frac{1}{2}$, -1 from $m_1 = -\frac{1}{2}$, $m_2 = -\frac{1}{2}$, while 0 can result from *either* $m_1 = +\frac{1}{2}$, $m_2 = -\frac{1}{2}$ or $m_1 = -\frac{1}{2}$, $m_2 = +\frac{1}{2}$, where subscripts 1, 2 label the two methylene protons. It would then be expected that the single resonance line which would result from the chemically equivalent methyl protons in the absence of spin coupling would split into three lines, with the central component twice as intense as either of the other two. A similar line of reasoning will predict for the methylene resonance a splitting into four lines in a $1:3:3:1$ intensity ratio.

These first-order arguments succeed reasonably well in this case because the spin-coupling constant is fairly small ($J = 7.15$ cps) compared with the chemical shift of $\Delta\nu = 147.4$ cps. The second-order corrections to the energies of the levels concerned are proportional to $J^2/\Delta\nu$ and are not negligible even in the present case, as is obvious from the spectrum. If the spectrometer operating frequency is reduced to 12 Mc, the methylene group spectrum as shown in Fig. 87*b* in no way resembles the first-order prediction. The second-order effects obviously constitute a complication which it is desirable to minimize, and this is a further reason for use of as high an operating frequency (and field) as possible, to increase the chemical shift between the interacting groups. Since the coupling constants are independent of the field, the result is a reduction of the second-order splittings.

In the high-resolution proton NMR spectrum, then, the positions of the line groups relative to the selected reference permit the identification of the various types of structural environments of the protons present. Determination of the integrated line intensities leads to a calculation of the relative numbers of protons in the several environments. Further structural information can be obtained from analysis of the fine structure due to spin-spin interaction, since a particular postulated structure will necessarily imply particular spin-coupling effects in the spectrum.

In order to obtain results characteristic of a given molecule, the experimental conditions must provide for the elimination of the effects of interactions between different molecules. For this reason high-resolution studies are made on liquids or solutions, where the reorientation of molecules due to random brownian motion proceeds rapidly enough to average out the effects of all such interaction. A molecule in a liquid phase is present essentially in a cavity in a medium having magnetic properties which must be considered in a comparison of chemical shifts as measured, for example, on different pure liquids or on solutions in different solvents. Such comparisons are best made in terms of data obtained with solutions in a common solvent and for dilutions high enough for solute-solute interactions to be negligible. Solvents commonly used include CCl_4, $CDCl_3$, and a variety of other deuterated species.

····→ *Apparatus.* NMR spectrometer; precision sample tubes; tetramethyl silane; acetaldehyde, benzene; 2,3-dibromothiophene; methyl ethyl ketone.

PROCEDURE. The operating instructions for the NMR spectrometer should be carefully read and understood before any experimental work is attempted. Recommended preliminary reading also includes Chaps. 4 and 6 of Ref. 5, and Chaps. 6 and 7 of Ref. 2.

The spectrometer system is kept in continuous operation in order to obtain improved stability. Adequate performance usually can be obtained by adjustment of the field homogeneity controls, which govern the magnitudes and directions of flat coils of copper wire inside the pole cap covers and of configurations appropriate for creation of small field contributions in the magnet gap to cancel out gradients otherwise present. Care should be taken to ensure smooth spinning of the sample tube, as it is essential to take full advantage of the increase in effective field homogeneity to be obtained thereby. The ringing of lines in the low-field quartet in the spectrum of acetaldehyde is suggested as a criterion of field homogeneity.

Spectrometer records are then obtained to illustrate:

1. Comparison of line or multiplet spectra with spinning and nonspinning sample
2. Effect of H_1 field intensity of typical line or multiplet
3. Effect of sweep rate on spectrum obtained
4. Calculation of chemical shift of typical line, such as benzene line, relative to TMS by sideband method
5. Determination of T_1 for air-saturated and degassed benzene samples
6. Effect of solvent on AB pattern for 2,3-dibromothiophene (Fig. 84a)
7. Methyl ethyl ketone (MEK) spectrum, with calibration sidebands, for MEK as solute in CCl_4 (Fig. 88)

CALCULATIONS. The values of T_1 for the benzene samples are calculated as described previously (page 302). The line positions in the various spectra are calculated, relative to TMS or otherwise as appropriate, by use of the sideband data. The chemical shift and coupling-constant values are then determined for the dibromothiophene for each solution studied. For this purpose the relations of Eqs. (10) and (11) can be applied as shown in Fig. 85. Calculations based on Fig. 84b are also made to verify that the coupling constant is independent of the intensity of the H_0 field, while the chemical shift is proportional to the H_0 field intensity.

Estimates of the chemical shifts and coupling constants for MEK are derived from the observed line spacings, which are compared with those indicated in Fig. 88. An explanation of the structure of the spectrum is advanced.

A qualitative discussion is given of the results obtained in procedures 1 to 3.

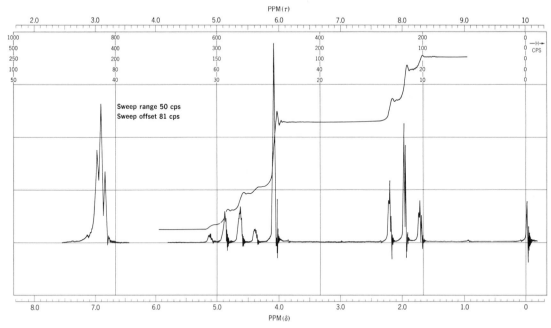

figure 88. NMR spectrum of methyl ethyl ketone in CCl₄ solution at 60 Mc.

Practical applications. High-resolution NMR spectroscopy is the most important new tool for the qualitative study of molecular structure introduced in the last several decades. Studies are not restricted to proton spectra but commonly are extended to fluorine, boron, phosphorus, and carbon 13 as well. Other applications of NMR spectroscopy include studies of crystal structure[3] (notably the location of protons, which may be difficult to establish by x-ray methods), rate processes,[17] etc.

Suggestions for further work. Spectra of compounds containing fluorine as well as hydrogen may be studied, through both fluorine and proton spectroscopy, to further illustrate the reciprocal nature of spin couplings. The consequences of magnetic nonequivalence for chemically equivalent nuclei may be shown in the spectrum of 1,1-difluoroethylene. Rate processes may be studied.[17] Interpretation may be attempted for spectra of other compounds, on the basis either of experimental determination of the spectra or the use of records, such as those available in spectrum catalogs.[8]

References

1. J. A. Pople, W. G. Schneider, and H. J. Bernstein, "High-resolution Nuclear Magnetic Resonance," McGraw-Hill Book Company., New York, 1959.
2. J. W. Emsley, J. Feeney, and L. H. Sutcliffe, "High Resolution Nuclear Magnetic Resonance Spectroscopy," Pergamon Press, New York 1965.

3. E. R. Andrew, "Nuclear Magnetic Resonance," Cambridge University Press, New York, 1955.
4. E. D. Becker, "High Resolution NMR," Academic Press Inc., New York, 1968.
5. Staff, Varian Associates Instrument Division, "NMR and EPR Spectroscopy," Pergamon Press, New York, 1960.
6. A. L. Van Geet and D. N. Hume, *Anal. Chem.*, **37:** 979 (1965); **37:** 983 (1965).
7. W. Brugel, "Nuclear Magnetic Resonance Spectroscopy and Chemical Structures," Academic Press Inc., New York, 1967.
8. Varian Spectrum Catalogs, Varian Associates, Palo Alto, Calif.
9. O. Haworth and R. E. Richards, in J. W. Emsley, J. Feeney, and L. H. Sutcliffe (eds.), "Progress in Nuclear Magnetic Resonance Spectroscopy," vol. I, Pergamon Press, New York, 1966.
10. G. A. Williams and H. S. Gutowsky, *Phys. Rev.*, **104:** 278 (1956).
11. F. Bloch, *Phys. Rev.*, **94:** 496 (1954).
12. W. A. Anderson and J. T. Arnold, *Phys. Rev.*, **94:** 497 (1954).
13. P. J. Paulsen and W. D. Cooke, *Anal. Chem.*, **36:** 1713 (1964).
14. R. R. Ernst in J. S. Waugh (ed.), "Advances in Magnetic Resonance," vol. 2, Academic Press Inc., New York, 1966.
15. A. A. Bothnerby in J. S. Waugh (ed.), "Advances in Magnetic Resonance," vol. 1, Academic Press Inc., New York, 1965.
16. S. Sternhell, *Rev. Pure Appl. Chem.*, **14:** 15 (1964).
17. L. W. Reeves in V. Gold (ed.), "Advances in Physical-Organic Chemistry," Academic Press Inc., New York, 1965.

Chapter 14

Diffraction

This experiment illustrates the determination of the lattice type for cubic crystals and the calculation of the size and mass of the unit cell.

THEORY.[1-7]   In the interpretation of the x-ray diffraction patterns of crystals, ideas of symmetry are of basic importance. A *symmetry operation* is a geometrical manipulation (such as rotation about an axis or reflection in a plane) such that an object assumes an aspect indistinguishable from its appearance before the operation. The individual symmetry operations are summarized in Table 1. In the operation of inversion every point in the crystal is displaced in a straight line through a point, called the center of inversion, to a position an equal distance on the other side of the point. A twofold rotation-inversion about a given axis is equivalent to a reflection.

An *element of symmetry* is the line, point, or plane about which the operation is performed. For example, among the elements of symmetry of a cube are fourfold axes (which are perpendicular to the centers of faces), threefold axes (which pass through diagonally opposite corners), reflection planes, and a center of symmetry. For any object the symmetry elements must pass through a common point. Thus the symmetry operations leave one point unchanged.

A set comprising symmetry operations of an object constitutes a *point group*. Only axes of orders 1, 2, 3, 4, and 6 can occur in crystals; considering all compatible combinations of symmetry elements that pass through a common point, it then follows that there are only 32 for crystal point groups. A crystal can be classified according to symmetry by stating to which of these 32 point groups it belongs.

A *space lattice* is an array of points repeated indefinitely through space so that the environment of each point is identical. In 1848 Bravais showed that there are only 14 different ways of arranging points in a space lattice. Three of these are cubic space lattices: primitive, body-centered, and face-centered. If in any space lattice the distance from one point to another is measured, another point will be found at twice this distance in the same direction, a third at three times the distance, etc. A *unit cell* is defined by three of these unit displacements in three definite directions, all originating from the same point. There is an infinite number of ways in which a unit cell may be chosen, but it is advantageous to select a unit cell so that it has the smallest possible volume and the maximum symmetry of the lattice. In

Table 1. Symmetry Operations

Operation	Element	Symbols
Rotation	Axis	1, 2, 3, 4, 6
Reflection	Mirror plane	m
Inversion	Center of symmetry	$\bar{1}$
Rotation followed by inversion	Axis of rotatory inversion	$\bar{1}, \bar{2}, \bar{3}, \bar{4}, \bar{6}$

the primitive cubic unit cell there is a lattice point at each corner of the cube, in the body-centered unit cell there is an additional lattice point in the center of the unit cell, and in the face-centered unit cell there are six additional lattice points, one in the center of each face.

A crystal structure based on a lattice can be compared to a wallpaper pattern; the design of the wallpaper consists of a pattern or motif placed at each point of the two-dimensional lattice. In an analogous fashion the structure of a crystal consists of an atom or group of atoms as a motif situated about each lattice point. The lattice points themselves may be occupied by atoms, as in the crystal structures of many common metals, or they may serve as imaginary points about which a group of atoms is clustered. When the various symmetry operations are combined with the various space lattices, it is found that there are only 230 different possible combinations. These combinations are referred to as *space groups*, and each consists of a collection of symmetry elements, each with its own location in the unit cell defined. It is most helpful to know the space-group symmetry since the atoms, molecules, or ions in the unit cell must conform to give the configuration of the space group. As a result of symmetry, one need only specify the positions of a certain number of atoms in the cell; the positions of the other atoms are generated by the symmetry operations of the crystal. In this experiment we shall be concerned only with the space lattices of the cubic system.

Scattering Theory. X-rays are scattered by the electrons of the atoms, and the superposition of waves scattered by the individual atoms results in diffraction, the intensity in any direction depending on whether the individual scattered waves are in phase. Hence the *intensities* of diffracted beams are determined by the *distribution of atoms* within the unit cell. One can think of x-rays as being reflected from a given family of planes as specified by Bragg's law: the *angles of reflection* of x-rays are determined by the *geometry* of the lattice, i.e., the size and shape of the unit cell.

In cubic crystals the interplanar spacing d_{hkl} of a family of planes with Miller indices *hkl* is given, from strictly geometrical considerations, by

$$d_{hkl} = \frac{a}{\sqrt{h^2 + k^2 + l^2}} \tag{1}$$

where a is the lattice constant, the length of the edge of the unit cell. The Miller indices of a given type of plane may be obtained by counting the number of planes crossed in moving one lattice spacing along the three axes of the crystal; h, k, and l are the resulting integers for the A, B, and C axes, respectively.

The angles θ which the incident and diffracted beams make with the planes having Miller indices *hkl* depend upon the wavelength of the x-rays and d_{hkl} according to Bragg's law,

$$\lambda = 2d_{hkl} \sin \theta \tag{2}$$

Equations (1) and (2) may be combined to obtain

$$\sin^2 \theta = \frac{\lambda^2}{4a^2} (h^2 + k^2 + l^2)$$

$$= \frac{\lambda^2 N}{4a^2} \tag{3}$$

where $N = h^2 + k^2 + l^2$. The characteristic values of N possible for the various types of cubic crystals are summarized in Table 2. For primitive cubic crystals there is a repeating unit (ion, atom, molecule, or group of atoms) at each corner of the cubic unit cell. All lattice planes that can be drawn through the corners of the unit cells can lead to the reflection of x-rays. As can be seen from Table 2, $N = 7, 15,$ 23 and certain larger values are missing; this is the distinguishing characteristic of the primitive cubic lattice. These values of N cannot be obtained as the sum of the squares of three integers.

For face-centered cubic crystals there is a repeating unit (ion, atom, molecule, or group of atoms) in the center of each face of the cubic unit cell, as well as at each corner. Consideration of the various types of reflecting planes shows that these extra lattice points lie on the planes for which the Miller indices are either all even or all odd: 111, 200, 220, 311, etc. Reflections are obtained from these planes, but reflections are not obtained from the planes for which the Miller indices are not all even or all odd, because for these cases there are lattice points between these planes which destroy the reflections by interference. Such systematic absences make it possible to determine the nature of the crystal lattice from x-ray diffraction.

For body-centered cubic crystals there is a repeating unit (ion, atom, molecule, or group of atoms) at the center of each unit cell, as well as at each corner. Inspection of the various types of reflecting planes shows that these extra lattice points lie on the planes for which the sum of the Miller indices $h + k + l$ is even: 110, 200, 211, 220, etc. The reflections from planes with $h + k + l$ odd are destroyed by interference.

As a result of certain arrangements of atoms within the unit cell, certain reflections may be very weak. However, the correct structure can be obtained even when there are accidental absences of this sort.

The density ρ of a crystal is equal to the mass of a unit cell divided by the unit-cell volume. For a cubic crystal,

$$\rho = \frac{n(M/N_0)}{a^3} \tag{4}$$

where n is the number of units of molecular weight M in the unit cell. If a single atom or molecule is associated with each lattice point, $n = 1$ for primitive cubic, $n = 2$ for body-centered cubic, and $n = 4$ for face-centered cubic.

Table 2. Miller Indices of Cubic Crystals

N	Primitive	Face-centered	Body-centered	N	Primitive	Face-centered	Body-centered
1	100			36	600, 442	600, 442	600, 442
2	110		110	37	610		
3	111	111		38	611		611
4	200	200	200	39			
5	210			40	620	620	620
6	211		211	41	621, 540, 443		
7				42	541		541
8	220	220	220	43	533	533	
9	300, 221			44	622	622	622
10	310		310	45	630, 542		
11	311	311		46	631		631
12	222	222	222	47			
13	320			48	444	444	444
14	321		321	49	700, 632		
15				50	710, 543		710, 543
16	400	400	400	51	711, 511	711, 511	
17	410, 322			52	640	640	640
18	411, 330		411, 330	53	720, 641		
19	331	331		54	721, 633, 552		721, 633, 552
20	420	420	420	55			
21	421			56	642	642	642
22	332		332	57	722, 544		
23				58	730		730
24	422	422	422	59	731, 553	731, 553	
25	500, 430			60			
26	510, 431		510, 431	61	650, 643		
27	333, 511	333, 511		62	732, 651		732, 651
28							
29	520, 432						
30	521		521				
31							
32	440	440	440				
33	441, 522						
34	530, 433		530, 433				
35	531	531					

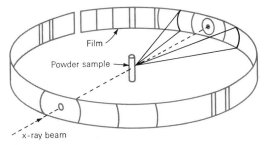

figure 89. X-ray powder camera.

In the powder method, originated by Hull and by Debye and Scherrer, a monochromatic x-ray beam is passed through a large number of fine crystals oriented in random directions. The arrangement of the film is shown in Fig. 89. Since crystals are oriented in all directions with respect to the incoming x-ray beam, all possible reflections are obtained. The rays reflected at a given angle form a cone.

There are a number of other experimental arrangements for studying x-ray diffraction. In the Laue method a single oriented crystal is irradiated with a collimated beam of polychromatic x-rays. In other methods the crystal is rotated or rocked in a particular way during the exposure, and the film may also be moved. The interpretation of such experiments is more complicated than for powder patterns, but much more detailed information about the crystal structure may be obtained.

Apparatus. X-ray apparatus equipped with a powder camera; fine powders of cubic crystals, for example, KCl, NaCl, NaBr; x-ray film and supplies for developing it; accurate millimeter scale or microcomparator.

PROCEDURE. It is important that the crystalline sample be pure, since small amounts of impurities will give rise to additional lines that may make the interpretation of the x-ray pattern difficult. If the crystals are not sufficiently fine, the sample is ground further with a mortar and pestle, so that the diffraction pattern will not contain large spots due to single crystals. The sample is either mounted in a thin-walled capillary or is attached to the outside of a thin glass rod, using a cement, as prescribed by the instructor.

The sample is mounted in the powder camera so that it is precisely lined up with the collimator for the x-ray beam. The alignment of the sample is checked with the motor rotating the sample, as it will during the exposure. The specific instructions for using the particular x-ray powder camera employed should be followed in detail.

After the exposure the photographic film is developed, washed, and dried. Information about photography is to be found on page 566.

CALCULATIONS.   The first step is to find the θ values (definition, page 322) for the observed reflections. The diffraction angles in degrees will be calculated from

$$2\theta = 180\frac{X}{R}$$

where X = distance of a line from center of exit hole

R = distance from center of entrance hole to center of exit hole

The distances X and R are lengths of arc measured along the film. The best procedure is to first tabulate the positions of the lines relative to an arbitrary reference point. This may be done by means of a steel scale which is laid along a line between the points where the x-ray beam would have passed through the film in entering the camera to the position where it would have passed through the film in leaving the camera. Note that the lines occur in pairs placed symmetrically about the exit hole. The positions of the holes are found by averaging the positions of several pairs of corresponding lines.

Next, the lattice type is to be identified by finding which column of N values (Table 2) is consistent with the values of θ found. To do this, first tabulate values of $\sin^2 \theta$, since these should be proportional to N values. (Tables of $\sin^2 \theta$ are available in certain handbooks.) The largest common factor of the values of $\sin^2 \theta$ is found by subtracting each value from the next higher and examining the differences. The values of $\sin^2 \theta$ are divided by this common factor to obtain the apparent N values, which are then compared with Table 2. These values should be sufficiently close to integers so that the corresponding integer can be recognized, provided that N is not too large. Impurities in the crystalline sample will produce extra lines which cannot be accounted for in this way.

In order to obtain the lattice parameter a for the crystal, $\lambda^2 N/4 \sin^2 \theta$ is computed for a number of lines and a is calculated using Eq. (3).

The average wavelength of copper $K\alpha$ radiation is 1.542 Å. Actually, this radiation consists of two lines, $K\alpha_1$ and $K\alpha_2$. For low values of θ these two lines are not resolved, but at large values of θ they may be visible. When they are, the average position is taken.

From the value of a obtained, the density of the crystal is calculated from Eq. (4). This value is compared with the literature value.

Practical applications. Diffraction measurements of x-rays are important in the identification of crystalline materials, including biological materials, and in the interpretation of the structure of crystals.

Suggestions for further work. Unknown powder samples may be identified by determining the d spacings of the three most intense lines and looking these distances up in the ASTM x-ray Diffraction Pattern Card Index. If there are several crystals which have nearly the same strong lines, these crystals may be distinguished from each other on the basis of further lines in the pattern.

The next level of complexity in powder patterns is encountered with tetragonal crystals. The procedure for analyzing such patterns is discussed in textbooks of x-ray diffraction.

Other types of x-ray diffraction experiments may be used. For example, Laue patterns may be taken of a single crystal which is oriented so that a beam of "white" x-rays passes down various axes of the crystal.

References

1. L. V. Azároff and M. J. Buerger, "The Powder Method in X-ray Crystallography," McGraw-Hill Book Company, New York, 1958.
2. B. D. Cullity, "Elements of X-ray Diffraction," Addison-Wesley Publishing Company, Inc., Reading, Mass., 1956.
3. F. Daniels and R. A. Alberty, "Physical Chemistry," John Wiley & Sons, Inc., New York, 1966.
4. R. W. M. D'Eye and E. Wait, "X-ray Powder Photography," Butterworth & Co. (Publishers) Ltd., London, 1960.
5. H. P. Klug and L. E. Alexander, "X-ray Diffraction Procedures," John Wiley & Sons, Inc., New York, 1954.
6. W. N. Lipscomb in A. Weissberger (ed.), "Technique of Organic Chemistry," vol. 1, "Physical Methods of Organic Chemistry," 3d ed., pt. II, Interscience Publishers, Inc., New York, 1960.
7. D. F. Eggers, N. W. Gregory, G. D. Halsey, Jr., and B. S. Rabinovitch, "Physical Chemistry," John Wiley & Sons, Inc., New York, 1964.

Chapter
15

Macromolecular
Chemistry

Viscosity determinations are very important in the study of high polymers. Using simple viscosity measurements, an average molecular weight of the polymer may be determined and certain qualitative conclusions may be made as to the general form of the macromolecules in solution.

THEORY.   One is confronted with a number of special problems in the characterization of macromolecules. Sometimes such molecules assume the form of "random coils" in solution, but in other cases they may exist in helical or partial helical array. One differentiation of certain biological macromolecules from the synthetic organic high polymers is that the former may assume several different helical forms. Such secondary structures ordinarily do not appear with the synthetic materials of commercial value. It is not surprising then that to make a study of the size and shape of the macromolecular solutes different approaches may be required in the two cases.

In this experiment the objective is to make an estimate of the size of an average organic high-polymer molecule from observations of the viscosity behavior of its very dilute solutions. The viscosity of such solutions is probably their most widely measured property; it is very useful with molecules of the random-coil type. In the viscosity tube the flow arises from an externally applied gradient, and depending on its size and shape, the macromolecule is subjected to a torque which produces a small effective orientation in the field. In this way a dissipation of energy is produced to give rise to an increase of the viscosity of the solution as compared to that of the solvent.

Before providing definitions of the several viscosity functions for use, it perhaps should be noted that transport of this kind is not as simple as the one found in diffusion and in sedimentation and that the frictional factor which measures the resistance to flow is now a more complicated quantity. Furthermore, although the viscosity of a polymer solution does depend in part on the size of the polymer molecules, the viscosity experiment does not "count" or "weigh" them and the molecular-weight average for a polydisperse polymer so obtained does not often conform to either the simple number- or weight-average quantity; it is called the *viscosity average*. In general it is much closer to the weight-average molecular weight than to the number-average quantity. We thus have a simple experiment from the point of view of requirements of apparatus but one which is complicated in theory and interpretation of the data.

The viscosity experiment may provide not only mass and size information about the polymer molecule but also certain thermodynamic data for the solution. For the former purpose the interpretation of the experimental data may be said to have begun with Einstein,[1] who showed that the coefficient of viscosity η of a dilute suspension of small unsolvated and uncharged rigid spheres is given by the expression

$$\lim_{\Phi \to 0} \frac{\eta/\eta_0 - 1}{\Phi} = 2.5 \tag{1}$$

where η_0 = coefficient of viscosity of solvent

η/η_0 = viscosity ratio

Φ = fraction of total volume occupied by the spheres themselves = phase volume

Later Simha extended the formula to apply to ellipsoidal particles, making use of what is termed the viscosity increment ν

$$\lim_{\Phi \to 0} \frac{\eta/\eta_0 - 1}{\Phi} = \nu \tag{1a}$$

However, in the case of polymer solutions it is impossible to calculate the volume occupied by the polymer in an unambiguous way, and the equations were so written that the concentration scale was changed from volume fraction to weight per volume unit. Thus, Kraemer[2] defined a quantity called the *intrinsic viscosity* $[\hat{\eta}]$ as follows:

$$\lim_{\hat{c} \to 0} \frac{\eta/\eta_0 - 1}{\hat{c}} = [\hat{\eta}] \tag{2}$$

The quantity $\hat{c}$ is the concentration in grams of solute per 100 ml of solution. It has since seemed advisable to use the concentration c in grams of solute per milliliter, so that

$$\lim_{c \to 0} \frac{\eta/\eta_0 - 1}{c} = [\eta] \tag{2a}$$

and to designate the quantity $[\eta]$ as the *limiting viscosity number*. Thus, the viscosity number, $(\eta/\eta_0 - 1)/c$, measures the contribution to the viscosity of the polymer molecules at concentration c, with the limiting viscosity number being a measure of the contribution per macromolecule, the really significant quantity. With the different concentration scale it has the dimension milliliters per gram. The quantity $\eta/\eta_0 - 1$ also occurs frequently in the theory of the viscosity of solutions; it is designated as η_{sp}. In passing, it will be noted that the equivalent of the Einstein equation in terms of the limiting viscosity number is

$$[\eta] = \lim_{c \to 0} \frac{\eta/\eta_0 - 1}{c} = 2.5\,\bar{v} \tag{2b}$$

where $\bar{v}$ is the partial specific volume of the polymer in the solution.

Plots of $(\eta/\eta_0 - 1)/c$ versus c are generally linear at low solute concentrations so that the extrapolated value $[\eta]$ may be determined with precision.

Another plot, often preferred, is $(1/c) \ln (\eta/\eta_0)$ versus c to give the limiting viscosity number. That the two plots have the same intercept is shown by expansion of the logarithm.

$$\ln \frac{\eta}{\eta_0} = \ln (1 + \eta_{sp}) = \eta_{sp} - \frac{\eta_{sp}^2}{2} + \cdots \tag{2c}$$

and noting that second- and higher-order terms in η_{sp} become negligible as compared with the first as the concentration approaches zero.

In contrast with the case for spheres, the limiting viscosity number for a linear soluble high polymer is found to be a function of its molecular weight. For example, there exists an empirical relationship of the form

$$[\eta] = KM^a \tag{3}$$

where K and a are constants which depend on the solvent used and the temperature. The exponent a usually varies between 0.5 and 1.0 for different solvents, polymers, and temperatures. For the so-called random-coil molecule in a "poor" solvent (little or no solvent binding) the value 0.5 can be expected. The extensive analysis has been presented by Flory and Fox,[3,4] by Kirkwood and Riseman,[5] and others.

A much simplified argument suggests that this value is not unexpected. For a flexible chainlike molecule the average end-to-end distance $\bar{r}$ will be $\bar{r} = l\sqrt{n}$, where l is the length per monomer unit and n is their total number in the molecule. For carbon-carbon backbone, with tetrahedral angle structure, this expression is modified, so that

$$\bar{r} = l\sqrt{n}\sqrt{2}$$

Now, if a vector $\mathbf{R}$ is drawn from the center of mass to each segment of the chain, it can be shown that its rms average length is proportional to $\sqrt{n}$.

The volume occupied by the polymer molecule is proportional to $(\overline{R^2})^{\frac{3}{2}}$ so that it is proportional to $n^{\frac{3}{2}}$ or $M^{\frac{3}{2}}$. The actual volume occupied by the segments is proportional to n or M. Thus the effective volume per unit mass, that is $[\eta]$, is proportional to $M^{\frac{3}{2}}/M$, or $M^{\frac{1}{2}}$. The more detailed theories are clearly indications of the reason why a is usually somewhat greater; there is an expansion of the molecule as a whole due to solvent permeation.

The values of K and a listed in Table 1 have been determined for fractionated polymers.

In addition to yielding molecular weights, viscosity measurements give us some insight into the general form of polymer molecules in solution.[4,5,7-11] A long-chain molecule in solution takes on a somewhat kinked or curled shape, intermediate between a tightly rolled-up mass and a rigid linear configuration. Presumably, all possible degrees of curling are represented, owing to the internal brownian move-

Table 1. Parameters for Eq. (3)[6]

Polymer	Solvent	T, °C	$K \times 10^4$	a
Cellulose acetate	Acetone	25	1.49	0.82
Polyisoprene	Toluene	25	5.02	0.67
Polystyrene	Toluene	25	3.7	0.62
GR-S copolymer	Toluene	30	5.4	0.66
Methyl methacrylate	Benzene	25	0.94	0.76
Polyisobutylene	Toluene	20	3.6	0.64

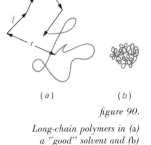

(a)   (b)

figure 90.

Long-chain polymers in (a) a "good" solvent and (b) a "poor" solvent.

ment of the flexible chains. In a "good" solvent, i.e., one which shows a zero or negative heat of mixing with the polymer, the polymer molecule is rather loosely extended, as represented in Fig. 90a, and the intrinsic viscosity is high. In a "poor" solvent, i.e., one in which the polymer dissolves with positive heat of mixing (absorption of heat), the segments of the polymer molecule attract each other in solution more strongly than they attract solvent molecules, and the result is that the molecule assumes a more compact shape as illustrated in Fig. 90b. Consequently, in a poor solvent the intrinsic viscosity will be lower than in a good solvent.

Apparatus. Ostwald viscometer; stopwatch; pipettes; 25-ml volumetric flasks; polymer sample;† toluene; methanol.

PROCEDURE.   Solutions of the polystyrene sample of unknown molecular weight are prepared in a good solvent (toluene) and a poor solvent (a mixture of toluene and methanol). Since the polymer may dissolve rather slowly, warming in a water bath may be used to accelerate solution. When this is done, the solution should be cooled to 25° before adding solvent to bring the meniscus up to the mark on the volumetric flask. The following solutions are needed:

1. Five hundred milligrams of polystyrene is dissolved in toluene and diluted quantitatively to 25 ml in a volumetric flask.
2. One hundred milliliters of a solution containing 15 percent methanol and 85 percent toluene by volume is prepared. Again 500 mg of polystyrene is dissolved in this solvent and diluted quantitatively to 25 ml. The remaining solvent is required for dilutions and a flow-time determination.

After the viscometer has been thoroughly cleaned with cleaning solution, it is rinsed and dried by aspirating clean air from the laboratory through it. It is important for the viscometer to be perfectly dry inside before organic solvents are added.

† Dow polystyrenes.

The flow time in the viscometer is determined for toluene and for the methanol-toluene solvent as described in Experiment 26. The flow time for the solution of polystyrene in toluene is determined. The sample is then diluted by a factor of 2 and the flow time determined. The accuracy of the dilutions may be improved by use of two calibrated pipettes, one calibrated for withdrawal of solution and the other for delivery of solvent. This dilution is repeated until the viscosity ratio of the polymer solution becomes so close to 1 that there is a large error in the specific viscosity. Since the densities of the dilute polymer solutions are not significantly different from that of the solvent, it is unnecessary for our present purpose to determine the densities.

The same procedure is now repeated with the solution of polystyrene in methanol-toluene, in this case diluting with the mixed solvent.

Since methanol is a nonsolvent for polystyrene, the addition of further quantities will cause precipitation of the polymer. The percent methanol by volume required to cause the first turbidity is determined by titrating a few milliliters of the solution of polystyrene in toluene.

Notice: Before the viscometer is allowed to dry inside, it should be rinsed thoroughly with toluene so that a film of polymer will not be left in the capillary.

CALCULATIONS. The limiting viscosity number is obtained by plotting η_{sp}/c and $(1/c) \ln (\eta/\eta_0)$ versus c (in grams of polymer per milliliter of solution) for the solution of polystyrene in toluene. The limiting viscosity number in methanol-toluene is determined by a similar graph. The advantage of the double extrapolation is that the intercept may be determined more precisely than by using only one straight line. Furthermore, there is a simple relationship between the slopes of the two lines which may be verified. For the dependence of specific viscosity with concentration, Huggins has written

$$\eta_{sp} = [\eta]c + k_1[\eta]^2c^2$$

Similarly,

$$\ln \frac{\eta}{\eta_0} = [\eta]c - k_2[\eta]^2c^2$$

To a good approximation it can be shown by application of Eq. (2c) that $k_1 + k_2 = 0.5$.

Using the values of K and a given in Table 1, the molecular weight of the sample of polystyrene is calculated from its limiting viscosity number in toluene. It should be realized that this is an average molecular weight, some of the molecules being larger and some smaller; even the more homogeneous of the so-called living polymers show some heterogeneity.[10]

An estimate as to the size of the average polystyrene molecule in the sample is obtained by calculating the volume of one molecule from the molecular weight, using 0.903 g ml^{-1} for the density of polystyrene. The radius of the molecule is calculated, assuming the molecule to be spherical. The length of the extended polymer molecule

$$\rightarrow| \quad 2.5 \text{ Å} \quad |\leftarrow$$

is calculated, assuming the length per monomer unit is 2.5 Å. The actual shape of the polystyrene molecule in solution is intermediate between the spherical and stretched-out forms and depends upon the solvent and temperature.

Practical applications. Determination of the molecular weight of the soluble organic high polymers such as polyisobutylene, polymethyl methacrylate, and polyvinyl chloride is important because the physical properties of these materials depend markedly on molecular weight.

Suggestions for further work. The general subject of the viscosity of polymer solutions has been well reviewed in the literature, where suggestions for further work may be found.

Viscosity determinations at higher concentrations are used to show that the linear relation between η_{sp}/c and c does not hold at higher concentrations. Ordinary polystyrene may be separated into fractions by precipitating part of it from toluene solution by adding methanol. The average molecular weight of each fraction is estimated by means of viscosity determinations.

For a given polymer, $[\eta]$ may be determined for various concentrations of nonsolvent and plotted against percent nonsolvent.

References

1. A. Einstein, *Ann. Physik*, **19:** 259 (1906); **34:** 591 (1911).
2. E. O. Kraemer, *Ind. Eng. Chem.*, **30:** 1200 (1938).
3. P. J. Flory and T. G. Fox, Jr., *J. Am. Chem. Soc.*, **73:** 1904, 1909, 1915 (1951).
4. P. J. Flory, "Principles of Polymer Chemistry," Cornell University Press, Ithaca, N.Y., 1953.
5. J. G. Kirkwood and J. Riseman, *J. Chem. Phys.*, **16:** 565 (1948).
6. A. I. Goldberg, W. P. Hohenstein, and H. Mark, *J. Polymer Sci.*, **2:** 502 (1947).
7. C. Tanford, "Physical Chemistry of Macromolecules," John Wiley & Sons, Inc., New York, 1961.
8. H. Mark and A. V. Tobolsky, "Physical Chemistry of High Polymeric Systems," 2d ed., Interscience Publishers, Inc., New York, 1950.

9. P. F. Onyon in P. W. Allen (ed.), "Techniques of Polymer Characterization," Academic Press Inc., New York, 1959.
10. M. Szwarc, M. Levy, and R. Milkovich, *J. Am. Chem. Soc.*, **78:** 2656 (1956).
11. H. Tompa, "Polymer Solutions," Academic Press, Inc., New York, 1956.

47 DETERMINATION OF THE OSMOTIC PRESSURE OF A SOLUTION OF HIGH POLYMER

The number-average molecular weight of a sample of polystyrene and the second virial coefficient of its solution in methyl ethyl ketone are calculated from osmotic pressure data taken in dilute systems.

THEORY.   The organic-high-polymer physical chemist uses the osmotic-pressure experiment to obtain a number-average molecular weight of a dissolved polymer and to acquire definitive information about polymer-solvent interactions.[1-4] In the performance of an experiment it is required that the solution come to equilibrium with pure solvent across the semipermeable membrane, one which is permeable only to the solvent. The chemical potential of the solvent in the solution has been rendered less than that of the pure solvent by the introduction of solute. An effect of pressure on the solution is to increase the chemical potential of the solvent component. By the osmotic pressure is understood that excess pressure which must be applied to the solution to establish equilibrium, the condition for which is equality of chemical potential for the diffusible component (or components) in the two phases. Since $(\partial \mu_i / \partial P)_{T,n_i} = \overline{V}_i$, the increase in μ_1 at equilibrium due to the additional pressure π is

$$\Delta \mu_1 = -\pi \overline{V}_1$$

Here, in a two-component system the subscript 1 may be used for the solvent and 2 for the solute. The quantity $\overline{V}_1$ is the partial molal volume of the solvent, and π, as the experimental arrangement indicates, is the osmotic pressure.

In the classical experiment with the two-component system, then, a solution of polymer is placed on one side of a semipermeable membrane, the β phase, and the solvent on the other side, the α phase. With the establishment of equilibrium the chemical potentials of the solvent, the component which can pass the membrane, in the two phases become equal, thus $\mu_1^\alpha = \mu_1^\beta$. The osmotic pressure is defined as the difference in pressure, at equilibrium, on the two sides of the membrane, that is, $P^\beta - P^\alpha$. The experiment is conducted at constant temperature, and the pressure on the solvent, or α phase, is that of the atmosphere, so that μ_1^α and P^α are constants.

The traditional problem of the basic texts in physical chemistry is that of the two-component system of neutral molecules, with the assumption of ideal-solution behavior. However, the description of nonideality enters in a simple and direct way, and it will be included in the derivation of the equation upon which the interpretation of the experimental data depends. We make use of the Gibbs-Duhem equation

$$S\,dT - V\,dP + n_1\,d\mu_1 + n_2\,d\mu_2 = 0$$

For the β phase, with constant temperature,

$$n_1^\beta\,d\mu_1^\beta + n_2^\beta\,d\mu_2^\beta = V^\beta\,dP$$

At equilibrium μ_1^β is constant, and

$$\frac{n_2^\beta}{V^\beta}\,d\mu_2^\beta = dP^\beta = \frac{C_2}{1000}\,d\mu_2^\beta \tag{1}$$

where V, the volume of the β phase, is measured in cgs units and C_2 is the solute concentration in moles per liter. The quantities n represent the numbers of moles of the indicated components.

Now, remembering that $\mu_2 = (\mu_2^\circ) + RT\ln y_2 C_2$, where y_2 is the activity coefficient on the C scale, $\overline{V}_2 = M_2\bar{v}_2$, and that

$$d\mu_2 = \left(\frac{\partial\mu_2}{\partial P}\right)_{C_2}dP + \left(\frac{\partial\mu_2}{\partial C_2}\right)_P dC_2$$

$$= \overline{V}_2\,dP + \frac{RT}{C_2}\left(1 + C_2\,\frac{\partial\ln y_2}{\partial C_2}\right)$$

we have

$$\frac{dP}{dC_2} = \frac{RT}{1000}\left(\frac{1 + C_2\,\partial\ln y_2/\partial C_2}{1 - C_2 M_2\bar{v}_2/1000}\right) \tag{2}$$

The superscripts β have not been continued, but the statements are still restricted to the β phase. When the solute molecular weights are unknown, their concentrations are ordinarily expressed in grams solute per 100 ml solution, $C_2 = (10/M_2)c_2$, and

$$\frac{dP}{dc_2} = \frac{RT}{100 M_2}\left(1 + c_2\,\frac{\partial\ln y_2}{\partial c_2}\right)\left(1 + \frac{c_2\bar{v}_2}{100} + \cdots\right) \tag{3}$$

On integrating,

$$P^\beta - P^\alpha = \pi = \frac{RTc_2}{100 M_2}\left[1 + \frac{c_2}{2}\left(\frac{\bar{v}_2}{100} + \frac{\partial\ln y_2}{\partial c_2}\right)\right] \tag{3a}$$

The quantity $c_2\bar{v}_2/100$ is the volume fraction of component 2, a small quantity in dilute solutions, and it can be neglected. Since we treat a system of neutral molecules, the quantity $\ln y_2$ can be expanded in a power series in c_2, so that

$$\frac{\partial \ln y_2}{\partial c_2} = \text{constant} = B$$

and

$$\pi = \frac{RTc_2}{100M_2} \left(1 + \frac{B}{2} c_2 + \cdots\right) \tag{4}$$

If the concentration of solute is expressed in terms of grams per milliliter of solution, the factor 100 in the denominator at the right disappears, and the equation takes another common form.

This equation indicates that a plot of the quantity π/c_2 versus c_2 will be nearly linear at low concentrations and that on extrapolation to $c_2 = 0$ the intercept will permit the calculation of M_2. The limiting slope of the line makes possible an estimate of the value of the second virial coefficient B.

Ordinarily, the molecules in a sample of an organic high polymer do not all have the same molecular weight. The molecular-weight distribution may be represented by a plot such as that shown in Fig. 91. The fraction of molecules having molecular weights in the interval M to $M + dM$ is represented by $f\,dM$. The dM appears here because the fraction of molecules in a given range of width dM is proportional to the magnitude of dM. In the figure, f is plotted versus M.

In such cases, it can be readily shown that the osmotic-pressure experiment gives what is called a *number-average molecular weight* M_n. By definition,

$$M_n = \frac{\displaystyle\sum_i n_i m_i}{\displaystyle\sum_i n_i} = \frac{\displaystyle\sum_i c_i}{\displaystyle\sum_i \frac{c_i}{M_i}}$$

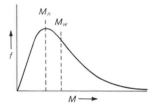

figure 91.

Distribution of molecular weight for a high polymer. The values of M_n and M_w for this molecular-weight distribution are shown.

where n_i is the number of molecules of molecular weight M_i; and c_i, their concentration, is $n_i M_i/V$. Starting with the limiting osmotic-pressure law and making use of the facts that $c = \displaystyle\sum_i c_i$ and $\pi = \displaystyle\sum_i \pi_i$, we have

$$\pi = \frac{RT}{100} \sum_i \frac{c_i}{M_i}$$

Dividing through by c,

$$\frac{\pi}{c} = \frac{RT}{100} \frac{\displaystyle\sum_i \frac{c_i}{M_i}}{\displaystyle\sum_i c_i} = \frac{RT}{100 M_n} \tag{5}$$

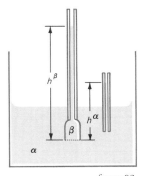

figure 92.

Principle of the osmometer.

From the equations that have been presented it can be demonstrated that the osmotic pressure is proportional to the number of molecules of polymer per molecule of solvent, no matter what their weight may be; in other words, the solute molecules are being counted and not weighed.

The osmometer to be used is of the Schulz-Wagner type,[5,6] illustrated in Fig. 92. It consists of a graduated capillary (0.75 to 1.0 mm inside diameter) attached to a short section of larger tubing (12 mm inside diameter) which has a ground lower surface against which the membrane is held by a brass clamp. The purpose of the lower plate is to hold the membrane tightly against the glass tubing and to support the membrane so that it will not bulge out and stretch during the experiment. The osmometer is filled with polymer solution by use of a syringe with a long stainless-steel needle. This method of filling the osmometer has the advantage that ground-glass joints and valves are avoided, reducing possibilities for leaks to a minimum. The capillary tube provides a correction for capillary rise.

Since the permeability of a membrane to various types of solvents may be very different, osmotic-pressure measurements may be made more rapidly with one solvent than with others.

Apparatus. Two Schulz-Wagner osmometers; polystyrene; methyl ethyl ketone (MEK); 5-ml syringe with long stainless-steel needle; volumetric flasks; cellophane membranes.

PROCEDURE. Du Pont cellophane 600 is soaked overnight in 30 percent sodium hydroxide and washed a few minutes in progressively more dilute solutions of sodium hydroxide and then water. This treatment is necessary to increase the porosity of the membrane. The progressive dilutions of alkali are essential to avoid wrinkling of the membrane. The membranes are then washed in progressively more concentrated solutions of MEK in water. Since they may be stored in MEK, it is unnecessary for each student to carry out the sodium hydroxide treatment.

A weighed sample of polystyrene is dissolved in MEK, and the solution is diluted to the desired volume in a volumetric flask. A suggestion as to the concentration to use may be obtained from an instructor. Since 2 to 4 days are required for equilibration, two or more osmometers should be used. The osmotic pressure for at least three and preferably four concentrations of the same polymer are measured.

In assembling the osmometer, it is important to keep the membrane moistened with MEK, since the effect of the sodium hydroxide treatment will be lost if the membrane is allowed to dry out.

Two disks of a smooth hard filter paper are moistened with MEK and placed on the bottom brass plate. The filter paper serves as a support for the membrane so that it cannot sag into the holes and facilitates a good seal between the membrane and the ground-glass surface of the osmometer bulb. The membrane is placed on the filter paper, and the glass osmometer tube is attached. The knurled nuts are tightened with the fingers. The polystyrene solution is placed in the osmometer by use of a syringe with a long stainless-steel needle, taking care to avoid trapping bubbles at any point.

Precaution: The syringe and needle should be rinsed out with MEK after being used to transfer the polymer solution so that no polystyrene residue will be left in them. Since MEK is a solvent for many of the plastics used in fountain pens and pencils, carelessness may result in damage to them.

Enough MEK is placed in the outer glass tube so that the meniscus comes to the lower part of the graduated capillary when the osmometer is hung in the tube.

The height of the liquid column in the capillary is set at a value several centimeters above, or below, that expected at equilibrium. If it is possible to use two osmometers for each solution, the height is set higher than the expected pressure in one and lower in the other. The osmometer is then suspended in MEK so that the membrane is completely immersed.

The osmometers are placed in a well-regulated thermostat. After allowing some time for temperature equilibration, an initial reading is taken. Readings are then taken at intervals over a period of 2 or more days.

An unfractionated polymer sample contains molecules of a wide range of molecular weights, and some of the lowest-molecular-weight material may diffuse through the membrane. If there is appreciable leakage of low-molecular-weight material, a constant osmotic pressure will not be obtained. It is necessary to correct the observed meniscus height for the capillary rise of the solution. It is satisfactory to determine the capillary rise of the solvent and to assume, for the purpose of this experiment, that the capillary rise for the polymer solution will be the same as that determined for the solvent by measurement of h^α (Fig. 92).

Data required for making another correction should also be obtained with the osmometer used. The need for this further correction arises since the equilibrium height in the capillary is, in general, different from the initial height. If the equilibrium height of the meniscus is greater than the initial height, solvent has passed into the polymer solution, and so the equilibrium polymer solution is more dilute than the original solution. In order to compute the equilibrium concentration it is necessary to know the volume of the osmometer bulb and the radius of the capillary. The volume of the bulb may be determined with sufficient accuracy by measuring the volume of water required to fill it, and the radius of the capillary may be calculated from the capillary rise of MEK (cf. Experiment 51). The surface tension of pure MEK is 23.9 dynes cm^{-1} at 25°, and its density at 25° is 0.803 g cm^{-3}.

CALCULATIONS.   Plots of height h^β versus time are used in determining the osmotic pressure. The osmotic pressure π in atmospheres may be calculated from the equilibrium height difference $h = h^\beta - h^\alpha$ from the equation

$$P^\beta - P^\alpha = \pi = (h^\beta - h^\alpha)\rho_s g \qquad (6)$$

where h^α and h^β = heights indicated in Fig. 92
   ρ_s = density of high-polymer solution (assumed to be the same as for solvent)

If h is in centimeters, ρ_s is grams per cubic centimeter, and g in centimeters per second per second, π will be in dynes per square centimeter. In order to obtain the pressure in atmospheres, this value is divided by 1.013×10^6 dynes cm^{-2} atm^{-1}.

The concentrations c in grams per 100 ml are calculated by correcting the initial concentrations as described above for the flow of solvent, in or out. A plot of π/c versus c is extrapolated to zero concentration. Such plots are expected to be linear in the concentration range suggested, but if the experimental points do indicate a slight curvature, a straight line which is tangent to the curve at zero concentration may be used.   The number-average molecular weight is calculated from the extrapolated value of π/c, and the second virial coefficient B is obtained from the slope of the linear limiting tangent.

Practical applications. Since the properties of a solid plastic or synthetic rubber will depend upon the molecular weight of the polymer, the measurement of osmotic pressure is widely used in industry. The viscosity method, which is also used, is illustrated in Exp. 46.

Suggestions for further work. The molecular weight of a sample of polystyrene may be determined in another solvent. Although the number-average molecular weight obtained from the intercept of a plot of π/c versus c should be the same, the slope may be quite different. On the other hand, the slopes for a series of polystyrene samples of different molecular weight in a given solvent will be very nearly the same.

If the polymer sample contains low-molecular-weight material which leaks through the membrane, this material may be eliminated by precipitating about 75 percent of the polymer by the addition of methyl alcohol to a solution in benzene. The lowest-molecular-weight fraction will remain in the supernatant liquid. The precipitate may be dissolved in benzene and dried by vacuum sublimation of the benzene to obtain a porous preparation of polystyrene which can be readily redissolved.

The polymer sample may be fractionated into samples of different number-average molecular weight by dissolving it in a good solvent and partially precipitating it by the addition of a poor solvent, e.g., methyl alcohol. The molecular weights of the polymer in the precipitate and in the solution phase may be shown to be different by osmotic-pressure measurements.

The osmotic pressure of gelatin or serum albumin in aqueous solution may be determined, and the number-average molecular weights calculated. To reduce the Donnan effect, $0.2\ M$ sodium chloride is used as solvent and the pH is adjusted to the isoelectric point[2] of the protein. The Donnan effect is the contribution to the osmotic pressure due to the unequal distribution of the electrolyte ions across the membrane at equilibrium.

References

1. G. Scatchard, *J. Am. Chem Soc.*, **68:** 2315 (1946).
2. R. L. Baldwin in M. Florkin and E. H. Stotz (eds.), "Comprehensive Biochemistry," vol. VII, pt. 1., American Elsevier Publishing Company, New York, 1963.
3. P. J. Flory, "Principles of Polymer Chemistry," Cornell University Press, Ithaca, N.Y., 1953.
4. E. F. Casassa and H. Eisenberg, *Advan. Protein Chem.*, **19:** 287 (1964).
5. R. H. Wagner, *Ind. Eng. Chem. Anal. Ed.*, **16:** 520 (1944).
6. R. H. Wagner and L. D. Moore, Jr. in A. Weissberger (ed.), "Technique of Organic Chemistry," vol. 1, "Physical Methods of Organic Chemistry," 3d ed., pt. I, Interscience Publishers, Inc., New York, 1959.

48 ION-EXCHANGE CHROMATOGRAPHY

The separation of cations by means of elution from a cation-exchange resin with a complexing reagent is illustrated. The relation of an idealized chromatographic experiment to the adsorption isotherm is discussed.

THEORY.   An ion-exchange resin is made up of insoluble macromolecules with ionizable groups attached. A cation-exchange resin has sulfonic, phenolic, or carboxylic groups, which provide negatively charged binding sites for cations. An anion-exchange resin has basic binding sites. A cation-exchange resin, treated with acid, is said to be in the hydrogen form because the negatively charged groups are neutralized by protons. If metal ions are added to the solutions, protons are displaced. For example, if Ni^{++} and Co^{++} ions are present, the following equilibria are established:

$$Ni^{++} + H_2R \rightleftharpoons NiR + 2H^+ \tag{1}$$

$$Co^{++} + H_2R \rightleftharpoons CoR + 2H^+ \tag{2}$$

where R represents the part of the insoluble macromolecular ion that binds one divalent metal ion. Since these equilibria are reversible, the metal ions may be displaced by raising the acid concentration.

The affinity of the resin for a metal ion is affected by two factors, the radius of the hydrated ion and the valence of the ion. Thus two different metal ions will be held by the resin to different extents. A separation of the two ions might then be achieved by taking advantage of this difference in the affinity of the two ions for the resin.

If a bed of resin is arranged in column form and metal ions are adsorbed in the top layer, they may be washed through the column by a flow of solvent, the rate of transport of the ions being dependent upon their equilibrium concentration in the solution. As the equilibrium solution is washed down the column, the adsorbed ions at the upper edge of the band dissociate into solution and a certain portion of the ions, carried to the layer of fresh resin, becomes adsorbed by the resin.

The separation of different metal ions obtainable in this way may be enhanced by use of a complexing agent (such as citrate buffer or tartrate buffer) which complexes with metal ions and competes with the resin for the cations. The speed with which a given ion moves down the resin column is then dependent upon the affinity of the resin for the ion and the extent to which the ion is complexed by the complexing agent.

When a solution of ammonium citrate (of a pH at which nickel citrate complex ions are formed) is passed through a resin column on which nickel ion has been adsorbed at the top, the following reversible reaction is set up:

$$2NH_4^+ + NiR \rightleftharpoons (NH_4)_2R + Ni^{++} \tag{3}$$

At the same time, the citrate anion of average charge $-m$ (represented by A^{-m}) reacts with the Ni^{++} ion as indicated by the reversible reaction

$$Ni^{++} + xA^{-m} \rightleftharpoons NiA_x^{2-xm} \tag{4}$$

where x is the number of anions which complex with each nickel ion. The fraction of the total nickel found in each form, i.e., as Ni^{++}, NiR, and NiA_x^{2-xm}, is determined by the equilibrium constants for reactions (3) and (4). A similar set of two reactions occurs for the Co^{++} ion. Now if these two equilibrium constants for the nickel reactions are sufficiently different from those for the cobalt reactions, a separation may be achieved. By using a complexing agent, there are two equilibrium constants instead of one with which to work.

When ammonium citrate buffer is passed into the column, the Ni^{++} ion is complexed, thereby shifting the equilibrium of reaction (3) to the right. The nickel in solution moves down the column to a region of "fresh" resin $(NH_4)_2R$, where the Ni^{++} ion is readsorbed. Thus, as the solution passes down the column, the Ni^{++} is successively adsorbed and desorbed in a process similar to fractional distillation in which a substance is successively vaporized and condensed. The same process is going on with the Co^{++} ions. Thus the small differences in equilibrium constants are made use of many times, with the result that a better separation is achieved by column operation than in a batch experiment. When citrate buffers of higher pH are used, reaction (4) is displaced to the right and the metal ions are eluted more rapidly from the column.

For the case in which the adsorption isotherm is linear, as in the low-concentration range, the slope of the isotherm may be calculated from the velocity of the zone of adsorbed substance down the chromatographic column. The slope of the adsorption isotherm K (distribution coefficient) is the ratio of the weight of solute adsorbed by 1 g of adsorbent to the equilibrium concentration in grams per milliliter of solution. Although the theory[1-3] of chromatographic separation is complicated, the relation between the position of the zone and the distribution coefficient is quite simple provided that the adsorbed solute is in equilibrium with dissolved solute at every step in the process.

$$K = \frac{V}{Fm}$$

where V = volume of elutant which has flowed into column, ml

$\quad m$ = mass of adsorbent in column, g

$\quad F$ = fraction of length of column swept through by center of zone

····> **Apparatus.** Ion-exchange column filled with Dowex-50; citrate buffer (50 g citric acid mono-hydrate and 1 g phenol per liter of water adjusted to pH 3.30 to 3.40 with concentrated aqueous ammonia); $NiCl_2$; $CoCl_2$; two 25-ml graduates; spectrophotometer; spectrophotometer cells; lens tissue; long tube with bulb.

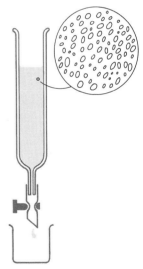

PROCEDURE.   The experimental arrangement for this experiment is illustrated in Fig. 93. If it is not certain that the resin has been completely freed of nickel and cobalt and washed with distilled water, the column is washed with about 300 ml of 2 N HCl, followed by about 300 ml of distilled water. Washing may be stopped when the effluent is basic to methyl orange (about pH 4.5).

Quantities of nickel and cobalt chlorides containing 200 mg of the metal are weighed out, mixed, and dissolved in a small quantity of distilled water (not over 20 ml). In order to save time, it is advisable to suck out the water above the resin bed by using a long glass tube and rubber bulb before pouring on the solution of the metals. The solution containing metal ions is poured on the top of the resin bed and allowed to flow into the column. If distilled water is then poured carefully into the tube above the resin bed, it will form a layer above the salt solution and there will be an adequate head to force the solution through the column. A siphon is connected to the column, and about 150 ml of distilled water is allowed to flow through it. If the washings are colorless, they are discarded.

The distilled water above the column is then withdrawn with the long glass tube and bulb, and citrate buffer is poured on the column. A large container of citrate buffer is connected by means of a siphon. The rate of elution is controlled by means of a pinch clamp so that the flow rate is 2.5 to 3.0 ml min⁻¹. Twenty-five-milliliter samples of the effluent are collected until the first nickel appears, and from that point on, 10-ml samples are collected. For each fraction the volume and time are recorded (the flow rate may be checked from these data), and a test tube full of the solution is corked for analysis.

Since the cobalt tends to tail out through a large volume, the last bit may be

figure 93.

Ion-exchange column.

removed by washing the column with 2 *N* HCl and then with distilled water (at least 300 ml).

The solutions are analyzed directly with a spectrophotometer. It has been found convenient to analyze for cobalt at a wavelength of 510 and for nickel at 650 mμ. A small amount of nickel does not interfere with the analysis for cobalt, and vice versa. The concentrations are read directly from plots of absorbancy, log (I_0/I) (Experiment 16) versus concentration of nickel or cobalt; these are provided in the laboratory or determined separately.

CALCULATIONS.   The concentrations of nickel and cobalt in the various fractions are plotted versus volume of effluent. The quantities of nickel and cobalt recovered from the column are calculated from the concentrations and volumes of the fractions. The percent recovery is computed.

Practical applications. Chromatographic adsorption is finding many applications. It is widely used for qualitative analysis of organic compounds and for the separation of different compounds in a mixture of biological materials. It has been used for the separation of isotopes.[4,5]

Suggestions for further work. The influence of pH, rate of elution, or ratio of weight of metal to weight of resin may be investigated.

A vertical tube packed with powdered sugar under the proper conditions may be used for separating by chromatographic adsorption the various plant pigments obtained by crushing leaves and treating them with petroleum ether.

References

1. J. C. Giddings, Dynamics of Chromatography, I: Principles and Theory, M. Dekker, Inc., New York, 1965.
2. F. H. Spedding and J. E. Powell in E. Heftmann (ed.), "Chromatography," chap. 25, Reinhold Publishing Corporation, New York, 1963.
3. D. DeVault, *J. Am. Chem. Soc.*, **65:** 532 (1943).
4. F. H. Spedding, J. E. Powell, and H. J. Svec., *J. Am. Chem. Soc.*, **77:** 6125 (1955).
5. Series of papers on the separation of rare earths and radioisotopes by ion exchange, *J. Am. Chem. Soc.*, **69:** 2769–2881 (1947).

49   *SEDIMENTATION RATE AND PARTICLE SIZE DISTRIBUTION*

The measurement of the velocity of sedimentation of particles in the earth's gravitational field or a centrifugal field gives valuable information concerning their size. When the dispersed particles are so large that they exceed the limit of colloidal dimensions, and when the density of the particle relative to that of the suspension medium is sufficiently great, they settle out under the force of gravitation. By measuring rates of sedimentation or, as in this experiment, by making observations of the rate of accumulation of sediment, particle size and size-distribution determinations can be made with finely divided solids for which other sizing methods would be impractical or impossible.

THEORY. The constant velocity with which a spherical particle falls in a liquid may be expressed by a relatively simple law. The force of friction resisting the fall of the particle is $6\pi\eta r\, dx/dt$, and the force of gravity acting on the particle is mg, or $\frac{4}{3}\pi r^3(\rho_p - \rho)g$. In these expressions η is the coefficient of viscosity of the liquid, r is the radius of the particle, dx/dt is the velocity of fall of the particle, ρ_p and ρ are the densities of the particle and of the suspension medium, m is the effective mass of a particle, and g is the acceleration due to the earth's gravitational field. The gravitational force and the force of friction are exactly opposed and equal when the system reaches a steady state; i.e., the particle falls with constant velocity, and

$$6\pi\eta r\,\frac{dx}{dt} = \frac{4}{3}\pi r^3(\rho_p - \rho)g$$

or

$$r = \sqrt{\frac{9}{2}\eta\frac{dx/dt}{(\rho_p - \rho)g}} \tag{1}$$

This equation is known in the literature as *Stokes' law*. Thus, if the sedimentation rate or rate of accumulation of spherical particles is measured and the coefficient of viscosity of the liquid, the difference in density between the particle and the liquid, and the constant of the sedimentation field are known, the particle size can be calculated.

In actual particulate systems, heterogeneity of size is the rule rather than the exception. These systems have usually been prepared by grinding, milling, precipitation from solution, etc., and may be characterized in two ways, either by an average size or weight or by a size frequency distribution or tabulation. If the sample contains a broad range of sizes, the values found for the average size by various methods may be greatly different. The methods which give number averages are strongly influenced by the particles of the smaller sizes, which make less important contributions to the weight of the sample; in turn, these particles have relatively little effect on the weight-average values.[1,2] The size frequency distributions give more complete information. They may be integral or differential distributions.

Sedimentation methods are useful for the determination of the size distribution of particles with diameters below 50 μ. These methods, which use gravitational force, may be applicable to particles as small as 1 μ provided their density is sufficiently different from that of the suspension medium. There are accumulation methods and incremental methods. The one illustrated in this experiment is a cumulative method which makes use of a graphical analysis of tangents to correct for the continually separating particles and thus to produce a differential distribution curve.

Descriptions of Particle Size Distribution. The object of this experiment is to describe in detail the number of different sizes of particles in a dispersion and their relative amounts. In order to simplify matters consideration is restricted to spherical particles. The most obvious method would be to measure in some way the diameters of a thousand or more particles and then construct a table showing the numbers of each of the various sizes present. For obvious reasons such data are more useful in graphical form. One plot of the information is to arrange the table in order of increasing (or decreasing) size and then to construct a supplementary table showing in percent the total number of particles (or total weight of particles) that are smaller (or larger) than the several sizes. The two quantities are then graphed to produce what is termed the *cumulative curve* (Fig. 94).

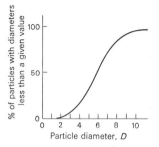

figure 94.

Cumulative distribution curve.

In another analysis, these same data could have been arranged into a relatively small number of size classes of uniform breadth, with subsequent counting (or weighing) of the particles in each size class; then a different type of plot is indicated. Thus, on a number basis the resultant curve is the same as would have been obtained if the slopes of the cumulative curve had been plotted at various points against the corresponding sizes, say diameters D. On this form of curve, shown in Fig. 95, the total area under the curve represents 100 percent of the material, and the ratio of areas under the curve between any two vertical lines to the total area represents the fraction of the material the sizes of which fall between the corresponding abscissas. Thus,

$$\left(\frac{1}{n}\frac{\Delta n}{\Delta D}\right)\Delta D = \frac{\Delta n}{n}$$

Considering some property P, the total number of particles is

$$n = \sum_i n_i = \int_0^\infty f_n(P)\,dP$$

where $f_n(P)$, the number distribution function, is $\dfrac{1}{n}\dfrac{dn}{dP}$. On a weight basis another distribution function is involved, namely, $f_w(P)$, and

$$w = \sum_i n_i m_i = \int_0^\infty f_w(P)\,dP = \int_0^\infty \frac{1}{w}\frac{dw}{dP}\,dP$$

where m_i is the mass of the ith particle.

Thus, in general with statistical methods one requires knowledge of the distri-

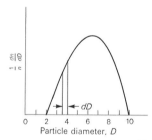

figure 95.

Differential distribution (or size-frequency) curve. (Same distribution as in Fig. 94.)

bution of some observable quantity with respect to the variable or variables on which it depends. In this experiment, the determination of a distribution function from data taken in a simple sedimentation tube is sought. The external force is that of the earth's gravity field. Throughout, we consider the radius r to be the fundamental variable. For the fraction, a dimensionless quantity, of the total mass of a sample contributed by particles whose radii lie between r and $r + dr$ we write

$$f_w(r)\ dr = \frac{m(r)\ dr}{\int_0^\infty m(r)\ dr} \tag{2}$$

It is noted that the quantity $m(r)$, a mass corresponding to the radius r, has been multiplied by the radius interval dr which is involved.

In the sedimentation tube all particles fall through the same distance L_0, and all particles start their fall at time $t = 0$. At time t all particles whose radii exceed $r(t)$ will have reached the bottom of the tube. From Stokes' law, eq. (1),

$$r(t) = \left[\frac{9\eta L_0}{2(\rho_p - \rho)gt}\right]^{\frac{1}{2}} \tag{3}$$

If the height of the column of suspension in the tube is assumed to be proportional to the mass of material sedimented out, then

$$\frac{h(t) - h(0)}{h(\infty) - h(0)} = \frac{M_t}{M_\infty} = \frac{\int_{r(t)}^\infty m(r)\ dr}{\int_0^\infty m(r)\ dr} \tag{4}$$

In this equation the several heights of accumulated sediment at the bottom of the tube are measured at time t, 0, and ∞, as indicated, and M_t is the total mass sedimented by time t. The integral or cumulative mass fraction as it depends on t, Eq. (2), can be made to provide the differential distribution function for mass, which is

$$\frac{m(r)}{\int_0^\infty m(r)\ dr}$$

a quantity which now has the dimension cm^{-1}.

····▸ *Apparatus.* Sedimentation-tube assembly (Fig. 96); finely divided slurry of sulfur; appropriate spherical glass beads, etc.; cathetometer; stopwatch; dispersing agent Daxad 11.†

PROCEDURE.   The sedimentation column consists of a long tube with a stopcock at the top and an outer ground-glass joint at the bottom, to permit insertion of a

† Dewey and Almy Chemical Division, W. R. Grace Co., Cambridge, Mass.

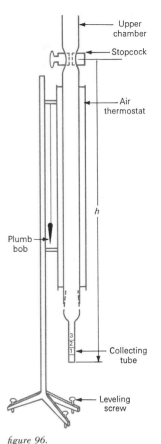

figure 96.

Palo-Travis apparatus.

collecting tube. The assembly is mounted on a wall bracket for the observations with the cathetometer of the sedimentation accumulation in the collecting tube.

A solution of approximately 1 g Daxad 11 dispersing agent in 200 ml water is prepared. The collecting tube is filled with this solution and fastened to the outer ground-glass joint using rubber bands (wrapped around *twice*). The sedimentation tube is then filled with this solution up to the stopcock. The stopcock is closed.

The suspension is prepared in the following manner. Some 5 g of sulfur is ground in a mortar for from 3 to 5 min. A slurry is formed by the addition of about 20 ml of the dispersing-agent solution. It is placed above the stopcock. The sedimentation-tube assembly is placed on the wall mount and the cathetometer is focused on the bottom of the collecting tube. The cathetometer reading is noted. The slurry is stirred and the stopcock is then opened and the timer started. The increasing level $h(t)$ of sedimented material in the collecting tube is determined at 100-sec intervals. The sedimentation is allowed to proceed until all material has reached the bottom of the tube (about $1\frac{1}{2}$ to 2 hr). It may be necessary to tap the collecting tube gently (with a pencil) to keep the sedimented material level.

When the sedimentation is complete, the apparatus is removed from the wall mount and *inverted* in the three-legged holder provided. Needless to say, the stopcock should be closed for this procedure and the small amount of material remaining above the stopcock poured into a beaker. A beaker is placed under the stopcock and the stopcock opened. Finally the collecting tube is removed. The apparatus should be rinsed with distilled water and returned to the wall mount.

CALCULATIONS.   A graph of the quantity

$$\frac{h(t) - h(0)}{h(\infty) - h(0)}$$

versus time is prepared from the data. It is the slope of this experimental curve which provides the key to the computation of the mass-differential distribution function. Equation (4) is now differentiated with respect to time, using Leibniz' rule, to give

$$\frac{d}{dt}\left[\frac{M_t}{M_\infty}\right] = -\frac{m[r(t)]}{\int_0^\infty m(r)\ dr}\frac{d}{dt}r(t)$$

The differential distribution function is then determined by dividing the slope of the experimental curve by $-dr/dt$,

$$\frac{m(r)}{\int_0^\infty m(r)\ dr} = -\frac{\text{slope}}{dr/dt} = \text{slope}\ \frac{2t}{r}$$

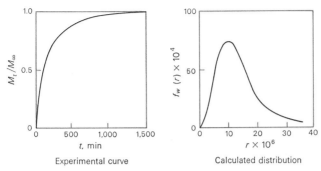

figure 97. Idealized plots of M_t/M_∞ versus t and of $f_w(r)$ versus r.

Slopes are taken at 500-sec intervals between the times 1000 and 4000 sec. In doing this, the slope is best evaluated by taking tangents to a smooth curve drawn through the data points, using two x and y coordinates on the extremities of each tangent. The critical radii $r(t)$ corresponding to these times are calculated by using Eq. (3).

The overall results of the experiment are presented by the inclusion in the report of graphs M_t/M_∞ versus time and of the function $f_w(r)$ versus r. They should appear as indicated in Fig. 97.

Practical applications. Knowledge of particle size and size distribution is important in cement, ceramic, ore flotation, photographic-emulsion, and paint-pigment technologies. The rate of settling of precipitates is often important in analytical chemistry. Relationships between the size and the behavior of soils, the stability of pastes, etc., have been recognized for a long time.

Suggestions for further work. Sedimentation studies with typical soils, paint pigments, or cements may be carried out, and more complete mathematical analyses of the distribution curves may be attempted.

The weight of particles settling from a suspension may be determined directly by suspending a pan from one arm of a balance in the medium. When carefully done, this method is capable of considerable precision.[3,4]

References

1. G. Herdan and M. L. Smith, "Small Particle Statistics," Elsevier Press, Inc., Houston, Tex., 1953.
2. C. Orr, Jr., and J. M. Dalla Valle, "Fine Particle Measurement," The Macmillan Company, New York, 1959.
3. S. Odén, *Proc. Roy. Soc. Edinburgh,* **36:** 219 (1916).
4. T. Svedberg, "Colloid Chemistry," 2d ed., pp. 167–182, Reinhold Publishing Corporation, New York, 1928.

50   THERMODYNAMIC ANALYSIS OF RUBBERLIKE ELASTICITY

From measurements on an elastic rubber sample of the tension force as a function of temperature and length, the force is separated into contributions from energy and entropy changes. The results are compared with the predictions of a statistical theory.

THEORY.[1-3]   When an elastic body is deformed by an external force, work is done in addition to the usual work of volume expansion, and in fact the latter may usually be neglected. Thus, for a rubber sample in the form of a strip of length l subject to stretching force or tension f (dynes), the element of work done *on* the sample is

$$dw = f \, dl \tag{1}$$

The change in internal energy dU for a reversible process is then

$$dU = T \, dS + f \, dl \tag{2}$$

Defining the Helmholtz free energy A as usual by

$$A = U - TS \tag{3}$$

one has

$$dA = -S \, dT + f \, dl \tag{4}$$

From (4),

$$f = \left(\frac{\partial A}{\partial l}\right)_T$$
$$= \left(\frac{\partial U}{\partial l}\right)_T - T\left(\frac{\partial S}{\partial l}\right)_T \tag{5}$$

Thus, reversible stretching of the sample by an amount dl at constant temperature increases the free energy by an amount $f \, dl$. Equation (5) allows one to separate the force into two contributions associated, respectively, with the internal energy and entropy changes which occur on stretching.

By equating the second-order cross derivatives from (4), one obtains

$$\left(\frac{\partial S}{\partial l}\right)_T = -\left(\frac{\partial f}{\partial T}\right)_l \tag{6}$$

and putting this with (5) gives

$$\left(\frac{\partial U}{\partial l}\right)_T = f - T\left(\frac{\partial f}{\partial T}\right)_l \tag{7}$$

Thus, the two contributions to the force can be separately determined by measuring the force as a function of temperature at a fixed length.

One should note that Eq. (5) does not mean that the force is a linear function of temperature, since the two derivatives are implicit functions of T.

It is interesting to consider the molecular-scale processes which can account for these thermodynamic properties. Soft vulcanized rubber consists of a network of flexible threadlike molecular strands which are in constant agitation because of their thermal energy. On stretching, these strands assume a partial alignment in the direction of stretch. The second term in (5) can then be understood as the contribution to the force from the tendency of the system to go toward a condition of greater randomness, i.e., shorter length. The first term, due directly to variation of internal energy with length, is often relatively small.

The molecular theory of rubber elasticity has been developed by a number of authors. Excellent treatments are given by Treloar[1] and Wall.[2] The model assumed is that of a cross-linked network, with N strands in the sample. (A segment between cross-linked points is counted as one strand.) The deformation of a rectangular parallelepiped sample is specified by three extension ratios,

$$\lambda_1 = \frac{l_1}{l_{01}} \qquad \lambda_2 = \frac{l_2}{l_{02}} \qquad \lambda_3 = \frac{l_3}{l_{03}}$$

where l_1, l_2, l_3 are the dimensions of the deformed sample and l_{01}, l_{02}, l_{03} are those of the undeformed sample. By a statistical-mechanical derivation, the entropy of the deformed sample relative to that $S_0(T,V)$ of the undeformed sample at the same temperature and volume is given by

$$S(T,V,\lambda_1,\lambda_2,\lambda_3) = S_0(T,V) - \tfrac{1}{2}Nk(\lambda_1^2 + \lambda_2^2 + \lambda_3^2 - 3) \tag{8}$$

In deriving this, it is assumed that all configurations at this volume have the same energy, so that

$$U(T,V,\lambda_1,\lambda_2,\lambda_3) = U_0(T,V) \tag{9}$$

For a pure elongation along axis 1, at constant volume,

$$\lambda_1\lambda_2\lambda_3 = 1 \qquad \text{and} \qquad \lambda_2 = \lambda_3 \tag{10}$$

Such a deformation can be specified by the single ratio λ_1, since $\lambda_2^2 = \lambda_3^2 = \lambda_1^{-1}$. Writing λ henceforth for λ_1, we get

$$S(T,V,\lambda) = S_0(T,V) - \tfrac{1}{2}Nk(\lambda^2 + 2\lambda^{-1} - 3) \tag{11}$$

$$U(T,V,\lambda) = U_0(T,V) \tag{12}$$

$$\begin{aligned} A(T,V,\lambda) &= U(T,V,\lambda) - TS(T,V,\lambda) \\ &= A_0(T,V) + \tfrac{1}{2}NkT(\lambda^2 + 2\lambda^{-1} - 3) \end{aligned} \tag{13}$$

Then by Eq. (5), the tension force for elongation λ, with volume assumed constant, is

$$f = \frac{1}{l_0} \left(\frac{\partial A}{\partial \lambda} \right)_T = \frac{NkT}{l_0} (\lambda - \lambda^{-2})$$

$$= \frac{NkT}{l_0} \left[\frac{l}{l_0} - \left(\frac{l_0}{l} \right)^2 \right] \tag{14}$$

where l and l_0 are the stretched and unstretched lengths of the sample.

In an actual experiment, stretching causes a slight change in volume as well as shape, and Eq. (14) becomes

$$f = \left(\frac{\partial A}{\partial l} \right)_T = \left(\frac{\partial A}{\partial V} \right)_{T,\lambda} \left(\frac{\partial V}{\partial l} \right)_T + \left(\frac{\partial A}{\partial \lambda} \right)_{T,V} \left(\frac{\partial \lambda}{\partial l} \right)_T$$

$$= -P \left(\frac{\partial V}{\partial l} \right)_T + \frac{NkT}{l_0} (\lambda - \lambda^{-2}) \tag{14a}$$

A simple calculation (required data given below) will show that the term due to the volume change is negligible under conditions of most experiments. The more general approach is desirable, however, because the previous assumption of a zero volume change on stretching would imply a zero bulk compressibility, and this in turn would imply a vanishing thermal coefficient of expansion, requiring l_0 to be independent of temperature. But the temperature dependence of l_0 *is* important; in particular, it affects the temperature dependence of f as given by Eq. (14).

Equation (14) can be tested by measuring the tension force on a sample as a function of length and temperature. If it is found to represent the behavior of the sample satisfactorily, then N, the number of network strands in the sample, can be calculated from the data.

The average molecular weight M_c of a network strand, i.e., of the chain between two cross links, may then be calculated from N and the density ρ (in grams per cubic centimeter), since

$$\frac{M_c N}{N_0} = \rho V \tag{15}$$

where V is the sample volume in cubic centimeters and N_0 is Avogadro's number.

Apparatus. Force transducer (e.g., Hewlett-Packard model FTA-100-1) and exciter-detector unit (model HP-311A); special micrometer with nonrotating shaft, e.g., Hewlett-Packard model FTA-1011; recording potentiometer (optional); mounting assembly; temperature control system; bead thermistor and Wheatstone bridge; rubber samples, e.g., pure latex rubber tubing, or gum rubber bands; cement for holding sample;† precision calipers; thermometer and water bath for calibrating thermistor.

† Quick set adhesive 404, Loctite Corp., Newington, Conn., is very convenient and effective for this purpose.

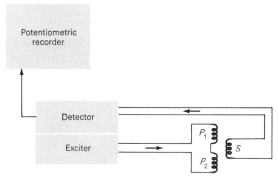

figure 98. Diagram of force transducer and associated units. The primary windings, P_1 and P_2, are fixed to the case of the unit. The secondary S is fastened to the movable shaft, which in turn is supported by a stiff spring mounted in the case.

PROCEDURE.   The force measurements are to be made by means of a force transducer. The LVDT type (linear variable differential transformer) is recommended for this experiment because of its excellent temperature stability and linearity. The principle is illustrated in Fig. 98.

The basic element is a differential transformer. The exciter unit delivers an ac voltage to the primary windings. The secondary is mounted on a movable shaft, which in turn is mounted on a rather stiff spring. When the shaft is in its equilibrium position, the net voltage induced in the secondary winding is zero, as the effects of the two primaries then cancel. When the secondary is displaced, a voltage is induced which is very accurately linear with the displacement; since the spring is designed to follow Hooke's law closely, the voltage is accurately linear with force. The force range appropriate for this experiment is about 0 to 100 g, the corresponding displacement being only 0.010 in.

The ac voltage from the secondary is converted into direct current by a phase-sensitive detector (page 634). The output of this detector is a dc voltage proportional to the force.

The dc voltage from the detector unit is measured with a precision voltmeter, or potentiometer. The use of a recording potentiometer is recommended because it facilitates determining when equilibrium has been reached.

The assembly in which the transducer is mounted is sketched in Fig. 99. Length can be read directly from the micrometer scale. The range of this is such as to allow a change of length of 1.000 in. Therefore, the length of the sample, unstretched, should be a little less than 1 in., so that it can be stretched to about double its initial length. The movable shaft of the force transducer is displaced very slightly downward when the sample is stretched. The platform at the top is mounted

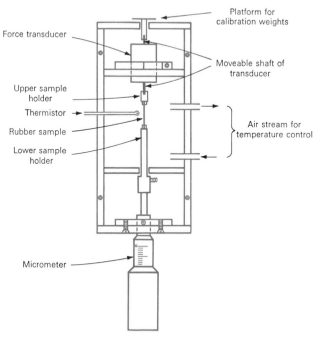

figure 99. Transducer assembly for measurement of tension as a function of length and temperature. The rubber sample is clamped by the two sample holders, which can be removed by loosening setscrews. The micrometer is of a special design such that the shaft undergoes a pure translation without rotation. Rectangular blocks of polyurethane foam fit behind, on either side, and in front of the sample to provide thermal insulation.

on the same shaft. The system is calibrated by placing known weights on this platform and noting the recorder deflection produced.

Temperature is controlled by passing a stream of air through a heat exchanger, either heating or cooling it, and then through the sample. With polyurethane foam insulation surrounding the sample, temperatures ranging from 0 to 80°C can readily be achieved.

Temperature is measured by means of a small bead thermistor (page 609) placed close to the sample. The thermistor resistance is measured by means of a Wheatstone bridge, Fig. 8 (page 29). It is appropriate to calibrate the thermistor by measuring the resistance R at several known temperatures over the range of interest. It is convenient to make a graph of these data, with log R plotted versus $1/T$, as this is approximately linear [see Eq. (74) on page 609].

By means of a razor, the rubber sample may be cut to the desired size, a little less than 1 in. long and about 1 mm square in cross section, the latter being chosen such that the force required to produce the maximum desired extension ratio is somewhat less than the full-scale range of the force transducer, about 100 g. If the sample is to be cut from a flat sheet, a good technique is to mount the sheet between two metal plates in a vise for cutting. The cross-sectional dimensions are measured with a micrometer at several points or, alternatively, the average cross section determined from the weight and length. The length can be measured accurately if the sample is placed between two metal bars to hold it straight with negligible tension or pressure. The ends of the sample may then be cemented or clamped to the two sample holders (Fig. 99). If cement is used, it should not cover any more of the sides of the sample than necessary for holding, as this reduces the effective length of the sample. A clamp with smooth jaws tightened with a setscrew is more convenient if a large number of samples is to be used.

The exciter unit is connected to the transducer and the detector output to the recorder. A shunt is used across the recorder input to make the full-scale range correspond to the voltage obtained for approximately 100 g force. The recorder reading should be noted with no weights placed on the platform and then with various known weights placed upon it, to provide a calibration for the force transducer. The upper sample holder is first attached to the shaft of the transducer and the setscrew carefully tightened. *The transducer can be damaged by excessive upward force or twisting.*

With the upper sample holder in place, it is necessary to find the recorder deflection corresponding to zero tension force. This can be done with the lower sample holder held off to the side so that the sample itself is bent in a wide quarter circle. While doing this, one should also estimate the magnitude of error due to any vertical force from the rubber in this position.

Next the lower sample holder may be put into place, in such a way that there is no appreciable twist or tension, but also a minimum of slack, and the setscrew then tightened.

The micrometer setting d_0, corresponding to the unstretched sample length l_0, is determined as the setting for which the recorder deflection is the same as that previously noted for zero tension force.

In subsequent measurements, the sample length (in centimeters) corresponding to micrometer reading d (in inches) is given by

$$l = l_0 + 2.54(d - d_0)$$

The length l_0 can be measured with the sample in place by adjusting the micrometer to give the recorder deflection corresponding to zero force, and then measuring the length directly with precision calipers.

Force-elongation data at a single temperature can be taken by setting the micrometer at various points and allowing the sample to equilibrate at each length. It is important to take two cycles of measurements, e.g., increasing, decreasing, increasing, decreasing lengths, so that the extent of hysteresis can be gauged.

If there is evidence of enough hysteresis in the data to be troublesome, it may be helpful to prestress the sample.[4] This involves stretching it to a length greater than the maximum length to be used in subsequent studies, taking it to a temperature at the upper end of the range to be covered, and holding it at these conditions for one or several hours. After prestressing, l_0 should be remeasured.

Force-temperature data can then be taken with the length at the desired fixed value (nominally corresponding to double the unstretched length). Since time is needed for adjustment and equilibration at each temperature, it is better not to try to get too many points but rather to get good data for just enough temperatures to define the shape of the curve. Again, it is important to take the sample through at least one complete cycle of measurements and preferably through a second. If necessary, the temperature intervals can be made larger for the second cycle. The thermistor resistance is measured at the time of each force measurement.

CALCULATIONS.   For the determination of sample temperatures, a graph of $\log R$ versus $1/T$ may be prepared. A plot of potentiometer deflection versus force may be used as calibration curve for the transducer.

The isothermal force-versus-length data are plotted on a graph.

Next, the force data are plotted as a function of temperature, and the two contributions $(\partial U/\partial l)_T$ and $T(\partial S/\partial l)_T$ to the force f evaluated graphically at several temperatures. These results are valid, to within the accuracy of the data, without regard to any molecular theory. For reporting, they should be reduced to correspond to a sample of unit cross section and length when unstrained.

The equation of state, Eq. (14), predicted by the network theory can be tested graphically, by plotting the force data against $\lambda - \lambda^{-2}$.

If the statistical equation is in reasonable accord with the data, M_c can be obtained by use of Eq. (15).

Some properties of rubber required for thermodynamic calculations are[5-7]

$$\rho = \text{density} = 0.95 \text{ g cm}^{-3}$$

$$\alpha \equiv \frac{1}{V}\left(\frac{\partial V}{\partial T}\right)_P = 6.6 \times 10^{-4} \text{ deg}^{-1}$$

$$\kappa \equiv -\frac{1}{V}\left(\frac{\partial V}{\partial P}\right)_T = 5.3 \times 10^{-11} \text{ cm}^2 \text{ dyne}^{-1}$$

$$\eta \equiv \frac{l}{V}\left(\frac{\partial V}{\partial l}\right)_{T,P} \approx 2 \times 10^{-4}$$

The linear coefficient of thermal expansion is equal to $\alpha/3$. The dilation coefficient η varies considerably according to the nature of the sample,[7] and so the above value should be used only for purposes of estimation.

Practical applications. Most of the uses to which natural and synthetic rubbers are put depend on the remarkable mechanical properties which have been here illustrated. The modulus of elasticity, or stiffness, depends on the degree of cross-linking as measured by the number of network strands per unit volume. The molecular structure of the rubber will determine the relative magnitudes of the entropy and energy contributions to the retractive force, and hence the degree to which the force depends on temperature. In many applications, such as in automobile tires, the operating temperature may be considerably different from that of the surroundings.

Suggestions for further work. The measurements may be repeated for a different type of rubber, such as silicone, GR-S, or a heavily loaded stock such as is used in tire casings. In the latter, the presence of a large proportion of finely divided carbon alters the mechanical properties. Data for force-temperature plots may be obtained at higher and lower elongations than 100 percent. The thermodynamic analogies between stretching a "perfect" rubber [for which $(\partial U/\partial l)_T = 0$] and compressing a perfect gas [for which $(\partial U/\partial V)_T = 0$] may be explored by comparing equations in which f corresponds to pressure and l corresponds to volume.

References

1. L. R. G. Treloar, "The Physics of Rubber Elasticity," 2d ed., Oxford University Press, Fair Lawn, N.J., 1958.
2. F. T. Wall, "Chemical Thermodynamics," 2d ed., chap. 15, W. H. Freeman and Company, San Francisco, 1965.
3. P. J. Flory, "Principles of Polymer Chemistry," Cornell University Press, Ithaca, N.Y., 1953.
4. L. A. Wood and F. L. Roth, *J. Appl. Phys.*, **15:** 781 (1944).
5. J. Brandrup and E. H. Immergut, "Polymer Handbook," Interscience Publishers, John Wiley & Sons, Inc., New York, 1966.
6. A. H. Scott, *J. Res. U.S. Natl. Bur. Std.*, **14:** 99 (1935).
7. F. G. Hewitt and R. L. Anthony, *J. Appl. Phys.*, **29:** 1411 (1958).

Chapter
16

Surface
Chemistry

Three different methods are used for determining surface tensions of liquids. The effect of temperature on surface tension is investigated.

THEORY.[1,2]   The molecules at the surface of a liquid are subject to the strong attractive forces of the interior molecules. A resultant force, whose direction is in a plane tangent to the surface at a particular point, acts to make the liquid surface as small as possible. The magnitude of this force acting perpendicular to a unit length of a line in the surface is called the *surface tension* γ. The surface, or interface, where the tension exists is between the liquid and its saturated vapor in air, usually at atmospheric pressure. A tension may also exist at the interface between immiscible liquids; this is commonly called the *interfacial tension*. The dimensions of surface tension are force per unit length and are commonly expressed in the cgs system as dynes per centimeter.

In order to illustrate the above definition of surface tension, we shall consider the principles involved in the three methods of measurement discussed in this experiment and shown in Fig. 100. Surface tension is a property of an interface. Usually it is measured by balancing it along a certain boundary line against an equal force which can be measured.

In the *capillary-rise* method, the liquid rises, because of its surface tension, in a capillary tube of small internal diameter immersed in it (provided that the liquid wets the tube). The circular boundary line is located at some point above the meniscus and has a length $2\pi r$, where r is the inside tube radius. The force which causes the liquid to rise in the tube is $2\pi r\gamma \cos\theta$, where θ is the contact angle shown in Fig. 100a. At equilibrium the downward force mg due to gravity prevents the liquid from rising higher in the capillary. Here m is the mass of liquid in the capillary and g is the acceleration of gravity. For a liquid of density ρ rising to a height h in the capillary, this downward force is $\pi r^2 h \rho g$. At equilibrium this force

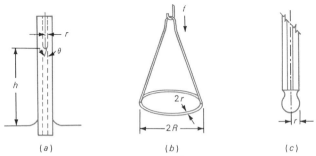

(a)   (b)   (c)

figure 100. Principles of three surface-tension methods: (a) capillary rise; (b) ring; (c) drop weight.

is just balanced by the vertical force due to the surface tension, $2\pi r \gamma \cos \theta$. For water and most organic liquids this contact angle is practically zero; this means that the surface of the liquid at the boundary is parallel with the wall of the capillary. Setting the two forces equal, we have, for zero contact angle,

$$2\pi r \gamma = \pi r^2 h \rho g$$
$$\gamma = \tfrac{1}{2} h r \rho g \tag{1}$$

In this derivation it has been assumed that the vapor does not have an appreciable density as compared with that of the liquid.

In the *ring* method, a platinum-iridium ring in the surface of the liquid is supported by a stirrup attached to the beam of a torsion balance. The ring is pulled upward from the liquid by turning the torsion wire, thus applying a force which is known from calibration of the instrument. For an idealized system, the force just necessary to break the liquid film is equal to $4\pi R \gamma$, where R is the mean radius of the ring. Doubling of the perimeter $2\pi R$ arises from the fact that there are two boundary lines between liquid and wire, one on the outside and one on the inside of the ring. This treatment holds for liquids with zero contact angle, a condition usually met, and for an ideal situation where the ring holds up a thin cylindrical shell of liquid before the break occurs, a condition which is not met. Actually, the shape of the liquid held up influences the force necessary for breaking away. The shape is a function of R^3/V and R/r, where V is the volume of liquid held up and r is the radius of the wire. This volume V is computed from the force equation, $f = mg = \rho V g$. The surface tension is thus given by the equation

$$\gamma = \frac{f}{4\pi R} F_r \tag{2}$$

where f = maximum force registered on the torsion-balance scale
$\quad\quad F_r$ = correction factor due to shape of liquid held up and ring dimensions (values in Table 1)

Table 1. Correction Factors for the Ring Method
$\dfrac{R}{r} = 40$

R^3/V	F_r	R^3/V	F_r
0.30	1.038	0.80	0.923
0.40	0.996	0.90	0.913
0.50	0.969	1.00	0.905
0.60	0.950	1.10	0.897
0.70	0.935	1.20	0.890

These factors have been determined experimentally by Harkins and Jordan.[3] Over extreme variations of R^3/V and R/r, F_r varies between about 0.75 and 1.02. In ordinary cases it is close to 1.

In the *drop-weight* method, a drop forms at the end of a tube, and the boundary line is the outside perimeter of the tube, $2\pi r$. When the drop just detaches itself, the downward force on the drop, mg, is equal to the force acting upward, $2\pi r\gamma$. Actually, only a portion of a drop falls, and Harkins and Brown[4] propose the equation

$$m_i g = 2\pi r\gamma \tag{3}$$

where m_i is the mass of an ideal drop. An equation which is equivalent to Eq. (3) and more convenient to use is $\gamma = (mg/r)F_d$, where F_d is an empirically determined function of V/r^3 and V is the actual volume of the drop.

Correction factors for the drop-weight calculations are given in Table 2. Values of V/r^3 between 1 and 3 give the best results. For values of F_d outside the limits of Table 2, the tables of Harkins and Brown[1,4,5] are to be consulted.

Table 2. Experimental Values of Drop-weight Corrections

V/r^3	F_d
2.995	0.261
2.637	0.262
2.341	0.264
2.093	0.265
1.706	0.266
1.424	0.265
1.211	0.264
1.124	0.263
1.048	0.262

Surface tension decreases as the temperature rises and is practically unaffected by changes in total area, pressure, or volume. The surface tension vanishes at the critical point. The temperature coefficient of surface tension, $d\gamma/dT$, is of importance in the thermodynamic treatment of surfaces. The total surface energy per unit area U_A of a film is given by the equation[2]

$$U_A = \gamma - T\frac{d\gamma}{dT} \tag{4}$$

where γ = work done on increasing surface by 1 cm^2 = free surface energy
$-T(d\gamma/dT)$ = heat absorbed during the process

The negative of the temperature coefficient of surface tension, $-(d\gamma/dT)$, is the surface entropy S_A per unit area. Equation (4) is the two-dimensional analog of the Gibbs-Helmholtz equation.

····➤ *Apparatus.* Assemblies for surface-tension measurements by capillary-rise, ring, and drop-weight methods; organic liquid such as acetone or absolute ethyl alcohol.

PROCEDURE.   The three methods for measuring surface tension described above are to be compared by careful measurements on a single organic liquid. Acetone or absolute ethyl alcohol is suggested. Distilled water is used for calibration purposes.

The *capillary-rise* apparatus is shown in Fig. 100a. The capillary tube of radius about 0.02 cm is provided with engraved millimeter graduations.† It is cleaned with *hot* cleaning solution and rinsed with distilled water and then with the liquid to be used. In case the liquid involved is immiscible with water, an intermediate rinsing with acetone is necessary. Air from pressure lines should not be blown through the capillary, since it is contaminated with oily substances. The test tube must be cleaned in the same manner.

The assembled apparatus is placed in a 25° thermostat, and pure liquid is poured into the test tube to a depth of several centimeters. The stopper is replaced; a clean dust-free rubber tube is fitted with a loose wad of cotton to keep out dirt and spray and is attached to the projecting tube as shown. After coming to the temperature of the thermostat, the liquid in the capillary is raised slightly by gently blowing into the rubber tube and allowed to fall back to its equilibrium level. Then it is depressed by slight suction and again allowed to come to equilibrium. If the capillary is clean, the reading on the scale after equilibrium is attained should be the same after raising the level as after depressing it.

The difference in the level of the liquid in the capillary tube and in the test tube is read on the scale, the bottom of the meniscus being read in each case. A wide test tube gives a more nearly flat meniscus. Four or five measurements are made on the liquid, and the results are averaged.

The radius of the tube is obtained by observing the capillary rise with pure water, the surface tension of which is known to be 71.8 dynes/cm^{-1} at 25°C. At least five observations are made. The average value of h is obtained from the scale readings, and r is obtained by solving Eq. (1), using the known value of γ.

The surface tension of acetone is measured at 0°C by using an ice bath and at three other temperatures, 25, 35, and 50°, using thermostats if they are available. A hand-regulated bath may be used for less accurate work, with a thermometer immersed directly in the liquid.

---

† Another convenient form of apparatus makes use of a U tube, with one arm a capillary of this radius and the other a tube of radius sufficiently large to give a flat meniscus. This rise in the capillary would then be measured with a cathetometer to give data of higher precision.

The DuNouy tensiometer, the principle of which is illustrated in Fig. 100*b*, is a widely used example of the *ring method.*

A ring of platinum-iridium wire is cleaned with *hot* cleaning solution, thoroughly rinsed with warm distilled water, and dried by carefully touching with a clean filter paper or cloth and allowing to stand in the air. Occasionally, the ring may be heated momentarily in a bunsen flame if further cleaning is necessary. It must not be touched with the fingers, and care should be used not to bend it. The apparatus is adjusted after hanging the ring on a hook attached to a torsion balance.

Several readings are taken of the force required to pull the ring away from the surface of the liquid, and the average is computed. The temperature is determined by immersing a thermometer in the liquid. Unfortunately, close temperature control is not convenient with this instrument.

After the determinations have been made, the instrument is calibrated over the range of scale readings involved. The ring is dried, a weighed square of paper is inserted, and a fractional gram weight is added. The pointer is turned until the lever just barely clears its support and lies in its standard horizontal position. The mass of the paper and fractional gram weight divided by the scale reading and multiplied by *g* gives the value in dynes of one scale division. The procedure is repeated once or twice, using more weights.

A useful form of *drop-weight* apparatus is shown in Fig. 101. The end of the capillary is ground flat and polished. It must be free from chips in order to obtain reproducible results, and to this end a permanent guard tube surrounds it.

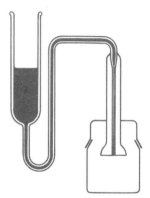

figure 101.

Drop-weight apparatus for measuring surface tension.

Before the performance of the experiment the bulb and capillary should be scrupulously cleaned with hot cleaning solution, followed by rinsings with distilled water and the liquid to be studied. After the apparatus has been conditioned and assembled, a dried weighing bottle is weighed and placed in the protecting bottle, which is then screwed into the cap. A very thin layer of stopcock grease applied to the ground joint of the weighing bottle will prevent loss by evaporation in subsequent weighings. A small air vent is provided in the cap to keep pressure from building up inside. The assembly is now introduced into the thermostat.

The liquid whose surface tension is to be measured is added through the side tube and capillary, and the liquid levels are adjusted until the time of formation of a drop is of the order of 5 min. The difference in liquid level to meet this requirement is ordinarily less than 1 cm. The apparatus may be tipped at first to get the liquid started in the capillary.

The first drop is allowed to form over this relatively long period of time in order to saturate the space within the container. After the detachment of this first drop, additional liquid is added to the side tube, or slight pressure is carefully applied to increase the drop rate. Care must be taken that each drop falls only under the influence of gravity. After the first drop, 30 sec should suffice for the formation of each of the others. A total of 20 to 25 drops should be adequate for the determination. The bottle is then weighed again.

If the radius of the capillary is not known, it may be measured with a comparator microscope.

CALCULATIONS.   The surface tension of the liquid as determined by the different methods is computed, using Eqs. (1) to (3). The density of the liquid may be found in tables.

The value of R/r for the ring of 4-cm perimeter usually supplied with the commercial-type student ring-method instrument may be taken as 40 unless a different value is specifically given. The volume V of liquid held up is calculated from the density and the reading on the dial scale, which can be converted to mass of liquid. Correction factors to be applied in Eq. (2) are given in Table 1 for $R/r = 40$ and for different values of R^3/V. For other values of these parameters the work of Harkins and Jordan[3] must be consulted. If the temperature of the liquid differs much from $25°$, a correction to this temperature may be estimated from the temperature coefficient of surface tension.

The surface tensions determined by the three methods are compared with the literature values, and the precentage deviations are computed.

The surface tensions of acetone at the several temperatures are plotted as ordinates against temperature as abscissa. The slope of the line is determined, and the total surface energy computed [Eq. (4)] in ergs per square centimeter at some representative temperature.

Practical applications. Surface tension is an important phenomenon in the study of macromolecular chemistry. It is an important factor in the concentration of ores by the flotation process. Surface-tension measurements find valuable applications in the biological sciences, particularly in bacteriology; the movement of the moisture of the soil and the passage of sap in plants are only two of the many agricultural phenomena that involve surface tension.

Suggestions for further work. Measurements are made so rapidly with the DuNouy apparatus that the surface tension of a large number of liquids and solutions may be determined. If a solute lowers the surface tension, it concentrates in the outer layers of the solution, but if it increases the surface tension of the solution, it is driven away from the surface. Surface tension, then, is never increased very much by the addition of a solute, but it may be decreased by a considerable amount. This theory may be checked by a number of determinations.

Another convenient apparatus to measure the change in surface tension when solute is added to a solvent or solution is the Wilhelmy balance.[6] It consists of a wetted plate, conveniently a microscope cover glass, which dips into the solution while hanging from the arm of a balance. The force on the plate corresponds to the surface tension of the liquid. Some of the data obtained with the DuNouy ring may be verified in this way.[7]

References

1. W. D. Harkins and A. E. Alexander, in A. Weissberger (ed.), "Technique of Organic Chemistry," vol. 1, "Physical Methods in Organic Chemistry," 3d ed., pt. 1, chap. 14, Interscience Publishers, Inc., New York, 1959.
2. E. O. Kraemer, J. W. Williams, and R. A. Alberty in H. S. Taylor and S. Glasstone (eds.), "Treatise on Physical Chemistry," 3d ed., vol. II, D. Van Nostrand Company, Inc., Princeton, N.J., 1951.
3. W. D. Harkins and H. F. Jordan, *J. Am. Chem. Soc.*, **52:** 1751 (1930).
4. W. D. Harkins and F. E. Brown, *J. Am. Chem. Soc.*, **41:** 499 (1919).
5. "International Critical Tables," vol. IV, p. 435, McGraw-Hill Book Company, New York, 1928.
6. J. F. Padday and D. R. Russell, *J. Colloid Sci.*, **15:** 503 (1960).
7. K. J. Mysels, "Introduction to Colloid Chemistry," Interscience Publishers, Inc., New York, 1959.

52 ADSORPTION FROM SOLUTION

Quantitative measurements of adsorption by an ion-exchange resin are made. The isothermal data are expressed by means of an equation, and the regeneration of the adsorbent is illustrated.

THEORY. Solids have the property of holding molecules at their surfaces, and this property is quite marked in the case of porous and finely divided material. Various forces are involved, ranging from those which are definitely physical in nature to those which are referred to as chemical. Adsorption is frequently quite specific, so that one solute may be adsorbed selectively from a mixture. Such differences in adsorbability are required for the success of the several chromatographic separations.

In the case of so-called *exchange adsorbents,* the adsorption is actually a chemical reaction in which an ion is liberated from the adsorbent as another is adsorbed. The softening of water, using naturally occurring or synthetic zeolites, is an example of this phenomenon. An important advance in the field of exchange adsorbents was made by Adams and Holmes[1] in 1935, when they discovered that phenol-formaldehyde resins exhibited ion-exchange properties. The advantage of synthetic resins is that the exchange-adsorption properties may be varied at will by the selection of the reactants for the polymerization reaction.[2-4] For example, the condensation of polyhydric phenols with formaldehyde yields resins which adsorb calcium ions, liberating hydrogen ions or sodium ions.

$$2NaR + Ca^{++} \rightleftharpoons CaR_2 + 2Na^+$$
$$2HR + Ca^{++} \rightleftharpoons CaR_2 + 2H^+$$

where NaR represents the sodium salt of the cation-exchange resin and HR represents the acid form. By treating the exhausted resin with an excess of Na^+ or H^+, the reactions may be reversed, regenerating the resin.

Synthetic resins prepared by condensing aromatic amines with formaldehyde exhibit anion-exchange, or acid-adsorbent, properties as illustrated by the following equation:

$$RX + HCl \rightleftharpoons RX \cdot HCl$$

where RX represents the anion-exchange resin. The capacity of a resin ranges from about 4 to 9 mequiv of acid per gram of dry resin. In this case the regeneration is accomplished by treating the exhausted resin with a solution of sodium carbonate.

$$RX \cdot HCl + Na_2CO_3 \longrightarrow NaCl + NaHCO_3 + RX$$

By passing water successively through columns of base-exchange (hydrogen form) and acid-binding resins, deionized water comparable in quality with distilled water may be obtained.

The amount of solute adsorbed by a given quantity of adsorbent increases with the concentration of the solution. In some cases the layer of adsorbed molecules is only one molecule deep, and further adsorption ceases when the surface of the crystal lattice is covered. The equilibrium between the dissolved solute and the material adsorbed also depends upon the nature of the solvent and the temperature, the amount adsorbed increasing at lower temperatures. In many cases the equilibrium may be rapidly established; in any event one must ascertain that the system is at equilibrium for a satisfactory representation of the data.

The relation between the amount adsorbed and concentration in a limited concentration range may be represented by the adsorption isotherm of Freundlich,

$$\frac{x}{m} = kc^n \tag{1}$$

where x = weight of material, g, adsorbed by m grams of adsorbing material
$\qquad c$ = concentration in solution, g liter^{-1}
$\qquad n$ = constant ranging from 0.1 to 0.5
$\qquad k$ = another constant

Although k varies considerably with the temperature and nature of the adsorbent, the ratio of k values for two different adsorbents is constant for different solutions. By taking the logarithm of Eq. (1) we obtain

$$\log \frac{x}{m} = n \log c + \log k \tag{2}$$

According to this equation a plot of $\log(x/m)$ versus $\log c$ is a straight line, and the constants may be evaluated from the slope n and the intercept $\log k$.

The adsorption equation of Langmuir[5] is based upon a theoretical consideration of the process of adsorption. This equation may be written

$$\frac{x}{m} = \frac{\alpha c}{1 + \beta c} \qquad \text{or} \qquad \frac{c}{x/m} = \frac{1}{\alpha} + \frac{\beta}{\alpha} c \tag{3}$$

where α and β are constants. For cases in which this equation represents the data, $c/(x/m)$ may be plotted as a linear function of c, and the constants evaluated from the slope β/α and intercept $1/\alpha$. The Langmuir equation differs from the Freundlich equation in that the adsorption approaches a finite limit as the concentration is increased.

The Langmuir equation is actually one description of the law of mass action, which may be seen in the following argument. Suppose we use the symbol R for the part of the resin which combines with a hydrogen ion according to the reaction

$$R + H^+ \rightleftharpoons RH^+$$

The constant for the association is $K = c_{RH^+}/c_{H^+}c_R$; or $c_{RH^+} = Kc_{H^+}c_R$. The total active resin concentration is $c_{R,t} = c_{RH^+} + c_R$, so that

$$c_{RH^+} = Kc_{H^+}(c_{R,t} - c_{RH^+})$$

Thus

$$\frac{c_{RH^+}}{c_{R,t}} = \frac{Kc_{H^+}}{1 + Kc_{H^+}}$$

The ratio $c_{RH^+}/c_{R,t}$ represents the number of moles of bound hydrogen per mole of resin combining sites, and the expression for it formally resembles the Langmuir function.

····> **Apparatus.** Twelve 250-ml Erlenmeyer flasks; two burettes; 1 N acetic acid; 0.1 N sodium hydroxide; anion-exchange resin† or, as an alternative, highly activated adsorbent charcoal; 100-ml volumetric flask; weighing bottle.

PROCEDURE. The exchange resin consists of particles (approximately 20 to 50 mesh) which are kept in a regenerating solution of 8 percent sodium carbonate when not in use. The carbonate solution is decanted, and the resin rinsed with three portions of distilled water, allowing the resin to settle before each decantation. The loss of finer particles during decantation is not serious.

† Amberlite IR-4B is satisfactory and is obtainable from chemical supply houses.

The resin is now collected on a filter and pressed to remove excess moisture. Twelve samples of damp resin of approximately 2 g in weight are weighed to ± 10 mg. Two additional samples of the resin weighed at the same time are placed in the oven and dried at $110°C$ to constant weight. In the calculations, the adsorption x/m is calculated on the basis of the dry resin.

Acetic acid solutions of different concentrations are made by running out $1\ N$ acetic acid from a burette and diluting to 100 ml with water; 50, 25, 10, 5, 2.5, and 1 ml are diluted with distilled water to 100 ml. Each solution is transferred to a 250-ml Erlenmeyer flask, and 2 g of wet exchange resin (or 1 g of adsorbent charcoal) is then added to each flask. The solutions are set up in duplicate. The solutions are agitated and allowed to stand overnight or longer. A thermostat is unnecessary if the room temperature is fairly constant.

After equilibrium has been reached and the resin has settled, a suitable volume (5 or 25 ml, depending on the concentration) is pipetted from the clear supernatant solution in each flask and titrated with $0.1\ N$ sodium hydroxide. Whenever the titration volume falls below 25 ml, the base may be diluted by a known ratio, such as $1:1$. Blank determinations should be made on all distilled water used for this purpose.

The experiment may be repeated with oxalic acid instead of acetic acid.

After the experiment has been completed, the resin is regenerated by the addition of 10 ml of 8 percent by weight Na_2CO_3 per gram of resin and returned to the stockroom.

CALCULATIONS.   The total weight of acetic acid in each solution is calculated from the data of the original solutions, and titration gives the weight remaining in 100 ml of the solution in equilibrium with adsorbent. The difference gives directly the weight of acetic acid adsorbed by the m grams of adsorbent.

The concentration of the solution in equilibrium with the adsorbed acetic acid is calculated from the sodium hydroxide titrations, and the values of x/m are plotted against these equilibrium concentrations.

The data are examined graphically with the object of learning which of Eqs. (1) and (3) better represents the behavior of the system studied.

Practical applications. The adsorption isotherm or equivalent form is important in the quantitative expression of the adsorption process, and as such it finds use in some dyeing and in various purification operations. It is useful also to characterize the adsorption of gases. In the theories descriptive of chromatography, the Langmuir equation has been very useful in accounting for the shape of the band passing through the column, the explanation of "tailing," etc.[6]

Ion-exchange resins are important in softening water, recovering ions from solutions of low concentration, and separating the rare earths.[7,8]

Suggestions for further work. If the system is in equilibrium, the same results should be obtained whether approached from concentrated or from more dilute solutions. Equilibrium may be tested by repeating the adsorption experiments and then diluting the solution with water after it has stood with the adsorbent. Acetic acid should be released by the adsorbent, and the final values of x/m and c should still fall on the same curve.

Various other materials may be adsorbed—weak acids, or bases. Ammonium hydroxide is suitable if a cation-exchange resin is used.

The adsorption isotherm may be tested nicely with dyes adsorbed from solutions (not necessarily in water) on charcoal, silicic acid, certain fibers,[9] etc. The initial and final concentrations of dye in solution may be obtained with either colorimeter or spectrophotometer.

The adsorption experiments may be carried out at 0°C and at elevated temperatures.

The values of the constants k and n in Eq. (1) are compared for the different materials and temperatures.

References

1. B. A. Adams and E. L. Holmes, *J. Soc. Chem. Ind. London*, **54:** 1 (1935).
2. M. C. Schwartz, W. R. Edwards, Jr., and G. Boudreaux, *Ind. Eng. Chem.*, **32:** 1462 (1940).
3. R. J. Myers and J. W. Eastes, *Ind. Eng. Chem.*, **33:** 1203 (1941).
4. R. J. Myers, J. W. Eastes, and F. J. Myers, *Ind. Eng. Chem.*, **33:** 697 (1941).
5. I. Langmuir, *Trans. Faraday Soc.*, **17:** 621 (1921).
6. J. C. Giddings, "Dynamics of Chromatography, I: Principles and Theory," M. Dekker, Inc., New York, 1965.
7. F. H. Spedding et al., *J. Am. Chem. Soc.*, **69:** 2777, 2786, 2812 (1947).
8. G. E. Boyd et al., *J. Am. Chem. Soc.*, **69:** 2818, 2836, 2849 (1947).
9. H. E. Schroeder et al., *Discussions Faraday Soc.*, **16:** 210 (1954).

53   ADSORPTION OF GASES

In this experiment the amounts of gas adsorbed at various pressures on activated charcoal or silica gel are determined by a volumetric method.

THEORY.   Adsorption is a process whereby gases or solutes are attracted and held to the surface of a solid. The material adsorbed is called the *adsorbate,* and the material on which it is adsorbed is called the *adsorbent.* Often the force of attraction is physical in nature, involving an interaction between dipoles or induced dipoles, but sometimes the force of attraction involves chemical bonds, as when oxygen is adsorbed on charcoal. In many cases the layer of adsorbed molecules is only one molecule deep, an atom of adsorbent at the surface being unable to extend its attractive force beyond the first molecule of adsorbate. Since the amount of gas which is adsorbed is proportional to the amount of surface exposed, the good adsorbents are those which have enormous surface areas, such as activated charcoal or silica gel, but adsorption of gases at low temperatures has been measured on clean glass surfaces, mercury surfaces, and metallic wires. Sometimes a large surface is produced by a cellular structure originally present in the plant, as in the case of charcoal.

If the adsorbent is porous on a submicroscopic scale, so-called *capillary condensation* may take place below the normal saturation pressure. This type of adsorption can be distinguished from the unimolecular-layer type of adsorption by heats of adsorption and by other criteria.

The experimental data for adsorption are plotted as adsorption isotherms, in which the quantity of gas adsorbed (expressed as milliliters at $0°$ and 760 mm) per gram of adsorbing material is plotted against the equilibrium pressure.

In many cases of adsorption it is possible to relate the amount of adsorbed material to the equilibrium pressure, using the empirical equation of Freundlich,

$$V = kP^n \tag{1}$$

where V = number of milliliters of gas, corrected to $0°$ and 760 mm, adsorbed per gram of adsorbing material

P = pressure

The constants k and n may be evaluated from the slope and intercept of the line obtained when $\log V$ is plotted against $\log P$.

One of the most successful theoretical interpretations of gas adsorption is that of Langmuir,[1] who considered adsorption to distribute molecules over the surface of the adsorbent in the form of a unimolecular layer. Consideration of the dynamic equilibrium between adsorbed and free molecules leads to the following relation:

$$\frac{P}{V} = \frac{P}{V_u} + \frac{1}{kV_u} \tag{2}$$

where P = gas pressure

V = volume of gas ($0°C$, atm) adsorbed per gram of adsorbent

V_u = volume of gas ($0°C$, atm) adsorbed per gram of adsorbent when unimolecular layer is complete

k = constant characteristic of adsorbent-adsorbate pair

Thus, if P/V is plotted against P, a straight line will be obtained if the Langmuir equation (2) applies. The slope of the line is equal to $1/V_u$; when the line is extrapolated to low pressures, as $P \to 0$, P/V approaches the finite limit $1/kV_u$. The values of the constants in the Langmuir equation may also be obtained by plotting $1/V$ versus $1/P$.

By postulating the building up of multimolecular adsorption layers on a surface, Brunauer, Emmett, and Teller[2,3] have extended the Langmuir derivation for unimolecular layer adsorption to obtain an isotherm equation for the more complicated case. The extension is not without some empiricism, yet it is often useful. Thus, knowing the volume of gas required to form a complete unimolecular layer over the surface of the adsorbent, it is possible to compute the surface area of the adsorbent if it is assumed that each molecule of the adsorbate occupies the volume that it would occupy if the density of the unimolecular film is the same as that of the liquid adsorbate at the same temperature.

A graph of adsorption data giving the pressure as a function of temperature for a specific volume of gas adsorbed is referred to as an *isostere*. Such graphs bear a close resemblance to graphs in which the vapor pressure is plotted against the temperature; the heat of adsorption $q_{\text{isosteric}}$ may be calculated in a manner similar to that used for calculating the heat of vaporization of a liquid, using the following modification of the Clausius-Clapeyron equation:

$$\ln \frac{P_2}{P_1} = \frac{q_{\text{isosteric}}(T_2 - T_1)}{RT_1 T_2} \tag{3}$$

where P_1 = equilibrium pressure for a given amount of gas adsorbed at T_1

$\quad\;\; P_2$ = equilibrium pressure of same amount of gas adsorbed at T_2

By calculating $q_{\text{isosteric}}$ for different volumes of gas adsorbed, the variation of heat of adsorption with volume adsorbed may be found. A variation of heat of adsorption with the volume adsorbed indicates changes in the magnitude of the forces between the adsorbent and the adsorbate.

····> ***Apparatus.*** Apparatus as shown in Fig. 102, consisting of a mercury manometer, a mercury-filled gas burette, six stopcocks, and a ground-glass joint; vacuum pumps; tanks containing methyl chloride or other suitable gas; adsorbent charcoal or silica gel.

PROCEDURE.   The flexible Tygon connecting tube from the gas tank attached to *E* is flushed out by removing the ground-glass tube *J* and opening the stopcocks *E, D, C,* and *B*. Stopcock *E* is closed; the tube *J* is replaced; the leveling bulb of mercury is lowered as far as possible, and the whole system is evacuated with a vacuum pump through stopcock *A*. Stopcock *D* is closed, and methyl chloride is admitted through *E* until the burette *H* is filled at atmospheric pressure. Stopcock *E* is then closed, and the leveling bulb *G* is adjusted carefully until the gas in the burette is at precisely atmospheric pressure. The volume of gas in the burette is then recorded.

The amount of gas which is adsorbed by the adsorbent is to be determined by measuring the volume of gas which must be admitted from the burette in order to give a specified pressure. Part of the gas from the burette is adsorbed, but an additional part, known as the *dead volume,* is needed for filling the apparatus up to the given pressure at which the adsorption measurement is made. The dead volume of the apparatus between the manometer *K* and the stopcock *D* is determined by finding the volume of gas admitted from the burette which is necessary to produce a given pressure in the absence of any adsorbing material in *J*.

The apparatus is first tested for leakage by observing the manometer. There should be no detectable change in the levels of the manometer *K* over a period of at

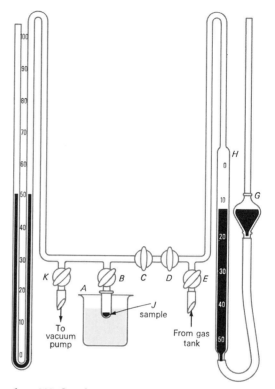

figure 102. Gas-adsorption apparatus.

least 5 min when the apparatus is evacuated. Then gas from the burette is admitted into the system through stopcocks C and D until the pressure in the apparatus is about 50 mm, and the exact manometer reading, burette readings, and temperature of the room are recorded. These data are used for calculating the dead volume of the gas under standard conditions. The dead volume is determined in a similar manner at pressure increments of about 100 mm. The burette is refilled when necessary by closing stopcock D, opening stopcock E, and lowering the leveling bulb G.

The volume of methyl chloride adsorbed by activated charcoal at a given pressure is determined by placing 0.080 to 0.100 g of the charcoal in the ground-glass tube J, inserting it at B, and placing a beaker of water at room temperature around it. Stopcocks C and D are closed, and the charcoal is outgassed for 5 min by evacuating through A. Methyl chloride is then admitted from the burette until the pressure is about 50 mm. The pressure will decrease as the gas becomes adsorbed on the charcoal, but after 5 min or so there is no tendency toward further change if the pressure is kept constant. The two stopcocks C and D with the intervening space of about 0.05 ml volume provide a convenient means for introducing gas into the system in small amounts so as to maintain the pressure constant. (A much longer time is required to reach equilibrium if a given quantity of gas is introduced and allowed to decrease in pressure until reaching an equilibrium.) When equilibrium has been reached and no more additions of gas are necessary to maintain the pressure, the volume of gas introduced from the burette is recorded.

This operation is repeated by using pressure increments of 30 to 50 mm up to about 250 mm and then using increments of about 100 mm up nearly to atmospheric pressure.

The adsorption measurements on charcoal are then repeated at $0°$ by using an ice bath in the beaker which surrounds the adsorbent.

CALCULATIONS. Part of the gas introduced from the burette is adsorbed on the charcoal, and part remains in the manometer and connecting tubes. The volume of gas adsorbed at a given equilibrium pressure is obtained by subtracting the dead-space volume in the apparatus from the total volume of gas introduced from the burette. The dead volume in milliliters at a given pressure is obtained by interpolation on a line obtained by plotting the corrected dead volume against the pressure measured in the absence of any adsorbent. All these observed volumes are reduced by calculations to the volumes of gas at $0°$ and 760 mm. When several additions of gas have been made to the adsorbent, giving a specified equilibrium pressure, they are all added together to obtain the total volume. The volume V adsorbed per gram of charcoal is then determined by dividing the corrected volume by the weight of the adsorbent.

Three graphs are drawn to interpret the adsorption. In the first graph, the corrected volume of gas V adsorbed per gram of adsorbent is plotted vertically, and the equilibrium pressures are plotted horizontally.

In the second graph, $\log V$ is plotted against $\log P$ in accordance with the Freundlich equation (1), and the constants k and n are calculated for the equation $V = kP^n$ as discussed on page 366.

In the third graph, the values of P/V are plotted against P to evaluate the constants k and V_u of the Langmuir equation (2); and the constant V_u is used for calculating the surface area of the adsorbent.

The volume of one molecule in the layer of adsorbate is calculated on the assumption that the molecule has the same volume as a molecule in the liquid state. The number of molecules in the unimolecular layer is calculated by dividing the volume V_u of adsorbate in the unimolecular layer by the volume of a mole of the gas, 22.4 liters, and multiplying by the Avogadro number 6.02×10^{23}. Then the volume of a molecule of adsorbate is calculated by dividing the volume of a mole of the material in the liquid state (obtained from tables of liquid densities) by the Avogadro number. The surface area covered by a single molecule is equal to the two-thirds power of the volume of the molecule. The surface area of a gram of adsorbent is equal, then, to the number of molecules in the saturated unimolecular layer multiplied by the cross-sectional area of a molecule.

The isosteric heat of adsorption is computed by Eq. (3) for several different volumes of gas adsorbed, and these values are plotted against the volume adsorbed. This plot is then used to interpret qualitatively the forces which exist between the adsorbent material and the gas being adsorbed.

Practical Applications. The adsorption of gases is used for purification and recovery of vapors. Solvent vapors are adsorbed from a stream of gas by adsorption in activated charcoal or silica gel and then recovered in concentrated form by heating the adsorbent to drive out the adsorbed vapors.

The drying of air is carried out on a large scale by adsorbing the water vapor with silica gel. When the silica gel becomes saturated, it is reactivated by heating to about $150°$ to expel the water and prepare the adsorbent for another cycle of adsorption.

High vacuum is conveniently produced in a vessel by connecting to a tube containing an adsorbent at liquid-nitrogen temperatures.

The effective surface of powders and catalysts is determined by measuring the amount of gas adsorbed. The adsorption of nitrogen on material at the temperature of liquid air has been used in such determinations.

Measurements of the Brunauer, Emmett, and Teller[3] (BET) constants are carried out on a routine basis in some catalyst testing programs.

Suggestions for further work.[4,5] The experiments and calculations may be repeated, using 0.5 g of silica gel instead of 0.1 g of activated charcoal. The silica gel should be activated by evacuating the system while gently heating the sample at J to about $150°$ with a bunsen flame.

Other gases and vapors may be used instead of the methyl chloride, e.g., ammonia, sulfur dioxide, or Freon, with suitable weights of adsorbents.

In general, the higher the boiling point of a liquid, the higher the temperature at which the adsorption experiments should be carried out.

Different grades of silica gel and activated charcoal may be used for adsorption experiments.

The rate of adsorption may be studied by maintaining the pressure as nearly constant as possible and measuring the volume adsorbed at different times. The rate of adsorption at different temperatures may be studied.

Instead of computing the amount of gaseous adsorption from pressure-volume data the weight of an adsorbed gas may be determined by direct gravimetric procedures, using a quartz spiral-spring balance.[6,7]

References

1. I. Langmuir, *J. Am. Chem. Soc.*, **38:** 2267 (1916); **40:** 1361 (1918).
2. S. Brunauer, "The Adsorption of Gases and Vapors," Princeton University Press, Princeton, N.J., 1943.
3. S. Brunauer, P. H. Emmett, and E. Teller, *J. Am. Chem. Soc.*, **60:** 310 (1938).
4. A. W. Adamson, "Physical Chemistry of Surface," Interscience Publishers, Inc., New York, 1959.
5. J. H. de Boer, "The Dynamical Character of Adsorption," Oxford University Press, Fair Lawn, N.J., 1953.
6. J. W. McBain and A. M. Baker, *J. Am. Chem. Soc.*, **48:** 690 (1926).
7. S. J. Gregg, *J. Chem. Soc.*, **1946:** 561, 564.

Chapter 17

Photochemistry

54   *PHOTOHYDROLYSIS OF MONOCHLOROACETIC ACID*

This experiment illustrates the use of an actinometer and the calculation of the quantum yield.[1,2]

THEORY.   Monochloroacetic acid hydrolyzes according to the reaction

$$CH_2ClCOOH + H_2O = CH_2OHCOOH + HCl$$

At room temperature the reaction rate is relatively small, but under irradiation by ultraviolet light, the reaction proceeds much more rapidly. Not every molecule that absorbs a photon undergoes reaction, since some molecules lose their excess energy before they react. The *quantum yield* for a photochemical reaction is defined as the number of molecules reacted per photon *absorbed.*

For determination of the number of photons absorbed, an actinometer is used. This consists of a solution of uranyl nitrate and oxalic acid. Oxalic acid alone in solution is nearly transparent to the radiation employed. With the addition of uranyl ions, the solution absorbance increases to a large value, and the reaction

$$UO_2^{++} + H_2C_2O_4 + light \longrightarrow UO_2^{++} + H_2O + CO + CO_2$$

takes place. It is to be noted that the uranyl ion, though essential to the process, is finally left unchanged. The amount of oxalic acid decomposed is readily found by titration. From the known quantum yield for this reaction and the amount reacted, the number of photons absorbed by the actinometer solution can be calculated.

The actinometer samples are so placed as to receive the same *incident* intensity as the monochloroacetic acid samples. In order to calculate the number of quanta *absorbed* by the monochloroacetic acid solution, it is necessary to know the relative absorbancies of the two solutions. With the concentrations and thicknesses used, the actinometer solution absorbs practically all the incident light for $\lambda < 4360$ Å. The monochloroacetic acid solution absorbs completely the light at 2537 Å but not that of the longer wavelengths present in the radiation. The high-voltage "cold" mercury lamp used emits radiation of which approximately 85 percent has a wavelength of 2537 Å. It may therefore be assumed for the present experiment that the mono-chloroacetic acid solution absorbs 85 percent of the incident light from these lamps, practically all of which is in the ultraviolet.

····⟶ ***Apparatus.*** Germicidal mercury-vapor lamp available at electric-fixture stores; four small rectangular plastic dishes or small crystallizing dishes; 50-ml pipette; two burettes; 200 ml 0.2 M uranyl nitrate; 200 ml 0.02 M oxalic acid; 500 ml 0.025 N potassium permanganate; 200 ml 0.1 M monochloroacetic acid, freshly prepared; 100 ml 0.01 M mercuric nitrate and 10 percent sodium nitroprusside *or* 200 ml 0.005 M silver nitrate; potentiometer; silver electrodes.

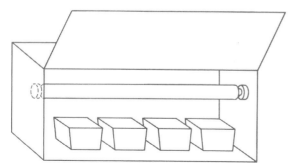

figure 103. Apparatus for photohydrolysis of monochloroacetic acid.

PROCEDURE.   The oxalic acid is weighed out accurately so that this solution may be used to standardize the potassium permanganate solution. An equal volume of the uranyl nitrate solution is mixed with the oxalic acid solution. The monochloro-acetic acid need not be made up accurately because the determination of the chloride ion, produced by the ultraviolet light, is the only analytical measurement to be made.

The apparatus is shown in Fig. 103. Four rectangular polyethylene dishes, commercially available for household uses, are filled with 50 ml of solution and placed in line under two or more 8-watt germicidal ultraviolet lamps. These are mounted parallel in the top of a wooden box provided with a hanging hinged door in front to protect the eyes from the ultraviolet light. To prevent overheating, air at about 15 ft^3 min^{-1} is drawn through the box past the lamps by an exhaust fan.

Caution: The lamp should not be viewed directly.

Since the light intensity varies with position, the actinometer solution is placed in the first and third positions and the monochloroacetic acid solution in the second and fourth. The time of exposure is recorded. It should be long enough (over 2 hr) to give a satisfactory titration.

The uranyl oxalate is titrated with 0.02 N potassium permanganate after adding an excess of sulfuric acid (15 ml of 5 N H$_2$SO$_4$ is sufficient for 50 ml of oxalate solution) and heating to about 60°C.

The two monochloroacetic acid solutions are titrated with 0.01 M mercuric nitrate. The titration follows the reaction Hg^{++} + 2Cl$^-$ = HgCl$_2$; the HgCl$_2$ produced is soluble but undissociated. About 0.2 ml of 10 percent sodium nitroprusside is added as indicator. The end point is recognized by the appearance of a faint, permanent turbidity due to the formation of a precipitate of mercuric nitroprusside. The amount of reaction is small, and the solutions are dilute, so that sharp end points cannot be expected. For this titration, it is helpful to use a darkened room and to arrange side lighting of the sample flask.

In a control experiment, the extent of any *dark reaction*, i.e., reaction taking place by a mechanism not involving absorption of light, is obtained by titration of unirradiated samples of the monochloroacetic acid solution.

The mercuric nitrate solution is standardized against a solution of HCl or NaCl of a known concentration similar to that of the Cl^- in the samples. A blank experiment is performed with nitroprusside in water alone.

Alternatively, the chloride ion may be titrated with 0.005 M $AgNO_3$ and the titration followed potentiometrically. Two silver wire electrodes are used, one being placed in a 0.01 M $AgNO_3$ solution; the latter is connected through a salt bridge of ammonium nitrate to a beaker containing the monochloroacetic acid solution, in which is immersed the second silver electrode. The voltage is plotted as a function of the volume of $AgNO_3$ added. The voltage remains nearly constant until the chloride ion is nearly gone. Thereafter, it changes rapidly with further addition of silver nitrate. The end point may be taken as the point of greatest slope in the titration curve.

In a check experiment, the uranyl oxalate solution may be placed in the first and second positions and the monochloroacetic acid in the third and fourth.

CALCULATIONS.   The number of photons absorbed by the actinometer is calculated from the permanganate titration data and the known quantum yield, 0.57 molecule per photon, for the actinometer reaction.[3] The number of photons absorbed by the two monochloroacetic acid samples is assumed to be 15 percent less than the number absorbed by the actinometer solutions. The quantum yield for the monochloroacetic acid samples is calculated, and the result compared with literature values.

Practical applications. The determination of quantum yield is an essential first step in seeking to understand the mechanism for a photochemical reaction. A great deal of effort has been devoted to the study of photochemical processes in nature such as photosynthesis.

Suggestions for further work. A better method[4] for measuring the energy absorption consists in making a determination with a flat quartz dish containing a concentrated solution of monochloroacetic acid (which absorbs light of wavelength 2537 Å) placed over the uranyl oxalate dish. In a second experiment the upper quartz dish contains water. The titration a obtained for uranyl oxalate with the monochloroacetic acid filter gives a measure of the number of quanta of wavelength longer than 2537 Å, and the titration b with the water gives a measure of all the radiation absorbed by uranyl oxalate, together with a correction for losses of light from the filter due to reflection—amounting to about 4 percent each at the air-quartz interface and at the air-solution interface. If the titration obtained in the absence of an upper filter solution is denoted by c, the corrected titration for light of 2537 Å alone is $c - (c/b)a$.

The bleaching of dyes such as methylene blue or malachite green may be followed colorimetrically, using as standards various concentrations of the unbleached dyes.

The photodecomposition of hydrogen peroxide may be followed by titration with potassium permanganate. This is a chain reaction subject to catalytic influences. Chain reactions give large quantum yields and do not require micro methods for chemical analysis.

The photobromination of cinnamic acid in carbon tetrachloride provides a good experiment.[5] The quantum yield shows that the reaction is a chain reaction. It increases when dissolved oxygen is removed by degassing the solution by boiling under reduced pressure. Equal portions of 0.005 M bromine in carbon tetrachloride and 0.01 M cinnamic acid in carbon tetrachloride are mixed and placed in sunlight. The bromine adds to the double bond when exposed to light, and the decrease in free bromine is determined by adding potassium iodide solution and titrating with standard sodium thiosulfate. The bromine removal may be determined also spectrophotometrically.

Additional suggestions for further work may be found in Ref. 6.

References

1. R. N. Smith, P. A. Leighton, and W. G. Leighton, *J. Am. Chem. Soc.*, **61:** 2299 (1939).
2. L. B. Thomas, *J. Am. Chem. Soc.*, **62:** 1879 (1940).
3. P. A. Leighton and G. S. Forbes, *J. Am. Chem. Soc.*, **52:** 3139 (1930).
4. W. A. Noyes, Jr., and P. A. Leighton, "The Photochemistry of Gases," pp. 82–85, Reinhold Publishing Corporation, New York, 1941.
5. W. A. Bauer and F. Daniels, *J. Am. Chem. Soc.*, **56:** 387, 2014 (1934).
6. J. G. Calvert and J. N. Pitts, "Photochemistry," John Wiley & Sons, Inc., New York, 1966.

Chapter 18

Radioactive Isotopes and Tracers

55 THE SZILARD-CHALMERS PROCESS AND THE HALF-LIFE OF RADIOIODINE

In this experiment a radioisotope is prepared by the neutron-bombardment process, and its half-life is determined. The Szilard-Chalmers technique is illustrated.

THEORY.[1-5] When a slow neutron is captured by a target nucleus, the resulting nucleus, which is one unit larger in mass, is produced initially in an unstable state and becomes stabilized through the emission of one or more γ-ray photons. This process is called the (n,γ) reaction. The resulting nuclear state may exhibit radioactivity associated with subsequent nuclear transformations.

The present experiment illustrates the production of radioactive nuclei by an (n,γ) process and the measurement of their decay rate. A liquid sample of ethyl iodide (^{127}I) is irradiated with neutrons to produce ^{128}I by an (n,γ) process,

$$n + {}^{127}I \longrightarrow {}^{128}I + \gamma$$

The subsequent decay

$$^{128}I \longrightarrow {}^{128}Xe + \beta^- + \bar{\nu}$$

is studied by counting the emitted β^- particles. The antineutrino $\bar{\nu}$ is not observed in this experiment.

The radioactive nuclei are produced only in trace amounts, and it is desirable to concentrate them in order to improve the accuracy of the decay-rate measurement. Although it might seem impossible to achieve appreciable isotopic enrichment of the radioactive species (^{128}I relative to ^{127}I) by simple chemical techniques, the possibility of doing so arises as a consequence of the chemical events following the (n,γ) reaction, as shown by Szilard and Chalmers.[6]

Because the law of conservation of momentum must be satisfied, the recoiling nucleus of mass M, in the emission of a γ-ray photon of energy E_γ, acquires a momentum E_γ/c and therefore a large kinetic energy $(E_\gamma/c)^2/2M \sim 100$ ev. This is very much greater than chemical-bond energies, which are typically only a few electron volts. The result is not only the rupture of the bond originally holding the atom involved in the molecule but also the breaking of numerous other bonds as the recoiling atom expends its excess energy. The atoms and molecular fragments so produced recombine in various ways to form stable molecular species, and many of the radioactive nuclei become bound up in molecular species different from the original target substance. Because of this, chemical techniques can be effective in concentrating them.

Thus, when ethyl iodide is irradiated, a large fraction of the radioiodine produced ends up in inorganic form, as $H^{128}I$ and $^{127}I^{128}I$, which may be separated from the organic medium by extraction with aqueous sodium hydroxide solution. Some of the ^{128}I ends up in organic form, and this is not extracted. Since the proportion of organic to inorganic yield depends on the competition among various parallel reaction paths, such as

$$^{128}I + CH_3 = CH_3^{128}I$$
$$^{128}I + CH_3CH_2^{127}I = CH_3CH_2 + {}^{127}I^{128}I$$

the inorganic yield can be improved by adding a small amount of I_2 to the solution before irradiation.[7] This is effective through processes such as

$$^{128}I + {}^{127}I_2 = {}^{127}I^{128}I + {}^{127}I$$
$$CH_3 + {}^{127}I_2 = CH_3^{127}I + {}^{127}I$$

which enhance the inorganic yield. This effect of added I_2 is called the *scavenger* effect.

For success of the Szilard-Chalmers technique, it is essential that isotopic exchange not take place to any appreciable extent between the bulk target material and the species in which the radioactive isotopes are to be concentrated.

For a bulk sample containing a large number of radioactive nuclei, the average rate of disintegration, $-dN/dt$, is found to be proportional to the number N of radioactive atoms present:

$$-\frac{dN}{dt} = \lambda N \tag{1}$$

Integration from $t = 0$, when the number of atoms present is N_0, to time t, when the number of atoms present is N, gives

$$2.303 \log \frac{N}{N_0} = -\lambda t \tag{2}$$

or

$$N = N_0 e^{-\lambda t} \tag{3}$$

For characterizing a radioisotope, the half-life $t_{\frac{1}{2}}$ is more commonly used than the rate constant λ. The half-life is the time required for the number of radioactive atoms to decrease to one-half the number originally present; by setting N_t/N_0 equal to $\frac{1}{2}$ in Eq. (2), one obtains

$$t_{\frac{1}{2}} = \frac{2.303 \log 2}{\lambda} = \frac{0.693}{\lambda} \tag{4}$$

The activity R of a sample is defined as the number of disintegrations per second, $-dN/dt$, and through Eq. (1) this is proportional to N. Equations (2) and (3) clearly remain valid when N is replaced by R or any other quantity proportional to N.

For laboratory use an adequate neutron flux can be obtained from a small radium-beryllium source

$$^{226}_{88}\text{Ra} \longrightarrow {}^{222}_{86}\text{Rn} + {}^{4}_{2}\text{He}$$
$$\alpha \text{ particle}$$

$$^{4}_{2}\text{He} + {}^{9}_{4}\text{Be} \longrightarrow {}^{12}_{6}\text{C} + {}^{1}_{0}n$$

Approximately 10^7 neutrons per second are produced per gram of radium; a typical laboratory source containing 10 mg of radium thus provides 10^5 neutrons per second. The neutrons produced may have kinetic energies up to 8 Mev and must be slowed down to be effective in the (n,γ) process. This is achieved by elastic collisions with nuclei. Hydrogen-containing substances make good moderators since the transfer of energy from a neutron to a proton in an elastic collision is very efficient because of the similarity in mass. Neutrons which have been slowed down to energies of the magnitude of that of thermal agitation, about RT per mole, are called *thermal neutrons*. Such neutrons are not monoenergetic but are characterized by a maxwellian distribution of velocities; their average energy depends on the temperature of the medium in which they were slowed down.

If a neutron beam having a flux ϕ (neutrons per second per square centimeter) is incident upon a thin layer of material with $\mathfrak{N}$ target nuclei per cubic centimeter, the number of neutron captures occurring per second per cubic centimeter is $\mathfrak{N}\phi\sigma$, where σ is a proportionality factor called the *cross section* for neutron capture. The cross section is commonly reported in terms of barns (1 barn $= 10^{-24}$ cm^2), and its value depends on the energy of the incident neutron as well as the identity of the nucleus involved. For ^{127}I the cross section σ is 6.4 barns for thermal neutrons.[8]

With a small neutron source centered in a flask containing an organic iodide the neutron flux is not constant throughout the medium. The number of radioiodine atoms produced per unit time, however, is constant at a value $\mathfrak{P}$ determined by the neutron-capture cross section and the radial depth and ^{127}I concentration in the medium. The net rate of increase of radioactive atoms is this rate of production minus the rate of disintegration λN, as given by Eq. (1):

$$\frac{dN}{dt} = \mathfrak{P} - \lambda N \tag{5}$$

Integration from $t = 0$, $N = 0$, to a later time t, when the number of radioactive atoms is N, gives

$$N = \frac{\mathfrak{P}}{\lambda}(1 - e^{-\lambda t}) \tag{6}$$

Thus, N follows the standard exponential growth curve for a first-order process, approaching the limiting value

$$N_\infty = \frac{\mathscr{P}}{\lambda} = 0.693\mathscr{P}t_{\frac{1}{2}} \tag{7}$$

which corresponds to the steady state, where dN/dt of Eq. (5) has become zero. The half-life for the growth of N during the irradiation period is the same as that for the decay in the absence of irradiation. From the (n,γ) cross section and incident neutron flux, the actual value of N_∞ can be estimated.

When the disintegration is observed by counting the emitted particles, fluctuations in the rate become obvious. Since the emission process appears to be strictly random, a statistical analysis is appropriate. Assume that the probability that a given radioactive nucleus will disintegrate during a small time interval Δt is $k\,\Delta t$. The actual number n of disintegrations which will occur in a time interval Δt for a sample of N radioactive nuclei cannot be predicted, but statistical theory can be used to predict the relative probabilities $P(n)$ of occurrence of various values of n if the nuclei are considered to be independent. The result of such an analysis is [1,9,10]

$$P(n) = \frac{(kN\,\Delta t)^n}{n!} e^{-kN\Delta t} \qquad \text{Poisson distribution} \tag{8}$$

with

$$\sum_{n=0}^{\infty} P(n) = 1 \tag{9}$$

The predicted average $\bar{n}$ for a large number of trials is

$$\bar{n} \equiv \sum_{n=0}^{\infty} nP(n) = kN\,\Delta t \tag{10}$$

Since the average rate is given by Eq. (1), we can identify k with λ. Of interest is the predicted standard deviation σ, which is given by

$$\sigma^2 = \sum_{n=0}^{\infty} (n - \bar{n})^2 P(n) = \bar{n} \tag{11}$$

$$\sigma = \sqrt{\bar{n}} \tag{12}$$

The last part of Eq. (11) results from substituting Eq. (8) into the sum. The quantity σ provides an estimate of the rms fluctuation to be expected in n. From actual measurement, $\bar{n}$ is not known, but σ may be approximated satisfactorily (provided n is not too small) by

$$\sigma \approx \sqrt{n} \tag{13}$$

It should be noted that the absolute value of σ increases with n, but the fractional value σ/n decreases.

The estimated standard deviation is useful for assessing the reliability of a measurement. If an actual measurement yields n counts and the object was to measure the true average $\bar{n}$, the result may be stated as

$$\bar{n} = n \pm \sqrt{n} \tag{14}$$

with a confidence level of about 70 percent. If the range is increased to $\pm 2\sqrt{n}$, the confidence level becomes 95 percent. Again, as an approximation, we have replaced $\sqrt{\bar{n}}$ by $\sqrt{n}$.

An individual measurement of λ will then be subject to error due to statistical fluctuations in the values of n. Up to a point, the accuracy can be improved by extending the measurements over a longer period of time t, but a practical limitation is imposed by the existence of a background of counts due to cosmic rays and stray radioactive nuclei in the laboratory. A correction based on the average background rate R_b is subtracted from the observed count n to obtain a corrected count n', for time interval Δt,

$$n' = n - R_b \, \Delta t \tag{15}$$

To estimate the error in n' due to statistical fluctuations, we need to consider how standard deviations are compounded in simple arithmetic operations. Let $\sigma(x)$ be the standard deviation for the quantity x and $\sigma(z)$ that for z. Now if

$$z = ax$$

where a is a constant, then

$$\sigma(z) = a\sigma(x) \tag{16}$$

If

$$z = x + y$$

where x and y are independent quantities (errors uncorrelated), then

$$\sigma(z) = \sqrt{\sigma(x)^2 + \sigma(y)^2} \tag{17}$$

Equations (16) and (17) follow from the definition of σ, Eq. (11). The rule for combining probabilities of x and y as uncorrelated quantities is $P(z) = P(x)P(y)$.

Suppose the average background counting rate in the laboratory R_b is measured by counting for a period $(\Delta t)_b$. The number of counts for that measurement then is $R_b(\Delta t)_b$. Application of Eqs. (12), (16), and (17), with the assumption of a Poisson distribution for the background gives

$$\sigma[R_b(\Delta t)_b] = \sqrt{R_b(\Delta t)_b}$$

$$\sigma[R_b] = \sqrt{\frac{R_b}{(\Delta t)_b}}$$

$$\sigma[R_b(\Delta t)] = \sqrt{\frac{R_b}{(\Delta t)_b}}\,\Delta t$$

$$\sigma(n') = \sqrt{\sigma^2(n) + \sigma^2(R_b\,\Delta t)}$$

$$= \sqrt{n + \frac{R_b(\Delta t)^2}{(\Delta t)_b}} \tag{18}$$

On the assumption that R_b is constant, i.e., that the measured background shows only random fluctuations, the contribution of the second term in Eq. (18) can be made small by choosing $(\Delta t)_b$ large enough. But even if this is done, the background still contributes to the error in n', as the standard deviation in n' is $\sqrt{n}$ rather than $\sqrt{n'}$. To understand this result, one should recognize that even if the mean background rate R_b is accurately known, one still does not know how many of the n counts obtained in the time interval Δt were actually due to the background.

An alternative form of Eq. (18) is useful which gives standard deviations in the rates rather than numbers of counts. Let

$R_s =$ counting rate due to sample
$R = R_s + R_b =$ total counting rate

Since $n = R\,\Delta t$ and $n' = R_s\,\Delta t$, division of both sides of Eq. (18) by Δt gives

$$\sigma(R_s) = \sqrt{\frac{R_s}{\Delta t} + \frac{R_b}{\Delta t} + \frac{R_b}{(\Delta t)_b}} \tag{18a}$$

····> ***Apparatus.*** Neutron source† and lead storage shield; remote-handling device; irradiation-flask assembly; irradiation-flask shield; 1-liter separatory funnel; 50-ml pipette; two 150-ml beakers; two 125-ml Erlenmeyer flasks; 25-ml graduate; special sintered-glass filter crucible or small Büchner funnel and filter paper; filter-flask assembly; bunsen burner and ring stand; counting tube, scaler, and timer; ethyl iodide; 0.5 N sodium hydroxide; nitric acid; 0.01 N silver nitrate solution; carbon tetrachloride; iodine.

PROCEDURE.[11]   About 850 ml of ethyl iodide is placed in a 1-liter Pyrex flask. Approximately 10 mg of iodine is weighed out and dissolved in the ethyl iodide to

---

† Such sources may be obtained from Atomic Energy of Canada, Limited, Commercial Products Division, P.O. Box 379, Ottawa, Canada, or the Oak Ridge National Laboratory, Oak Ridge, Tenn.

act both as a scavenger and as a carrier for the iodine released by the neutrons from chemical combination in the organic molecule, ethyl iodide. The neutron source is then removed from its lead protective housing *by the instructor,* using a remote-handling device,† and transferred to the irradiation flask of ethyl iodide as shown in Fig. 104. The irradiation is allowed to proceed for 1 hr.

The test-tube holder for the neutron source is of soft glass because Pyrex glass contains boron, which has a high capture cross section for neutrons.

Caution: The source must always be kept at a safe distance from all personnel.§ The radium-beryllium neutron source represents a **radiation hazard;** γ rays as well as neutrons are emitted. It should be stored in a lead housing and must be handled by means of a remote-control device.

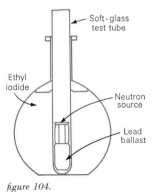

figure 104.

Irradiation-flask assembly.

Lead shielding is also provided for the irradiation flask during the irradiation period. The ^{128}I is not produced in hazardous amounts, so that the special precautions are required only with the operations involving the neutron source itself.

Because of the relatively short half-life of ^{128}I, rapid processing of the irradiated material is essential. During the irradiation period preparations should be carefully made for the subsequent operations. By means of a pipette, 50 ml of 0.5 N sodium hydroxide is placed in a 1-liter separatory funnel supported on a ring stand. A 1 N nitric acid solution is prepared by dilution of concentrated acid. A quantity of this acid slightly in excess of that required to neutralize the 50 ml of the sodium hydroxide solution is placed in a 250-ml beaker. In separate flasks are placed 15 ml of carbon tetrachloride and 25 ml of a 0.01 N solution of silver nitrate. The auxiliary equipment (filter flask, burner, etc.) is set up, and the counting equipment prepared for use at this time.

The Geiger-Müller tube, shown in Fig. 105, the impulse register, and the electric timer are connected to the scaling unit. With all switches in the "off" position and the high-voltage control in the "low" position, the unit is connected to the 110-volt ac line. The main power switch is then turned on, and a warm-up period of 2 min allowed. The high-voltage switch is then turned on, and when the applied potential has registered on the voltmeter, the high-voltage control is turned up slowly, until the operating potential for the particular tube in use is reached.

The number of counts in a 10-min period is determined with no sample in the holder or near the Geiger-Müller tube. The counts registered constitute the background of the counter, which contributes to the observed counting rate in any

† Obtainable from suppliers of nuclear apparatus.

§ Permissible radiation exposures are outlined in Ref. 12, available from the Superintendent of Documents, Washington, D.C. 20402. A survey meter should be used to check radiation levels.

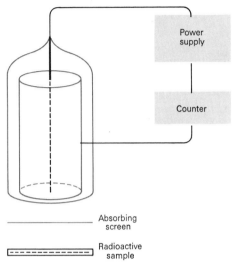

figure 105. Geiger-Müller tube.

measurement. Radioactive materials, including radium-activated luminous watch dials, should be kept remote from the counter tube. (When maximum accuracy is required, as with weakly radioactive samples, heavy lead shields may be placed around the counting tube to reduce the background.)

At the end of the irradiation period, the neutron source is removed by the instructor and returned to storage. The time is noted, so that the period of irradiation as well as the time required for the subsequent chemical operations will be known.

The ethyl iodide with dissolved iodine is transferred to the separatory funnel containing the sodium hydroxide solution, and the mixture shaken vigorously. The principal reaction is

$$3I_2(org) + 6\ OH^-(aq) = 5I^-(aq) + IO_3^-(aq) + 3H_2O$$

The rapid disappearance of the iodine color from the organic layer indicates the extraction of the iodine into the aqueous phase. The two phases are allowed to separate, and the heavier ethyl iodide layer is collected in the irradiation flask and saved. When the experiment has been completed, the ethyl iodide should be returned to the instructor since it can be used again.

The aqueous phase is then extracted in a separatory funnel with the carbon tetrachloride to remove any residual ethyl iodide. The carbon tetrachloride extract is withdrawn and discarded. The aqueous layer is then drained into the dilute nitric acid; the stirred solution is tested with methyl orange indicator to make sure it is slightly acid. The silver nitrate solution is added, and the mixture heated rapidly to boiling to coagulate the silver iodide formed. The precipitate is collected by use of a glass-filter crucible, the body of which has been cut off close to the sintered-glass disk to permit the precipitate to be mounted close to the window of the radiation counter tube. Alternatively, filter paper and a small Büchner funnel may be used. The precipitate is washed with distilled water, then with acetone, and air is drawn through to dry it. The crucible (or filter paper) is mounted on an aluminum or cardboard plate for counting. The number of counts obtained in 1-min counting periods is determined at 2-min intervals for a period of at least 50 min. The time elapsed between the removal of the irradiation source from the sample and the beginning of the counting should be recorded.

If the initial counting rate is low, the accuracy of the data will be poor. Low initial counting rates are usually the result of spending too much time on the chemical separation procedure.

The scavenger effect of the iodine may be demonstrated by repeating the experiment, with the same procedure as before, except that the I_2 is added *after* the irradiation. (If ethyl iodide from the previous experiment is reused, it may need to be distilled in order to remove any iodine remaining.) The iodine added in this case still serves a purpose, as a *carrier*. The amount of radioactive iodine formed is so small that it could not be precipitated alone. The ordinary iodine exchanges with the radioactive iodine, and when it is precipitated as iodide, most of the radioactive iodide is precipitated with it.

CALCULATIONS. A plot is made of the logarithm of the number of counts per minute from the radioiodine samples against time of observation. The counting rate which is measured is proportional to the number of radioactive atoms present in the sample. Probable errors due to statistical fluctuations (including those in the background) are calculated for the various points and indicated on the graph. At high counting rates the *dead time* due to lag in recovery of the Geiger-Müller counter and associated amplifier circuit may cause an error. If the dead time is τ and the observed counting rate r, the true rate R can be estimated as $R = r/(1 - r\tau)$. The rate constant for disintegration of ^{128}I is determined from the slope of the straight line considered to give the best representation of the experimental points. The scatter of the data points relative to this line should be examined in the light of the predicted statistical fluctuations. The half-life of ^{128}I is calculated and compared with the accepted value.

From the measured half-life, the irradiation time, and the time elapsed between the irradiation and the start of the counting period, one may easily compute the ratio of the sample activity reached at the end of the irradiation period to the limiting value for the given source, and the ratio of the sample activity at the start of the counting to that at the end of the irradiation period.

Practical applications. Neutron bombardment, particularly in the high neutron fluxes furnished by nuclear reactors, is the most important method for the production of radioisotopes.

Suggestions for further work.[9,11,13] The (n,γ) reaction on bromine may also be studied. The modes of production of the three radioactive species formed and their individual decay schemes may be represented as follows:

The ^{80}Br and ^{80m}Br constitute a pair of nuclear isomers; the latter is a relatively long-lived metastable excited nuclear state of the former. The decay of ^{80m}Br to ^{80}Br is termed an *isomeric transition*. The 0.049-Mev transition occurs almost exclusively by the process of *internal conversion*, in which a direct transfer of energy from the nucleus to an adjacent orbital electron takes place, with the resultant ejection of a *conversion electron*. The 0.037-Mev transition occurs by conversion electron emission in about half of the events and by γ-ray emission in the other half.

Ethylene dibromide is irradiated for at least 4 hr, and preferably overnight, with the neutron source. The irradiated mixture is extracted with a solution of about 25 mg of potassium bromide in 50 ml of distilled water. (A small amount of a reducing agent such as sodium sulfite may be helpful also.) The aqueous layer is separated and acidified to methyl orange indicator with nitric acid, and silver bromide precipitated by addition of silver nitrate solution. The precipitate is collected as described previously for silver iodide, and the number of counts per minute from the radiobromine sample determined at 2-min intervals for the first half hour and at longer intervals thereafter. An absorber of thickness about 100 mg cm^{-2} is placed between the sample and the counting tube to absorb the relatively weak β emission from the ^{82}Br present. A total counting period of 4 hr or more is recommended. The counting rate initially decreases rapidly as the ^{80}Br present disintegrates. Ultimately the rate of decay of ^{80}Br becomes equal to its rate of production from ^{80m}Br; the activity thereafter decays with a half-life of 4.4 hr. It should be noted that the counts registered here are due to the β-ray emission from the daughter ^{80}Br in equilibrium with the ^{80m}Br, since the counting efficiency for the accompanying γ rays is negligible relative to that for the β particles. The γ rays are so penetrating that only a few are counted.

A plot is made of the logarithm of the number of counts per minute from the radiobromine sample versus the time of observation. A straight line is drawn through the points corresponding to the later observations, and the half-life of ^{80m}Br is calculated from its slope. Data for a corresponding plot for the 18-min activity can be obtained by subtracting from the total numbers of counts per minute at the early times of observation the contributions of the 4.4-hr activity evaluated from the extrapolation of the straight line referred to above.

After the irradiated ethylene dibromide has stood for about 2 hr following the first extraction, a second extraction may be made. The activity thus isolated will decay with a half-life of 18 min. It arises from ^{80}Br released from organic combination by bond rupture accompanying the isomeric transition from ^{80m}Br by the internal conversion process.

Radiomanganese, ^{56}Mn, whose half-life is 2.6 hr, may be isolated by the Szilard-Chalmers process by irradiation of concentrated aqueous potassium permanganate solution.[13,14] The efficiency of the separation depends on the pH of the solution and becomes low when the pH is high.[5]

The nonequivalence of the two sulfur atoms in the thiosulfate ion may be demonstrated[15] by the use of radiosulfur obtained from the U.S. Atomic Energy Commission. The maximum β-ray energy for ^{35}S is low, 0.165 Mev, so that a counting tube with a very thin mica window must be used.

References

1. G. Friedlander, J. W. Kennedy, and J. M. Miller, "Nuclear and Radiochemistry," 2d ed., John Wiley & Sons, Inc., New York, 1964.
2. I. G. Campbell, *Advan. Inorg. Chem. Radiochem.,* **5:** 135 (1963).
3. J. E. Willard, *Ann. Rev. Nucl. Sci.,* **3:** 193 (1953); *Nucleonics,* **19:** 61 (1961).
4. R. R. Williams, Jr., W. H. Hamill, and R. H. Schuler, *J. Chem. Educ.,* **26:** 210 (1949).
5. W. H. Hamill, R. R. Williams, Jr., and R. H. Schuler, *J. Chem. Educ.,* **26:** 310 (1949).
6. L. Szilard and T. A. Chalmers, *Nature,* **134:** 462 (1934).
7. G. Levey and J. E. Willard, *J. Am. Chem. Soc.,* **74:** 6161 (1952).
8. C. M. Lederer, J. M. Hollander, and I. Perlman, "Table of Isotopes," 6th ed., John Wiley & Sons, Inc., New York, 1967.
9. R. D. Evans, "The Atomic Nucleus," chap. 26, McGraw-Hill Book Company, New York, 1955.
10. E. B. Wilson, Jr., "An Introduction to Scientific Research," pp. 191–195, McGraw-Hill Book Company, New York, 1952.
11. R. H. Schuler, R. R. Williams, Jr., and W. H. Hamill, *J. Chem. Educ.,* **26:** 667 (1949).
12. Maximum Permissible Body Burdens and Maximum Permissible Concentrations of Radio-nuclides in Air and in Water for Occupational Exposure, *Natl. Bur. Stand. U.S. Handbook* 69, 1959; see also ref. 1. p. 127.
13. G. B. Cook and J. F. Duncan, "Modern Radiochemical Practice," Oxford University Press, Fair Lawn, N.J., 1952.
14. W. F. Libby, *J. Am. Chem. Soc.,* **62:** 1930 (1940).
15. W. J. McCool and R. R. Hentz, *J. Chem. Educ.,* **32:** 329 (1955).

56 EXCHANGE REACTIONS WITH DEUTERIUM OXIDE

By the use of concentrated deuterium oxide it is possible to determine the number of hydrogen atoms in a hydrogen-containing compound which may be exchanged with hydrogen atoms of water. A dry sample of the compound is dissolved in heavy water, and then the heavy water is sublimed out of the sample by use of a vacuum pump and Dry Ice trap. The increase in weight of the compound is due to the substitution of deuterium for hydrogen. Some hydrogen atoms exchange so rapidly that it has not been possible to determine rates, while hydrogen atoms in other positions may exchange very slowly.

THEORY.   Although the chemical reactions of isotopes are nearly identical, there are very minor differences which show up most prominently in the elements of low atomic weight, and especially in hydrogen.[1,2] If a mixture of water containing equal parts of the light isotope H and the heavy isotope D is electrolyzed, it is found that H is liberated by electrolysis about five times as fast as D. Exchange reactions and highly efficient fractional distillation as well as electrolysis are used on a large scale to produce an enrichment in deuterium oxide.

Apparatus. Water containing a known percentage of D_2O;† analytical balance; 15-ml weighing bottle; vacuum pump (0.3 mtorr); drying trap (Fig. 106); Dewar flask; Dry Ice; hot plate; ammonium sulfate; trichloroethylene.

PROCEDURE.   Ammonium sulfate is satisfactory for this experiment because it is nonvolatile and 6 percent of its mass is due to hydrogen. A convenient amount of ammonium sulfate to use is 0.7 g. The ammonium sulfate is dried by placing it in a weighing bottle in an apparatus such as that illustrated in Fig. 106. The apparatus

† Heavy water containing 99.8 percent D_2O or higher may be purchased from the Stuart Oxygen Company of San Francisco.

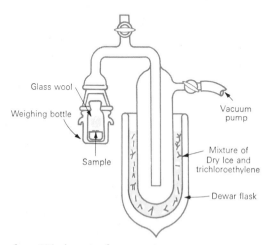

figure 106. Apparatus for evaporating water.

is evacuated by use of a mechanical vacuum pump. A trap surrounded by Dry Ice in trichloroethylene is placed between the drying chamber and the pump so that moisture does not get into the oil of the vacuum pump or oil vapor into the sample. The Dewar flask is about one-third filled with liquid, and then crushed Dry Ice is added (slowly at first). The ammonium sulfate may be dried more quickly if the drying chamber is immersed in a beaker of boiling water.

After the ammonium sulfate has been dried, it is weighed and about 1 g of heavy water is placed on the sample. The weighing bottle and contents are then weighed again. The supply of heavy water should be kept stoppered so that D_2O will not exchange with H_2O in the air. The sample must be completely dissolved in the heavy water (by heating if necessary) so that the entire sample has an opportunity to exchange. It is necessary to evaporate the D_2O out of contact with room air so that exchange with ordinary H_2O will not occur. To avoid bumping, the water is sublimed from the frozen solution. To avoid losses, a small wad of glass wool is weighed and is placed in the top of the weighing bottle before the solution is frozen in the Dry Ice—trichloroethylene freezing mixture.

The weighing bottle containing frozen solution is placed in the drying apparatus and evacuated. The evaporation of D_2O molecules may not be sufficiently rapid to keep the solution frozen. It is essential that the sample not be allowed to become so warm that bumping occurs, as some of the sample will be lost in this way. However, if the chamber containing the weighing bottle is kept immersed in ice water, the sample will generally remain sufficiently cold. When the liquid content has been reduced to the point that there is no longer danger of bumping, the ice bath is replaced by a beaker of room-temperature water, and finally by a beaker of boiling water. By raising the temperature in this way, the drying time is considerably reduced.

The dried salt is weighed and then placed in the drying apparatus for another hour at 100° and reweighed to be sure constant weight has been reached.

If a vacuum pump is not available, the experiment may be carried out by passing air dried with Ascarite or other desiccant over the sample. If air is drawn through the sample chamber by use of a water aspirator, a second drying tube should be used to prevent diffusion of ordinary water vapor into the sample chamber.

CALCULATIONS.   It is desired to determine how many of the hydrogens in the ammonium sulfate molecule can exchange with deuterium in the solvent water. Only hydrogen atoms in certain sites in organic compounds do exchange with the hydrogen atoms in water. If the ammonium sulfate were in equilibrium with pure deuterium oxide, all the readily dissociable hydrogens would be replaced by deuterium. However, the deuterium oxide becomes diluted with ordinary water by the exchange with hydrogen from the ammonium sulfate. Since the ammonium sulfate and heavy water are in equilibrium, the mole fraction of replaceable hydrogens which are replaced by deuterium will be equal to the mole fraction of the deuterium oxide in the water which is in equilibrium with ammonium sulfate.

The mole fraction of D_2O in the water at equilibrium is calculated from the original weight w of heavy water taken, the weight fraction f of D_2O in the original deuterium oxide, and the gain in weight of the ammonium sulfate, $g_2 - g_1$. The total number of moles of water is

$$\frac{fw}{20} + \frac{(1-f)w}{18} \tag{1}$$

This is the number of moles of water added originally, and it is also the total number after exchange, because there is no change in the total number of molecules of D_2O and H_2O, merely a change in the relative proportions of each. The decrease in the number of moles of D_2O as a result of exchange with the ammonium sulfate, leaving deuterium in the ammonium sulfate instead of hydrogen and forming H_2O, is $(g_2 - g_1)/2$, where the factor 2 enters because there are 2 atoms of hydrogen per molecule of water and a difference in atomic weights of D and H of 1. Thus the mole fraction of D_2O at equilibrium is

$$X_{D_2O} = \frac{(fw/20) - (g_2 - g_1)/2}{(fw/20) + (1-f)w/18} \tag{2}$$

The gain in weight per mole of ammonium sulfate would be AX_{D_2O}, where A is the number of gram atoms of hydrogen involved in the exchange equilibrium per gram molecule of ammonium sulfate.

The gram-molecular weight of ammonium sulfate is designated by M. If g_1 is the initial dry weight of the ammonium sulfate and g_2 is the weight after exchange, the gain in weight for g_1/M moles of ammonium sulfate is

$$g_2 - g_1 = AX_{D_2O}\frac{g_1}{M} \tag{3}$$

Solving for A,

$$A = \frac{(g_2 - g_1)M}{g_1 X_{D_2O}} \tag{4}$$

Practical applications. The interest in isotopic hydrogen lies chiefly in the fact that hydrogen atoms can be labeled by their greater weight and followed through various chemical reactions and physical processes. From the final distribution of the heavy and light atoms, much information can be obtained concerning the nature of the process. The applications in biology have been particularly important.

Suggestions for further work. Other exchange reactions may be studied, such as acetone with water,[3] carbohydrates with water,[4] or urea with water.

The concentration of deuterium in D_2O-H_2O mixtures can be determined by mass spectrometry, or by the thermal conductivity of H_2 and D_2.

The main application of this method is to nonvolatile substances of high solubility in water which have a reasonably large percentage change in mass when D is substituted for H. Several organic compounds of interest for this type of work are malonic acid, succinic acid, and malic acid. The hydrogens of —COOH and —OH are expected to exchange very rapidly with the solvent. According to the literature, the α hydrogens of malonic acid are 100 percent exchanged in 760 hr at 50°. The α hydrogens in sodium malate are 100 percent exchanged in 5 hr at 100°. There is no exchange of α hydrogens of succinic acid in 160 hr at 100°.

References

1. A. H. Kimball, "Bibliography of Research on Heavy Hydrogen Compounds," McGraw-Hill Book Company, New York, 1949.
2. J. Kirshenbaum, G. M. Murphy, and H. C. Urey, "Physical Properties and Analysis of Heavy Water," McGraw-Hill Book Company, New York, 1951.
3. J. O. Halford, L. C. Anderson and J. R. Bates, *J. Am. Chem. Soc.*, **56:** 491 (1934).
4. W. H. Hamill and W. Freudenberg, *J. Am. Chem. Soc.*, **57:** 1427 (1935).

Chapter 19

General Experimental Techniques

This experiment offers the opportunity for learning simple techniques used in constructing glass apparatus. Practice is obtained in performing some basic operations in glassblowing.

THEORY.   Pyrex glass is almost universally used for bench-blown glassware and will be used in this experiment. The disadvantage of the higher softening point of Pyrex is more than compensated for by its lower coefficient of thermal expansion and high strength. The characteristics of Pyrex and soft glass shown in Table 1 are of importance to the glassblower.

Table 1. Characteristics of Pyrex and Soft Glass

Characteristic	Pyrex (No. 774)	Soft glass
Softening point	820°C	626°C
Strain point	510°C	389°C
Annealing point	560°C	425°C
Linear coefficient of expansion	32×10^{-7}	90×10^{-7}

Pyrex glass requires an oxygen-air-gas or oxygen-gas flame. Manipulation of Pyrex at temperatures above the strain point may introduce harmful strains; however, these can be eliminated by annealing. Annealing is best done in a furnace which may be heated to 580 to 585°C, but in the laboratory a small piece of glass apparatus may be annealed in a flame. The glass is heated in a soft bushy flame until it is uniformly softened. The working temperature is slowly decreased by manipulating the glass in the cooler parts of the flame and by lowering the flame temperature until a layer of soot has been deposited from the smoky flame which is finally used.

A more detailed description of glassblowing than can be given here is to be found in Refs. 1 to 5.

····➤ *Apparatus.* Burner; hand torch; oxygen tank; Pyrex glass tubing 10 to 12 mm outside diameter; file or glass knife; 4-mm Pyrex cane (rod); corks; forceps; rubber tubing; didymium eyeglasses.

PROCEDURE.   Practice is obtained in drawing "points" and making bends, straight seals, and T seals. It is essential that all tubing used for glassblowing be clean and dry.

Small tubes and rods are easily cut by making a single file scratch with a sharp file, placing the two thumbs toward each other on either side of the scratch but on the other side of the tubing, and breaking with a combined bending and pulling force. It will be found helpful to moisten the scratch before the break is attempted.

In the case of large tubes, a small point of heated glass is touched against one end of the file scratch, and a crack is produced under the file scratch which extends for a short distance beyond. The crack may be extended if necessary by touching the heated glass point to the tube just beyond the end of the crack.

The tube may be cracked more neatly and faster by wrapping nichrome or other heating wire around the tube over the file scratch. The wire is heated red hot with an electric current, using a suitable resistance in series.

The proper procedure for lighting a burner is to ignite the gas first, then turn on the air, and lastly turn on the oxygen. The optimum temperature for the flame depends upon the size of the piece being worked and the ability of the operator. A more skillful operator may use a higher temperature, which makes possible faster working. The hottest part of a flame is at the tip of the inner cone. Didymium glasses should be worn to protect the eyes from the sodium light produced by sodium vaporized from the glass.

The first operation to be learned is uniform rotation of the glass (Fig. 107). A suitable length of tubing 10 to 12 mm outside diameter is selected as a convenient size for practice in this and subsequent operations. The tubing is held by the last three fingers of the left hand, which act as a bearing; the thumb and forefinger are used to rotate the glass. The right hand supports the other end of the tubing. (Left-handed persons may reverse the order given.) Here again, the thumb and forefinger are used to rotate the tube while the other fingers are used mainly for support. The palm of the left hand is downward, while that of the right hand is upward. These positions permit the glassblower to blow into the right-hand end of the tubing, which should be the shorter end. Rotation is in such a direction that the top of the tubing moves away from the glassblower. The right- and left-hand movements are synchronized to prevent twisting of the tubing. The tube is held in a straight line, and bending, pushing, or pulling the glass is avoided except when required for a specific purpose. The student will find that a considerable amount of practice will be required.

The importance of mastering this rotation technique cannot be too strongly emphasized; it is essential for obtaining uniform wall thickness and symmetrical shapes. The student should practice this operation until he gets the "feel" before any glassblowing is tried.

figure 107. Glassblowing: rotating the tube.

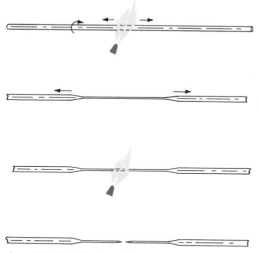

figure 108. Glassblowing: pulling a point.

The next operation is pulling a point (Fig. 108). This is an elongation on the end of a tube formed by pulling the tube to a small diameter. Points form convenient handles for holding short pieces of tubing and provide a means for closing the tube and for cutting the glass with the flame. To pull a point, the tube is rotated in the flame so as to heat a length of about 1 cm. When the glass is pliable, it is removed from the flame, and while still rotating, it is pulled slowly to a length of about 20 cm. The drawn-out portion is melted apart at the center, thus closing both points. If the points do not have the same axis as the tube, it will be necessary to heat at the shoulders, where they join the tube, to straighten them. The position of the point with respect to the tube is a test of the student's mastery of the rotation technique.

The straight seal is tried next (Fig. 109). One end of the tube to be held in the left hand is stoppered, the tube in the right hand being left open for blowing. The ends of the tubes which are to be sealed together are heated in the flame, with rotation, until softened. The two ends are then pushed together carefully on the same axis, and as soon as the contact is effected, the joint is pulled slightly. The joint is now rotated in the flame until the diameter of the tubing is somewhat decreased and the wall thickness is increased at the point of juncture. While it is still being rotated, the tube is removed from the flame and the joint is blown to a somewhat greater size than the original tubes. The tubing should not be pulled at this point, since this will decrease the wall thickness at the seal. The tubing is now reheated at the enlarged portion until its diameter is decreased; it is then removed from the flame and blown to a slight bulge. Before the tube has cooled appreciably, the joint is

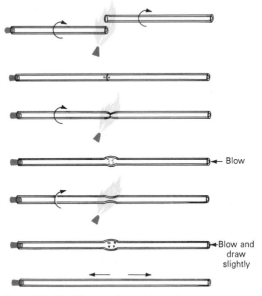

figure 109. Glassblowing: the straight seal.

pulled sufficiently to reduce the diameter to that of the tube. Note that the tubing is continuously rotated in all these operations. If the rotation technique has been mastered, seals which are all but indistinguishable from the remainder of the tubing can be made very quickly. If two pieces of tubing of different diameter are to be joined, the larger is first drawn down and cut off to give an end of the same diameter and wall thickness as the smaller tube.

The T seal (Fig. 110) presents a different problem, since the tubing cannot be rotated easily except by using a special glassblower's clamp. One end of a tube is closed by a cork, and with a sharp flame a spot on the side of the tube is heated. The heated spot is blown to form a bulge, which is then reheated and blown to a small bulb having thin walls. This bulb is broken, and the excess glass chips removed by scraping with wire gauze. The size of the hole thus formed should be about the same size as or slightly smaller than the tube which is to be attached. The other end of the tube with the side hole is now sealed with another cork. The side opening and the end of the tube to be sealed are heated until soft, removed from the flame, and brought quickly together and given a slight pull as soon as contact has been made. The joint is blown slightly to expand it and to remove any irregularities. If the glass was sufficiently softened when joined, this procedure results in a good seal; however, should it not appear uniform, small parts can be heated with a sharp flame and then

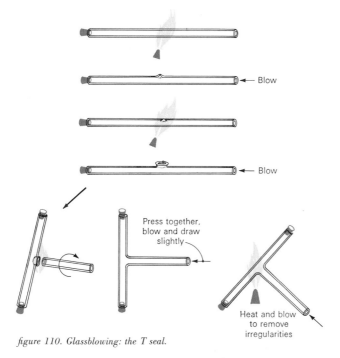

Blow

Blow

Press together,
blow and draw
slightly

Heat and blow
to remove
irregularities

figure 110. Glassblowing: the T seal.

blown to proper size. The entire seal is reheated to remove stresses and to adjust the angle between the tubes.

A bend may be made after the tubing has been heated until it is pliable and it has been removed from the fire. In order to obtain a smooth uniform bend, one end of the tube is closed with a cork, and as soon as the bend is completed, the open end of the tube is blown into with sufficient pressure to eliminate any irregularities in the bend. In order to prevent sagging of a bend, the ends of the tube are bent upward with the heated portion downward.

Practice should be continued on each of the techniques described above until at least one satisfactory specimen of each of the following types has been produced:

1. Straight seal, 10-mm tubing
2. Straight seal, 7-mm tubing
3. T, 10-mm tubing, two ends sealed off, open end polished
4. T, 7-mm tubing, two ends sealed off, open end polished
5. Straight seal, 10- to 7-mm tubing, large end sealed, open end polished
6. 135° bend, 10-mm tubing
7. 90° bend, 7-mm tubing

Items 4 and 5 are tested for pinholes. The test is made by attaching the unit to a vacuum system, which is then pumped down to a vacuum of the order of 0.1 mm Hg, and exploring the tip with a Tesla coil (page 416) held 5 to 10 mm from the surface. A bright streamer of ionized air points to any small hole in the glass.

The completed items should be assembled together for inspection and approval by the instructor.

Suggestions for further work. As soon as the student has acquired reasonable proficiency in the above operations, he may proceed to more difficult operations such as joining capillary tubing or small-diameter tubing and making ring seals and closed circuits of tubing. Further directions will be found in the reference books. Flaring the end of the tubing and blowing small bulbs are also good exercises.

Metal-to-glass seals are required in certain types of work. Platinum wire can be sealed into soft glass and also into Pyrex if the diameter is small; these seals, however, are not recommended for vacuum work. Tungsten in small diameters can be sealed directly into Pyrex, and larger diameters can be sealed if an intermediate grading glass is used. Special alloys are available for sealing to low-expansion glasses like Pyrex. One or more grading glasses are usually required for the Pyrex-to-metal seal. Kovar† and Fernico§ are examples and can be obtained in the form of tubing, wire, and various fabricated shapes, either alone or already sealed to glass. The latter is the preferred way to obtain these materials, since the sealing operation is an art which requires considerable practice. Copper can be sealed into either soft glass or Pyrex by the Housekeeper[4] method, which requires that the copper be very thin where it is sealed to the glass.

References

1. F. C. Frary, C. S. Taylor, and J. D. Edwards, "Laboratory Glass Blowing," 2d ed., McGraw-Hill Book Company, New York, 1928.
2. J. D. Heldman, "Techniques of Glass Manipulation in Scientific Research," Prentice-Hall, Inc., Englewood Cliffs, N.J., 1946.
3. "Laboratory Glass Blowing with Pyrex Brand Glasses," Corning Glass Works, Corning, N.Y., 1952.
4. J. Strong, "Procedures in Experimental Physics," chap. 1, Prentice-Hall, Inc., Englewood Cliffs, N.J., 1938.
5. R. H. Wright, "Manual of Laboratory Glass-blowing," Chemical Publishing Company, Inc., New York, 1943.

58   HIGH VACUUM

This experiment illustrates some of the elements of importance in the production and measurement of low pressure.

THEORY.[1,2]   In vacuum technology, pressures have been expressed in *microns* of mercury. The cgs unit of pressure, 1 dyne cm^{-2}, is called the *microbar*. Pres-

† Manufactured by Latronics Corporation, Latrobe, Pa.

§ Manufactured by General Electric Co., Schenectady, N.Y.

sures are now often given in torrs or millitorrs (1 standard mm Hg = 1 torr). The relations between these units can be summarized as follows:

$$1 \text{ micron} = 1 \mu = 10^{-3} \text{ mm Hg} = 10^{-6} \text{ m Hg}$$
$$= 1.3332 \mu\text{bar} = 1 \text{ mtorr} = 10^{-3} \text{ torr}$$

The volume unit used commonly is the liter. Quantity of gas, at a particular temperature, is conveniently expressed as liter-microns, or the equivalent in other pressure and volume units.

The speed S of a pumping system which is removing gas from a vessel of fixed volume V may be defined† by the relation

$$S = -\frac{dv}{dt} = -\frac{V}{P}\frac{dP}{dt} \tag{1}$$

where dv = volume of gas, measured at temperature and pressure P of vessel, removed in time dt

dP = corresponding change in pressure

The pumping speed can be regarded as a *conductance*, since it has the dimensions of volume of gas removed per unit time, and is customarily expressed in liters per second.

The speed of the pumping system determines the time required to reduce the pressure in the volume V from one specified level to another. This problem is considered in detail by Dushman,[1] but for the case where S is *constant* over the range from the initial pressure P_1 to the final pressure P_2, integration of Eq. (1) gives for the required pumping time

$$t_2 - t_1 = \frac{V}{S} \ln \frac{P_1}{P_2} \tag{2}$$

Part A. Vacuum Pumps

In laboratory practice two types of vacuum pumps are commonly employed, mechanical and diffusion pumps. The principle of operation of a typical rotary oil pump for producing a vacuum is indicated in Fig. 111. The vane V is kept in close contact with the eccentric cylinder C by the spring-loaded rocker arm A. As the cylinder rotates in the oil-filled chamber, air is drawn in at the inlet tube I and driven around to the outlet at O. The vane V and the close fit of the eccentric rotor with the stator produce an efficient pumping action. The entire pump unit is oil-immersed; the ball check shown at O prevents backflow of oil when the rotor is

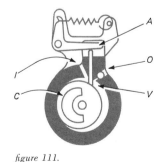

figure 111.

Principle of operation of a rotary oil pump for producing a vacuum.

† A more realistic definition takes into account the finite low-pressure limit the pumping system can reach, P_s, and gives $S(P - P_s) = -V dP/dt$. If it is assumed that P_s is negligibly small compared with P, Eq. (1) follows directly.

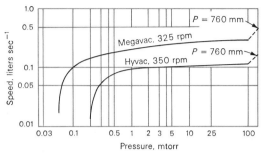

figure 112. Performance curves for typical rotary oil pumps.

stopped. Ordinarily, two such units, with rotors on a common shaft, are connected in series to form a single compound pump. The pumping speed of such a rotary oil pump depends on its size, but in any case drops off rapidly as the millitorr range is approached. Performance curves for typical laboratory pumps are shown in Fig. 112.

For efficient pumping at low pressures the diffusion pump is employed. A typical mercury-vapor pump is illustrated in Fig. 113. The pressure in the pump

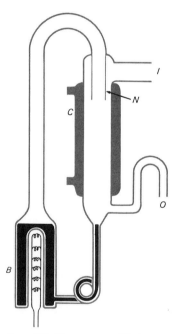

figure 113. Mercury-vapor diffusion pump.

and system is first reduced by means of a rotary oil pump, the *forepump*, connected to the outlet *O*. A high-velocity stream of mercury vapor generated in the boiler *B* by electrical heating is then driven through nozzle *N*. Air molecules diffuse through the inlet tube *I* into the vapor stream and are driven downward by collisions with mercury atoms. The latter, because of their relatively high mass, are only slightly deflected and proceed to the water-cooled wall *C*, where they are condensed and returned to the boiler. The gas molecules are driven down to the outlet, where they are removed by the forepump. Very high pumping speeds and low ultimate pressures can be obtained by this pump. The pressure maintained by the forepump must be below a critical value, determined by the diffusion-pump design and the power input to the boiler, if the diffusion pump is to work. This necessary forepressure may be increased by the use of multistage diffusion pumps.

Mercury as the working fluid has one disadvantage: its vapor pressure at room temperature is relatively high, about 1 mtorr, so that an efficient cold trap must be used between a mercury-vapor pump and a system in which mercury cannot be tolerated. High-molecular-weight organic pumping fluids, such as butyl sebacate, are commonly used instead of mercury, since they have vapor pressures at room temperature which are negligible for practically all purposes. They are, however, subject to cracking if overheated and must be heated only under vacuum because of susceptibility to air oxidation. The use of silicone pumping fluids offers a solution to these particular problems.

It should be noted, however, that a particular application may restrict the choice of pump fluid. Silicone oils, for example, are not suitable for use with mass-spectrometer inlet systems because they can result in the deposition on source elements of a tenacious, electrically insulating oxide coating, with consequent impairment of instrument performance. Polyphenyl ether pump fluids, which are of particularly low volatility at room temperature, are useful.

Modern oil-diffusion pumps can produce pressures below 10^{-6} mm Hg at room temperature without cold traps, and pumping speeds as high as 20,000 liters sec^{-1} at 10^{-4} mm Hg are available with large-scale commercial units. In Fig. 114 is given a

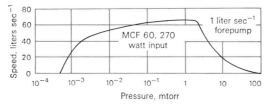

figure 114. *Performance curve for a small three-stage diffusion pump.* (Consolidated Vacuum Corp.)

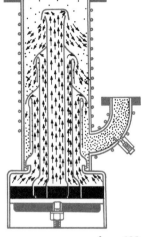

figure 115.

Three-stage oil-diffusion pump. (Consolidated Vacuum Corp.)

performance curve for the small three-stage commercial pump illustrated in Fig. 115. On the low-pressure side the measured pumping speed drops to zero at a pressure fixed by the vapor pressure of the pumping fluid; the actual pumping speed for *air* remains high at this same pressure, as may be shown by the use of a pressure gauge with a liquid-air trap.

A diffusion pump can remove vapors (water, organic liquids) which would condense into the oil of the forepump and seriously impair its performance. Where such a possibility exists, the forepump should be protected by a suitable trap. Drying towers filled with solid desiccants, etc., are of limited utility when it is desired to conserve the speed of the pump. Liquid-cooled traps can provide very efficient performance. For routine forepump protection, cooling with solid carbon dioxide is in general adequate; a mixture of solid carbon dioxide and trichloroethylene is used. Trichloroethylene is now readily available commercially and should always be used instead of the inflammable acetone which was commonly used in the past. It should be remembered that the vapor pressure of ice at the temperature of solid carbon dioxide is about 0.1 mtorr.

Liquid air is a very effective trap refrigerant but involves an explosion hazard if organic materials are brought into contact with it, through breakage of a glass trap, for example. The problem is intensified by the fractionation that takes place on evaporation of the liquid air, which leaves the residual liquid progressively richer in oxygen. Liquid-air traps are preferably made of metal, or else a metal jacket should be used around a glass trap so that the latter does not come in direct contact with the liquid air.

Liquid nitrogen has marked advantages over liquid air as a refrigerant because of its chemical inertness. It may be prepared by fractionation of liquid air; if only a moderate quantity is required, it may be produced by expansion of tank nitrogen through a throttle valve after precooling by passage through a copper tubing coil immersed in liquid air. Liquid nitrogen is commercially available in many areas.

Of increasing importance is the use of ion pumping to attain high vacuum. A typical ion pump contains an anode and a titanium cathode within a strong magnetic field produced by an external permanent magnet. A high-voltage power supply is connected between the anode and cathode. Ionization of gas molecules is produced by collisions with electrons; the intense magnetic field results in electron travel in a long spiral path which increases ionization efficiency. The positive gas ions are accelerated toward the cathode, where they sputter titanium from the cathode surface. This sputtered titanium is largely monatomic and highly reactive. The pumping action achieved results from chemical combination of the metal with reactive gas species to give nonvolatile products and the burying of nonreactive species under layers of sputtered metal on the pump walls and other surfaces. Typically, pressures as low as 10^{-9} mm Hg can be produced without the use of cold traps or baffle systems. Disadvantages include inability to operate except at low pressure and inability to handle large quantities of gas.

The ion pump has a particular advantage in cases where contamination of a system by organic material must be avoided. The pumping speed is relatively high for hydrogen, low-molecular-weight hydrocarbons, nitrogen, and oxygen and relatively low for helium, argon, etc. The pump can be damaged by operation for any length of time at pressures greater than about 2 to 3×10^{-5} mm Hg, and an auxiliary pump must be used to reduce the pressure to the low millitorr range before the ion pump can be started.

Such roughing down may be done with an ordinary mechanical pump. To avoid possible system contamination with pump oil, a sorption pump may be used instead. Such a pump contains an adsorbent (alumina, activated charcoal, molecular sieve) which is cooled by liquid nitrogen. Adsorption of gas then lowers the pressure to the point where the sorption pump can be valved off and the ion pump started.

Part B. Conductance of a Pumping System

The speed of the pumping system depends not only on the intrinsic speed of the pump proper S_0 but also on the conductance of the connection between the pump and the vessel to be evacuated. For the connecting tube this conductance F relates the quantity Q of gas transferred per unit time to the pressure drop ΔP across the tube:

$$Q = F \Delta P \tag{3}$$

In vacuum work the experimental conditions are in general such that turbulent flow of gas through the connecting tubes is not encountered. At the higher pressures of interest Poiseuille's law of viscous flow applies:

$$n = \frac{\pi}{256\eta} \frac{d^4}{l} \frac{P_2^2 - P_1^2}{RT} \tag{4}$$

where n = number of moles of gas (assumed to obey ideal-gas law) which flow per second through a cylindrical tube

d = diameter of tube, cm

l = length of tube, cm

P_2, P_1 = inlet and outlet pressures, μbars, i.e., dynes cm^{-2}

η = coefficient of viscosity of gas, poises, at $T\,^\circ$K

Since nRT can be set equal to PV, the quantity of gas Q in microbar cubic centimeters transferred per second can be written

$$Q = \frac{\pi d^4}{256\eta l} (P_2^2 - P_1^2) = \frac{\pi d^4}{128\eta l} \frac{P_2 + P_1}{2} (P_2 - P_1) \tag{5}$$

As $(P_2 + P_1)/2$ is the average pressure in the tube in millitorrs, $\overline{P}_{mt}$, Eq. (3) shows that the conductance F_v of the tube in the viscous-flow region is given by

$$F_v = \frac{\pi d^4}{128\eta l}\overline{P}_{mt} \qquad \text{cm}^3 \text{ sec}^{-1} \tag{6}$$

At very low pressures a different relation obtains because the mechanism of flow changes. As the pressure is reduced, the mean free path of the molecules increases; for air at 25°C, the mean free path λ in centimeters is given by $\lambda = 5.09/P_{mt}$, where P_{mt} is the pressure in millitorrs. When the mean free path becomes larger than the diameter of the tube through which the gas is moving, collisions between molecules become of negligible importance compared with collisions of the molecules with the tube walls. The gas is then referred to as a *Knudsen gas*, because of the contributions made by the physicist Knudsen to the kinetic theory of gases for such conditions. Each molecule then moves in essential independence of the others. The wall, to an incident molecule, is an extremely rough surface. It is hence legitimate to assume that the direction in which a molecule will bounce off the wall will be independent of the angle of incidence calculated on the assumption of a perfectly smooth surface. The resulting transfer of gas down the tube, termed *molecular flow* by Knudsen, can then be treated by statistical methods to give the following result for the conductance F_m for a cylindrical tube of diameter d centimeters and length l centimeters:

$$F_m = \frac{1}{3}\sqrt{\frac{\pi}{2}}\frac{1}{\sqrt{\rho_1}}\frac{d^3}{l} \qquad \text{cm}^3 \text{ sec}^{-1} \tag{7}$$

Here ρ_1 is the gas density in grams per cubic centimeter at a pressure of 1 μbar. It should be noted that in the molecular-flow region the conductance is independent of pressure.

At intermediate pressures the flow is partly viscous and partly molecular. The general expression given by Knudsen must be used for this range:

$$F = F_v + F_m Z = F_v + F_m \frac{1 + [(d\sqrt{\rho_1})/\eta]\overline{P}_{mt}}{1 + [(2.47d\sqrt{\rho_1})/2\eta]\overline{P}_{mt}} \qquad \text{cm}^3 \text{ sec}^{-1} \tag{8}$$

For air at 25°C, with the average pressure expressed in millitorrs, d and l in centimeters, and F in liters per second,

$$F = 0.177\frac{d^4}{l}\overline{P}_{mt} + 12.2\frac{d^3}{l}Z \tag{9}$$

$$Z = \frac{1 + 0.246\overline{P}_{mt}d}{1 + 0.304\overline{P}_{mt}d} \tag{10}$$

The factor Z varies between 0.81 at high pressures and 1 at low pressures. As pointed out by Hecker,[2] if $\overline{P}_{mt}d$ is greater than 1000, the flow is over 95 percent viscous, while for $\overline{P}_{mt}d$ less than 2, it is over 95 percent molecular. An alternative criterion for molecular flow given by Knudsen states that the flow is over 95 percent molecular if d/λ is less than 0.4. The molecular-flow range is of primary concern in high-vacuum work, but conductance calculations for the intermediate pressure range are often required in the design of connecting lines between diffusion pumps and forepumps (see below).

The quantity of gas moved per second through two tubes in series is the same for each tube. If the conductances of the tubes are F_1 and F_2, F that of the series combination, $P_3 - P_2$ the pressure drop across the first tube, and $P_2 - P_1$ that across the second,

$$Q = F_1(P_3 - P_2) = F_2(P_2 - P_1) = F(P_3 - P_1) \tag{11}$$

so that

$$\frac{1}{F} = \frac{1}{F_1} + \frac{1}{F_2} \tag{12}$$

This result may also be obtained by thinking of the conductance as the reciprocal of a resistance to flow. The additivity of resistances in series again leads to Eq. (12).

Bends and elbows in a tube of constant diameter and axial length have relatively little effect on the conductance at low pressures. As the mean free path becomes large compared with the diameter of the tube, however, molecules experience a difficulty getting into the tube from an adjacent region of relatively larger cross section. This difficulty may be characterized quantitatively by a conductance for the tube *entrance*, which for a circular aperture of diameter d can be written approximately as

$$F_0 = \frac{1}{4}\sqrt{\frac{\pi}{2}}\frac{d^2}{\sqrt{\rho_1}} \tag{13}$$

For molecular flow the resultant conductance for a tube plus the entrance is obtained by combining the relations of Eqs. (7) and (13) in accordance with Eq. (12). The result for air at 25°C is

$$F = \frac{9.17d^2}{1 + \frac{3}{4}l/d} \qquad \text{liters sec}^{-1} \tag{14}$$

Consider a composite pumping system of speed S consisting of a pump of speed S_0 in series with a connecting tube of conductance F. The rate Q at which this system removes gas from a vessel where the pressure is P must be equal to SP, from the definition of pumping speed. Similarly, $Q = S_0 P_0$, where P_0 is the pressure at the pump entrance proper. Also, since the same gas is driven through the conductance F by the pressure drop $P - P_0$, $Q = F(P - P_0)$. Hence

$$Q = SP = S_0 P_0 = F(P - P_0) \tag{15}$$

and

$$\frac{1}{S} = \frac{1}{S_0} + \frac{1}{F} \tag{16}$$

This relation is of basic importance in vacuum technique. It must be utilized in the design of high-vacuum systems to obtain efficient use of the pump employed.

Part C.   Measurement of Speed of Vacuum Pumps

A number of methods are available for measuring the speed of vacuum pumps. The procedure suggested by Howard[3] is convenient for pumps of moderate speed. The apparatus used is shown in Fig. 116. The capillary tube and the 1-mm-bore stop-

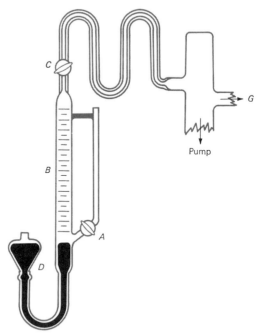

figure 116. Apparatus for Howard's method for measuring the speed of a pump.

cock C, whose plug is grooved slightly at the edge of the hole to facilitate adjustment of the leak rate, permit a slow leakage of air from the 50-ml gas burette B. Initially the stopcock C is closed and the pump set in operation. The mercury leveling tube D is lowered so that stopcock A can be opened to the atmosphere. Stopcock C is then opened carefully until the pressure maintained by the pump, as measured by a McLeod gauge connected at G, is approximately equal to the pressure for which the pumping speed is to be determined. The pressure in the system is checked periodically, and when it has become constant, the mercury level in the burette is raised slowly past the stopcock A, which is then closed. The position of the mercury level in the burette, the time, and the barometric pressure are recorded. The leveling bulb is progressively raised as required to keep the pressure in the burette constant as gas is removed by the pump. When an appropriate volume of air has been pumped out, stopcock C is closed and the time and burette reading are recorded. The pumping speed is calculated by multiplying the volume of air removed per second by the ratio of the burette pressure to the pump working pressure.

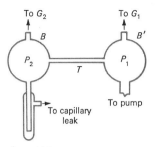

figure 117.

Apparatus for flow-tube method for pumping-speed determination.

In another useful method the gas removed by the pump is delivered by a flow tube whose conductance can be calculated; the experimental arrangement is shown in Fig. 117. The bulbs B, B' should be of about 1 liter volume. The appropriate dimensions for the flow tube T depend on the kind of pump under test,[4] but for the smaller laboratory diffusion pumps a tube of 1 cm diameter and 50 cm length can be used for the low-pressure range. Initially the capillary leak (a sensitive high-vacuum needle valve should be used) is closed and the pump started. The leak is then opened gradually until the pressure P_1 has reached the level for which the speed is desired. When the pressures P_2 and P_1 have become constant (as measured by the vacuum gauges G_2 and G_1), the rate of flow of gas through the tube can be evaluated by use of the calculated conductance F of the flow tube.

$$Q = F(P_2 - P_1) \tag{17}$$

The pump speed is then given by Q/P_1, provided the pump is connected directly to B'. Otherwise the conductance of the connecting tube must be taken into account [cf. Eq. (16)].

Part D. Vacuum Gauges

The McLeod gauge, one design of which is shown in Fig. 118, is a basic instrument for low-pressure measurements. A large volume of gas V at the unknown pressure p is compressed to a small volume v, and the corresponding pressure P is measured; p is then calculated from the relation $p = Pv/V$, which holds quite accurately for the low pressures involved. In use, the high-vacuum stopcock C is gradually opened, and as the pressure in the gauge is reduced, the two-way stopcock T is opened care-

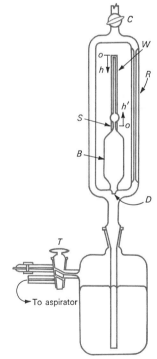

figure 118.

McLeod gauge.

fully to the auxiliary vacuum line to keep the mercury from rising appreciably in the connecting tube. To make a pressure measurement the control stopcock T is opened *slightly* to the atmosphere (a capillary leak here may be used to prevent an undesirable rush of air into the vessel) to produce a slow rise of mercury into the gauge. When the rising mercury reaches the cutoff D, the gas in the gauge bulb B is trapped at the pressure to be determined. For highest sensitivity in the pressure measurements, this gas is compressed until the mercury meniscus in the reference capillary R has reached the level of the top of the gauge capillary W. The difference of h mm in the levels of the mercury in the two capillaries represents the final pressure of the compressed gas. The reference capillary is made of the same tubing as the gauge capillary in order to eliminate the effect of the capillary depression.

Previous calibration measurements give the total volume V of the bulb B and the volume per millimeter length v_0 of the capillary W. The gas volume at the unknown pressure p is V, while at the final pressure h mm Hg it is $v_0 h$, so that

$$p = \frac{v_0 h}{V} h = \frac{v_0}{V} h^2 \text{ mm Hg} \tag{18}$$

The volumes V and v_0 must be expressed in the same units. The various levels h corresponding to initial gas pressures of 10^{-6}, 10^{-3}, and intermediate values, in millimeters of mercury, are calculated, and the gauge calibration scale is constructed. This scale is termed the *quadratic scale* or the *logarithmic scale*.

A different method of operation can be used to extend the range of the gauge to higher pressures than can be measured on the quadratic scale. The gas in the bulb B is compressed until the meniscus *in the bulb* has reached the reference mark S in the lower capillary section. The final volume of the gas is then always equal to the volume V_f of the section above S. The final pressure is read as the height h' mm above the level S of the mercury meniscus in the reference capillary R. Then

$$p = \frac{V_f}{V} h' \text{ mm Hg} \tag{19}$$

This method thus yields a linear scale. A typical laboratory McLeod gauge will have a range of 10^{-1} to 10^{-6} mm on the quadratic scale and of 10^{-1} to 2 mm on the linear scale.

After the pressure reading has been made, the stopcock T is opened to the vacuum line and the mercury drawn out of the gauge. When the gauge pressure is raised to the atmospheric level, the reservoir pressure must be increased at the same rate. The care required in this process is one of the main disadvantages of this particular design. In another version the gauge bulb is connected to a tube at least 760 mm long, connected to a mercury well provided with a plunger with a threaded top. When the plunger is forced down and the threads engaged, the mercury rises in the gauge. The threaded section facilitates the fine adjustment of the mercury level. The tube and well can be made of metal to provide a very rugged construction.

Small McLeod gauges are available in which a flexible or rotating connection to the system permits the whole gauge to be tipped or rotated as required to force mercury into the gauge bulb.

The sensitivity of the McLeod gauge is governed by the ratio v_0/V. It is not practical to make V larger than 500 ml, because of the weight of such a large volume of mercury. Similarly, the capillary diameter should not be less than 0.5 mm because of the tendency of mercury to stick in small capillaries. The mercury used must be quite pure, and the gauge itself carefully cleaned before filling. An efficient cold trap is required when a McLeod gauge is used with a system from which mercury must be excluded. This gauge cannot be degassed, nor can it be used to measure the pressure of a vapor which condenses to a liquid when compressed into the gauge capillary.

The basic importance of the McLeod gauge is that it is an *absolute* pressure gauge because the calibration scale can be calculated directly from the measured physical characteristics of the instrument. As such it can be used as a reference standard for the calibration of other types of gauges commonly employed in vacuum work. In such applications, or in any case in which quantitatively significant measurements of low pressures are to be made with a McLeod gauge, it is necessary to recognize that the gauge can exert a pumping action even though the vapor pressure of mercury at room temperature is about 1 mtorr.[5,6] For this reason, when accurate results are required the gauge is cooled, normally to about 0°C. For the corresponding lower vapor pressure of the mercury, the pumping effect is negligible.

At low pressures, the thermal conductivity of a gas becomes directly proportional to the pressure when the distance between the hot and cold surfaces becomes smaller than the mean free path. This property is exploited in the *thermocouple gauge* and the *Pirani gauge*. In the thermocouple gauge a constant current is passed through a resistance wire, to the center of which a thermocouple is attached. The temperature of the wire changes as the pressure changes because of the variation of the thermal conductivity of the gas, and this change is indicated by a microammeter connected in series with the thermocouple. The thermocouple gauge is rugged and relatively inexpensive and has a useful range extending from 1 to 500 mtorr. In the Pirani gauge the resistance wire is connected as one arm of a Wheatstone bridge; the extent of the bridge unbalance, as registered by a microammeter connected as the bridge detector, indicates the pressure in the gauge. A second resistance filament, identical with the first but sealed off in a tube under high vacuum, is used in the opposite arm of the bridge to compensate for room-temperature changes, etc., to obtain better gauge performance. The useful range of a typical commercial Pirani gauge extends from 1 to 500 mtorr. The Pirani gauge is subject to zero shift, and its calibration must be checked regularly. A separate calibration is required for every different gas for both these thermal conductivity gauges.

For measurements of very low pressures *ionization* gauges are employed. The electrode arrangement in the conventional type is similar to that of the triode in Fig. 191 (page 600), except that a filamentary cathode and more rugged grid construction are used. Electrons emitted from the heated cathode are accelerated toward the concentric helical grid, which is maintained *positive* with respect to the cathode by a suitable power supply. In collisions between these electrons and gas molecules present, ionization of molecules takes place. The positive ions so produced are attracted to the cylindrical plate, which is made *negative* with respect to the cathode. The resulting current in the plate-cathode circuit is proportional to the pressure of the gas in the gauge provided the electron current from cathode to grid is held constant. The plate current may be measured directly with a microammeter in the less sensitive instruments, but for low-pressure work the potential drop caused by the flow of the current through a series resistance is measured after amplification. The gauge sensitivity is only slightly dependent on the plate potential provided the latter is at least -10 volts relative to the cathode.

This particular electrode arrangement, i.e., plate negative and grid positive, is utilized to provide increased sensitivity. The grid has a relatively small area, so that most of the electrons accelerated toward it pass through the helix and continue on toward the plate. As they approach the negative plate, they are repelled and move back toward and through the grid. This process continues and increases considerably the average distance traveled in the gauge by the electrons before they are captured by the grid. The lengthened path means an increased number of ionizing collisions with gas molecules, and hence increased gauge sensitivity.

The standard ionization gauge can measure gas pressures as low as 10^{-8} mm Hg. The upper pressure limit is about 5 mtorr; it is not possible to operate this gauge at higher pressures without harming it. Commonly, a thermocouple or Pirani gauge is used to show when the pressure is low enough to permit the operation of the ionization gauge.

The type of ion gauge described above can measure pressure down to about 10^{-8} mm Hg. This limit, as first suggested by Nottingham,[7] can be explained as follows. Electrons striking the grid cause emission of soft (low-energy) x-rays. The resultant x-ray bombardment of the large-area collector plate in turn causes emission from it of photoelectrons. For fixed filament current and grid potential, this photoemission current is constant and, of course, independent of gas pressure. A photoelectron emitted from the plate has the same effect on the collector current as a positive gas ion arriving at the plate; a lower limit to useful pressure measurements with the gauge is thus reached when the positive-ion current drops to the neighborhood of the photoelectron current, which happens at about 10^{-8} mm Hg pressure.

Bayard and Alpert[8] altered the gauge design to obtain useful results at lower pressures. They placed the filament *outside* the grid, and instead of the conventional cylindrical plate employed a fine wire collector centered *inside* the grid. The resultant reduction in photoemission current permits the gauge, due to its inverted geometry, to provide useful indications of pressures as low as 10^{-11} mm Hg.

A second type of ionization gauge is the *Phillips gauge*. In this unit a high-voltage ($\sim$2000 volts) discharge is created between a cold cathode and a ring-shaped anode. The resulting current depends on the gas pressure and is measured directly by a microammeter in the grounded side of the high-voltage supply. Small permanent magnets are suitably arranged in the gauge to cause the electrons to move to the anode in spiral paths to obtain increased sensitivity. The Phillips gauge has a range from about 10^{-2} to 25 mtorr. It is a rugged instrument and is unharmed when subjected to atmospheric pressure. The discharge aids in the removal of adsorbed surface gases within the gauge, making it unnecessary to degas this gauge for low-pressure measurements. The response of all ionization gauges depends on the nature of the gas present, so that separate calibrations are required for different gases. It should be emphasized also that in the use of any vacuum gauge to determine the pressure in a system, particularly if that pressure is not constant, the possible effect of the conductance of the connection to the gauge must be considered.

The general ranges of applicability for the several vacuum gauges are summarized in Table 1.

Table 1. Ranges of Vacuum Gauges

Gauge	Millimeters Hg
Mercury manometer	1–1000
Butyl phthalate manometer	0.01–10
Pirani gauge	10^{-4}–0.5
Thermocouple gauge	10^{-3}–0.5
McLeod gauge	10^{-6}–2
Phillips gauge	10^{-5}–10^{-2}
Knudsen gauge	10^{-6}–10^{-2}
Ionization gauge	10^{-11}–10^{-3}

One of the basic problems in the attainment of very low pressures is the outgassing of surfaces, particularly of metals, as the pressure is reduced, due to the release of adsorbed gases or vapors. The best method for elimination of this interference is a degassing procedure in which the system is heated to drive off these materials. Adsorbed water comes off at about 200°C, and other gases, mainly carbon monoxide, at higher temperatures. Obviously, the whole system must be degassed at once for efficient performance, and no materials containing volatile components, such as vacuum waxes, Glyptal, rubber, brass (zinc volatilizes easily), etc., can be present. Without degassing, the best vacuum that can be expected is about 1×10^{-6} mm, and protracted pumping may be required to do this well. For good results vacuum gauges, such as ionization gauges, must be degassed. Ordinarily, provision is made in the gauge-control unit for degassing by electrical heating of the gauge elements.

In any vacuum system, connections between the various sections present a problem. In metal systems, flange joints with neoprene gaskets are often used; for a demountable joint in a section which must be degassed, copper gaskets can be employed. Great convenience can be obtained by use of O-ring gaskets† with appropriately machined flanges; with these units, connections between metal sections, or between metal and glass sections, are easily made, as are effective seals on rotating or sliding rods or tubes which enter the system. For glass systems and moderately high vacua, standard taper joints can be used with vacuum wax or a good grade of vacuum grease. Stopcocks used should be of the specially processed high-vacuum type and of a suitably large bore for the particular application.

Part E. Leak Detection

The detection and elimination of leaks is an inevitable step in the setting up of a vacuum system. All leaks must obviously be located, and very small ones may be extremely difficult to find. A complication in the latter case is the so-called *virtual leaks*, which actually are due to the continuous evolution of gas inside the system. If it is possible to build up an excess air pressure within the system, leaks may be found by painting the outside in the suspected areas with a soap solution. In glass sections moderate leaks may be found with a Tesla coil; the high-frequency discharge will jump to a pinhole or crack in the glass, illuminating it brightly. Precautions must be taken to avoid puncturing thin-walled areas by use of too violent a discharge. Neither of these first two methods is useful for small leaks.

Because their response depends upon the nature of the gas present as well as its pressure, the thermal conductivity and ionization gauges can be used in leak hunting in appropriate pressure ranges. When the area including the leak is sprayed with acetone, for example, the acetone is drawn into the system and causes a change in the reading of the gauge. The Pirani gauge is considered more sensitive than the thermocouple gauge for this work. The ionization gauge can be used when the leaks are small enough so that an appropriately low pressure can be maintained in the system.

† Viton O rings, formed from a fluoroelastomer, can withstand system temperatures as high as 250°C and permit attainment of lower pressures than can be reached with neoprene O rings.

Special leak-detecting units are available commercially. The halogen-sensitive type utilizes the increase in the emission of positive ions (presumably originating from alkali-metal impurities) from hot platinum which occurs when halogen-containing molecules strike the electrode surface; carbon tetrachloride, chloroform, or Freon 1-1 is commonly used with this gauge. Extremely high sensitivity is obtained in the helium-leak detector by use of the mass-spectrometer principle to obtain a specific response to helium. Leaks are located by virtue of the response when a stream of helium is played over the outside of the system.

It is obviously desirable in any case to know the composition of residual gas in a vacuum system, and a knowledge of the species present can facilitate distinguishing between a real leak and a virtual leak due to outgassing. A recent development of major importance in high-vacuum technology has been the introduction of the *residual-gas analyzer,* a compact, low-resolution mass spectrometer, commonly of the quadrupole type, which permits both quantitative and qualitative detection, with high sensitivity, of components in the mass range up to about 300 atomic mass units.

The detailed treatment of problems and procedures in vacuum technology given by Dushman[1] is a basic reference in this field. A very useful general coverage, with extensive references, has been provided by Roberts and Vanderslice.[9]

····→ ***Apparatus.*** Rotary oil pump(s); mercury diffusion pump; oil diffusion pump; Howard pumping-speed apparatus; flow-tube pumping-speed apparatus; McLeod gauge(s); Tesla coil; thermocouple gauge, Pirani gauge, ionization gauge, and control units; acetone.

PROCEDURE.   *Part A.* The pumping speed for a rotary oil pump at a pressure near atmospheric is determined by the flow-tube technique. In this case a flow tube 1 m long and of 3 mm internal diameter is suitable; one end is open to the atmosphere, the other is connected to a short section of larger-diameter tubing which leads directly to the pump. A differential manometer is used to measure the pressure drop across the flow tube. One side of this manometer is connected through a stopcock to the low-pressure side of the flow tube; the other is open to the atmosphere. With the stopcock closed, the pump is started. The stopcock is then opened slowly, and when the pressure drop indicated by the manometer has become constant, its value is recorded, together with the barometric pressure and the dimensions of the flow tube. It is convenient to use an oil as the manometer fluid; its density will be specified by the instructor.

Part B. The rotary pump is then connected as the forepump for a mercury diffusion pump. The latter is connected to a manifold to which a McLeod gauge and a Howard pumping-speed assembly are attached. The stopcocks leading to these two units are closed, and the forepump is started. After a few minutes the stopcock connecting the McLeod gauge is opened, and a pressure measurement is made with the gauge, as described above. The pressure is rechecked until it has dropped below 100 mtorr. The water line to the diffusion-pump condenser jacket is then turned on, and power applied to the pump heater. As the pump goes into operation, the pressure in the system will drop rapidly, and the McLeod gauge is used to measure the resulting low pressure, which soon reaches 10^{-2} mtorr. The capillary leak of the Howard apparatus is then opened slowly until a pressure of about 1 mtorr is found with the McLeod gauge. The pumping speed of the diffusion pump is then determined at this pressure as described under Theory. Finally, the mercury is drawn down out of the gauge, and the gauge stopcock C is closed. The diffusion-pump heater is turned off, then the forepump, and the system slowly opened to the atmosphere by means of a stopcock connected to the forepump line. The gauge is then returned to atmospheric pressure also. After the diffusion-pump boiler has cooled, the condenser water supply is shut off.

Part C. The calibration of a McLeod gauge is undertaken next. A gauge-bulb assembly will be supplied by the instructor, together with a piece of the tubing used for the gauge capillary. The clean and dry gauge bulb is weighed empty. It is then filled completely with water to the cutoff at D (Fig. 118) and reweighed. A hypodermic syringe with a long needle will facilitate the filling of the capillary. The temperature of the water is recorded. The bulb is emptied with the help of an aspirator, placed in a drying oven, and reevacuated, with the aspirator, while warm, to assist in drying it. A thread of mercury is drawn into the capillary tube supplied, and its length is measured. The mercury is emptied into a tared weighing bottle, and its weight determined.

Part D. Measurements are made with an oil diffusion pump attached to a flow-tube apparatus like that shown in Fig. 117. It is convenient to use two thermocouple gauges, one to measure the forepressure and the other on the high-vacuum side of the diffusion pump to indicate when it is safe to turn on the ionization gauge connected at G_1. The gauge G_2 may be a Pirani gauge or a Phillips gauge.

All stopcocks connected to the system are closed, the forepump is turned on, and several minutes allowed for the pressure in the system to be reduced below the millimeter level. The thermocouple forepressure gauge is turned on, and when the pressure drops below 200 mtorr, the diffusion-pump heater can be connected. For an air-cooled pump the air supply is turned on, or the water in the case of a water-cooled pump; the manufacturer's operating instructions for the particular pump used should be carefully followed.

While the system pressure is being reduced, the Pirani or Phillips gauge is put into operation. When the pressure drops below 1 mtorr, the ionization gauge may be turned on. A pressure of the order of 10^{-2} mtorr is reached after the diffusion pump has been in operation for some time. The speed of the latter is then determined for a pressure of about 0.2 mtorr, following the procedure given in the Theory section. The capillary leak connected to bulb B (Fig. 117) is slowly opened until the pressure P_1 has reached the desired level. The pump is then allowed to operate until the pressures P_2 and P_1 become constant; their values are recorded, together with the dimensions given for the flow tube. The capillary leak must be opened very carefully for the protection of the ionization gauge. It is possible to dispense with the leak by connecting bulb B to the forepump, but the high-vacuum needle valve provides greater versatility.

Part E. The ionization gauge is turned off, and leak detection is illustrated by use of a tube with a pinhole leak in it connected through a high-vacuum stopcock to bulb B. This stopcock is now opened slowly, and the Tesla coil turned on and the discharge played over the tube surface to locate the leak. The readings of the various gauges still in operation are noted. The area around the pinhole is then sprayed with acetone, and the effect on the gauge response noted.

The diffusion-pump heater and the vacuum gauges are turned off. After the pump fluid has cooled, the forepump is stopped and the system pressure raised to 1 atm. This is conveniently done by opening a stopcock connected to the forepump line.

CALCULATIONS.   *Part A.* The pressure drop across the flow tube used with the rotary oil pump is converted to millimeters of mercury, and the mean pressure in the tube and the pressure at the pump are calculated. The tube conductance in liters per second is calculated by use of Eq. (6), which for air at 25°C may be written as

$$F_v = 0.177 \frac{d^4}{l} \overline{P}_{mt} \tag{20}$$

where $\overline{P}_{mt}$ is the mean pressure in millitorrs and d and l are in centimeters. The quantity of gas Q delivered per second to the pump by the flow tube is calculated by use of Eq. (17). Division of Q by the pressure at the pump gives the pumping speed.

Part B. The determination of pumping speed by the Howard method has been described in the Theory section.

Part C. The volume V of the McLeod gauge bulb is calculated from the weight of water it holds and the density of the water (page 653). The volume per millimeter length, v_0, of the gauge capillary is calculated from the weight of the mercury thread, its length, and the density of the mercury. A calibration scale for the gauge bulb is drawn for pressure in decade steps from 10^{-6} to 10^{-2} mm Hg, in accordance with Eq. (18).

Part D. The conductance of the flow tube used with the oil diffusion pump is calculated by use of the molecular-flow formula, Eq. (14). The speed of the pump is then calculated as described for the other flow-tube experiment. For this pumping speed there is calculated the diameter for a 20-cm connecting tube necessary to give a conductance which would not reduce the pump speed by more than 10 percent.

Practical applications. High-vacuum techniques, fundamental in many research fields in natural science, are finding many practical industrial applications, such as in vacuum furnaces in metallurgy, the application of surface coatings by the vacuum evaporation process, freeze-drying of various materials, vacuum distillation of high-molecular-weight compounds, etc., as well as the more familiar examples of light bulbs, electron tubes, and Dewar flasks.

Suggestions for further work. The volume of the McLeod gauge capillary above the reference mark S (Fig. 118) may be determined, and a linear calibration drawn up. The formula for the conductance of an orifice, Eq. (13), may be derived from the relation $n_s = \frac{1}{4}n\sqrt{8kT/m\pi}$ for the number of molecules striking a surface per square centimeter per second; n represents the number of molecules per cubic centimeter of gas, m is the mass per molecule, k the Boltzmann constant, and T the absolute temperature. The effect on the speed of a diffusion pump of variation in the heater-power dissipation or the forepressure may be studied. The conductance formula of Eq. (9) may be checked with tubes of various diameters and lengths. The vacuum evaporation of aluminum may be attempted; directions are given by Strong.[10]

References

1. S. Dushman, "Vacuum Techniques," 2d ed., John Wiley & Sons, Inc., New York, 1962.
2. J. C. Hecker in A. Weissberger (ed.), "Technique of Organic Chemistry," vol. 4, E. S. Perry and A. Weissberger (eds.), "Distillation," 2d ed., pt. 2, chap. 6, Interscience Publishers, Inc., New York, 1951.
3. H. C. Howard, *Rev. Sci. Instr.*, **6:** 327 (1935).
4. I. Backhurst and G. W. C. Kaye, *Phil. Mag.*, **47:** 918 (1924).
5. H. Ishii and K. Nakayama, *Trans. Natl. Vacuum Symposium, 8th*, **1:** 519 (1962).
6. E. W. Rothe, *J. Vacuum Sci. Tech.*, **1:** 66 (1964).
7. W. B. Nottingham, unpublished remarks, 7th Annual Conference on Physical Electronics, M.I.T. (1947).
8. R. T. Bayard and D. Alpert, *Rev. Sci. Instr.*, **21:** 572 (1950).
9. R. W. Roberts and T. A. Vanderslice, "Ultra-high Vacuum and Its Applications," Prentice-Hall, Inc., Englewood Cliffs, N.J., 1963.
10. J. Strong, "Procedures in Experimental Physics," Prentice-Hall, Inc., Englewood Cliffs, N.J., 1938.

59   ELECTRONICS

The properties of several types of electronic circuits are illustrated, and experience is provided with the application of the oscilloscope and other test instruments.

THEORY.   Chapter 28 includes an introduction to the principles of operation of the circuits considered here.

····▸ *Apparatus.* Experimental chassis and components for the construction of circuits of Figs. 119 and 214; oscilloscope; electronic voltmeter; power supply (0 to 250 volts dc, 6.3 volts ac outputs); tube manual; soldering iron or gun; needle-nose pliers; diagonal cutter; hookup wire; screwdriver with insulated handle.

PROCEDURE.   Two of the circuits discussed in Chap. 28 are to be wired up and studied quantitatively. For this purpose units will be furnished with the heavier components mounted on a metal plate or plywood board, which serves as experimental chassis.

The schematic symbols and color-code conventions for electronic components are given in the Appendix. Tube-socket terminals (pins) are numbered clockwise as viewed from *below*, starting at the key or gap. The manufacturer's tube manual gives the pin numbers for the various electrodes. Any unused transformer leads should be doubled back and taped carefully, or otherwise insulated, to avoid accidental short circuits. All circuit wiring should be carefully rechecked *before* the power switch is turned on. Circuit changes should be made *only* when the power is *off*. Safety precautions in electronics work are emphasized on page 636.

The principles of operation of the auxiliary equipment, especially the oscilloscope and voltmeter, should be studied in advance. In practice, the ground terminal of the oscilloscope should be connected to the ground of the circuit under study and an *insulated* test prod employed to pick up the voltage whose waveform is to be viewed.

Part A. The dc power supply of Fig. 214 (page 625) is constructed, with

L = 15 henrys, rated at 50 ma dc
C = 10 μf, 450 volts dc (VDC) voltage rating
R_b = 10,000 ohms, 25 watts, fixed-power resistor
T = power transformer, 470 volts center-tapped (VCT) at 40 ma; 5.0 VCT at 2.0 amp; 6.3 VCT at 2.0 amp

The experimental chassis is furnished with these units and the tube socket already mounted and the various component leads connected to binding posts or barrier-type terminal strips. Interconnections are conveniently made by means of leads consisting of lengths of test prod wire with insulated banana plugs or spade lugs, as appropriate, attached at the ends. Banana plugs and binding posts of the best quality should be used to ensure that secure connections are made. The power cord for connection to the 115-volt 60-Hz line is fused and is terminated at an outlet box which is fastened to the experimental chassis to provide a switch-controlled power outlet.

When the connections have been made and carefully checked, the power is switched on and the dc voltage measured at points A, B, and C with a vacuum-tube voltmeter. Next, the ac voltage waveforms at points A, B, and C are observed with the oscilloscope. This ac ripple voltage is shown to be primarily 120-Hz frequency by comparison of its waveform with that of a 60-Hz signal obtained from the 6.3-volt filament winding of the power transformer; the center tap of this winding should be grounded.

The effectiveness of the filtering action may be illustrated by comparing the ac ripple voltages at points *A*, *B*, and *C* as measured with a voltmeter. The accuracy of this measurement is limited because as a rule an electronic voltmeter nominally calibrated to indicate rms voltage indicates this correctly only for a sinusoidal waveform.

When the power has been turned off, the filter condensers should be shorted by means of a screwdriver with an *insulated handle* to ensure their complete discharge.

Part B. The properties of the triode amplifier circuit of Fig. 198 (page 605) are to be studied from the point of view of the equivalent circuit, Fig. 197 (page 605). In particular, the gain is to be measured as a function of load resistance to illustrate the theoretical relationship (page 604)

$$A = \frac{e_p}{e_g} = -\mu \frac{R_L}{r_p + R_L} \tag{1}$$

where A = gain of triode stage

e_p = ac component of plate voltage (relative to cathode)

e_g = ac component of grid voltage (relative to cathode)

R_L = load resistance

r_p = plate resistance of triode, at given operating point

μ = amplification factor of triode, at given operating point

The full definitions of r_p and μ are given on page 603. Since the cathode resistor is bypassed with capacitor C_1, the cathode is held practically at ac ground potential. The load resistance $R_L = R_1 R_2 / (R_1 + R_2)$ is the resistance equivalent to that in the actual ac circuit between triode plate and ground.

The experimental arrangement is shown in Fig. 119. The precision potentiometer, tube socket, switch assembly with precision resistors, and a terminal board (solder type) are supplied already mounted on a sturdy aluminum plate. The latter is supported in a horizontal position and held secure by two wooden blocks, in such a fashion that the tube itself is below the plate and all terminals are on top for easy access.† The remaining components of the triode circuit are soldered into place by the student. To minimize damage to the tube-socket pins from repeated soldering, each pin is connected to a terminal of the terminal board, so that all connections to the rest of the circuit can be made at the terminal board.

---

† An alternative type of experimental chassis which is satisfactory for this work can be constructed easily from chassis kits or punched insulating boards, both available from Vector Electronics, Glendale, Calif.

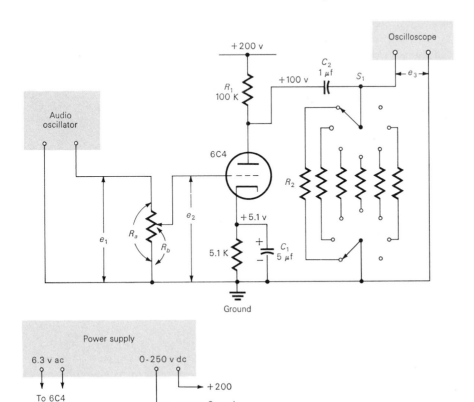

figure 119. Experimental arrangement for study of triode amplifier circuit. The numerical voltages given are nominal dc values, while e_1, e_2, and e_3 represent ac components. Tables of electronic symbols and abbreviations are given in the Appendix.

R_a, R_b = Precision potentiometer, 0.1% linearity accuracy
  R_a = 1000 Ω
  R_1 = 100 K, 2 W, 1%, carbon film type
  R_2 = Switch-selected precision resistors, 2 W, 1%, carbon film type, in the following resistances:
        100 K, 50 K, 25 K, 12 K, 6 K, 2 K; one position open
  C_1 = 5 μf, 150 VDC, electrolytic
  C_2 = 1 μf, 450 Mylar
  S_1 = Two-deck, 2-pole, 7-position, nonshorting steatite wafer switch
   K = Kilohms
  μf = Microfarads

Individuals who have not had previous experience in the soldering of electronic circuits should learn the proper technique from an instructor. Failure of circuits constructed by inexperienced persons is often attributable to poorly soldered connections. It is a good idea to place a check mark near the corresponding symbol in the diagram when a component is wired into place, as an aid to avoiding mistakes.

The precision potentiometer has three terminals, one at each end and one connected to a variable tap. The end-to-end resistance is R_a; the resistance between the variable tap and ground is R_b. The potentiometer should be connected into the circuit in such a way that the scale reading *increases* as the variable tap moves *away* from the grounded terminal; the scale readings will then be proportional to R_b. Commercially available precision potentiometers usually bear labels showing the internal connections.

Lastly, the leads to the audio generator,† oscilloscope, and power supply are connected to the triode circuit, but these units are not turned on until the entire assembly has been checked again to see that the connections and part sizes are correct and that each junction has been properly soldered.

When the assembly has been completed and checked, the power supply is turned on; the filament power is supplied to the triode first, and then, after about 30 sec, the plate voltage is applied. The latter should initially be set at a low value, and then gradually increased to 200 volts.

The dc voltages to be expected at the plate and cathode are indicated on the schematic. As a preliminary test, these voltages should be checked with an electronic voltmeter; variations of the order of 10 percent from the values shown are to be expected. If an abnormal discrepancy is found, the tube should be tested or replaced and the wiring checked again. If the difficulty persists, the instructor should be consulted. When proper voltages have been obtained, the measured values are recorded and used to calculate the tube current, the dc plate-to-cathode voltage, and the plate dissipation, all of which should be within the maximum ratings specified for this tube type in the tube manual.

The amplifying action of the circuit is studied qualitatively by displaying the input- and output-voltage waveforms successively on the oscilloscope; these are the ac components labeled on the schematic as e_2 and e_3, respectively. While the output voltage is being observed, the input-voltage amplitude is increased gradually from a very low value to illustrate the result of overdriving the tube.

Next, the gain of the circuit is measured as a function of R_2. In the procedure to be described, the precision potentiometer serves to provide a variable and accurately known attenuation (voltage loss) which is adjusted to compensate exactly for the gain of the amplifier, so that e_1 equals e_3. The oscilloscope is used only as an indicating device for matching the amplitudes of these two voltages.

---

† If an audio generator is not available, a suitable input voltage may be derived from the 6.3-volt filament supply. A voltage of proper amplitude (about 1 volt rms) may be obtained by connecting one side of the 6.3-volt supply (or the center tap) to the experimental circuit ground, and inserting a 5.1-kilohm (or 2.7-kilohm) resistor between the ungrounded 6.3-volt lead and the top of R_a. In this case, however, C_1 should be increased to 25 μf, C_2 to 2 μf, and the lowest value of R_2 omitted.

The oscilloscope is connected initially to display e_1. The audio generator is set at a nominal frequency of 1000 Hz and at an amplitude somewhat below the level at which appreciable distortion in the amplifier will occur; a value of about 3 volts peak to peak (p-p) is satisfactory. The oscilloscope gain or vertical amplifier control is adjusted carefully until the peak-to-peak deflection has an arbitrary, conveniently reproducible value. The oscilloscope gain is then left unchanged until the measurements have been completed.

The oscilloscope is next connected to the output to indicate the amplitude of e_3. With the switch for selecting R_2 set at the open position, the precision potentiometer is adjusted to make the amplitude of e_3 the same as that of e_1. Actually, the condition achieved is

$$e_3 = -e_1 \tag{2}$$

but the fact that e_3 is inverted relative to e_1 will not be observed unless the oscilloscope sweep is synchronized directly with the audio generator. The gain of the amplifier itself is now

$$A = \frac{e_3}{e_2} = -\frac{e_1}{e_2} = -\frac{R_a}{R_b} \tag{3}$$

so that A can be calculated from the known ratio R_a/R_b.

A series of measurements is made in which R_2 is set successively at the available values; for each value of R_2, the precision potentiometer is adjusted to match e_3 with e_1 and the values of R_b and R_2 are recorded.

When the experimental work has been completed, the components and leads which were soldered into place by the student are removed and the experimental chassis restored to its original condition.

CALCULATIONS.   The observed ripple-reduction factors for the LC filter sections used with the full-wave rectifier are compared with the values predicted by the equation

$$\alpha = \frac{\text{alternating voltage at output of section}}{\text{alternating voltage at input of section}} = \frac{1}{(2\pi f)^2 LC - 1} \tag{4}$$

which applies to each of the two LC sections. For the full-wave rectifier the frequency f is to be taken as 120 Hz. The higher ripple frequencies present (higher multiples of 60 Hz) will be attenuated more than the 120-Hz component.

The measured values of A are plotted versus R_L to illustrate the general character of their relationship. For the purpose of testing the form of Eq. (1) more critically and deriving the values of the tube parameters μ and r_p for the operating point used, a graph of $1/A$ versus $1/R_L$ is drawn; this procedure is based on a rearranged form of Eq. (1).

Practical applications. The circuits studied are so widely used in research apparatus that a thorough grasp of the principles involved will prove invaluable in almost any sort of physical-chemical experimental work. The concept of the equivalent circuit for a triode is very useful for design purposes and can be quite helpful as a means to understanding the behavior of triodes under a wide variety of circumstances.

Suggestions for further work. With the circuit of Part B, the effect of the cathode bypass capacitor on the gain is easily and strikingly demonstrated. With R_2 set at any convenient value, the peak-to-peak amplitude at the grid is adjusted for about 1 volt, as measured on the oscilloscope, and the output amplitude is noted. The cathode capacitor C_1 is then removed, and the amplitude at the output is again noted. The equivalent circuit of Fig. 199 (page 606) covers this case.

With the addition of a suitable feedback circuit, the amplifier of Fig. 119 becomes an audio oscillator (page 632).

It is instructive to examine the variation of gain of the triode circuit with frequency. The frequencies at which the gain drops off can be estimated theoretically with reasonable accuracy with the aid of the equivalent circuit including bypass, coupling, stray, and tube electrode capacitances (page 606).

References

Pertinent references are given in Chapter 28.

PART
2

APPARATUS
AND
METHODS

Chapter 20

Treatment of Experimental Data[1,2]

KINDS AND QUALITY OF DATA

Many of the quantities measured in the physical chemistry laboratory, like temperature, pressure, or voltage, can assume continuously varying values. Instruments serve to transform one quantity of this type into another: a barometer converts pressure to the length of a column of mercury; a thermocouple converts a temperature difference to a voltage. The quantities thus represented are known as *analog* data.

Other data represent essentially discrete counting processes, like the number of drops of liquid falling from the end of a capillary or the number of radioactive disintegrations in a sample in a given length of time. A datum represented by a number is said to be *digital*. Analog data can be converted to digital form by comparison with some standard scale. In an undergraduate laboratory this comparison is usually made by the student, who compares the length of the mercury column in a barometer with marks on the brass scale or the position of the needle of a meter with a printed scale on the meter face. He thus serves as an analog-to-digital converter.

Strictly speaking, we can never measure the true value of any quantity but only an approximation to it. The error of a measurement is the difference between the observed value and the true value of the quantity. If the error is small compared with the magnitude of the measured quantity, the measurement is said to be *accurate*. If successive measurements of the same quantity agree closely with each other, the measurement is said to be *precise*. A precise measurement is not necessarily accurate, however, as will be discussed below.

A precisely measured value contains more information than an imprecisely measured value. The amount of information in the value depends on the number of *significant figures* used in expressing it. It is becoming increasingly common to refer to the number of *bits* of information in a number. This is essentially the number of digits used in expressing the number in the binary system, or, approximately, that power of 2 which just exceeds the number. In the case of very large or small numbers, exponential notation should be used. The digits in the exponent are not counted as significant figures but are often included in counting the number of bits. Thus 0.0000053 would be written as 5.3×10^{-6} and has two significant figures. It has 6 bits ($2^6 = 64 > 53 > 2^5 = 32$) if the exponent is excluded and 3 more bits ($2^3 = 8 > 6 > 2^2 = 4$) in the exponent.

The exponent need not be considered as additional information in cases where the magnitude of the number is known in advance. The precision of a measurement should be reflected in the number of significant figures used in expressing its value.

ERRORS OF MEASUREMENT

Errors in experimental measurements may be divided into two classes: (*a*) *systematic errors* and (*b*) *random errors*. It is possible to correct for errors of the first type if the source of the error is known, and they are therefore frequently designated as *corrigible* or *determinate* errors. Random errors are indicated by fluctuations in successive measurements and lead to imprecise measurements. Systematic errors are reproduced in successive measurements, made under the same conditions.

Many systematic errors can be eliminated by the application of familiar corrections. For example, in the determination of atmospheric pressure using a mercury barometer, corrections must be applied to allow for the difference between the thermal expansion of mercury and of the brass scale (page 652). This is required because 1 standard atm (page 485) corresponds to the pressure exerted by a column of mercury 760 mm in height in an evacuated glass tube at a temperature of $0°C$ at sea level at a latitude of $45°$. In very precise work it is necessary, in addition, to correct for the capillary depression of the mercury and for the difference between the acceleration of gravity where the barometer is being used and the defined acceleration at sea level, $45°$ latitude.

In other cases where the theory has not been as well developed it is necessary to determine corrections experimentally. For example, in the drop-weight method for the determination of surface tension (Experiment 51), corrections which have been determined by using substances of known surface tension are applied. The correction factor to be applied is a function only of V/r^3, where r is the outer radius of the tube and V is the volume of the drop, and therefore does not depend on the nature of the liquid studied. A calibration of the scales of many instruments can best be obtained by making measurements on standard materials with well-known properties. This procedure tends to eliminate systematic errors introduced by the instrument.

Systematic errors may not manifest themselves by fluctuations in measurements and cannot be eliminated by merely repeating the measurements. These errors are therefore especially serious and insidious, and can be avoided only by careful calibrations and consideration of all possible corrections. Sometimes systematic errors are indicated by the change in the measured value resulting from a change of experimental technique or when different values are obtained on different days.

Errors of the second type, random errors, are indicated by fluctuations in successive measurements. These random variations are due to small errors beyond the control of the observer. For example, if a barometer is read several times in succession, the values read from the vernier will be found to fluctuate about a mean value. Random errors are not necessarily of instrumental origin. There is sometimes an essential background noise superimposed on the signal being measured (cf. Experiment 55). If the fluctuations are in fact random, they can be treated by the methods of statistics.

STATISTICAL TREATMENT OF EXPERIMENTAL DATA[2]

The purpose of the statistical treatment of experimental data is to determine the most probable value of a measured quantity and to estimate its reliability. Small random errors occur more frequently than large ones, and for many experimental measurements the errors may be assumed to be adequately described by a gaussian (or normal) probability distribution. Then the probability that a given measured value will be in error by an amount between x and $x + dx$ is

$$G(x,h) \, dx = \frac{h}{\sqrt{\pi}} e^{-h^2 x^2} \, dx \tag{1}$$

(Note that x can be either positive or negative.) The constant h is a measure of the width of the distribution: if h has a large value, the probability decreases rapidly from its maximum value at $x = 0$, indicating that the probability of large errors is small; if h is small, the probability curve falls off slowly and larger errors occur more frequently.

  If a sufficiently large number of measurements are made, the arithmetic mean of the values will be a good approximation to the true value of the quantity being measured, in the absence of systematic error. In the limit of an infinite number of measurements, the *standard deviation* σ_x of the errors

$$\sigma_x = \left[\frac{1}{n} \left(\sum_{i=1}^{n} x_i^2 \right) \right]^{\frac{1}{2}} \tag{2}$$

is related simply to the width of the distribution

$$\sigma_x = \frac{1}{h\sqrt{2}} \tag{3}$$

Since only a finite number of measurements are in fact made, the true value is not determined exactly. The standard deviation is estimated in terms of the *residuals* r_i, or difference between the individual values and the mean of the whole set,

$$\sigma_x = \left[\frac{1}{n-1} \left(\sum_{i=1}^{n} r_i^2 \right) \right]^{\frac{1}{2}} \tag{4}$$

If enough values are available for a statistical treatment to be meaningful, 68.3 percent of measurements are predicted to be within one standard deviation of the mean. It is also common to use the probable error, P_x

$$P_x = C\sigma_x \qquad C = 0.6745 \cdots \tag{5}$$

where the constant C is chosen so that

$$\int_{-P_x}^{P_x} \frac{h}{\sqrt{\pi}} e^{-h^2 x^2} \, dx = \tfrac{1}{2} \tag{6}$$

Thus half the measurements will be within one probable error of the mean.

If several sets of measurements are made, the means of the different sets will in general differ. The *standard deviation of the means,* sometimes also called the *standard error,* will be

$$\sigma_x = \left[\frac{1}{n(n-1)} \left(\sum_{i=1}^{n} r_i^2 \right) \right]^{\frac{1}{2}} \tag{7}$$

and can thus be estimated from a single set of measurements. A corresponding probable error of the mean is given by a relationship similar to that of Eq. (5).

Precise statements about the probability distribution can be made when the standard deviation or probable error is known. For example, 82.2 percent of all measurements will give values within $2P_x$ and 95.7 percent of the measurements will give values within $3P_x$ of the correct value. Unfortunately, it is often not practical to make enough measurements for valid use of a statistical treatment. Systematic errors in measurement or interpretation also contribute.

ESTIMATION OF EXPERIMENTAL ERRORS

When it is not possible to repeat a measurement enough times for a statistical treatment, as is often the case in the physical chemistry laboratory, it is necessary to estimate the precision of a measurement. No fixed values can be given for the precisions of various types of measurements because the precision depends upon the apparatus, the conditions under which it is used, and the care taken by the operator. Therefore it is necessary to develop an awareness of various sources of error in order to make reliable estimates.

In the measurement of weight in the laboratory, the precision may vary over a wide range. An ordinary analytical balance may be used to obtain weights to ± 0.1 mg, but the precision will depend upon the sensitivity of the balance and the way in which it is used, as well as the quality of the weights. Large objects may be weighed on a trip balance with a precision of ± 0.1 g.

In the measurement of volume the precision will depend upon whether volumetric flasks, pipettes, or burettes are used and on the size of the volume to be measured. The National Bureau of Standards tolerances for volumetric equipment are given in textbooks on quantitative analysis. In brief, a 25-ml volumetric flask (to contain) should be reliable to 0.03 ml, or 0.12 percent, and a liter volumetric flask to 0.5 ml, or 0.05 percent. A 10-ml transfer pipette should be reliable to 0.01, ml, or 0.1 percent, and a 2-ml pipette to 0.006 ml, or 0.3 percent. Of course, the precision of a volume required to reach an end point in a titration depends upon the sharpness of the end point as well as the accuracy of the burette.

In the measurement of pressure an ordinary laboratory barometer can be read to ± 0.2 mm, and the pressure should be accurate within this uncertainty after the necessary corrections have been made (see Appendix). On the other hand, the pressure obtained with a simple mercury manometer without a special reading device will be uncertain by about ± 1 mm.

The uncertainty in a measurement of temperature will be quite different if the temperature is measured with a good mercury-in-glass thermometer near room temperature or by use of a thermocouple at a high temperature. In calculating the percentage error in the temperature it must be remembered that it is the uncertainty of the value which is used in the calculation that is significant. Thus an uncertainty of $1°$ at $25°$ would cause not a 4 percent error in a calculation of molecular weight from the ideal-gas law, but a 0.3 percent error. In other types of experiments it is the change in temperature that is significant, rather than the absolute temperature, and so it is important to estimate the precision with which this difference can be measured.

In other cases, a scale can be read to a greater precision than is warranted by other factors in the experimental arrangement. For example, if a low-sensitivity galvanometer is employed for detection of the null point in a potentiometric circuit, it may be observed that the potentiometer slide-wire can be adjusted several divisions before a detectable movement of the galvanometer occurs. In this case, the precision is determined by the galvanometer rather than the graduated slide-wire.

INFLUENCE OF EXPERIMENTAL ERRORS ON THE FINAL RESULT

A final physical-chemical result is usually obtained by combining the results of different kinds of measurements. The accuracy of any final result is influenced by the accuracy of the measurements of the several quantities involved. If it happens that one of the quantities involved is subject to a much greater error than the others, it will have the preponderant effect in determining the accuracy of the final result. For example, in the determination of molecular weight from the elevation of the boiling point (Experiment 12), the solvent and solute can be weighed more accurately than the boiling-point elevation can be determined. If, however, the relative errors in the various measured quantities are of the same order of magnitude, the errors introduced by all the measured quantities must be considered. In trying to improve the accuracy of a given experimental determination it is important to emphasize improvement of the least accurate measurement.

The result u, calculated from a set of experimentally determined quantities x, y, z, $\ldots$, constitutes a function dependent on the values assumed for these quantities as independent variables. Corresponding to differential changes dx, dy, dz, $\ldots$ in the independent variables, the differential change in the result u is given by the conventional expression for the exact differential of a function of several independent variables. Restricting the treatment to a basis of three independent variables (the extension to a larger number is obvious),

$$du = \left(\frac{\partial u}{\partial x}\right)_{y,z} dx + \left(\frac{\partial u}{\partial y}\right)_{x,z} dy + \left(\frac{\partial u}{\partial z}\right)_{x,y} dz \tag{8}$$

The partial derivatives are derived from the relation by means of which u itself is calculated. Equation (8) provides a simple basis for estimation of the possible range of uncertainty which must be assigned to the value calculated for u as a consequence of the acknowledged uncertainties in the experimental data. In this procedure it is assumed that the accuracy of the measurements is reasonably good (of the order of a few percent or better), so that to an adequate degree of approximation Eq. (8) may be replaced by

$$\Delta u \approx \left(\frac{\partial u}{\partial x}\right)_{y,z} \Delta x + \left(\frac{\partial u}{\partial y}\right)_{x,z} \Delta y + \left(\frac{\partial u}{\partial z}\right)_{x,y} \Delta z \tag{9}$$

Here Δu approximates the finite change $u(x + \Delta x, y + \Delta y, z + \Delta z) - u(x,y,z)$ in the calculated value of u, which results from the changes Δx, Δy, Δz in the independent variables away from the values x, y, z, for which u and the partial derivatives are evaluated.

In the present application, the variations in x, y, z correspond to the uncertainties in the experimental data, which are known with respect to *estimated magnitude* but not with respect to sign. It is thus possible to calculate with confidence only the magnitude of each term appearing on the right-hand side of Eq. (9); the sum of these quantities then gives the extreme value for the absolute magnitude of the change in the derived quantity u which is consistent with specified uncertainties in x, y, z. Thus, for

$$x = x_m \pm \Delta x$$
$$y = y_m \pm \Delta y$$
$$z = z_m \pm \Delta z$$

$$u = u_m \pm |\Delta u| = u(x_m, y_m, z_m) \pm |\Delta u|$$

$$|\Delta u| = \left|\left[\left(\frac{\partial u}{\partial x}\right)_{y,z}\right]_{\substack{x=x_m \\ y=y_m \\ z=z_m}}\right||\Delta x| + \left|\left[\left(\frac{\partial u}{\partial y}\right)_{x,z}\right]_{\substack{x=x_m \\ y=y_m \\ z=z_m}}\right||\Delta y| + \left|\left[\left(\frac{\partial u}{\partial z}\right)_{y,x}\right]_{\substack{x=x_m \\ y=y_m \\ z=z_m}}\right||\Delta z|$$

$$(10)$$

It must be emphasized that if the ranges of uncertainty in the experimentally measured quantities are not properly estimated, the above calculation loses its practical significance and that in any case the possible effects of systematic errors are not included. If a systematic error can be estimated, its effect can be approximated. Equation (9) is used rather than Eq. (10) since in this case the sign of the error would be known.

Example. The molar refraction for a fluid is defined by the relation

$$\mathcal{R} = \frac{n^2 - 1}{n^2 + 2} \frac{M}{\rho} \tag{11}$$

where M = molecular weight

n = refractive index

ρ = density of fluid

For 25°C, at 1 standard atmosphere pressure, the following results are specified for liquid benzene, for which the molecular weight may be considered as 78.114 g mole^{-1} with negligible error.

$$\rho = 0.8737 \pm 0.0002 \text{ g ml}^{-1}$$
$$n = 1.4979 \pm 0.0003$$

Then $\mathcal{R} = 26.20$ ml mole^{-1}.

From Eq. (11),

$$\left(\frac{\partial \mathcal{R}}{\partial n}\right)_\rho = \frac{M}{\rho}\left[\frac{2n}{n^2 + 2} - \frac{2n(n^2 - 1)}{(n^2 + 2)^2}\right] = \frac{M}{\rho}\left[\frac{6n}{(n^2 + 2)^2}\right]$$

$$= \mathcal{R}\left[\frac{6n}{(n^2 - 1)(n^2 + 2)}\right]$$

$$\left(\frac{\partial \mathcal{R}}{\partial \rho}\right)_n = \frac{n^2 - 1}{n^2 + 2}\left(\frac{-M}{\rho^2}\right) = -\frac{\mathcal{R}}{\rho}$$

$$d\mathcal{R} = \mathcal{R}\left[\frac{6n}{(n^2-1)(n^2+2)}\right]dn - \frac{\mathcal{R}}{\rho}d\rho$$

$$|\Delta\mathcal{R}| = (26.20)\left[\frac{(6)(1.4979)}{\{(1.4979)^2-1\}\{(1.4979)^2+2\}}\right](0.0003)$$

$$+ \frac{26.20}{0.8737}(0.0002)$$

$$= 0.013 + 0.006 = 0.019$$
$$\mathcal{R} = 26.20 \pm 0.02 \text{ ml mole}^{-1}$$

In an alternative approach which is often convenient, recognizing that $d \ln f = df/f$, one can write

$$d \ln \mathcal{R} = d \ln (n^2 - 1) - d \ln (n^2 + 2) - d \ln \rho$$

$$\frac{d\mathcal{R}}{\mathcal{R}} = \frac{d(n^2-1)}{n^2-1} - \frac{d(n^2+2)}{n^2+2} - \frac{d\rho}{\rho}$$

$$= \left(\frac{2n}{n^2-1} - \frac{2n}{n^2+2}\right)dn - \frac{d\rho}{\rho}$$

or, as above,

$$d\mathcal{R} = \mathcal{R}\left[\frac{6n}{(n^2-1)(n^2+2)}\right]dn - \mathcal{R}\frac{d\rho}{\rho}$$

SPECIAL SIMPLE CASES

For the special case of addition or subtraction, $u = x \pm y$,

$$du = dx \pm dy$$
$$|\Delta u| = |\Delta x| + |\Delta y| \tag{12}$$

Thus the possible uncertainty in a sum or difference is equal to the sum of the absolute values of the uncertainties in the terms. For multiplication or division, $u = x^{\pm n}y^{\pm m}$,

$$\ln u = \pm n \ln x \pm m \ln y$$

$$\frac{du}{u} = \pm n \frac{dx}{x} \pm m \frac{dy}{y}$$

$$\left|\frac{\Delta u}{u}\right| = |n|\left|\frac{\Delta x}{x}\right| + |m|\left|\frac{\Delta y}{y}\right| \tag{13}$$

Then the relative uncertainty in the product (or quotient) is the sum of the absolute values of the relative uncertainties in the factors.

PROPAGATION OF PROBABLE ERRORS

From the defining equations (4) and (5),

$$P_u^2 = C^2 \frac{1}{n-1} \sum_i (du_i)^2 \tag{14}$$

but

$$(du_i)^2 = \left[\left(\frac{\partial u}{\partial x} \right) dx_i + \left(\frac{\partial u}{\partial y} \right) dy_i \right]^2$$

$$= \left(\frac{\partial u}{\partial x} \right)^2 (dx_i)^2 + 2\left(\frac{\partial u}{\partial x} \right)\left(\frac{\partial u}{\partial y} \right) dx_i\, dy_i + \left(\frac{\partial u}{\partial y} \right)^2 (dy_i)^2$$

If the errors are in fact governed by a gaussian distribution, positive and negative deviations are equally likely. The cross-product terms will thus tend to cancel out in the sum provided that the errors in x and y are uncorrelated, so that

$$P_u^2 = C^2 \frac{1}{n-1} \sum_i \left[\left(\frac{\partial u}{\partial x} \right)^2 (dx_i)^2 + \left(\frac{\partial u}{\partial y} \right)^2 (dy_i)^2 \right]$$

$$= \left(\frac{\partial u}{\partial x} \right)^2 P_x^2 + \left(\frac{\partial u}{\partial y} \right)^2 P_y^2$$

Thus, finally,

$$P_u = \left[\left(\frac{\partial u}{\partial x} \right)^2 P_x^2 + \left(\frac{\partial u}{\partial y} \right)^2 P_y^2 \right]^{\frac{1}{2}} \tag{15}$$

SIGNIFICANT FIGURES

An excessive number of uncertain figures should not be retained. After the uncertainty estimate has been obtained, the value reported for the quantity should be rounded off so that it contains not more than one or two uncertain significant figures. If the figures to be dropped amount to more than half of the last figure retained, that figure should be rounded off upward; if less than half, downward. When the portion discarded is exactly one-half, it is customary to round off to the nearest even value.

The number of significant figures reported for the uncertainty should correspond to these in the value itself. These points are perhaps best clarified by the following examples.

Value obtained	Reported
543 ± 37	540 ± 40
2.76 ± 0.015	2.76 ± 0.02
$5.4176 \times 10^{-5} \pm 2.32 \times 10^{-7}$	$(5.42 \pm 0.02) \times 10^{-5}$

THE REPRESENTATION OF DATA[1]

A mere collection of data is of limited value. It should be treated in a way which establishes its significance. There are three principal ways by which data may be represented: tables, graphs, and equations.

TABULAR REPRESENTATION.   One form of table, the *statistical* table, presents a collection of facts in orderly form but does not try to establish functional relationships among them. The tables of molecular properties found in handbooks and Table 1 of Experiment 13 are of this type.

In a *functional* table, on the other hand, corresponding values of an independent variable and of one or more dependent variables are listed side by side. A further subdivision can be made on the basis of the use to which the table is to be put. If, as will usually be the case in the physical chemistry laboratory, the table is intended to present the results of one particular set of experiments, the entries should be the actual values observed. However, if the table is to summarize results of many experiments and is intended for future reference use, convenience becomes more important. One should choose rounded values of the independent variable, and the dependent variables should be given in terms of smoothed data.

Every table should have a title which is clear and complete, but brief, and each column in the table should have a heading identifying the quantity being tabulated and giving its units.

DATA SMOOTHING.   There are several ways of obtaining smoothed data, the most common one being to plot the primary data, draw a representative smooth curve, and read information directly from the curve at desired points. The equation of the curve may also be found (page 443), and data may be computed from the equation at any desired value of the independent variable.

Table 1. The Vapor Pressure of Acetone as a Function of Temperature Illustrating Different Means of Smoothing Data

t, °C	P, cm Hg		
	From curve, Fig. 120a (1)	From straight line, Fig. 120b (2)	From Eq. (16) with $m = -1.662 \times 10^3$, $b = 6.929$ (3)
5	8.91	9.02	8.98
10	11.39	11.6	11.47
15	14.50	14.1	14.49
20	18.17	18.2	18.16
25	22.53	22.8	22.67
30	28.00	28.0	27.98
35	34.57	34.3	34.28
40	42.09	42.0	41.82
45	50.76	50.9	50.64
50	61.07	61.5	61.09
55	72.30	73.8	73.40

These principles are illustrated in Table 1. Data represented are from a student's determination of the vapor pressure of acetone as a function of the temperature. In column 1 are listed data read directly from the curve of P versus T (Fig. 120a). The graph as originally prepared was of large scale and permitted readings to four significant figures. In column 2 are data obtained from the straight-line graph of Fig. 120b, which is plotted from the same experimental data. The scale of this graph did not permit reading beyond three significant figures. In column 3 are data computed from the equation

$$\log P = m\left(\frac{1}{T}\right) + b \tag{16}$$

where the constants have been evaluated by the method of least squares (page 446).

Actually, none of these methods yields smoothed data of the highest possible quality in this particular case, for the following reasons:

1. *From P versus T Curve.* Although the curve drawn for P as a function of T is the best approximation to the correct functional form, the decision as to how the curve should fit the points will be in slight error if the estimation is done purely by eye.
2. *From Eq. (16) with Constants Determined by Least Squares.* This would be the last word in data smoothing if $\log P$ versus $1/T$ were actually a straight line.

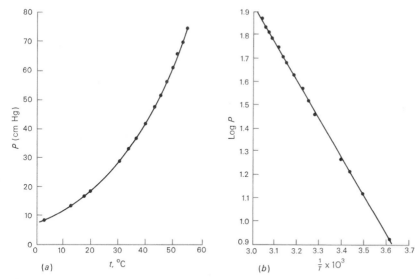

figure 120. (a) The vapor pressure of acetone as a function of temperature; (b) a plot of log P versus
1/T for acetone.

The best line through the points does, however, have a slight curvature. If the
precision of the data was not quite so good, it is unlikely that the curvature could be
assigned with certainty.

    If a three-constant equation of the form $\log P = (m/T) + a \log T + b$,
or better, $\log P = A/(T + C) + B$, were used in place of Eq. (16), the least-
squares method, or even an approximate method, for evaluating the constants
would yield smoothed data of very high quality. However, the labor of computation would be increased greatly.

3. *Directly from the Straight Line of Fig.* 120b. This is the usual procedure,
    yet it involves the errors encountered in both the other methods.

    Other topics concerned with functional tables, including several methods of
smoothing data by arbitrary numerical procedures, and the problems of interpolation
and extrapolation are treated fully by Worthing and Geffner.[1]

REPRESENTATION OF DATA BY GRAPHS. Graphs have many advantages
which favor their use in representing data. One of the most important of these
advantages is that a graph may reveal maxima, minima, inflection points, or other
significant features in data which might be overlooked in a tabular or formula
representation. Furthermore, direct differentiation may be performed by drawing
tangents to a curve, and integration may be accomplished by determining the

area under a curve; in many cases these operations would otherwise be tedious or impractical.

The steps to be followed in preparing a satisfactory graph have been extensively treated and illustrated with numerous examples by Worthing and Geffner.[1] We shall summarize here only some of the more important points.

Choosing the Graph Paper. Ordinary rectangular coordinate paper is satisfactory for a large majority of purposes. Semilogarithmic paper is convenient when one of the coordinates is to be the logarithm of an observed variable (Experiment 7). If both coordinates are to be logarithms of variables, log-log paper may be used (Experiment 53). Where an unknown functional relation is involved, these types of paper are sometimes used because it is found by trial and error that they yield a closer approximation to a straight line than rectangular coordinate paper. Another special-purpose paper which has triangular coordinates is used in Experiment 20.

It is practically always worth the extra expense to obtain high-quality paper made by a reputable manufacturer. Translucent paper is available for tracing or blueprinting. If the graph is to be reproduced in print, it is well to remember that blue coordinate lines will not show up; main coordinate lines may be traced over with India ink for satisfactory reproduction.

Choosing the Coordinate Scales. Five rules are listed by Worthing and Geffner. They are not altogether inflexible, and in case of doubt common sense should prevail.

1. The scale for the independent variable should be plotted along the X axis (abscissa).
2. The scales should be so chosen that the coordinates of any point on the plot may be determined quickly and easily.
3. The scales should be numbered so that the resultant curve is as extensive as the sheet permits, provided that the uncertainties of measurement are not made thereby to correspond to more than one or two of the smallest divisions.
4. Other things being equal, the variables should be chosen to give a resultant plot which approaches as nearly as practicable to a straight line.
5. Scales should be chosen such that the curve will, to the extent possible, have a geometrical slope approximating unity.

Labeling the Coordinate Scales. Main coordinate lines (or alternate ones) are labeled with the values they represent. The name of the quantity represented is given along each axis, together with the units in which it is measured.

Plotting the Data. Each point should be surrounded by a suitable symbol, such as a circle. If at all practicable, it is customary to have the size of the symbol correspond approximately to the precision of the determination.

It frequently happens that several curves are to be plotted on the same sheet of graph paper. This is the customary manner in which a third variable is treated on a two-dimensional graph. When this is done, a different type of symbol should be used for each set of data.

One should not carry the above considerations to the extreme. It is possible to have such a hodgepodge of points and curves that fundamental relations are obscured.

Fitting a Curve to the Plotted Points. If sufficient points are available and the functional relation between the two variables is well defined, a smooth curve is drawn through the points. These conditions prevail in practically all physical-chemical work. French curves, splines, or other devices should be used for maximum smoothness, unless a straight line is being graphed. Generally speaking, inflections or discontinuities will be absent; however, if such irregularities are greater than the experimental error, one must not ignore them. An inflection in the cooling curve of a molten alloy (Experiment 19) indicates the freezing point.

The curve should pass as close as reasonably possible to all the plotted points, though it need not pass through any single one. There is a natural tendency to overestimate the importance of the end points; often these are the least accurate points on the graph.

Preparing a Descriptive Caption. This should include a more or less complete description of what the graph is intended to show. The caption is usually included in an open region directly on the graph paper unless it is to be reproduced for printing. If the data have been taken from the work of another, the source should be acknowledged.

REPRESENTATION OF DATA BY EQUATIONS.   In order to obtain the maximum usefulness from a set of experimental data, it is frequently desirable to express the data by a mathematical equation. An advantage of this method is that data are represented in a compact fashion and in a form which is convenient for differentiation, integration, or interpolation. Frequently the form of the relationship between the dependent and independent variable is known, and it is desired to determine the values of the coefficients in the equation, since these coefficients correspond to physical quantities. Common examples of such equations follow.

Vapor-pressure equation:

$$\log P = \frac{-\Delta H}{2.303R} \frac{1}{T} + \text{constant} \qquad \text{Experiment 7}$$

Beer-Lambert law:

$$\log \frac{I}{I_0} = -acb \qquad \text{Experiment 16}$$

First-order reaction-rate equation:

$$\log c = \frac{-kt}{2.303} + \text{constant} \qquad \text{Page 138}$$

Langmuir adsorption equation:

$$\frac{c}{x/m} = \frac{1}{\alpha} + \frac{\beta}{\alpha} c \qquad \text{Experiment 52}$$

Radioactive-decay law:

$$\log \frac{N_t}{N_0} = -0.3010 \frac{t}{t_{\frac{1}{2}}} \qquad \text{Experiment 55}$$

In many cases the form of the relation between the independent and dependent variables is unknown and must be determined. This may be done by plotting the data and comparing the shape with that for known functions. Frequently the functional relationship is such that a straight-line graph may be obtained by changing the variables. For example, in the case of the Langmuir adsorption equation, a plot of amount adsorbed x/m versus concentration c is a curved line, while a plot of $c/(x/m)$ versus c yields a straight line. When the data or some function of the data can be plotted as a straight line, the constants can be determined simply from the slope and intercept. In many cases where a straight line is not obtained it is best to use a power series of the type

$$y = a + bx + cx^2 + dx^3 + \cdots$$

with as many empirical constants as necessary to represent the data to within the experimental uncertainty.

Three methods for the evaluation of the constants in a linear equation will be discussed. In order of increasing degree of objectivity, these methods are (1) graphical method, (2) method of averages, (3) method of least squares.

Graphical Method. This method is especially useful for the determination of the constants in a linear equation. If a given equation is not linear with respect to the variable, it may frequently be arranged in a linear form by making a simple substitution. For example, in the case of the vapor-pressure equation given above, the linear equation

$$y = mx + b$$

is obtained by substituting $\log P = y$, $1/T = x$, $m = -\Delta H/2.303R$, and $b = $ constant. Thus, when $\log P$ is plotted versus $1/T$, a straight line is obtained if this equation is correct. A plot of student data for the vapor pressure of acetone determined by the Ramsay-Young method is shown in Fig. 120*b*. The best straight line is drawn "through" these points with a transparent straightedge. The slope *m* of the line is calculated from the coordinates x_1, y_1 and x_2, y_2 of two points on the line:

$$m = \frac{y_2 - y_1}{x_2 - x_1}$$

These points are not selected from the original data and are taken as far apart as possible. The constant *b* is equal to the intercept on the *y* axis for $x = 0$. In this case, it is more convenient to calculate *b* from the slope *m* and the coordinates of one of the points on the line:

$$b = y_1 - mx_1$$

The values of *m* and *b* calculated graphically from Fig. 120*b* are

$$m = -1.662 \times 10^3 \text{ deg}$$
$$b = 6.929$$

It should be noted that the "best line" is not exactly determined. A visual estimate can often be made of a range of possible lines. From this, uncertainties in *m* and *b* can be estimated.

Method of Averages. The constants in a linear equation may be calculated from only two pairs of values for the variables. In general, more than two pairs of values are available, and different values for the constants will be obtained when different experimental points are used in the calculation. One method for determining the constants by using all the experimental data is the method of averages. This method is based on the assumption that the correct values of the constants *m* and *b* are those which make the sum of the residuals equal zero. The residuals v_i are the differences between the values of *y* calculated from the empirical equation and the experimentally determined values y_i. In the case we are discussing, the residuals are

$$v_i = mx_i + b - y_i \tag{17}$$

This assumption gives only one condition on the constants, and so it is further assumed that if there are *r* constants, the residuals may be divided into *r* groups and $\Sigma v_i = 0$ for each group. The groups are chosen to contain nearly the same number of experimental values, but it should be noted that different methods of choosing the groups will lead to different values for the constants. If the number of residuals in a group is *k*, the summation of Eq. (17) yields

$$\sum_1^k v_i = m \sum_1^k x_i + kb - \sum_1^k y_i = 0$$

If the data in Table 2 are divided into groups, 1 to 7 and 8 to 15, the two equations are

$$23.715m + 7b - 9.089 = 0$$
$$24.886m + 8b - 14.061 = 0$$

The values of m and b calculated from these simultaneous equations are

$$m = -1.6571 \times 10^3 \text{ deg}$$
$$b = 6.9125$$

The values of the residuals calculated from Eq. (17) by using these constants and the values obtained experimentally are given in Table 2 to indicate the precision with which the data are represented.

Method of Least Squares. The methods already described give different values of the constants, depending upon the judgment of the investigator. The method of least squares has the advantage of giving a unique set of values for the constants, and the

Table 2. *Application of a Linear Equation to Experimental Data*

Experiment number	$\frac{1}{T} \times 10^3 =$ $x \times 10^3$	$\log P$ $= y$	$x^2 \times 10^6$	$xy \times 10^3$	Residuals $(mx_i + b - y_i) \times 10^3$		
					Graphical method	Method of averages	Method of least squares
1	3.614	0.920	13.06100	3.324880	+5	+4	+2
2	3.493	1.121	12.20105	3.915653	+5	+3	+2
3	3.434	1.221	11.79236	4.192914	+3	+1	0
4	3.405	1.271	11.59402	4.327755	+1	−1	−2
5	3.288	1.463	10.80944	4.810344	+4	+1	0
6	3.255	1.522	10.59502	4.954110	−1	−3	−4
7	3.226	1.571	10.40708	5.068046	−1	−4	−4
8	3.194	1.623	10.20164	5.183862	0	−3	−3
9	3.160	1.679	9.98560	5.305640	0	−3	−3
10	3.140	1.711	9.85960	5.372540	+1	−2	−1
11	3.117	1.749	9.71569	5.451633	+2	−2	−1
12	3.095	1.783	9.56902	5.518385	+4	+1	+1
13	3.076	1.814	9.46178	5.579864	+5	+1	+2
14	3.060	1.838	9.36360	5.624280	+7	+4	+4
15	3.044	1.864	9.26594	5.674016	+8	+4	+4
Σ	48.601	23.150	157.89433	74.303922			

values of y calculated by using the constants determined by this method are the most probable values of the observations, it being assumed that the residuals follow the gaussian law of error. The principle of least squares asserts that the best representative curve is that for which the sum of the squares of the residuals is a minimum. In the case of the equation which we have been discussing, this sum S is

$$S = \sum_{i=1}^{n} (x_i m + b - y_i)^2$$

$$= m^2 \sum_{1}^{n} x_i^2 + 2bm \sum_{1}^{n} x_i - 2m \sum_{1}^{n} x_i y_i + nb^2 - 2b \sum_{1}^{n} y_i + \sum_{1}^{n} y_i^2$$

The necessary conditions for a minimum are

$$\frac{\partial S}{\partial m} = 0 = 2m \sum_{1}^{n} x_i^2 + 2b \sum_{1}^{n} x_i - 2 \sum_{1}^{n} y_i x_i$$

$$\frac{\partial S}{\partial b} = 0 = 2m \sum_{1}^{n} x_i + 2b(n) - 2 \sum_{1}^{n} y_i$$

These two equations may be solved simultaneously for m and b to yield

$$m = \frac{(n)\Sigma y_i x_i - \Sigma x_i \Sigma y_i}{(n)\Sigma x_i^2 - (\Sigma x_i)^2}$$

$$b = \frac{\Sigma x_i^2 \Sigma y_i - \Sigma x_i \Sigma x_i y_i}{(n)\Sigma x_i^2 - (\Sigma x_i)^2} \tag{18}$$

where the summations are to be carried out from 1 to n. Thus, in order to compute the constants by this method, it is necessary to calculate Σx_i, Σy_i, Σx_i^2, and $\Sigma x_i y_i$ as shown in Table 2. The calculations are carried out with more figures than the number of significant figures in the experimental data because the experimental values are assumed to be exact for purposes of the calculation.

The values of m and b obtained are

$m = -1.6601 \times 10^3$ deg
$b = 6.9221$

and the values of the residuals are given in the last column of Table 2 for comparison with those obtained by the other methods. Calculations by the least-squares method are time-consuming and are therefore carried out only for reliable data. The use of a computer greatly reduces the effort required.

When one is in doubt as to the form to be used in presenting graphs, tables, or equations, it is often a good idea to check recent issues of the journals of the American Chemical Society or the American Institute of Physics, where examples of the preferred style are usually easy to find.

THE USE OF COMPUTERS IN PROCESSING EXPERIMENTAL DATA

At the beginning of this chapter, a distinction was made between digital and analog data. Either type of data can be processed by electronic computers.

ANALOG COMPUTING.[3]   In an analog computer, operational amplifiers are combined with passive elements such as resistors, capacitors, and inductors to produce output voltages which are specified functions of input voltages. A computer is set up to solve one type of problem. Parameter values can be changed by resetting switches, but if a significantly different type of problem is to be considered, the computer must be rewired. In addition, analog computers are of limited accuracy. They are fast, however, and well suited to certain types of problems.

DIGITAL COMPUTING.   As the name implies, a digital computer deals with numbers in digital form. The binary system with only the two digits 0 and 1 is used, with all numbers being converted in the computer to that form. Mathematical and logical operations are carried out sequentially under the direction of a program which is itself stored in the computer. To change from one problem to another it is only necessary to store a new program. Accuracy can be improved by increasing the number of digits in each number, although more accurate calculations take more time.

Because digital computers tend to be expensive, it is important that they be used efficiently. It is not efficient to have a computer waiting for the relatively slow response of a chemical experiment or a human brain. The development of time-sharing techniques, which allow one computer to watch many experiments and communicate with many people at the same time, has alleviated this difficulty, and the use of computers in connection with experimental work may be expected to increase rapidly.

Another rapidly developing field is that of *hybrid* computation, in which analog computing elements are used under digital stored-program control.

Even if a computer is not connected directly to the experimental equipment, it may be used to process data. Calculations like the least-squares fit of equations become much more reasonable with respect to time required if a digital computer is available. In situations like this, the student serves as an analog-to-digital converter and then enters the data on punched cards, paper tape, or a teletype keyboard for entry into the computer.

One should not lose sight of the fact that a computer will do only what it is told to do and has no inherent sense for the physical reasonableness of a result. The use of a computer to process experimental data does not relieve the experimenter of his responsibility for determining the method by which the data are treated, estimating their value, and drawing meaningful conclusions from the results.

SOME COMPUTER PROGRAMS

It would not be practical or desirable to present programs for all the calculations arising in Part One. Three programs are presented here to serve as examples of what can be done and for use in cases where they are applicable. Two of these programs are for fitting data with a straight line determined by the method of least squares. The third is for the calculations arising in Experiment 9, using the van Laar equations.

These programs have been written in a simple version of FORTRAN,[4] which should be acceptable to nearly all present compilers.

A SIMPLE LEAST-SQUARES PROGRAM.   This program uses Eq. (18) to determine the slope and intercept of the best straight line through a set of data points. The input to the program consists of the following: N, GOOD, ANAME in a format I5, F10.4, 5A6 on one card, followed by the data points X(I), Y(I), four x, y pairs per card, in format 8F10.4. N is the number of data points to be read and used by the program. GOOD is a number used to establish a criterion for discarding bad points. After a fit has been made to all N points the residuals of the y values are calculated, assuming the x values to be exact, and the standard deviation of the y's determined. The program then discards those points with residuals greater than GOOD times the standard deviation and fits a new line to the remaining points.†
ANAME is an identification array of up to 30 characters, including blanks. It is not used in the program but is reproduced with the output. It might be the experiment number, student name, etc. Additional output consists of the slope and intercept of the line, the standard deviation, and a table of values of X and Y (the input data) and R, the Y residual. This information is repeated for the second line fit after discarding bad points, if there are any.

Note: The second table in the computer output includes the points which were not used in the fit as well as those which were.

```
    PRØGRAM LSTSQ
    DIMENSIØN X(100), Y(100), R(100), ANAME(5)
1   READ 30, N, GØØD, (ANAME(I),I=1,5)
    IF(N) 2,2,3
```

† This step is omitted if GOOD = 0. is read in.

```
2   STØP
3   READ 31, (X(I),Y(I),I=1,N)
    SUMX = 0.
    SUMY = 0.
    SUMXSQ = 0.
    SUMXY = 0.
    DØ 4 I=1,N
    SUMX = SUMX + X(I)
    SUMY = SUMY + Y(I)
    SUMXSQ = SUMXSQ + X(I)*X(I)
4   SUMXY = SUMXY + X(I)*Y(I)
    EN = N
    DEN = EN*SUMXSQ − SUMX*SUMX
    AM = (SUMXY*EN − SUMX*SUMY)/DEN
    AB = (SUMXSQ*SUMY − SUMX*SUMXY)/DEN
    SUMRSQ = 0.
    DØ 5 I=1,N
    R(I) = Y(I) − AM*X(I) − AB
5   SUMRSQ = SUMRSQ + R(I)*R(I)
    C = N − 1
    SIGMA = SQRTF(SUMRSQ/C)
    PRINT 32, (ANAME(I),I=1,5)
    PRINT 33, N, SIGMA, AM, AB
    PRINT 34, (X(I), Y(I), R(I), I=1,N)
    IF(GØØD)2,1,6
6   TEST = GØØD*SIGMA
    M = N
    DØ 8 I=1,N
    DEV = ABSF(R(I))
    IF(DEV−TEST) 8,8,7
7   M = M − 1
    SUMX = SUMX − X(I)
    SUMY = SUMY − Y(I)
    SUMXSQ = SUMXSQ − X(I)*X(I)
    SUMXY = SUMXY − X(I)*Y(I)
8   CØNTINUE
    IF(M−N) 9,1,1
9   EM = M
    DEN = SUMXSQ*EM − SUMX*SUMX
    BM = (SUMXY*EM − SUMX*SUMY)/DEN
    BB = (SUMXSQ*SUMY − SUMX*SUMXY)/DEN
    C = M − 1
    SUMRSQ = 0.
    DØ 11 I=1,N
    DEV = ABSF(R(I))
    R(I) = Y(I) − BM*X(I) − BB
    IF(DEV−TEST) 10,10,11
10  SUMRSQ = SUMRSQ + R(I)*R(I)
```

```
11  CØNTINUE
    SIGMA = SQRTF(SUMRSQ/C)
    PRINT 33, M, SIGMA, BM, BB
    PRINT 34, (X(I), Y(I), R(I), I=1,N)
    GØ TØ 1
30  FØRMAT(I5, F10.4, 5A6)
31  FØRMAT(8F10.4)
32  FØRMAT(22H1LEAST SQUARE FIT FØR , 5A6)
33  FØRMAT(1HO, 4X, I5, 25H PØINTS FIT WITH STD DEV , F10.4,
  1 10H BY SLØPE , F10.4, 12H, INTERCEPT , F10.4//1HO, 9X, 1HX, 14X,
  2 1HY, 14X, 1HR)
34  FØRMAT(3(5X, F10.4))
    END
```

The program is constructed so as to run through several sets of data in succession. After the last set to be used, a card should be inserted with 0 in the N position. This will stop the execution.

A WEIGHTED LEAST-SQUARES PROGRAM.   This program gives different weights to the points used to determine a straight line. It is again assumed that the x value for each point is accurate but that there is an uncertainty Δy in the y value. The contribution of each point in the least-squares calculation is then weighted by $1/(\Delta y)^2$.

The input consists of the number of points N, and identification ANAME, as described above, in format I5, 5A6, followed by the N values of X, Y, DY read in format 3F10.5, one triple per card. As in the previous program, any number of sets of data can be treated in succession, and the run should be terminated with a card having 0 in the N position. (Here DY is used for Δy.)

In addition to the best slope and intercept, the uncertainties in these quantities are estimated,[2] and the y residuals and the standard deviation of the actual y values from those predicted by the straight line are calculated.

```
    PRØGRAM WTLSQ,
    DIMENSIØN X(100), Y(100), DY(100), R(100), ANAME(5)
 1  READ 30, N, (ANAME(I),I=1,5)
    IF(N) 2,2,3
 2  STØP
 3  READ 31, (X(I), Y(I), DY(I), I=1,N)
    WSX=0.
    WSY=0.
    WSXSQ=0.
    WSXY=0.
    SUMW=0.
    DØ 4 I=1,N
```

```
      W = 1./(DY(I)*DY(I))
      WSX = WSX + W*X(I)
      WSY = WSY + W*Y(I)
      WSXSQ = WSXSQ + W*X(I)*X(I)
      WSXY = WSXY + W*X(I)*Y(I)
    4 SUMW = SUMW + W
      DEN = SUMW*WSXSQ − WSX*WSX
      A = (WSXSQ*WSY − WSX*WSXY)/DEN
      B = (SUMW*WSXY − WSX*WSY)/DEN
      XAVG = WSX/SUMW
      YAVG = WSY/SUMW
      SDXSQ = 0.
      SDA = 0.
      SDB = 0.
      SUMRSQ = 0.
      DØ 5 I = 1,N
      DX = X(I) − XAVG
      DXSQ = DX*DX
      SDXSQ = SDXSQ + DXSQ
      AI = (YAVG*X(I) − XAVG*Y(I))/DX
      BI = (Y(I) − YAVG)/DX
      DA = AI − A
      DB = BI − B
      SDA = SDA + DXSQ*DA*DA
      SDB = SDB + DXSQ*DB*DB
      R(I) = Y(I) − A − B*X(I)
    5 SUMRSQ = SUMRSQ + R(I)*R(I)
      C = N
      DEN = C*SDXSQ
      SDA = SQRTF(SDA/DEN)
      SDB = SQRTF(SDB/DEN)
      C = C − 1.
      SIGMA = SQRTF(SUMRSQ/C)
      PRINT 32, (ANAME(I), I = 1,5)
      PRINT 33, N, SIGMA, B, SDB, A, SDA
      PRINT 34, (X(I), Y(I), DY(I), R(I), I = 1,N)
      GØ TØ 1
   30 FØRMAT(I5,5A6)
   31 FØRMAT(3F10.5)
   32 FØRMAT(1H15X,5A6//26H0WEIGHTED LEAST SQUARE FIT)
   33 FØRMAT(1H04X,I5,23HPØINTS WITH STD DEV F10.5/20H BY LINE WITH
      1SLØPE F10.5,3H+ − F10.5, 11H INTERCEPT F10.5,3H+ − F10.5//7X,1HX
      2,12X,1HY,11X,2HDY,12X,1HR//)
   34 FØRMAT(4(3XF10.5))
      END
```

A PROGRAM FOR EXPERIMENT 9.   This program calculates liquid and vapor composition curves, as a function of temperature, for mixtures of ethanol and benzene. Both ideal behavior and nonideal behavior with activity coefficients determined from the van Laar equations are treated. The details of the calculation are described in Experiment 9. Equation (15), page 68, is solved by Newton's method.

The input consists of identification ANAME, as described for the simple least-squares program, the corrected barometric pressure P, in millimeters of mercury, the azeotropic composition as mole fractions X1AZ and X2AZ (ethanol is taken to be component 1), and the boiling point of the azeotropic mixture TAZ, in degrees Celsius (centigrade). The program will do any number of successive calculations, reading a new set of data for each. After the final set, a card should be included with P zero or negative. This will stop the run. The input format is 5A6, 4F8.4.

```
      PRØGRAM LIQVAP
      DIMENSIØN ANAME(5)
  1   READ 30, (ANAME(I), I=1,5), P, X1AZ, X2AZ, TAZ
      IF(P) 2,2,3
  2   STØP
  3   PRINT 31, (ANAME(I),I=1,5),P
      Y = 0.43429448 * LØGF(P)
      T2 = 1211.215/(6.90522 − Y) − 220.87
      T1 = 1595.76/(8.11576 − Y) − 226.5
      T=T1
      DT=(T2−T1)/20.
      DØ 10 I=1,21
      P1T=P1(T)
      P2T=P2(T)
      X2L=(P−P1T)/(P2T−P1T)
      X2V = X2L * P2T/P
      PRINT 32, T, X2L, X2V
 10   T=T+DT
      G1AZ=P/P1(TAZ)
      G2AZ=P/P2(TAZ)
      XX=X1AZ/X2AZ
      AC2 = LØGF(G2AZ)
      RATIØ=XX*XX*LØGF(G1AZ)/AC2
      A2=.43429448*AC2*(1.+RATIØ/XX)**2
      A1=A2/RATIØ
      PRINT 33, A1, A2
      CØNV = 2.3025851
      X2L=0.95
      DØ 20 I=1,19
      X1L=1.−X2L
      XX=X1L/X2L
      Y=XX/RATIØ
      G1 = EXPF(CØNV * A1/((1.+Y)**2))
      G2 = EXPF(CØNV * A2/((1.+1./Y)**2))
      C1=G1*X1L
```

```
       C2 = G2*X2L
15     F1 = C1*P1(T)
       F2 = C2*P2(T)
       DT = (P−F1−F2)/(F1*CØNV*1595.76/((226.5+T)**2) + F2*CØNV*1211.215/
  1       ((220.87+T)**2))
       T = T+DT
       IF(ABSF(DT)−.01) 16,15,15
16     X2V = F2/P
       PRINT 32,T,X2L,X2V
20     X2L = X2L−.05
       GØ TØ 1
30     FØRMAT(5A6, 4F8.3)
31     FØRMAT(1H15X,5A6/39H0BINARY EQUILIBRIUM ETHANØL BENZENE AT ,F8.2,
  1 6H MM HG //5X14HIDEAL BEHAVIØR //9X1HT,11X3HX2L,10X3HX2V//)
32     FØRMAT(3(3XF10.3))
33     FØRMAT(1H04X,24HVAN LAAR CALC WITH A1 = F10.3,2X,5HA2 = F10.3
  1       //9X1HT,11X3HX2L, 10X3HX2V//)
       END

       FUNCTIØN P1(T)
       P1 = EXPF(2.3025851*(8.11576−1595.76/(226.5+T)))
       END

       FUNCTIØN P2(T)
       P2 = EXPF(2.3025851*(6.90522−1211.215/(220.87+T)))
       END
```

References

1. A. G. Worthing and J. Geffner, "Treatment of Experimental Data," John Wiley & Sons, Inc., New York, 1943.
2. L. G. Parratt, "Probability and Experimental Errors in Science," John Wiley & Sons, Inc., New York, 1961.
3. H. V. Malmstad, C. G. Enke, and E. C. Toren, Jr., "Electronics for Scientists," W. A. Benjamin, Inc., New York, 1963.
4. E. I. Organick, "A FORTRAN Primer," Addison-Wesley Publishing Company, Inc., Reading, Mass., 1963.

Chapter
21

Opticochemical
Measurements

Many types of analyses may be made by optical methods. These methods have the advantage of being rapid and sensitive and leaving the sample unchanged. Also, the optical tools are popular as instruments of physical chemistry in the determination of the structure, size, and shape of both ordinary and large molecules in solution or of colloidal particles in suspension.

REFRACTOMETRY

REFRACTOMETERS MEASURING CRITICAL ANGLE.   The basic principles of operation and the construction of the Abbe and immersion refractometers have been described in a number of places.[1] The compensator (or Amici prism), which makes possible the use of white light in the Abbe and immersion refractometers, consists of two direct-vision prisms in the telescope barrel which can be rotated in opposite directions. The direct-vision prisms are made to give dispersion, with a minimum of deviation, by cementing a dense prism of flint glass between two prisms of crown glass. If the first direct-vision prism spreads out the light into a spectrum, and if the second prism is set at the same angle, the dispersion is doubled. However, if the prism is rotated through 180°, the second prism will subject the dispersed beam to an opposite dispersion, thus reproducing white light. The refractometer prism and the liquid produce dispersion, the amount varying with the refractive index of the liquid. The extent to which the prisms of the compensator must be rotated in order to offset the dispersion of the refractometer and liquid to produce white light must be determined each time by trial. The reading on the compensator drum is a measure of the dispersion of the liquid.

The Abbe refractometer is checked by placing against the upper prism a plate glass of known refractive index supplied with the instrument. The two surfaces are held together with a drop of liquid having a higher index of refraction, for example, α-bromonaphthalene. If the refractometer does not give the proper reading, it is adjusted by means of the small screw at the back of the telescope. The immersion refractometer is checked with distilled water. If the shadow edge does not fall at 15.0 for 17.5°C or at 13.25 for 25°C, adjustment is made by means of a screw inside the micrometer drum.

The Pulfrich refractometer is perhaps the oldest and the most accurate instrument of this kind. It requires a sodium or mercury lamp or other source of monochromatic light because it does not have a compensating Amici prism. The horizontal beam of monochromatic light goes through a cup of liquid cemented to the top of the prism and is refracted through an angle which is measured directly by rotating the eyepiece until it picks up the colored line of light, directly coincident with the point of intersection of the cross hairs. The angle of refraction is read directly in degrees on a circular scale, using a vernier and lens.

The zero setting of the instrument is obtained by holding a small light at arm's length in front of the small square window near the eyepiece and turning the graduated circular scale until the image of the cross hairs reflected from the face of the main prism coincides with the cross hairs themselves. This window has a right-angled prism which directs a light beam on to the prism surface, from which it is reflected back into the eyepiece. The zero setting is subtracted from the reading of the angle of refraction.

The refractive index n_λ at each wavelength λ is calculated by means of the formula

$$n_\lambda = \sqrt{N_\lambda^2 - \sin^2 i}$$

where N_λ = refractive index of glass prism against air for wavelength of light used
  $\sin i$ = sine of angle of emergence i measured on circular scale

The values of N_λ are furnished with the Pulfrich instrument, usually for the blue (4358 Å), green (5460 Å), and yellow (5780 Å) lines of the mercury arc.

The cup of the Pulfrich refractometer is cemented to the prism. For organic liquids, fish glue may be used. Still better is a concentrated gelatin solution containing potassium dichromate, which is exposed to bright sunlight after setting in place. When aqueous solutions are used, the glass cup must be attached with Canada balsam or other waterproof cement. A smaller metal cup with circulating thermostated water is set into the cup of liquid.

The Pulfrich and the immersion refractometers may be provided with interchangeable prisms which extend the range to different refractive indices. The immersion refractometer is provided with a small metal cap with a glass bottom which fits over the prism and permits measurements to be made with small amounts of liquid.

The refractive index of liquids changes considerably with temperature, and temperature control to 0.1 or 0.2° is necessary. The refractive index of glass against air also changes but to a considerably smaller extent. For example, in the Pulfrich refractometer, the refractive index of the prism is given for 20°, and an increase of 3° gives an increase in the refractive-index calculations of about 1 in the fifth decimal place. This temperature correction for the glass may be neglected in ordinary work at room temperature.

DIFFERENTIAL REFRACTOMETER.   Optical methods have become extremely useful in the study of the physical chemistry of solutions. At constant temperature and pressure, the refractive index of a solution n can be simply related to the concentration, with the volume-based scale being used. For a two-component system,

$$n = n_0 + R_1 c + R_2 c^2 + \cdots$$

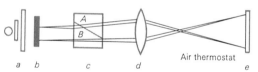

figure 121. Differential refractometer: a = light source, consisting of mercury-vapor lamp, light filter, and frosted glass screen; b = Lamm scale; c = prismatic cell, with two compartments; d = lens to focus scale on photographic plate; e = photographic plate.

where n_0 is the refractive index of the solvent. The higher terms are generally small, and the quantity R_1, the specific refractive increment of the solute, becomes $\lim_{c \to 0} (\partial n / \partial c)$. Measurements of n as a function of c thus provide means to determine the solute concentration. For multicomponent systems the interpretation is more difficult, but even in these cases it is sometimes possible to isolate the effects of a single solute for estimation.

The specific refractive increment for the solute may be determined by means of a differential refractometer. There are a number of varieties of instrument, but in all the refracting surfaces represent the equivalent of two adjacent prisms, one with solvent and the other with (dilute) solution (Fig. 121). With solvent in both prisms there is zero deflection of an image, but with solvent of refractive index n_0 in one and solution of refractive index n in the other, the displacement of the image is nearly proportional to $n - n_0$. There are many advantages of the differential method, especially that of very high accuracy. Temperature control is maintained by enclosing the cells in a jacket through which water can be pumped from a thermostat.

The differential refractometer is an especially important instrument for the macromolecular chemist. For example, in connection with the determination of molecular weight from light scattering, a knowledge of R_1 is required. The quantity often observed in the optical systems which are provided with ultracentrifuges is the refractive-index gradient, but solute concentrations may be required for proper interpretation of the data.

The construction and alignment of differential refractometers are described in a number of places in the literature.[2] The instruments are commercially available.

OPTICAL SYSTEMS

To observe both the changes in boundary form and position (if involved) which take place during the course of diffusion, velocity sedimentation, chromatography, and electrophoresis experiments and the redistribution of component concentrations at sedimentation equilibrium, optical methods are indispensable. They provide means of making the observations without in any way disturbing the course of the several processes. In general, three different properties of the dissolved molecules are used, those based on their power to absorb light of particular wavelengths, their ability to refract the light which passes through the solution, or their ability to modify the optical path length through the solution cell. Refractive-index methods are also described as being ray optical or wave optical in character. The several schlieren methods and the Lamm scale method belong to the first of these two kinds, while the interferometric procedures derive from wave optics. Svensson and Thompson's article contains an excellent general account of the subject, which is replete with references[3] and otherwise well documented.

LIGHT ABSORPTION METHOD.   The system used is shown in Fig. 122. For experiments with most biological macromolecules, quartz optical elements are required because the characteristic absorption takes place in the ultraviolet light regions. Illumination, with proper wavelengths, of the cell is provided by the lamp a and condenser lens and filter elements b. The camera lens d at the near right focuses the uniformly and constantly illuminated cell c, center of the diagram, in the plane of the photographic plate e. Thus, the solute concentration as a function of cell depth at any given time and position in the cell is given by a densitometer record of the blackening over the exposed and developed plate. The relationship between blackening and light intensity is defined by the statement

$$B = \log \frac{I_0}{I}$$

where I_0/I is the ratio of the light intensity measured through first an unexposed and then an exposed portion of the plate in the densitometer. For the evaluation of the experiment it is necessary to have available a measure of the dependence of blackening of the plate on solute concentration in the cell; it may be obtained in several ways.[4]

In the past difficulties in the control and calibration of the blackening of the photographic plate made the method unpopular. Now, with the availability of direct methods for the photoelectric recording of optical density (proportional to solute concentration) along a column and the realization that with proper and character-

a b c d e

figure 122. Diagram of optical system for light-absorption method: a = light source; b = condenser lens and filter system; c = cell; d = camera lens; e = photographic plate.

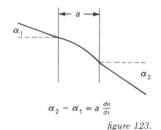

$$\alpha_2 - \alpha_1 = a\,\frac{dn}{dx}$$

figure 123.

Refraction of a beam of light through transparent media of variable refractive index.

istic wavelengths the behaviors of such solutes as proteins and nucleic acids in very dilute solutions can be studied, the situation has changed and there has been a marked resurgence in the use of the light-absorption method.

RAY OPTICAL METHODS.   Two methods which depend upon the bending of light by a refractive-index gradient, as indicated in Fig. 123, are now discussed. According to ray optics the angular deviation of the ray is directly proportional to the refractive-index gradient and to the thickness of the cell if the refractive index is constant in the region of the cell through which the ray passes. (The angles α are measured in air.)

LAMM SCALE METHOD.   A graduated scale is placed behind the cell and photographed through the cell with an ordinary camera. Refractive-index gradients in the cell cause displacements of the scale lines which are directly proportional to the refractive-index gradient. These displacements may be measured with a microscope comparator and plotted against the position of the line to yield a graph of refractive-index gradient versus distance in the cell. This is a laborious procedure, and it has become largely obsolete in practice. It is described in detail in the Svedberg-Pedersen monograph.[4]

SCHLIEREN METHODS.   These methods[5] depend upon the fact that light passing through the cell in the region of a gradient may be cut off by a blade so that the light passing through this part of the cell does not reach the photographic plate. The optical arrangement is shown in Fig. 124.

The camera lens focuses the illuminated cell *c* on the photographic plate *f*, but since the light passing through the gradient has been cut off by the blade, this portion of the cell appears dark on the viewing screen.

Schlieren Scanning Method. In this method the blade is moved vertically at a constant rate from a position below the most deviated rays to the normal slit position while at the same time the photographic plate is being moved in a horizontal

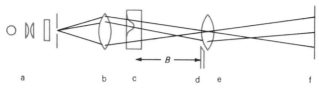

figure 124. Schlieren-scanning optical system: a = light source and illuminated horizontal slit; b = schlieren lens; c = cell with a refractive-index gradient in the middle; d = blade with a horizontal upper edge; e = camera lens; f = photographic plate or ground-glass screen.

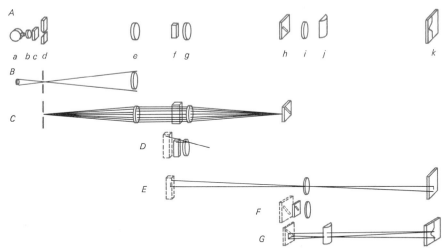

figure 125. Schlieren optical system for the observation of refractive-index gradient: a = light source; b = condensing lens; c = filter (77A for 5460-Å line); d = horizontal slit; e = schlieren lens 1; f = tiselius cell; g = schlieren lens 2; h = diagonal slit (or diagonal knife-edge, phase plate, etc.); i = camera lens; j = cylindrical lens; k = photographic plate.

direction at a constant rate. The resulting photograph is a plot of refractive-index gradient versus height in the cell.

   Cylindrical-lens Method. By the introduction of a cylindrical lens it is possible to obtain a plot of dn/dx versus x on the photographic plate or ground-glass screen without a scanning process. Such an optical system is illustrated in Fig. 125*A*. In this arrangement two schlieren lenses are used so that the monochromatic light will pass through the cell as a parallel bundle of rays, as illustrated in Fig. 125 *B* and *C*. The distance from the illuminated slit to the first schlieren lens is equal to the focal length of schlieren lens 1. The diagonal slit is mounted at a distance from schlieren lens 2 equal to the focal length of the latter. It is presently almost uniformly replaced by a phase plate.

   Figure 125*D* indicates that schlieren lens 2 forms a virtual erect image of the cell, which is focused on the photographic plate by the camera lens as indicated in Fig. 125*E*. The magnification of the cell on the plate is generally arranged to be close to unity.

   Figure 125*F* indicates that the camera lens forms a virtual image of the diagonal knife-edge, which is focused on the photographic plate by the cylindrical lens, as illustrated in Fig. 125*G*.

   If there is no refractive-index gradient in the cell, the only light which reaches the photographic plate is that which passes through the diagonal slit at the position of the normal slit image. This light appears on the photographic plate as a narrow vertical band (base line) in which the individual rays occupy positions corresponding to the levels at which they passed through the cell. It is to be noted that the cylindrical lens is without effect on the vertical positions of these rays.

If there is a refractive-index gradient in the cell, part of the light is deflected downward, as illustrated in Fig. 126*a*. The center of a diffuse boundary between two solutions of different refractive index is located at x_3. Plots of refractive index n and

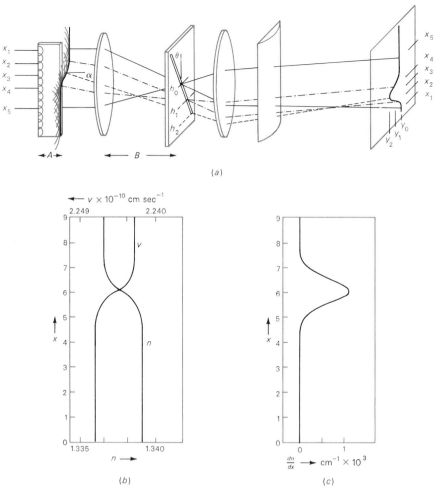

figure 126. *Complete optical system for the observation of refractive-index gradients: (a) light paths in schlieren optical system; (b) refractive index and light velocity versus height in cell; (c) refractive index gradient versus height in cell.*

velocity of light v as a function of height in the cell are given in Fig. 126b. Referring again to Fig. 126a, it is seen that the plane wavefront entering the cell is distorted by this variation in velocity of light. The rays passing through the center of the gradient are deflected downward so that they form an image of the horizontal slit at h_2. Since the diagonal slit is inclined at an angle θ from the vertical, the only rays from level x_3 which pass through the diagonal slit are at a distance $(h_2 - h_0) \times \tan \theta$ to the side of the optic axis. This sideways displacement is magnified by the cylindrical lens, and this light is focused on the opposite side of the optic axis and forms the top of the peak on the photographic plate. If other rays are traced through the cell, it will be seen that a plot of refractive-index gradient versus position in the cell will be obtained (see Fig. 126c).

WAVE OPTICAL METHODS

GOUY INTERFEROMETER.   If a photographic plate is placed in the plane of the diagonal knife-edge in the preceding optical system, instead of a continuous band of light below the normal slit image there is a series of interference fringes. These fringes result from interference between light from conjugate positions in the refractive-index gradient; i.e., the two positions at which dn/dx are equal. Figure 127 illustrates the important elements of the optical system when a single schlieren lens is used.

The total number of fringes j_m is equal to the number of wavelengths difference in path between the two homogeneous solutions on either side of the boundary

$$j_m = \frac{a(n_2 - n_1)}{\lambda}$$

It is noted that the Gouy method has had great success in the determination of diffusion coefficients D.[6,7] They are computed from the relation

$$D = \frac{(j_m \lambda b)^2}{4\pi C_t^2 t}$$

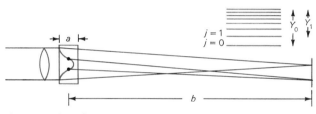

figure 127. Gouy fringes.

where t is the time after forming a sharp initial boundary and C_t is the greatest downward deflection of light at the plate that is predicted by ray optics; it is equal to $ab\,(dn/dx)_{max}$. This method for computing D is seen to be identical with the height-area method. Calculation of values of C_t is made from measured distances Y_j between the normal slit image and minima in the interference pattern, using the equation

$$C_t = \frac{Y_j}{e^{-z_j^2}}$$

where $e^{-z_j^2}$ is read from tables of the function

$$f(z_j) = \frac{2}{\sqrt{\pi}} \left(\int_0^{z_j} e^{-\beta^2}\,d\beta - z_j e^{-z_j^2} \right)$$

which in turn is obtained from the interference condition

$$f(z_j) = \frac{j + \frac{3}{4}}{jm}$$

The term $j + \frac{3}{4}$ in this interference condition is seen to differ by one-quarter of a wave from the ordinary ray optical interference condition, which states that two rays undergo destructive interference when their optical paths differ by *one-half* a wave, or $j + \frac{1}{2}$.

RAYLEIGH INTERFEROMETER.   Here the optical system utilizing Rayleigh fringes will be briefly described. The measurements in this case depend upon the difference in optical path through the cell and a reference optical path alongside it, rather than on deflection by the gradient. The simplest of several possible types of Rayleigh interferometer optical systems is illustrated in Fig. 128.

The cylindrical lens focuses the plane of the cell on the photographic plate. If the cell is filled with a homogeneous medium, a family of vertical fringes within a diffraction envelope will be obtained on the photographic plate. The separation of these fringes depends upon the distance d between the centers of the two slits, the optical distance b from the center of the cell to the photographic plate, and the wavelength λ of the monochromatic light used

$$s = \frac{\lambda b}{d}$$

If refractive index is a function of height in the cell, the fringes at any given level will be displaced a distance $ab(n - n_{\text{ref}})/d$ or $a(n - n_{\text{ref}})/\lambda$, expressed in numbers of fringes. Thus the fringes trace out the refractive index as a function of

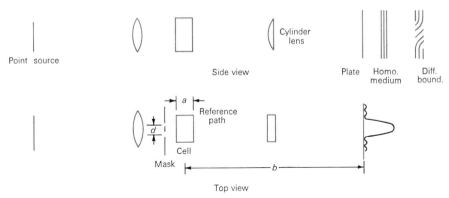

Point source

Side view

Plate Homo. medium Diff. bound.

Cylinder lens

a

Reference path

Cell

Mask

b

Top view

figure 128. Simple Rayleigh interferometer.

height in the cell. The principal advantages of this optical system are that refractive-index differences may be determined more accurately (about $10\times$) than by the schlieren method and it is not necessary to assume that ray optics is followed.[8,9]

LIGHT SCATTERING

A number of light-scattering phenomena have been studied, e.g., turbidimetry, nephelometry, Tyndall effects, dissymmetry of depolarization, etc. One effect of current interest for the physical chemist is the scattering from dilute solutions of macromolecules. Mathematical analyses and expressions for solute molecular weight are obtained by treating the scattering as resulting from statistical fluctuations in the density of the solvent and in the concentration of the solute, which produce momentary variations in dielectric constant and therefore in refractive index.

Some light is scattered even by a pure liquid because it is inhomogeneous on a microscopic scale. The scattering by a solution is correspondingly greater because of local differences in refractive index due to fluctuations in concentration of solute. The study of the intensity of light scattered by a solution of a high polymer (or protein) may be used to determine the molecular weight and radius of gyration of these molecules.[10] In the case of molecules having dimensions comparable with the wavelength of light or greater, measurements of light scattered at several angles are required; thus the apparatus is arranged so that the photomultiplier detector may be placed at several angles. The principle of the apparatus is illustrated by Fig. 129.

To use the equations, two optical measurements are required, a quantitative comparison of the two intensities of the scattered and incident beams and the determination of the specific-refractive-index increment of the solution. The measurement of the intensity of light scattered by a dilute solution is carried out in an apparatus

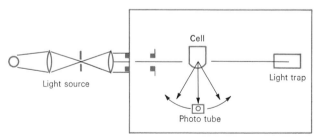

figure 129. Light-scattering apparatus.

(Fig. 129) which consists of a light source, a collimation system for the incident beam, a cell to hold the solution, a collimation system for the scattered light, a photomultiplier, and a measuring device for the impulse. The photomultiplier is so mounted that it may be moved in an arc around the solution cell in order that the quantity called the *Rayleigh ratio* may be determined at several angles. Some form of permanent working standard is used to check variations in the performance of the apparatus; molecular weights are given in reference to that of some standard substance.

Data on specific-refractive-index increments are obtained by using a differential refractometer.

Light scattering as applied to the study of large molecules has been reviewed by Stacey[11] and by Oster.[12]

MICROSCOPY

LIGHT MICROSCOPE.   The resolving power (RP) of the common microscope is a measure of its capacity to reproduce minute details of the structure of the object in the image. It is directly proportional to a quantity called the numerical aperture (NA) and inversely proportional to the wavelength λ of the light employed; thus

$$RP = \frac{NA}{\lambda}$$

This formula is valid only for central illumination; if sufficiently oblique illumination is used, the resolving power may be almost doubled. The resolving power also increases with an increase in the refractive index of the medium intervening between the front lens of the objective and the cover glass over the specimen. Thus, by using water-immersion and oil-immersion lenses, the resolving power can be further increased.

Eyepieces equipped with a measuring scale are known as *micrometer eyepieces*. The scale may be calibrated in absolute units by means of a stage micrometer, which is viewed in the same manner as any object. The scale must be calibrated for each objective used. The filar micrometer is almost indispensable for measurements of small lengths.

Binocular microscopes are advantageous from several points of view; they give a correct stereoscopic image, they prevent fatigue in prolonged studies, and with a properly constructed camera, very satisfactory stereophotomicrographs can be made.

Phase-contrast microscopy began with Zernicke.[13,14] It enhances the contrast of the image which is observed, making possible an observation of sections too thin or too transparent for the conventional techniques. It is also valuable in the microscopic determination of refractive index.

Microscopes for use in the chemical laboratory should be provided with polarizing and analyzing Nicol prisms. Very important information is often secured by examination of substances in polarized light. Special microscopes known as *chemical microscopes* have such prisms and other useful accessories.

ELECTRON MICROSCOPE.   The limiting factor in the resolution of small objects by the optical microscope is the wavelength of the light. The details of material under examination cannot be seen if they are smaller than the wavelength of the light used in the observation. A beam of monoenergetic electrons can exhibit interference effects characteristic of a wave motion with a wavelength λ of

$$\lambda = \frac{300hc}{Ee}$$

where E = accelerating potential, volts
$\quad\quad e$ = electron charge, esu
$\quad\quad h$ = Planck's constant
$\quad\quad c$ = velocity of light

It is readily calculated that a beam of electrons can have a wavelength of a fraction of an angstrom unit, as against 4000 to 8000 Å for the wavelength of visible light. Moreover, the electron beam can be bent by an electrostatic or an electromagnetic field so that it can be focused by simple adjustment of current and voltage. These principles have been applied in the design of electron microscopes,[15-17] and many excellent pictures have been taken with them. Bacteria, fine powders such as carbon black, fibers, and many things of biological interest have been examined with enormously greater magnification than was previously possible, and new details of structure have been revealed.

The principle of the electron microscope is shown in Fig. 130. Electrons are emitted from a hot filament accelerated by fields of 30,000 to 100,000 volts and

Electron
source

Magnetic condenser

Speciman

Magnetic objective

Intermediate image
projector

Magnified image

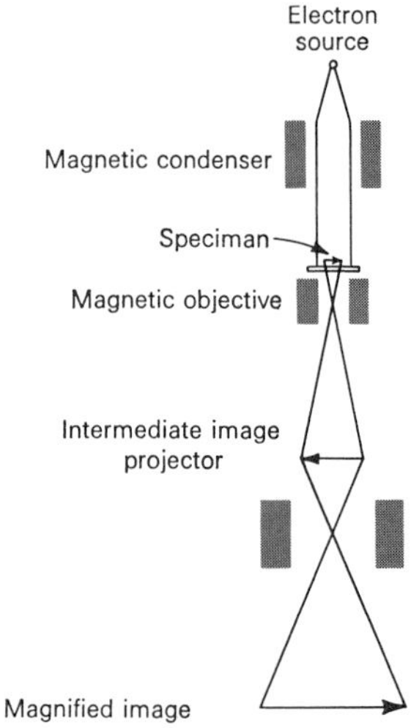

figure 130.

*Principle of the electron
microscope.*

focused with magnetic fields as indicated. The object is placed on a thin nitrocellu-lose film, which is fairly transparent to the beam and shows no structure of its own. The electrons pass through and are focused on a fluorescent plate, where the image is viewed by eye. The electrons are scattered by the denser parts of the object, but the transparent parts of the object show up more brightly. A photographic plate is then substituted for the fluorescent screen, and a short exposure gives a satisfactory picture. The monograph of Hall[15] and the chapter by Hamm[16] are but two sources of much information about the physical principles involved in the design of electron microscopes and the ways in which they are applied in the construction of the instrument.

The details in an electron-microscope picture sometimes may be brought out more clearly by producing "shadows" of the more prominent parts of an object. Vapors of a heavy element such as uranium may be directed across the object in an evacuated space, or the object may be embedded in a thin layer of an inorganic gel consisting of a compound of a heavy element. The less dense, or thinner, parts of the object then transmit the electron beam more readily, thus producing the shadows and revealing finer details of the structure.

POLARIMETRY

The development first of carbohydrate chemistry and later of protein chemistry and steroid chemistry is associated with polarimetry. When ordinary unpolarized light passes through a Nicol prism at the proper angle to the optic axis, only the plane-polarized extraordinary ray is transmitted. If this emergent ray is made to fall upon another Nicol, the ray will pass through if it has parallel orientation. If the second prism is rotated through 90°, the plane-polarized light is extinguished and the Nicols are said to be *crossed*. Such Nicols with light source, sample tube, and eyepiece form the basic elements of the polarimeter.

On passage through optically anisotropic liquids or solutions, a change in the direction of vibration of the linearly polarized light may be produced. Pasteur's principle that a compound will be optically active if its structure cannot be brought into coincidence with its mirror image is quite general; the presence of an asymmetric carbon atom is not *required*. This property will occur for fluids in which the individual molecules have neither a plane of symmetry nor a center of symmetry. Another case of optical activity, now of special interest, is the molecular dissymmetry due to the presence of a helical structure, producing an optical activity which is superimposed upon that due to the asymmetric carbon atoms in polypeptides, pro-teins, nucleic acids, etc.

A recent advance in instrumentation, the photoelectric spectropolarimeter of

Rudolph,[18] has made possible a very rapid extension of studies related to optical rotation and its dispersion. The original Rudolph instrument is shown in Fig. 131.

The function of the monochromator is to supply light of the desired wavelength from one of the sources. Beginning at the light source, upper left, the light (dotted line) traverses the monochromator and is reflected into the polarimeter and passes the quartz polarizer prism, the solution, and the quartz analyzer prism. It is then received at the phototube.

For researches in circular dichroism, the Roussel-Jouan dichrograph[19] permits measurements in the spectral range 220 to 600 mμ. A more flexible instrument, described by Grosjean and Tari,[20] covers the range 185 to 600 mμ and is equipped with an automatic recording device.

Several additional and more general discussions appear in the references.[21-23]

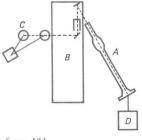

figure 131.

Rudolph photoelectric spectropolarimeter: A = circular-scale high-precision polarimeter; B = quartz monochromator; C = zirconium and xenon arcs; D = photovolt multiplier tubes, with photometer.

References

1. N. Bauer, K. Fajans, and S. Z. Lewin in A. Weissberger (ed.), "Technique of Organic Chemistry," vol. 1, "Physical Methods of Organic Chemistry," 3d ed., pt. 2, Interscience Publishers, Inc., New York, 1960.
2. P. P. Debye, *J. Appl. Phys.*, **17**: 392 (1946); B. A. Brice and M. Halwer, *J. Opt. Sci. Am.*, **41**: 1033 (1951); H. Svensson, *Anal. Chem.*, **25**: 913 (1953).
3. H. Svensson and T. E. Thompson, Translational Diffusion Methods in Protein Chemistry, in P. Alexander and R. J. Block (eds.), "Analytical Methods of Protein Chemistry," vol. 3, pp. 57–118, Pergamon Press, New York, 1961.
4. T. Svedberg and K. O. Pedersen, "The Ultracentrifuge," Oxford University Press, Fair Lawn, N.J., 1940.
5. L. G. Longsworth, *Ind. Eng. Chem. Anal. Ed.*, **18**: 219 (1946).
6. G. Kegeles and L. J. Gosting, *J. Am. Chem. Soc.*, **69**: 2516 (1947); L. J. Gosting and L. Onsager, *ibid.*, **74**: 6066 (1952).
7. C. A. Coulson, J. T. Cox, A. G. Ogston, and J. St.L. Philpot, *Proc. Roy. Soc. London Ser. A*, **192**: 382 (1948).
8. L. G. Longsworth, *Anal. Chem.*, **23**: 346 (1951); *Rev. Sci. Instr.*, **21**: 524 (1950).
9. J. M. Creeth, *J. Am. Chem. Soc.*, **77**: 6428 (1955); *J. Phys. Chem.*, **62**: 66 (1958).
10. P. Debye, *J. Phys. Colloid Chem.*, **51**: 18 (1947); B. Zimm, *J. Chem. Phys.*, **16**: 1099 (1948); P. Bender, *J. Chem. Educ.*, **29**: 15 (1952).
11. K. A. Stacey, "Light Scattering in Physical Chemistry," Academic Press Inc., New York, 1956.
12. G. Oster, *Chem. Rev.*, **43**: 319 (1948); in A. Weissberger (ed.), "Physical Methods of Organic Chemistry," 3d ed., pt. 3, Interscience Publishers, Inc., New York, 1960.
13. F. Zernicke in A. Bouwers (ed.), "Achievements in Optics," American Elsevier Publishing Company, New York, 1946.
14. A. H. Bennett, H. Jupnick, H. Osterberg, and O. W. Richards, "Phase Microscopy," John Wiley & Sons, Inc., New York, 1951.
15. C. E. Hall, "Introduction to Electron Microscopy," 2d ed., McGraw-Hill Book Company, New York, 1966.

16. F. A. Hamm in A. Weissberger (ed.), "Technique of Organic Chemistry," vol. 1, "Physical Methods of Organic Chemistry," 3d ed., pt. 2, Interscience Publishers, Inc., New York, 1960.
17. C. E. Hall and H. S. Slayter, *J. Biophys. Biochem. Cytol.*, **5:** 11 (1959).
18. H. C. Rudolph, *J. Opt. Soc. Am.*, **45:** 50 (1955).
19. L. Velluz and M. Legrand, *Angew. Chem.*, **73:** 603 (1961).
20. M. Grosjean and M. Tari, *Comp. Rend.*, **258:** 2034 (1964).
21. P. Crabbè, "Optical Rotary Dispersion and Circular Dichroism in Organic Chemistry," Holden-Day, San Francisco, 1965.
22. C. Djerassi, "Optical Rotatory Dispersion," McGraw-Hill Book Company, New York, 1960.
23. N. Klyne, *Advan. Org. Chem.*, **1:** (1960).

Chapter 22

Thermal Measurements

THERMOMETRY†

THE INTERNATIONAL TEMPERATURE SCALE.   Although the constant-volume hydrogen thermometer is the standard on which the Celsius (centigrade) and absolute scales are based, such a thermometer is inconvenient to use except as an ultimate standard. The international temperature scale is defined by several fixed points (Table 1), and means for interpolating between the fixed points are provided.

Table 1. *Fixed Points of the International Temperature Scale*
All at 1 atm pressure (760 mm Hg)

Substance	Designation	Temperature, °C
Oxygen	Boiling point	−182.97
Water (air-saturated)	Freezing point	0.000
Water	Boiling point	100.000
Sulfur	Boiling point	444.60
Antimony	Freezing point	630.50
Silver	Freezing point	960.5
Gold	Freezing point	1063

From the ice point to 660° the international Celsius temperature is computed from the resistance R_t of a standard platinum resistance thermometer by means of the equation

$$R_t = R_0(1 + At + Bt^2) \tag{1}$$

The constants R_0, A, and B are determined by calibration at the ice, steam, and sulfur points. From −190° to the ice point a platinum resistance thermometer is also used, with a modified interpolation formula:

$$R_t = R_0[1 + At + Bt^2 + C(t - 100)t^3] \tag{2}$$

R_0, A, and B are the same as before, and C is determined by resistance measurement at the oxygen point.

From 660 to 1063° the temperature is computed from the electromotive force E of a standard platinum versus platinum-rhodium thermocouple by means of the equation

$$E = a + bt + ct^2 \tag{3}$$

where a, b, and c are obtained from measurements at the antimony point, silver point, and gold point.

† A standard reference book on this subject is Ref. 1; see also Ref. 2.

Above the gold point an optical method, based on the Wien formula for black-body radiation, is used.

The international temperature scale assures laboratories throughout the world of an accurate and readily reproducible basis of temperature measurement. It should be pointed out, however, that the relationship between absolute Celsius and international scales is subject to such changes as are occasioned by production of new materials of higher purity or more refined experimental techniques in gas thermometry.

MERCURY THERMOMETERS.   The mercury thermometer is the simplest and most widely used instrument for measuring temperature. Mercury is particularly suitable because it has a very uniform coefficient of expansion, it does not wet glass, it is easily purified, and the thermometer is easily read. At atmospheric pressure it remains liquid from -40 to $+357°C$.

Thermometers of various grades and ranges are available, including (*a*) 0 to 100, 250, and 360°, graduated in degrees for general purposes; (*b*) sets of thermometers from -40 to $+400°$, each having a range of 50° and graduated to 0.1°; (*c*) 18 to 28°, graduated to 0.01°, or 17 to 31°, graduated to 0.02°, for calorimetric work; (*d*) -5.0 to $+0.5°$, graduated to 0.01° for freezing-point lowering; (*e*) Beckmann type thermometers with adjustable range, graduated to 0.01°; (*f*) high-temperature thermometers, in which special combustion glasses or quartz are used with nitrogen or argon under pressure to extend the upper temperature limit as high as 750°C.

The graduations should extend a little beyond the nominal limits, and a high-grade thermometer of any range has graduations for ice-point standardization if actual temperatures are to be measured. If only differences in temperature are required, the ice point is not necessary.

Reading. Exposed stem, parallax, and sticking mercury constitute three important sources of error in the reading of thermometers.

Thermometers are usually calibrated for total immersion of the mercury, and a correction is necessary when part of the stem is exposed. The thermometer will read too low if the air surrounding the stem is colder than the bath in which the bulb is immersed and too high if the air is warmer. A second thermometer is placed near the exposed stem, and the stem correction S is given by the formula

$$S = 0.00016n(t' - t)$$

where n = length of exposed mercury column in terms of scale degrees
   t' = temperature of bath
   t = average temperature of emergent stem
The factor 0.00016 is suitable for the glass used in most thermometers.

It should be emphasized that stem corrections are not accurate, and for very high or very low temperatures a considerable error may be introduced. It is better to avoid the exposed stem by improving the experimental conditions.

Parallax may cause an erroneous reading, depending upon the extent to which the eye is below or above the level of the top of the mercury thread. It may be eliminated completely by reading the thermometer from a distance with a cathetometer (telescope and cross hair), and it may be reduced considerably by carefully regulating the position of the eye. Special thermometer lenses sliding along the stem are helpful.

A thermometer should be read whenever possible with a rising thread rather than a falling thread, and in either case it is necessary in accurate work to tap the thermometer gently before reading, to prevent sticking.

Standardization. In the simplest method, the thermometer is compared with a standard thermometer, for which purpose a thermometer certified by the National Bureau of Standards is useful. The two thermometers are set side by side in a thermostat, vapor bath, or large, well-stirred body of liquid. The National Bureau of Standards thermometer is immersed nearly to the top of the thread, and the other thermometer is immersed to the depth at which it is to be used. The true temperature as given by the standard thermometer is obtained by adding or subtracting the correction indicated on the National Bureau of Standards certificate, *after correction for a possible change in the ice-point reading.*

The ice-point reading is always taken before a standardization, and it should check with the ice point given on the certificate, if the corrections as given in the certificate are to apply. In case the ice-point reading does not check, a constant adjustment of all the corrections is necessary.

In checking a thermometer at the ice point, it is necessary to have the ice very finely divided and intimately distributed throughout the whole bath. A large layer of water is not allowed to accumulate, and the spaces between the pieces of ice must be filled with water, not air. For ordinary work, finely shaved, close-packed ice in distilled water makes a satisfactory ice bath. For precision work, the recommendations of White[3] should be followed.

Fixed boiling points, freezing points, or transition temperatures of pure materials are also used for standardizing thermometers (Table 1). Calibration against a platinum resistance thermometer certified by the National Bureau of Standards is the best method.

The standardization of a mercury-in-glass thermometer should be rechecked frequently at one point, usually the ice point. Slow permanent changes in the glass result in changes in the volume of the bulb. Furthermore, temporary changes in the bulb volume are likely to result from heating; the bulb may not regain its original volume for several days.

Beckmann Thermometer. A Beckmann thermometer is shown in Fig. 132. This instrument reads directly to 0.01° and can be estimated to 0.001°. Its range is only 5 or 6°, but it can be set for any temperature by adjusting the mercury in the reservoir at the top of the scale. The thermometer is warmed until sufficient mercury has been driven over into the reservoir, and it is then given a sharp tap with the hand to break the thread at the entrance to the reservoir.

The thread is broken when the temperature is a little above the desired temperature, because a certain amount of cooling is necessary to bring the mercury back on the scale.

figure 132.

Beckmann thermometer.

OTHER LIQUID THERMOMETERS.   If thallium is added to mercury to give an 8.5 percent solution, the amalgam can be cooled to −60° before freezing. Liquid pentane can be used down to liquid-air temperatures, and toluene can be used to −100°, which is below the temperature of Dry Ice.

BIMETALLIC THERMOMETERS.   In this type of thermometer the temperature is indicated on a dial by a pointer actuated by the differential expansion of a bimetallic strip. These thermometers are usually accurate to only about 1 percent, but they are rugged and useful where high accuracy is not important.

GAS THERMOMETERS.   Gas thermometers are inconvenient to use for any purpose other than ultimate standardization. For such purposes the experimental arrangement and the calculations are quite complex.[4-6]

RESISTANCE THERMOMETERS.   The electrical resistance of a wire increases in a regular manner as the temperature rises, and since the resistance of a wire can be measured with great precision, this measurement offers an accurate method for determining temperatures.

Platinum wire is commonly used because of its chemical inertness and its high resistance. It must be of the highest grade, carefully purified, and annealed by heating to redness with an electric current. It is annealed again at a lower temperature after winding.

The wire is wound on mica supports in such a way that the metal is subjected to as slight a strain as possible when the thermometer is heated or cooled. Usually the coil is enclosed in a sealed glass or quartz tube; when it is desired to minimize the lag of the thermometer, the coil is enclosed in a flattened metal case. Platinum resistance thermometers are usually manufactured with a resistance of 25.5 or 2.55 ohms at 0°. The resistance then changes by about 0.1 or 0.01 ohm deg^{-1}, respectively.

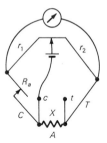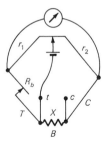

figure 133. Circuit diagram illustrating resistance measurements of a four-lead resistance thermometer.

In accurate resistance thermometry, it is essential that the resistance from which the temperature is computed be that of the thermometer element itself and contain no contribution from the leads. This is accomplished by using a four-lead thermometer and a suitable bridge, in which the lead resistance is effectively eliminated by a switching arrangement. In Fig. 133, r_1 and r_2 are equal ratio arms, C, c, T, and t are the four leads, X is the thermometer resistance, and R_a and R_b are the resistances required to balance the bridge in the two arrangements. It is evident that $R_a + C = X + T$ and $R_b + T = X + C$. Hence, $X = (R_a + R_b)/2$.

If temperatures are to be determined to a precision of $\pm 0.001°$, it is necessary to employ a carefully calibrated bridge in which the effect of switch contact resistance variation has been eliminated.[†]

For computing temperatures from observed resistances, the equation

$$t = \frac{100(R_t - R_0)}{R_{100} - R_0} + \delta\left(\frac{t}{100} - 1\right)\frac{t}{100} \tag{4}$$

is more convenient than Eq. (1). The constants R_0, $R_{100} - R_0$ (the *fundamental interval*), and δ are determined by calibration at the ice, steam, and sulfur points. This calibration is performed by the National Bureau of Standards for a reasonable charge. The value of t is calculated from R_t by successive approximations, a process which is not difficult because of the relatively small value of δ. Convenient tables for converting platinum-resistance values to degrees Celsius have been published.[8]

A carefully constructed and standardized platinum resistance thermometer is preeminent in the field of thermometry for its reliability in both accuracy and precision over long periods of time. Other resistance thermometers can be made of nickel, copper,[9] Hytemco,[§] and other metals and alloys in cases where extreme accuracy is subordinate to economy and ease of construction.

[†] For a discussion of this point and other factors in precision resistance thermometry, see Ref. 7.

[§] An alloy with an exceptionally high temperature coefficient of resistance, manufactured by the Driver-Harris Co., Harrison, N.J.

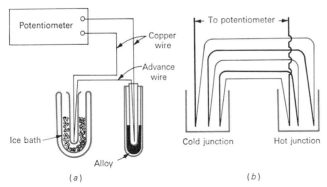

figure 134. (a) Use of the thermocouple in alloy-cooling-curve studies; (b) a three-junction thermel.

The resistance of a thermometer may also be determined by comparing with a potentiometer the potential drop across the thermometer with that across a standard resistor in series with the resistance thermometer, while a small current is flowing.

Resistance thermometers can also be made of semiconductors in which the resistance *decreases* very sharply as the temperature increases. Such thermometers are called *thermistors*. They permit the use of a less accurate bridge for temperature measurement and are suitable for small temperature *differences*.

THERMOCOUPLES.†   In a closed circuit of dissimilar metals, a current is generated when the junctions are at different temperatures. A simple thermocouple is shown in Fig. 134, together with a multiple-junction assembly which is called a *thermel*. The heavy line represents one metal, and the lighter line represents the other.

In Fig. 134 the use of the thermocouple in determining the freezing curves of alloys is illustrated. The potentiometer shown is a commercial instrument of moderate precision made expressly for use with thermocouples.

The common types of thermocouples and their important characteristics are illustrated in Table 2. Extended tables of electromotive force as a function of temperature for the different types are found in many of the standard handbooks and in a circular published by the U.S. National Bureau of Standards.[11]

Copper and constantan may be soldered together by using rosin or other non-corrosive flux, but the other metals are welded together in an oxy-gas flame or in an electric arc. The two wires are twisted together for a short distance, held in an insulated clamp, and connected to one pole of the electric circuit (110 volts). An insulated carbon rod is connected through a suitable resistance (15 ohms) to the

† For a comprehensive discussion, see Ref. 10.

Table 2. Types of Thermocouples

Type	Usual temperature range, °C	Maximum temperature, °C (for short periods)	Millivolts per degree at room temperature
Copper-constantan[a]	−190–300	600	0.0428
Iron-constantan[a]	−190–760	1000	0.0540
Chromel-P–alumel[b]	−190–1100	1350	0.0410
Platinum to platinum—10 percent rhodium	0–1450	1700	0.0064

[a] Constantan is a general name given to a group of copper-nickel alloys. It can be obtained under the trade name Advance from the Driver-Harris Co., Harrison, N.J., or from pyrometer manufacturers.

[b] Chromel-P and alumel are high-nickel alloys obtainable from the Hoskins Manufacturing Co., Detroit, or from pyrometer manufacturers.

other electrode. The rod is touched to the end of the thermocouple and pulled away slightly, giving an electric arc. As soon as the two wires are welded together, the electrode is pulled farther away to stop the arc.

For room temperature, insulation of cloth or enamel is sufficient; for high temperatures, wire can be obtained with Fiberglas insulation, or separate sleeving of the latter can be obtained. Porcelain tubes are also widely used. It is essential to protect the wires carefully from corrosion; in furnaces, long gastight tubes of glass, porcelain, or quartz are used. For work with solutions at room temperatures, the thermocouple is usually encased in a thin glass tube, frequently filled with oil to give better thermal contact. When several junctions are used, the exposed junctions come at higher and higher levels in the encasing tube to prevent short-circuiting of the uninsulated tips of the thermocouples.

For precision work, the wire must be carefully selected and tested. Full particulars for making and using these thermoelectric thermometers are available.[12] They can be made sensitive to 0.00001°.

A thermoelectric thermometer is used either with a potentiometer or with a millivoltmeter or galvanometer. The former is necessary in precision work, but the latter is so much more convenient that it is frequently used in industrial work. In the latter case, the current rather than the electromotive force is measured, and an error is introduced because the change in current with temperature is due not only to the thermocouple electromotive force but also to the change in resistance of the wires. To minimize this error, large wires are used, so that the changeable resistance of the thermocouple is small in comparison with the fixed resistance of the galvanometer. Galvanometer or millivoltmeter scales may be calibrated directly in terms of degrees. The resistance of the thermocouple should be about equal to the critical-damping resistance of the galvanometer.

Thermocouples are used extensively in many industrial operations. Recording potentiometers are commonly employed where a permanent record of temperature at all times is desired. Instruments are available which will record as many as 24 temperatures as a function of time on a single strip of paper.

For accurate work the potentiometer is specially designed to avoid spurious thermal electromotive forces or leakage. Reversing switches are particularly useful in this work. The White potentiometer,[†] Wenner potentiometer,[§] and microvolt potentiometer[§] are specially designed for use with thermocouples.

The cold junction is usually set into cracked ice in a vacuum-jacketed bottle as shown in Fig. 134, but in crude work at high temperatures it is sometimes left at room temperature. A fluctuation in the temperature of the cold end is, of course, just as effective as one in that of the furnace; the meter reading depends on the difference in temperature between the two junctions.

Thermocouples are calibrated with fixed temperatures, of which the most common are freezing mercury, $-38.87°C$; melting ice, $0°$; boiling water, $100°$ (with barometer correction); the transition temperature of sodium sulfate, $32.38°$; boiling sulfur, $444.6°$; freezing tin, $231.9°$; freezing cadmium, $320.9°$; freezing lead, $327.3°$; freezing zinc, $419.5°$; freezing silver, $960.8°$; freezing gold, $1063.0°$; freezing palladium, $1552°$; and freezing platinum, $1768°$. The National Bureau of Standards calibrates thermocouples, determines the thermal electromotive forces of thermocouple materials, and supplies a few metals with certified freezing points.

OPTICAL PYROMETERS.[13]   The operation of optical pyrometers depends on the fact that the radiation emitted by a hot body is a function of the temperature. They are very easy to use and are popular in industrial control operations. They are about the only instruments that can be used for the measurement of very high temperatures. They are not suitable for temperatures below 500 or $600°$, because the radiation is not sufficiently intense.

Several types are available, but the disappearing-filament type shown in Fig. 135 is one of the simplest and most practical.

The furnace, crucible, or other hot object is viewed through the telescope. An electric light bulb with a carbon filament is placed in the optical system so that the observer sees the filament across the field. A variable resistance changes the current through the lamp. The wire appears bright on a darker field when the wire is hotter than the hot object, and it appears dark on a light field when it is colder. When the two temperatures are exactly the same, the filament seems to disappear and the whole field becomes uniform. A red screen is used, and for very high temperatures the brightness of the radiation from the furnace is reduced with a thicker screen. The

† Leeds and Northrup Company, Philadelphia.
§ Honeywell, Inc., Minneapolis, Minn.

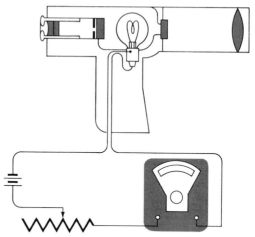

figure 135. Radiation pyrometer of the disappearing-filament type.

current required to make the filament disappear is noted on the milliammeter, and the corresponding temperature is obtained by reference to a table supplied by the manufacturers for use with the lamp.

The table is prepared by calibrating the lamp at known temperatures, reading the milliammeter when the rheostat is so adjusted as to make the filament disappear. A few readings are sufficient, the rest being obtained by interpolation. The temperature of the furnace may be determined with a thermocouple or resistance thermometer. The melting points of antimony, silver, and gold may be used for calibration temperatures.

The furnace should be nearly closed and viewed through a small opening so as to give a good approximation to true blackbody radiation. Otherwise, the calculated temperature may be considerably too low. Empirical corrections for radiation from the surface of exposed platinum and various other metals have been worked out.

CALORIMETRY

Calorimetric measurements, several examples of which are treated in Experiments 3 to 6, form an important part of experimental physical chemistry. A comprehensive survey of this field has been given by Sturtevant.[14] The specific details of the equipment and procedures employed in accurate work depend upon the particular type of measurement being made. Representative references on typical calorimetric problems include the following: heats of combustion (general;[15] liquids;[16] gases[17]); heat capacities of solutions;[18] heats of dilution;[19] heats of hydrogenation;[20] heat capacity of liquids;[21] heats of vaporization;[22] heat capacity of gases;[23] ice calorimeter;[24] low-temperature heat capacity; heat of fusion.[25]

Comprehensive collections of review articles on a variety of calorimetric techniques are now available.[26]

References

1. W. F. Forsythe, "Temperature: Its Measurement and Control in Science and Industry," Reinhold Publishing Corporation, New York, 1962.
2. E. Griffiths, "Methods of Measuring Temperature," 3d ed., Charles Griffin & Company, Ltd., London, 1947; J. M. Sturtevant in A. Weissberger (ed.), "Technique of Organic Chemistry," vol. 1, "Physical Methods of Organic Chemistry," 3d ed., pt. 1, chap. 6, Interscience Publishers, Inc., New York, 1959.
3. W. P. White, *J. Am. Chem. Soc.*, **56:** 20–24 (1934).
4. F. E. Keyes in Ref. 1, p. 45.
5. J. R. Roebuck and T. A. Murell in Ref. 1, p. 60.
6. C. S. Crego in Ref. 1, p. 89.
7. E. F. Mueller in Ref. 1, p. 162; "Precision Resistance Thermometry," and "Resistance Thermometers," Leeds and Northrup Co., Philadelphia.
8. F. D. Werner and A. C. Frazer, *Rev. Sci. Instr.*, **23:** 163 (1952).
9. C. G. Maier, *J. Phys. Chem.*, **34:** 2860 (1930).
10. W. F. Roeser in Ref. 1, p. 180; W. F. Roeser and S. T. Lonberger, Methods and Testing of Thermocouples and Thermocouple Material, *Natl. Bur. Std. U.S. Circ.* 590, 1958.
11. R. J. Coruccini and H. Shenker, *J. Res. Natl. Bur. Std. U.S.*, **50:** 229 (1953); *Res. Paper* 2415; *Natl. Bur. Std. U.S. Circ.* 561, 1955.
12. L. H. Adams, *J. Am. Chem. Soc.*, **37:** 481 (1915); "International Critical Tables." vol. I, p. 57, McGraw-Hill Book Company, New York, 1926.
13. W. E. Forsythe, Optical and Radiation Pyrometry in Ref. 1, pp. 1115ff.
14. J. M. Sturtevant in A. Weissberger (ed.), "Technique of Organic Chemistry," vol. 1, "Physical Methods of Organic Chemistry," 3d ed., pt. 1, chap. 10, Interscience Publishers, Inc., New York, 1959.
15. J. Coops et al., *Rec. Trav. Chim.*, **66:** 113–176 (1947); H. C. Dickinson, *Natl. Bur. Std. U.S. Bull.* 11, p. 189, 1915; R. S. Jessup and C. B. Green, *J. Res. Natl. Bur. Std. U.S.*, **13:** 469 (1934).
16. E. J. Prosen and F. D. Rossini, *J. Res. Natl. Bur. Std. U.S.*, **27:** 289 (1941).
17. E. J. Prosen, F. W. Maron, and F. D. Rossini, *J. Res. Natl. Bur. Std. U.S.*, **42:** 269 (1949).
18. F. T. Gucker, F. D. Ayres, and T. R. Rubin, *J. Am. Chem. Soc.*, **58:** 2118 (1936).
19. F. T. Gucker, H. B. Pickard, and R. W. Planck, *J. Am. Chem. Soc.*, **61:** 459 (1939); J. M. Sturtevant, *J. Phys. Chem.*, **45:** 127 (1941).
20. G. B. Kistiakowsky et al., *J. Am. Chem. Soc.*, **57:** 65 (1935).
21. D. W. Osborne and D. C. Ginnings, *J. Res. Natl. Bur. Std. U.S.*, **39:** 453 (1947).
22. G. Waddington, S. T. Todd, and H. M. Huffman, *J. Am. Chem. Soc.*, **69:** 22 (1947).
23. G. Masi and B. Petkof, *J. Res. Natl. Bur. Std. U.S.*, **48:** 179 (1952).
24. D. C. Ginnings and R. J. Corruccini, *J. Res. Natl. Bur. Std. U.S.*, **38:** 583 (1947).
25. J. G. Aston and M. L. Eidenoff, *J. Am. Chem. Soc.*, **61:** 1533 (1939); H. M. Huffman, *Chem. Rev.*, **40:** 1 (1947); R. A. Ruehrwein and H. M. Huffman, *J. Am. Chem. Soc.*, **65:** 1620 (1943).
26. IUPAC Committee on Chemical Thermodynamics, "Experimental Thermochemistry Techniques," F. D. Rossini (ed.), vol. 1, 1956; H. A. Skinner (ed.), vol. 2, 1962; Interscience Publishers, Inc., New York.

Chapter 23

Physical Properties of Fluids

PHYSICAL PROPERTIES OF GASES

VOLUME BY DISPLACEMENT.   A measured quantity of gas may be introduced into a system by displacing it with a measured quantity of liquid, in a flask provided with a two-holed rubber stopper. In an alternative method, the gas may be drawn over by running out a measured volume of liquid. The gas must be insoluble in the liquid. Mercury is the best liquid for most purposes, but it is too heavy for large volumes, and water, oil, nitrobenzene, and sulfuric acid have been used. Some of these liquids have a negligible vapor pressure, but a correction is necessary in the case of water. The temperature of the incoming liquid must be the same as that of the vessel.

Large volumes of gas are measured conveniently with commercial gas meters, in which cups or vanes rotate in a closed chamber containing a liquid at the bottom. The number of revolutions is recorded on one or more circular scales. The meters are calibrated with known volumes of gas.

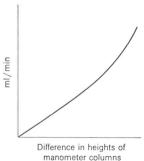

figure 136.

Flowmeter.

FLOWMETERS.   The rate of flow of a gas is conveniently measured with a flow-meter, as illustrated in Fig. 136. The difference in pressure on the two sides of a capillary tube, as indicated by the manometer, is a measure of the rate of flow of gas through the flowmeter. The flowmeter is calibrated at several different rates of flow, and a smooth curve is drawn showing the rate of flow as a function of the difference in levels (Fig. 137).

A convenient flowmeter is available which is made entirely of Pyrex glass. A graduated tube is mounted vertically in a larger concentric tube, and the displacement of the liquid due to the passage of gas is read directly on the tube. The rate of flow is nearly a straight-line function of the scale reading, and the rate of flow corresponding to any reading of the scale may be interpolated with accuracy. The factors involved in the theory and use of the flowmeter have been discussed by Benton.[1]

In calibrating a flowmeter, air or other gas is passed through the flowmeter by a displacement method, while the liquid used in the manometer is maintained continuously at a constant setting. The time taken for a given volume to pass through is determined accurately with a stopwatch. Capillary tubes of different bores may be used for different velocity ranges.

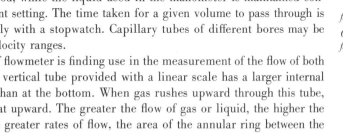

Difference in heights of manometer columns

figure 137.

Calibration curve for flowmeter.

Another type of flowmeter is finding use in the measurement of the flow of both gases and liquids. A vertical tube provided with a linear scale has a larger internal diameter at the top than at the bottom. When gas rushes upward through this tube, it carries a small float upward. The greater the flow of gas or liquid, the higher the float rises, but at the greater rates of flow, the area of the annular ring between the float and the containing tube becomes larger. These factors are so balanced that the height of the float is a linear function of the rate of flow. The Flowrator,† or

† Fischer and Porter Co., Hatboro, Pa.

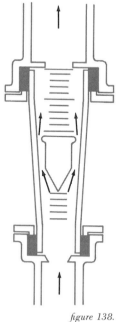

figure 138.

*Rotameter, or Flowrator,
for measuring the rate of
flow of gases or liquids.*

Rotameter,† diagramed in Fig. 138 is a convenient meter of this type. Laboratory kits with a set of interchangeable metering tubes and floats of different sizes are available with capacities from 0.06 to 220 ml of liquid water per minute. The meters can be used for fluids other than water and air, with the help of standard corrections, depending on the density of the fluid and the float and on viscosity and pressure.

MANOMETERS.   The mercury manometer is the instrument most commonly used for the measurement of pressure in the laboratory. It may be used for absolute measurements, by evacuation of the space over the liquid level in one arm, or for determination of the differential pressure between two levels. The range of pressure covered usually extends from a few centimeters to several meters of mercury for a single manometer; multiple manometers[2] connected in a cascade arrangement have been used for very accurate absolute-pressure measurements at as high as 200 atm.

An initially good vacuum over the mercury surface in the closed end of a mercury manometer can be obtained only by careful outgassing of the glass walls before the tube is filled, followed by vacuum distillation of mercury into the manometer while the tube is still fairly hot, i.e., over 100°C. The vacuum inevitably deteriorates with age, however, because of motion of gas along the interface between the tube wall and the mercury and further outgassing of the tube wall due to electric-discharge effects associated with the motion of the mercury resulting from changes in applied pressure. For best results it is desirable to evacuate the "closed end" continuously by means of a diffusion pump.

Because of the large surface tension of mercury, relatively large-internal-diameter tubes (about 20 mm) are required to give menisci adequately flat to permit accurate measurements and to minimize the uncertainty due to unequal surface-tension effects at the two menisci. It is often assumed that a common tube diameter for the two arms will provide a cancellation of such surface-tension effects. This is true only as a rough approximation, since variations in the conditions at the interface can result in a wide variation in the capillary depression for a given tube diameter. When the highest accuracy is required, the individual meniscus heights must be measured and the appropriate corrections applied; correction tables which facilitate this process are available.[3]

The accuracy of measurement of the difference in height of the two mercury levels is determined not only by the intrinsic accuracy of the scale used but also by the method by which the readings are obtained. The best procedure is to use a scale located in the (vertical) plane of the manometer which is read by means of a cathetometer telescope from a distance of several feet. This procedure is preferable to basing measurements on the cathetometer scale itself, which introduces the difficulty of assuring that the telescope axis is accurately horizontal.

† Brooks Rotameter Co., Lansdale, Pa.

In either case, the measured height difference h_t is obtained in terms of scale readings at the ambient temperature t. Since the scale can read correctly only at its calibration temperature t_c, the true height difference must be calculated from a knowledge of the linear coefficient of expansion s of the scale material:

$$h = h_t[1 + s(t - t_c)]$$

Application of the scale and capillary corrections gives the pressure in terms of millimeters of mercury at the manometer temperature. In terms of standard millimeters of mercury,

$$P = \frac{\rho \, g}{\rho_s \, g_s} h$$

Here ρ and g represent the density of mercury and acceleration of gravity for the local manometer conditions, while the standard values of these quantities are

$$\rho_s = 13.5951 \text{ g cm}^{-3} \qquad g_s = 980.665 \text{ cm sec}^{-2}$$

Extensive tables have been formulated to facilitate correction for the effect of temperature on mercury density and the scale material.[3] The correction for gravity is normally less important but is essential in accurate work.

A comprehensive discussion of mercury barometers and manometers has been made available by the National Bureau of Standards.[3]

Other manometric fluids than mercury can be used, of course. The essential properties include a known density, sufficiently low viscosity and vapor pressure, inertness toward the gas whose pressure is being measured, and an adequately low solvent power for the gas. The manometer readings can be expressed in terms of standard millimeters of mercury from a knowledge of the densities of the two fluids.

BOURDON GAUGE.[4]   The active element in a Bourdon gauge is a hollow tube of elliptical cross section bent into a circular or spiral form. Higher pressure inside the tube than outside causes a deflection of the end of the tube due to its tendency to straighten out. As shown in Fig. 139a, this motion of the tube end is coupled mechanically to the dial pointer of the gauge. Modern Bourdon gauges are stable, rugged, and reliable and are available in corrosion-resistant metals for use with reactive fluids. Accuracy as high as one-tenth of 1 percent of full scale and sensitivity several times as good as this can be obtained. Ranges as low as 0.1 to 20 mm Hg and as high as 0 to 50,000 psi are available.

DEADWEIGHT GAUGE.[5]   The principle of the deadweight is illustrated in Fig. 139b. The pressure measurement is made in terms of a balance of force at the bottom of the piston A, which has been carefully lapped to a close sliding fit in cylinder B,

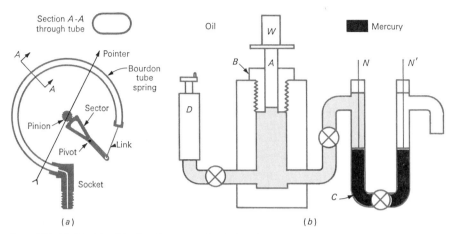

figure 139. (a) *Principle of the Bourdon gauge;* (b) *deadweight gauge assembly:* $A = piston;$ $B = cylinder;$ $C = steel$ U $tube;$ $D = oil$ $injector;$ $N, N' = indicator$ $contact$ $needles.$

from which it is separated by a very thin film of oil. The upward force exerted on the piston is given by the pressure acting there times the effective area of the piston. The downward force, in addition to that due to atmospheric pressure, is calculated as mass times acceleration, where the mass includes that of the piston, the weight pan, and the weights W required to achieve the balance condition and the acceleration is the local acceleration of gravity.

The effective area of the piston is determined by an effective diameter which is very nearly equal to the mean diameter for the piston and cylinder. On this basis the deadweight gauge can be used as an absolute instrument. More commonly, the effective piston area is determined experimentally by a gauge calibration against a known pressure. For this purpose the vapor pressure of liquid carbon dioxide at the ice point is used, measured very accurately by several workers.[6]

To prevent sticking of the piston in the cylinder, the piston may be oscillated back and forth by a yoke with eccentric drive. The yoke acts on a lever arm fastened to the piston; during part of each cycle the lever arm is not in contact with the yoke and the piston floats freely in the cylinder. Alternatively, the piston may be rotated continuously in the cylinder.

An unbalance in pressure will result in either a rise or fall of the piston, and this change in its position may be used as a balance indicator for work of moderate accuracy. Where maximum accuracy is required, the detection system may be based on the change in the volume below the piston resulting from its motion. Any such change will cause a displacement of the mercury in the steel U tube C, and thus provide a positive electrical connection between *one* of the insulated detector pins N, N'

and the steel tube, which may complete a relay actuating circuit. The mercury trap also serves to isolate the oil in the deadweight gauge from the medium whose pressure is to be measured.

The range of a deadweight gauge is determined by the nominal piston diameter, which usually is from $\frac{1}{2}$ to $\frac{1}{8}$ in. For accurate results the pressure range extends from about 2 atm to over 500 atm. The accuracy attained can be as high as 1 or 2 parts in 10,000, and the precision several times as good as that. Naturally, these figures refer to the pressure as measured at the gauge piston; for an arbitrary external system, all hydrostatic head contributions to the pressure between the system and the reference level at the bottom of the piston must be properly evaluated.

PUMPS.   Several types of vacuum pumps and certain vacuum techniques are described in Experiment 58.

The Toepler pump is used for transferring gas from one vessel to another under reduced pressure. It operates by alternately raising and lowering a mercury level as shown in Fig. 140. Each time the mercury level is lowered, the gas from A expands into B, and each time it is raised, the gas is forced from B into C or out into the room. The height of the mercury tube must be greater than the barometer height, or the mercury level may be raised by applying compressed air to the top of the mercury in the reservoir, thus permitting the reservoir to be just below the pump.

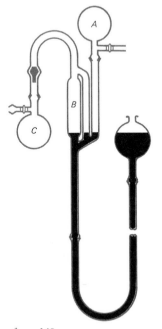

figure 140.

Toepler pump.

DETERMINATION OF THE BOILING POINT†

The boiling point of a liquid is a characteristic property which is often useful for purposes of identification and for the determination of purity. The boiling point is usually taken as the temperature established on the bulb of a thermometer on which a thin layer of the condensed liquid coexists with the vapor. Boiling liquids are generally superheated by the heating device used, and any boiling liquid is superheated because it boils under the pressure at the surface plus the hydrostatic pressure at the level in the liquid at which the temperature is measured. Special ebulliometers are required for the measurement of the boiling temperature of a liquid or of a solution. In the case of a pure liquid, the true boiling and condensation temperatures are equal.

If certain precautions are observed, the boiling point of a pure liquid may be determined in an ordinary distilling flask. The thermometer should be short, so that the whole column of mercury is surrounded by the vapor, or else a rather unsatisfactory stem correction is necessary (page 473). The thermometer bulb should be near the outlet tube so that it registers the temperature of the exit vapors. The boil-

† This subject has been reviewed in Ref. 7.

ing should not be so violent as to cause spray to reach the thermometer bulb or so rapid as to build up a pressure in the flask appreciably greater than atmospheric pressure. To avoid splashing, the flask should not be over half full.

One of the greatest sources of error comes from overheating the neck of the flask and heating the thermometer bulb by radiation. A large gas flame is particularly bad in this respect. A small gas flame without wire gauze or sand bath is better. It is a good plan to heat the distilling flask in a beaker of hot water (or oil, at higher temperatures) to a temperature only slightly above the boiling point of the liquid, as determined with an ordinary thermometer in the water bath. A thin-walled cylindrical tube placed in the vapor between the thermometer and the flask is sometimes used to minimize radiation.

Overheating may be minimized by using an electrical heating mantle covered with glass cloth or by using a heating coil immersed in the boiling liquid. The vapor rises from the electrically heated liquid and passes over into a side tube, completely enveloping the thermometer, which is suspended on a platinum wire attached to a glass hook on the stopper. A trap at the bottom may be used to return the liquid to the boiling flask. If the thermometer is completely within the vapor, a proper thermometer reading is ensured. Both stoppers may be made of glass if corks or rubber stoppers must be avoided.

This design gives good circulation of the liquid and allows the heating to be carried on until only a small volume of liquid remains. The heating coil of bare platinum wire is sealed through long glass tubes which pass down through the stopper of the flask. Copper wires leading to the source of current are welded to the platinum wires before the latter are sealed in glass.

SUPERHEATING.   The liquid should boil smoothly and steadily, but in some cases there is a tendency for the liquid to become heated above its boiling point. Superheating may be greatly reduced by any means of trapping small air bubbles in the liquid. Small chips of unglazed porcelain with air enclosed in the pores or pieces of platinum or platinized platinum are effective. The smoothest boiling is obtained with the internal electrical heating coil described above.

Distilling flasks covered with sintered glass powder are effective in preventing bumping.[8] Some of the same glass of which the flask is made is ground in a mortar and moved around inside while the flask is heated in a blast lamp to its softening temperature. Again, the surface may be coated with silica by evaporating a dilute solution of sodium silicate, heating to dull-red heat, cooling, and treating with dilute hydrochloric acid and rinsing.[9]

Bumping is particularly apt to be troublesome under reduced pressure, and it may be advisable to have a small stream of air bubbling through the liquid. A tube is inserted in the stopper, with its lower end drawn out into a capillary extending into the liquid. Its upper end is closed by a rubber tube and an adjustable pinchcock to control the rate of bubbling.

If only a small amount of liquid is available, the boiling point may be determined by the method of Smith and Menzies. The liquid is placed in an inverted bulb and fastened to the bulb of a thermometer and immersed in a bath of water or other transparent liquid which is immiscible with the liquid being studied. The temperature is raised gradually, and when the boiling point is reached, a stream of bubbles issues from the bulb.

EBULLIOMETERS.   An ebulliometer is a special apparatus for the measurement of the boiling temperature of a liquid or of a solution. With a differential ebulliometer, the boiling and condensation temperatures may be measured simultaneously. This apparatus may also be used for the determination of the degree of purity of liquid substances, of the molecular weight of a nonvolatile solute, or of the pressure coefficient of the boiling point.

Another differential boiling-point apparatus is described by Menzies and Wright.[10] A narrow graduated tube is closed at one end and bent up to give a U tube, with the closed arm about 2 cm long and the other about 12 cm long. It is partly filled with water or other liquid, sealed off at the ends, and placed vertically in a flask. There is thus an air pocket in the short arm and another one at the top of the long arm. A vapor pump pours the boiling solution over the pocket in the lower arm, whereas the pocket at the top of the long arm is surrounded only by vapor. Measured changes in the level of the enclosed liquid enable one to calculate the differences in vapor pressure of this liquid, and from the differences in vapor pressure, the corresponding differences in temperature can be calculated. This temperature difference then is equal to the difference in temperature between the solution boiling in the flask and its vapor. At 35° a difference in level of 1 mm of water corresponds to 0.0313°, and at 100° it corresponds to 0.0026°.

MEASUREMENT OF VAPOR PRESSURE

The measurement of vapor pressure is closely related to the measurement of boiling point, so that it is difficult to make a distinction between these two types of measurements. If the boiling point is determined at a series of pressures, the vapor-pressure–temperature relation is obtained. This is the essence of the *dynamic method* which is especially important. Three other general methods, some of which may be adapted as differential methods, will be described briefly.

DYNAMIC METHOD.   The construction of apparatus used at the National Bureau of Standards is described by Willingham and coworkers.[11] This apparatus consists of an electrically heated boiler, a vapor space with a vertical reentrant tube containing a platinum resistance thermometer, and a condenser. The pressure is controlled by an automatic device actuated by electrical contacts sealed through the barometer tube.

Special apparatus has been designed for the study of binary solutions.[12]

STATIC METHOD.   In the simplest method the liquid is contained in a bulb connected with a mercury manometer and a vacuum pump. The greatest source of error lies in the presence of air or other permanent gases which have been dissolved by the liquid or trapped by the mercury. Enough liquid is evaporated with the pump to sweep out all the gases. The evacuation is repeated until further evacuation gives no lowering of the vapor pressure. The whole apparatus should be thermostated. The method has been used by many investigators.

The isoteniscope[13] is useful for the determination of vapor pressure of a liquid or a solution.

Differential static methods have been developed by Lovelace et al.[14]

GAS-SATURATION METHOD (TRANSPIRATION METHOD).   In this method a measured volume of air or other inert gas is saturated by passing it through the liquid at a definite temperature. The quantity of liquid vaporized is obtained from the loss in weight of the liquid or by removal of the vapor from the gas stream in weighed absorbing tubes. Assuming Dalton's law of partial pressures and the ideal-gas law, the partial pressure of the vapor P is calculated by the formula

$$P = \frac{g}{MV} RT$$

where R = gas constant

T = absolute temperature

V = total volume of gas (including air and vapor) containing g g of vapor of molecular weight M

When V is expressed in liters and P in atmospheres, R is expressed in liter-atm deg^{-1} mole^{-1}. In case the vapor pressure is very low, it can be neglected in comparison with the atmospheric pressure in calculating the volume of the gas.

If a gas is passed first through pure solvent and then through a solution, the vapor pressure of the solution can be calculated from the vapor pressure of pure solvent, the total pressure at each saturator, and the gain in weight of each absorber, thus eliminating the need of measuring the volume of the gas.

Premature condensation must be avoided if the vapor is to be absorbed and weighed, and at higher temperatures the saturator, absorption tubes, and connecting tubes are all immersed in the thermostat. It is essential to saturate completely the air or other gas with the vapor of the liquid; but, on the other hand, there must be no stoppage in the apparatus which might build up a changing or an unknown hydrostatic pressure. If the air is passed through the saturator so slowly that a still slower rate gives no greater vapor pressure, it may be concluded that the air is completely saturated. The air-saturation method has been used in precision researches by Washburn[15] for the determination of vapor pressures of aqueous solutions and has been further developed by additional workers.[16] When both components of a solution are volatile, it is necessary to have a suitable means for analyzing the condensed vapor. A physical method such as refractometry or a chemical method such as titration may be used.

ISOPIESTIC METHOD.   This method is used in studies on solutions of nonvolatile solutes. Vessels containing solutions of two different solutes are placed side by side in a closed container. A net transfer of solvent from the solution of higher vapor pressure to the solution of lower vapor pressure proceeds until equilibrium is attained. The fugacity of the solvent (and hence the vapor pressure of each solution, since the solute is nonvolatile) is then the same for each solution. The solutions are analyzed, and if the vapor pressure of one of the solutions is known from other absolute measurements, the vapor pressure of the other has been determined.

It is important that the vessels be in good thermal contact. This is ordinarily achieved by using metal cups which fit snugly into holes bored in a large copper block. This block fits into a stainless-steel vessel with a cover and a lead gasket, so that the whole system can be evacuated and rotated in a large thermostat kept constant nearly to $0.001°$. After 24 hr, the little cups are covered and weighed. They are replaced in the vessel, the covers removed, and the determinations repeated again after 24 hr or until there is no further change in weight.

Robinson and Sinclair[17] compared the activities of water in solutions of inorganic halides with those in solutions of potassium chloride at different concentrations. Scatchard, Hamer, and Wood[18] determined the activities of water in solutions of potassium chloride, sulfuric acid, sucrose, urea, and glycerol as compared with solutions of sodium chloride. The results are very accurate and permit a check on the several different methods that have been used for the determination of the activity of water in solution.

FRACTIONAL DISTILLATION†

The separation of two liquids by distillation and the determination of the number of theoretical plates for laboratory fractionating columns have been described in Experiment 10. In addition to the number of theoretical plates, there are a number

† For further material see Ref. 19.

of factors, such as feed rate and operating holdup, to be considered in selecting or designing fractionating columns. The feed rate of the column is defined as the rate of entry of vapor into the bottom of the column, and, depending upon the design of the column, a certain feed rate may not be exceeded without flooding the packing. It is desirable to use as high a feed rate as possible so that a given distillation may be accomplished in the shortest possible time. The operating holdup of a column is the quantity of vapor and liquid in the column under operating conditions. It is desirable that the column have a small holdup so that a minimum amount of liquid is held in the column at any time. It may be shown that the sharpness of separation obtainable in a batch distillation is approximately a linear function of the ratio of charge to holdup. Thus decreasing the holdup relative to the charge enables a sharper separation to be obtained with the same number of theoretical plates and the same total distillation time. Actually, high feed rate and small holdup are not easily obtained in the same column and must be balanced against each other.

A number of different types of fractionating columns have been devised in an effort to improve the contact between the liquid and the ascending vapor. The ideal packing offers uniformly distributed interstices, a large surface for contact, and enough free space for a desirable feed rate. Packing of large units has a small area for contact and a tendency to channel, while packing of small units allows insufficient feed rate. If the ratio of column diameter to diameter of the individual packing units is greater than 8:1 and the ratio of column height to column diameter is greater than 15:1, the tendency to channel will be slight.

A fractionating column is most efficient when its operation is adiabatic throughout its length. An insulating jacket, a vacuum jacket, or an electrically heated jacket may be used. If an electrically heated jacket is used, it is important to keep the column somewhat hotter at the bottom than at the top.

Two types of still heads may be used to control the reflux ratio: (*a*) liquid-dividing heads and (*b*) vapor-dividing heads. Vapor-dividing heads have the advantage that the reflux ratio may be controlled more precisely. Automatic devices may be used with both types to control reflux ratios conveniently.

HIGH-VACUUM DISTILLATION.   With the development of efficient vacuum pumps (see Experiment 58), high-vacuum distillation of material of low volatility has become a common commercial and laboratory operation. This method has the advantage that high-molecular-weight organic molecules, such as vitamins, sterols, and synthetic polymers, which cannot be distilled at their normal boiling points without decomposition, may be distilled unharmed. Hickman[20] and others have developed new tools and special techniques for such distillations. When vacuum distillations are carried out with the usual flask–condenser–receiver–vacuum-pump apparatus, it is found that reducing the pressure in the receiver below about 5 mm Hg produces little increase in the rate of distillation or lowering of the temperature of the distillation. This is because of the resistance to the flow of the vapor exerted by the neck and sidearm of the distillation flask. In order to avoid this difficulty and to increase the amount of vapor which actually reaches the condenser, the condenser must be placed quite close to the surface where evaporation is taking place. If the distance of transfer is comparable with the mean free path of the vapor molecules, the process is known as *molecular* distillation.

There are important differences in principle between ordinary distillation and molecular distillation. In ordinary distillation, molecules from the vapor reenter the surface and tend to produce equilibrium between liquid and vapor phase. Under these conditions the quantities of the various constituents distilling are proportional to their partial pressures. In molecular distillation, on the other hand, molecules do not reenter the liquid phase and there is no equilibrium between liquid and vapor. The separation achieved depends only upon the differences in rates of evaporation of the various components.

The differences in the ordinary and molecular distillation processes lead to important differences in apparatus design. In molecular distillation it is clear that there is no generation of vapor bubbles below the surface because the vapor would have to exert a pressure of the order of a millimeter of mercury, while at the temperature of the distillation, the vapor pressure is actually less than 0.001 mm. It is important, therefore, for the liquid film to be thin and the surface constantly changed. This has been achieved by the falling-film still, in which a thin film of liquid flows down over the heater, and by the centrifugal still, in which a thin film of solution is spun out over the surface of a shallow conical evaporator which is rotated at a high speed. Another reason for using a very thin film is to reduce thermal decomposition by subjecting the sensitive organic compounds to a high temperature for only a very short time. In a centrifugal still, the liquid is heated for less than a tenth of a second. Thus thermal decomposition is greatly reduced by carrying out the distillation at a low pressure and by allowing only short exposure to the elevated temperature.

The measurement of the temperature of the vapor is not practical in short-path distillation, but information can be gained by distilling a number of known substances at different temperatures and plotting a curve of material condensed against temperature. In this way the relative temperature at which an unknown material comes over can be determined by reference to some known material.[21] These separations may be made very conveniently with a series of pilot dyes[22] of different volatilities which condense to give a colored deposit.

DENSITY†

Densities of liquids are most frequently expressed in grams per milliliter. Since the milliliter is defined as one one-thousandth of the volume of 1 kg of pure, ordinary water at its temperature of maximum density (3.98°C), the density in grams per

---

† Discussions of experimental methods for the determination of density are to be found in Refs. 23 and 24.

milliliter is numerically equal to the ratio of the absolute density of the liquid at $t°C$ to the absolute density of water at $3.98°$ and is frequently represented by ρ_4^t.

The densities of liquids may be determined by measurement of the weight of liquid occupying a known volume (pycnometric methods) and by buoyancy methods based on Archimedes' principle.

PYCNOMETERS.   Pycnometers are vessels with capillary necks in which a definite volume of liquid is weighed. The volume is determined by weighing the vessel filled with water at a definite temperature. A table giving the density of water as a function of temperature is given in the Appendix. Two types of pycnometers are illustrated in Fig. 24 (page 98).

In order to obtain fifth-place accuracy in density determinations, a number of precautions must be observed. The weights should be checked against each other to obtain their relative values. It is not necessary to compare the set with a certified standard mass. One of the largest errors is often due to the adsorption of an uncertain amount of moisture by the glass, and it is necessary to wipe the pycnometer with a damp cloth and allow it to stand in the balance case for several minutes before weighing. A similar pycnometer of approximately the same volume may be used to advantage as a counterpoise.

In order to obtain the true weight of the liquid in the pycnometer, it is necessary to correct for the buoyancy of the air. The volume occupied by the glass of the pycnometer can be left out of the calculation if the tare has very nearly the same weight and density. The true (vacuum-corrected) weight W_0 is calculated from the equation

$$W_0 = W\left(1 + \frac{\rho_{air}}{\rho_s} - \frac{\rho_{air}}{8.5}\right)$$

where W = apparent weight given by brass weights

ρ_{air} = density of air

ρ_s = density of substance being weighed

In the present case, ρ_s is the density of the liquid in the pycnometer. For fifth-place accuracy the density of the weights may be assumed to be 8.5 g ml^{-1} even when the small-denomination weights are made of material of different density. The density of the air may usually be taken to be 0.0012 g ml^{-1}, but for accurate work the variation of the density of air due to changes in room temperature, barometric pressure, and relative humidity must be considered. If the temperature ($t°C$), barometric pressure (P mm Hg), and the relative humidity (H, in percent) are measured at the time of the weighing, the density of air may be calculated from the equation

$$\rho_{air} = \frac{0.001293}{1 + 0.00367t}\frac{P - k}{760}$$

where $k = 0.0038HP_{H_2O}^t$

$P_{H_2O}^t$ = vapor pressure of water, mm Hg at $t°C$

For determining the density of a solid, a pycnometer with a wide mouth that will admit the solid is necessary. The ordinary type is a small bottle with a ground-glass stopper through which is bored a fine capillary.

The bottle is filled with water, and the stopper is inserted firmly, after which it is placed in a thermostat. The excess liquid is wiped off, and the pycnometer is dried and weighed. The pycnometer is weighed empty and again with the solid. After filling with water (plus the solid) it is weighed again, and all the data are then available for calculating the weight and volume of the solid and its density. In case the solid is soluble in water, some other inert liquid is used, and the density of the liquid is determined also.

The greatest source of error in determining the density of a solid is the adsorption of air by the solid. For this reason, the pycnometer containing the solid and some liquid is set in a larger bottle, which is connected to a vacuum pump, and evacuated until all air bubbles have ceased rising from the solid; then the pycnometer is filled completely.

BUOYANCY METHODS.   The Westphal balance is more accurate than the floating graduated hydrometer. It depends on the principle of Archimedes, according to which the buoyant effect is directly proportional to the weight of the liquid displaced. The sinker is suspended in pure water, with the unit weight in position, and a threaded counterpoise is turned until the pointer reads zero on the scale. The sinker is then dried and suspended in the liquid whose density is to be measured. The smaller weights are set at the proper places on the scale so as to restore the point of balance. Some balances are constructed with three riders, corresponding to 0.1, 0.01, and 0.001, and the scale is divided into 10 equal parts. The position on the scale gives the numerical value for each rider; e.g., if the 0.1 rider is at 9, the 0.01 at 8, and the 0.001 at 7, the specific gravity is 0.987.

The temperature is read directly on a thermometer which is enclosed in the sinker. A very fine platinum wire is used for suspending the sinker; the surface-tension effect on this wire is negligible for ordinary work, but for accurate work it may prove to be a source of error. The wire should be immersed to the same depth for each measurement.

The same principle of weighing a sinker which is suspended in a liquid is used in the chainomatic balance, shown in Fig. 141. With this more elaborate instrument, densities may be determined quickly with an accuracy of 1 part in 10,000. The value of the instrument is increased by the introduction of an electrical heating coil and switch, to maintain the liquid at a definite temperature.

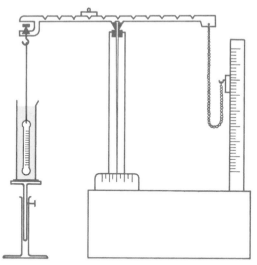

figure 141. Chainomatic balance for determining the density of liquids.

FLOATING EQUILIBRIUM.   The objection to the wire projecting through the surface of the liquid in the case of the Westphal balance can be eliminated by having the plummet so carefully adjusted that it neither sinks nor rises in the liquid.

Accurate density measurements may be made by placing a piece of iron in the float and measuring the strength of the electromagnetic field necessary to keep the bulb in floating equilibrium. Such an apparatus has been described by MacInnes et al.[25]

Gilfillan[26] adjusted the hydrostatic pressure on the liquid with a mercury column until equilibrium of the bulb was obtained. He calibrated the apparatus with dilute potassium chloride solutions of known densities.

The density of solids can be determined in a somewhat similar way by mixing two liquids of different density until the solid neither rises nor sinks. The system must be evacuated to remove air from the solid. A heavy liquid like methylene iodide is mixed with a light liquid like benzene, and the density of the final mixture is determined with a pycnometer. The density of the liquid is the same as the density of the solid with which it is in floating equilibrium. This method has been used by Hutchinson and Johnston[27] for the accurate determination of the density of lithium fluoride.

FALLING DROP.   The falling-drop method is especially useful when only small quantities of sample are available. The method requires measuring the velocity of fall of a drop of liquid through an immiscible fluid of known density.[28]

VISCOMETRY†

In the Ostwald viscometer (Experiment 26) the liquid is forced through a capillary with compressed air maintained at constant pressure. There are marks at the top and bottom of the bulb, and the time required for the liquid to flow first up and then down is recorded. Raaschow[30] gives a detailed discussion of the design and testing of Ostwald type viscometers.

The use of automatic timing for the transit of the meniscus of the liquid in the capillary tube, by means of a photoelectric cell, has been developed by Jones and Talley[31] and by Riley and Seymour[32]. Details of the necessary assembly have been given by these investigators in articles in which they also analyze sources of error in the use of viscometers.

The rotation-cylinder viscometer, which measures the torque that is required to rotate a cylinder at a given rate of speed in the solution, has certain advantages over the capillary-tube type. A source of error in a capillary-tube type of viscometer arises from the fact that the shearing stress exerted on the liquid is not uniform but varies with the distance from the center of the capillary. This effect is not important if the viscosity of the liquid is independent of the shearing stress (newtonian liquid), but in the case of nonnewtonian liquids, the viscosity obtained is an average value which depends on the dimensions of the instrument and the relation between viscosity and shearing stress. In the rotating-cylinder viscometer, this difficulty may be avoided.

One of the simplest methods for determining viscosity depends upon the determination of the velocity of a sphere falling through the liquid when it has reached a uniform velocity. This is based on Stokes' law relating the viscosity of the liquid to the frictional force which acts on a moving sphere. It is necessary to apply corrections for the effect of the wall of the tube on the velocity to avoid end effects. For a small sphere of radius r falling axially through a viscous liquid in a cylindrical tube, the complete expression is given on page 162. In measurements with spheres of equal radius in tubes of the same dimensions

$$\eta = K(\rho - \rho_0)t$$

The tube constant K may be evaluated experimentally by measuring the time of fall through a liquid whose density and viscosity are known or may be calculated from Eq. (22) on page 162. In determining relative viscosities, the tube constant cancels:

$$\frac{\eta}{\eta_s} = \frac{\rho - \rho_0}{\rho - \rho_s} \frac{t}{t_s}$$

The subscript s refers to the standard liquid. The sphere is discharged slowly into the tube, a few centimeters below the surface, through a glass tube slightly larger

---

† The general subject of viscometry is discussed in Ref. 29.

than the sphere; 1.5 mm is a suitable diameter for the sphere, and 20 cm is a satisfactory height for fall. The vessel should have a diameter at least 10 times that of the sphere. A steel ball such as those used in ball bearings is excellent.

The determination of absolute viscosities is rather difficult. It requires careful measurements of the apparatus, including the length and radius of the capillary, and some rather uncertain corrections. Further details concerning absolute measurements are given by Swindells, Coe, and Godfrey[33] and by Bingham, Schlesinger, and Coleman.[34]

SURFACE TENSION†

When the surface tension of a liquid is to be determined (Experiment 51), one should choose the particular method which will give the best results with the least effort. The realms of utility for three methods are outlined as follows:

1. *Single Liquids*
    a. *Capillary rise:* For highest accuracy, but not rapid.
    b. *Ring method:* Very fast and reasonably accurate if suitable apparatus is available. Can be used for interfacial tension.
    c. *Drop weight:* Best general method for both surface and interfacial tensions if both accuracy and speed are considered. Can be used with very small quantities of liquid.
2. *Solutions*
    a. *Drop weight:* Best method for surface and interfacial tension if long time effects are not involved.
    b. *Ring method:* Excellent for surface tension, even if time effects are involved.

DIFFUSION

An important transport process is that of molecular diffusion in solution, with any difference in concentration being reduced by a spontaneous transfer of matter. It is caused by the brownian motion of the dissolved molecules.

The flux of matter is defined as the amount of material which in unit time passes through a unit area of plane perpendicular to the direction of flow. It is the product of a concentration and a velocity. Thus, if the concentration is expressed in moles per cubic centimeter and the distance of displacement is given in centimeters, then the flux J is measured in moles per square centimeter per second.

† The measurement of surface tension has been reviewed in Ref. 35.

The diffusion coefficient D is defined by two laws of Fick. They are, for a two-component system,

$$J = -D \frac{\partial c}{\partial x} \tag{1}$$

$$\frac{\partial c}{\partial t} = \frac{\partial}{\partial x} \left(D \frac{\partial c}{\partial x} \right) \tag{2}$$

The coefficient which is defined by these equations is often referred to as a constant. This is only approximately true, and the study of the variation of D with concentration gives useful information. Equation (2) may be derived from Eq. (1) by introduction of a statement of the conservation of mass.

The diffusion of molecules in liquid media has been the subject of a number of reviews.[36] There are several types of experiment.

FREE DIFFUSION. In this method, which is based on the second law of Fick, an initially sharp boundary is formed between the solution and the solvent with the more dense phase in the bottom of the diffusion cell. The cell should be tall enough for the composition at the bottom and at the top of the column to remain unchanged during the period of observation. The gradual blurring of the boundary may be followed by a number of methods. In early work, samples of solution were taken from various levels in the cell and their concentrations used to calculate the diffusion coefficient. In some cases, it is possible to determine the concentration at every level in the diffusion cell by means of quantitative light-absorption measurements. The most generally applicable and accurate methods for the determination of the diffusion coefficient are optical methods which depend upon the fact that light is deflected upon passing through a refractive-index gradient such as that established in a diffusion column. In the Lamm scale method, the displacements of lines in the photograph of a linear scale placed behind the diffusion cell are measured with a microcomparator. The schlieren method, in which a plot of refractive-index gradient versus position in the cell is obtained directly, is described in Chapter 21. The diffusion coefficient may be calculated from the shape of the curve by several methods.

Interference methods developed by Kegeles and Gosting,[37] Longsworth,[38] and Coulson et al.[39] yield more accurate diffusion coefficients than any of the above methods.

A number of cells have been designed for the purpose of forming a sharp initial boundary between solution and solvent.[40]

RESTRICTED DIFFUSION. Harned[41] has used a conductivity method for the determination of concentration changes of salts in a diffusion cell. In this method, the difference in the conductances of the solution as measured between pairs of electrodes at the bottom and at the top of the cell is utilized in the calculation of the diffusion coefficient. This method is limited to the study of electrolytes, and an experimental precision of 0.1 percent may be obtained at concentrations less than 0.01 *N*.

STEADY-STATE DIFFUSION.   In this method the first law of Fick serves as the means for the computation of the result. Diffusion takes place through a region in which the concentration gradient is independent of time. An example of this method is the porous-plate method of Northrop and Anson.[42] One form of the diffusion cell consists of a bell-shaped glass vessel, closed at the narrow top end by a stopcock and at the wide-bottom end by a sealed-in sintered-glass disk. The cell is filled with a solution and is immersed in a beaker of water just touching its surface. Various sizes have been used, from 10 to 200 ml capacity. Diffusion is allowed to proceed until the concentrations in the pores of the sintered-glass disk are those for a steady state, which generally requires several hours. When steady-state concentrations have been attained, the cell is placed in contact with a fresh sample of solvent, and diffusion is allowed to proceed for a suitable length of time. From concentrations determined after various times in aliquots of solution and solvent, the diffusion coefficient may be calculated. It is necessary to calibrate the cell by an experiment with a substance of known diffusion coefficient. The advantage of this method is that it is useful for determining the diffusion coefficients of radioisotopes or biologically active substances at dilutions so high that optical methods cannot be used. In such experiments very careful control of temperature is required to prevent thermal convection.[43]

Micro methods for use with colored materials have been developed by Fürth[44] and Nistler.[45]

OSMOTIC PRESSURE†

Osmotic-pressure determinations may be divided into two classes according to the molecular weight of the solute. If the membrane is permeable only to solvent molecules, the so-called *total osmotic pressure* is obtained. In the case of solutions which contain solutes of both low and very high molecular weight, the total osmotic pressure would be due to both classes of solute molecules. However, if the osmotic-pressure determination is carried out with a membrane permeable to both solvent molecules and low-molecular-weight solute molecules, the osmotic pressure measured is due only to the large molecules and is referred to as *colloid osmotic pressure.*

† A general review of this subject is given in Ref. 46.

In the case of solutes of molecular weight less than 10,000, the great difficulties encountered in the preparation of membranes impermeable to solute molecules preclude the general use of the osmometric method. In the apparatus of Frazer and Myrick,[47] a porous clay cup contains an electrolytically deposited membrane of copper ferrocyanide which allows the passage of solvent but not of the low-molecular-weight solute. This clay cup, containing the solvent, is surrounded by a strong bronze cylinder containing the solution and the connection to the manometer. To measure the pressure, an electrical-resistance gauge or a deadweight gauge may be used in place of the customary mercury manometer. With this apparatus Frazer and his coworkers have measured osmotic pressures up to 273 atm.

Osmotic pressure may be successfully used to determine the molecular weights of macromolecules, such as high polymers, proteins, and polysaccharides. A simple calculation will show that an osmotic pressure of 1 mm of water corresponds to a freezing-point depression of roughly 0.0001°C. Such a freezing-point depression might be caused by a minute trace of salt in a protein solution, while the salt would not significantly affect the osmotic pressure of the isoelectric protein measured with a membrane permeable to small ions and water molecules.

Osmotic-pressure measurements may be carried out by dynamic or static methods. In the dynamic method,[48] the rate of movement of the meniscus in the capillary tube is measured at a number of heights and used to obtain the equilibrium height, whereas in the static method, the equilibrium height is determined directly. The construction of a standard type of static osmometer for high-polymer work[49] is illustrated in Fig. 142. The cell consists of two stainless-steel plates clamped

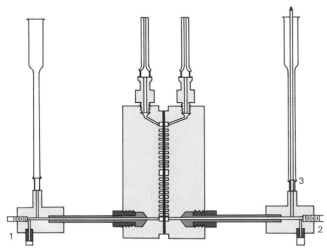

figure 142. Static osmometer.

together, with the membrane between the plates. The faces of the two plates are machined with a set of concentric cuts 2 mm wide and 2 mm deep. The solution is placed in one half-cell, and the solvent in the other, and the membrane simultaneously acts as a gasket. With valves 1 and 2 closed, and valve stem 3 removed, solution is poured into the left-hand tube, while solvent is simultaneously poured into the right one, so that the level rises at the same rate on both sides of the membrane. The stem is inserted, valve 3 is closed, and valve 1 is opened to drop the meniscus in the solution standpipe to its desired position. The left meniscus is maintained at constant height, while the rate of approach of the right meniscus to its equilibrium position is determined from both the high-pressure and low-pressure sides. Smaller glass osmometers[50] are also widely used.

Semipermeable membranes for use with polymer solutions are generally prepared by treating cellophane with 3 percent sodium hydroxide or by denitrating collodion (cellulose nitrate), using ammonium polysulfide.

Because of the nonideality of polymer solutions, it is necessary to extrapolate osmotic-pressure measurements at several concentrations to infinite dilution in order to calculate the molecular weight. A plot of π/c, where π is the osmotic pressure, versus c is quite linear in the low-concentration range and is used for the extrapolation.

A number of osmometers have been designed especially for protein work.[51] In these osmometers the protein solution is contained in a bag of collodion or regenerated cellulose (such as Visking sausage casing) and the pressure measured by means of a toluene manometer.

References

1. A. F. Benton, *Ind. Eng. Chem.*, **11:** 623 (1919).
2. J. R. Roebuck and H. W. Ibser, *Rev. Sci. Instr.*, **25:** 46 (1954).
3. Mercury Barometers and Manometers, *Natl. Bur. Std. U.S. Monograph* 8, 1960.
4. *Instr. Control Systems*, **34:** 1057 (1961).
5. C. H. Myers and R. S. Jessup, *J. Res. Natl. Bur. Std. U.S.*, **6:** 1061 (1931).
6. O. C. Bridgman, *J. Am. Chem. Soc.*, **49:** 1174 (1927); J. R. Roebuck and W. Cram, *Rev. Sci. Instr.*, **8:** 215 (1937).
7. W. Swietoslawski and J. R. Anderson in A. Weissberger (ed.), "Technique of Organic Chemistry," vol. 1, "Physical Methods of Organic Chemistry," 3d ed., pt. 1, chap. 8, Interscience Publishers, Inc., New York, 1959.
8. A. A. Morton, *Ind. Eng. Chem. Anal. Ed.*, **6:** 384 (1934).
9. W. Swietoslawski, *J. Chem. Educ.*, **5:** 469 (1928).
10. A. W. C. Menzies and S. L. Wright, *J. Am. Chem. Soc.*, **43:** 2314 (1921).
11. C. B. Willingham et al., *J. Res. Natl. Bur. Std. U.S.*, **35:** 219 (1945).
12. G. Scatchard, S. E. Wood, and J. M. Mochel, *J. Am. Chem. Soc.*, **61:** 3206 (1939); **62:** 712 (1940); D. F. Othmer and S. Josefowitz, *Anal. Chem.*, **39:** 1175 (1947); G. W. Thomson

in A. Weissberger (ed.), "Technique of Organic Chemistry," vol. 1, "Physical Methods of Organic Chemistry," 3d ed., pt. 1, chap. 9, Interscience Publishers, Inc., New York, 1959.

13. A. Smith and A. W. C. Menzies, *J. Am. Chem. Soc.*, **32:** 1412 (1910).
14. B. F. Lovelace, W. H. Bahlke, and J. C. W. Frazer, *J. Am. Chem. Soc.*, **45:** 2930 (1923).
15. E. W. Washburn and E. O. Heuse, *J. Am. Chem. Soc.*, **37:** 309 (1915).
16. J. N. Pearce and H. C. Eckstrom, *J. Am. Chem. Soc.*, **59:** 2689 (1937); M. F. Bechtold and R. F. Newton, *ibid.*, **62:** 1390 (1940).
17. R. A. Robinson and D. A. Sinclair, *J. Am. Chem. Soc.*, **56:** 1830 (1934).
18. G. Scatchard, W. J. Hamer, and S. E. Wood, *J. Am. Chem. Soc.*, **60:** 3061 (1938).
19. O. S. Robinson and E. R. Gilliland, "Elements of Fractional Distillation," 4th ed., McGraw-Hill Book Company, New York, 1950; A. A. Morton, "Laboratory Techniques in Organic Chemistry," chap. 4, McGraw-Hill Book Company, New York, 1938; T. P. Carney. "Laboratory Fractional Distillation," The Macmillan Company, New York, 1949; C. C. Ward, Review of the Literature on the Construction, Testing, and Operation of Laboratory Fractionating Columns, *U.S. Bur. Mines Tech. Paper* 600, 1939; A. L. Glasebrook and F. E. Williams in A. Weissberger (ed.), "Technique of Organic Chemistry," vol. 4, E. S. Perry and A. Weissberger (eds.), "Distillation," 2d ed., Interscience Publishers, Inc., New York, 1965.
20. K. C. D. Hickman, *Chem. Rev.*, **34:** 51 (1944); *Am. Scientist*, **33:** 205 (1945).
21. N. D. Embree, *Ind. Eng. Chem.*, **29:** 975 (1937).
22. K. C. D. Hickman, *Ind. Chem.*, **29:** 968, 1107 (1937).
23. N. Bauer and S. Z. Lewin in A. Weissberger (ed.), "Technique of Organic Chemistry," vol. 1, "Physical Methods of Organic Chemistry," 3d ed., pt. 1, chap. 4, Interscience Publishers, Inc., New York, 1959.
24. J. Reilly and W. N. Rae, "Physico-chemical Methods," 5th ed., vol. I, pp. 577–628, D. Van Nostrand Company, Inc., Princeton, N.J., 1953.
25. D. A. MacInnes and M. O. Dayhoff, *J. Am. Chem. Soc.*, **75:** 5219 (1953).
26. E. S. Gilfillan, *J. Am. Chem. Soc.*, **56:** 406 (1934).
27. C. A. Hutchinson and H. L. Johnston, *J. Am. Chem. Soc.*, **62:** 3165 (1940).
28. A. S. Keston, D. Rittenberg, and R. Schoenheimer, *J. Biol. Chem.*, **122:** 227 (1937).
29. E. Hatchek, "The Viscosity of Liquids," George Bell & Sons, Ltd., London, 1928; G. Barr, "Viscometry," Oxford University Press, Fair Lawn, N.J., 1931; W. Philipoff, "Viscosität der Kolloide," Theodor Steinkopff, Verlagsbuchhandlung, Dresden, 1942; C. M. Blair, "A Survey of General and Applied Rheology," Pitman Publishing Corporation, New York, 1944; J. F. Swindells and R. Ullman in A. Weissberger (ed.), "Technique of Organic Chemistry," vol. 1, "Physical Methods of Organic Chemistry," 3d ed., pt. 1, chap. 12, Interscience Publishers, Inc., New York, 1959.
30. P. E. Raaschow, *Ind. Eng. Chem. Anal. Ed.*, **10:** 35 (1938).
31. G. Jones and S. K. Talley, *J. Am. Chem. Soc.*, **55:** 624 (1933).
32. J. L. Riley and G. W. Seymour, *Ind. Eng. Chem. Anal. Ed.*, **18:** 387 (1946).
33. J. F. Swindells, J. R. Coe, and T. B. Godfrey, *J. Res. Natl. Bur. Std. U.S.*, **48:** 1 (1952).
34. E. C. Bingham, H. I. Schlesinger, and A. B. Coleman, *J. Am. Chem. Soc.*, **38:** 27 (1916).
35. W. D. Harkins and A. E. Alexander in A. Weissberger (ed.), "Technique of Organic Chemistry," vol. 1, "Physical Methods of Organic Chemistry," 3d ed., pt. 1, chap. 4, Interscience Publishers, Inc., New York, 1959; N. K. Adam, "Physics and Chemistry of Surfaces," Oxford University Press, Fair Lawn, N.J., 1942; Ref. 24, pp. 629–659.

36. J. W. Williams and L. C. Cady, *Chem. Rev.*, **14:** 171 (1934); H. Neurath, *Chem. Rev.*, **30:** 257 (1942); H. S. Harned, *Chem. Rev.*, **40:** 462 (1947); A. L. Geddes in A. Weissberger (ed.), "Technique of Organic Chemistry," vol. 1, "Physical Methods of Organic Chemistry," 3d ed., pt. 1, pp. 551–619, Interscience Publishers, Inc., New York, 1959; L. J. Gosting, *Advan. Protein Chem.*, **11:** 429 (1956); H. Svensson and T. E. Thompson in P. Alexander and R. J. Block (eds.), "Analytical Methods of Protein Chemistry," vol. 3, Pergamon Press, New York, 1961.

37. G. Kegeles and L. J. Gosting, *J. Am. Chem. Soc.*, **69:** 2516 (1947); L. J. Gosting et al., *Rev. Sci. Instr.*, **20:** 209 (1949).

38. L. G. Longsworth, *J. Am. Chem. Soc.*, **74:** 4155 (1952); **75:** 5705 (1953).

39. C. A. Coulson et al., *Proc. Roy. Soc. London Ser. A*, **192:** 382 (1948).

40. D. L. Loughborough and A. J. Stamm, *J. Phys. Chem.*, **40:** 1113 (1936); S. Claesson, *Nature*, **158:** 834 (1946).

41. H. S. Harned, *Chem. Rev.*, **40:** 462 (1947).

42. J. H. Northrup and M. L. Anson, *J. Gen. Physiol.*, **12:** 543 (1929).

43. M. Mouquin and W. H. Cathcart, *J. Am. Chem. Soc.*, **57:** 1791 (1935).

44. R. Fürth, *Kolloid-Z.*, **41:** 300 (1927).

45. A. Nistler, *Kolloid Chem. Beih.*, **28:** 296 (1929).

46. R. H. Wagner and L. P. Moore in A. Weissberger (ed.), "Technique of Organic Chemistry," vol. 1, "Physical Methods of Organic Chemistry," 3d ed., pt. 1, chap. 15, Interscience Publishers, Inc., New York, 1959.

47. J. C. W. Frazer and R. T. Myrick, *J. Am. Chem. Soc.*, **38:** 1907 (1916); J. C. W. Frazer and P. Lotz, *ibid.*, **43:** 2501 (1921).

48. R. E. Montanna and L. T. Jilk, *J. Phys. Chem.*, **45:** 1374 (1941).

49. R. M. Fuoss and D. J. Mead, *J. Phys. Chem.*, **47:** 59 (1943); P. J. Flory, *J. Am. Chem. Soc.*, **65:** 372 (1943).

50. B. H. Zimm and I. Myerson, *J. Am. Chem. Soc.*, **68:** 911 (1946).

51. H. B. Bull and B. T. Currie, *J. Am. Chem. Soc.*, **68:** 742 (1946); G. Scatchard, A. C. Batchelder, and A. Brown, *ibid.*, **68:** 2320 (1946).

Chapter 24

Electrical Measurements

A wide variety of electrical measurements are met with in physical chemistry. These include observations of the electromotive forces of electrochemical cells, uses of thermocouples and thermopiles, measurements of resistance in the determination of electrolytic conductance and of temperature, measurement of quantity of electricity in the determination of transference numbers and ionic mobilities, measurements of electrical energy in calorimetry, and determinations of dielectric constant.

The absolute electrical units are based upon the fundamental mechanical units of mass, length, and time by the use of accepted principles of electromagnetism. These units are maintained, as were the international units used before 1948, by groups of standard resistors and of standard cells. The international ohm and volt are slightly larger than the corresponding absolute units. The conversion factors for adjusting values of standards in this country are as follows.[1]

1 int ohm = 1.000495 abs ohms
1 int volt = 1.00033 abs volts
1 int amp = 0.999835 abs amp
1 int coulomb = 0.999835 abs coulomb
1 int henry = 1.000495 abs henrys
1 int farad = 0.999505 abs farad
1 int watt = 1.000165 abs watts
1 int joule = 1.000165 abs joules

GALVANOMETERS

In these days of automatic recording instruments for voltage (potentiometers), resistance, capacitance, and frequency, the electronic circuits have made galvanometers less common items of the physical chemistry laboratory as compared with a decade ago. However, galvanometers are still useful in many assemblies for electrical measurements; furthermore, ammeters and voltmeters are essentially portable galvanometers of low sensitivity, so that they warrant some description.

The d'Arsonval galvanometer consists of a rectangular coil of wire suspended from a fine wire in the field of a permanent magnet. The bottom part of the coil is made steady by a loosely coiled metallic spring, which also serves as a lead. The current in the moving coil flows perpendicular to the lines of magnetic force, producing a torque on the coil. The coil turns until the restoring moment due to the twist in the suspension is equal to the torque due to the current. The motion of the coil is observed and magnified by means of a beam of light reflected from a small mirror mounted on the coil. A wide variety of galvanometers is available to provide the sensitivity, period, and ruggedness desired. The current sensitivity of a galvanometer is defined as the current in microamperes required in the galvanometer coil to produce a standard deflection, usually 1 mm on a scale placed perpendicular to the reflected light beam at a distance of 1 m. Sometimes the sensitivity is expressed in microvolts, i.e., the electromotive force which produces the standard deflection when it is introduced into the series circuit consisting of the galvanometer and its external critical-damping resistance. The microvolt sensitivity is therefore the product of the microampere sensitivity and the sum of the galvanometer resistance and its external critical-damping resistance. The sensitivity of a galvanometer is less frequently given in terms of the ballistic sensitivity, which is the quantity of electricity which must be discharged through a galvanometer in a time which is short compared with its free period to produce the standard deflection.

The period of a galvanometer is the time in seconds required for one complete undamped oscillation of the galvanometer. In most galvanometer applications, a period as short as possible, consistent with other necessary requirements, is desired. Short periodicity conserves the time of the observer and makes possible precision of measurement of fluctuating phenomena that would otherwise be unobtainable. For other measurements, such as ballistic measurements, long periods are desirable to facilitate reading.

For most applications it is desirable to damp a galvanometer so that the final reading is obtained without oscillation. A critically damped galvanometer reaches its final reading without oscillation and in the shortest possible time. It is customary to take the period of a critically damped galvanometer as equal to its undamped period, for although the critically damped period is theoretically infinite, practically, a critically damped deflection is within about 1.5 percent of its final position in the undamped periodic time. If the resistance of the external circuit is too small, the galvanometer will be overdamped; i.e., the coil will rotate too slowly. In this case, more resistance is added in series. If the resistance of the external circuit is too great, the galvanometer will be underdamped, and critical damping is achieved by reducing the resistance in parallel.

For the most satisfactory operation of a galvanometer, one should select an instrument whose external critical-damping resistance is slightly lower than the resistance presented to the galvanometer by the circuit with which it is to be employed. An ordinary laboratory has use for low-resistance galvanometers (20 to 100 ohms) for thermopile work and calorimetry and for high-resistance galvanometers (over 1000 ohms) for measurements of electromotive force. In an ac galvanometer, sensitivity and period are important, but critical-damping resistance as such is of less significance because proper damping also depends on capacitance and inductance.

Portable box-type galvanometers are taut-suspension moving-coil instruments with built-in lamps and scales, as illustrated in Fig. 143. The light beam traverses the case five times after reflection from the moving mirror, thus considerably increasing the sensitivity. Such galvanometers are commonly available with sensitivities down to 2.5 μv mm^{-1} and 0.005 μa mm^{-1}.

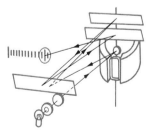

figure 143.

Portable galvanometer using multiple reflections.

For greater sensitivity, galvanometers with more delicate suspensions must be used. These galvanometers must be carefully leveled and mounted on vibrationless supports.[2] The beam of light from a small lamp is focused on a scale at 1 m distance, by means of a focusing mirror or a long-focus lens attached to the glass window of the galvanometer. High-sensitivity galvanometers are commonly available with sensitivities down to 0.1 μv mm^{-1} or 0.0001 μa mm^{-1}.

If a sensitivity much greater than 10^{-8} amp or 10^{-7} volt mm^{-1} is required with a low critical-damping resistance, it is necessary to use a special type of galvanometer or some type of amplification. Very small direct currents, as from thermopiles, may be amplified by electronic circuits if the current is subjected to mechanical interruption. Such methods are especially useful in infrared spectroscopy and have been reviewed by Williams.[3]

Another type of amplified galvanometer is the photoelectric galvanometer, which consists of a taut-suspension galvanometer and a double-cathode photocell in a balanced-bridge circuit. In the null position, the galvanometer light beam illuminates both photocell cathodes equally, under which condition the photocell constitutes a balanced bridge with zero potential difference across the photocell cathodes. Displacement of the light beam produces a signal in the photocell circuit whose magnitude and polarity are determined by the magnitude and polarity of the electromotive force applied to the galvanometer terminals. This method has been useful in amplifying thermopile currents.[4]

Ammeters for registering current are of very low resistance and are connected in series in the circuit. Voltmeters, for measuring the potential between two points in a circuit, on the other hand, are connected in parallel with the circuit, and their resistance must be so high that only a small fraction of the current flows through them. For example, if a meter with which the full-scale deflection is obtained with 1 ma is to be used as a 10-volt voltmeter, a series resistance S must be added. If the resistance of the meter is R, $R + S = 10/0.001$, and the resistance to be placed in series so that the meter will show a full-scale deflection for 10 volts is $(10/0.001) - R$. Similarly, the series resistance required to use the milliammeter as a 100-volt voltmeter is $(100/0.001) - R$. High-grade commercial ammeters and voltmeters are guaranteed to be correctly graduated to as close as $\frac{1}{4}$ percent of the full-scale reading, but cheaper instruments are generally accurate only to about 2 percent of the full-scale reading, irrespective of the actual reading.

MEASUREMENT OF ELECTROMOTIVE FORCE[5]

POTENTIOMETERS.   The principle of the potentiometer has been discussed in Experiment 29. Since the galvanometer acts only as a null-point indicator, the accuracy of measurement depends only on the accuracy of the standard-cell voltage

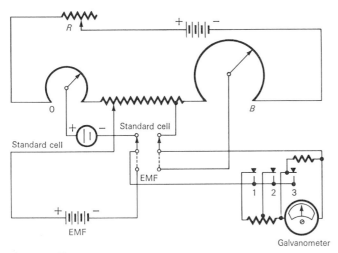

figure 144. Potentiometer circuit.

and the accuracy of the resistance coils, provided that the galvanometer is sufficiently sensitive and the resistance of the circuit is not too great. The resistance coils can be manufactured with exceedingly high precision, so the accuracy of the method depends principally on the constancy of the standard-cell voltage. Potentiometers for research purposes are designed in such a way that it is possible to check the potentiometer against the standard cell without disturbing the setting of the resistances in the measuring circuit. This is illustrated in Fig. 144. Fixed connections for the standard cell span a definite portion of the circuit *OB*, across which the fall in potential is adjusted to be equal to the voltage of the standard cell by varying *R*. By means of a double-pole double-throw switch, either the standard cell or the unknown electromotive force can be put in the circuit through the galvanometer. In making the first trial balance, the galvanometer keys 1 and 2 are used to include a high resistance in the galvanometer circuit and prevent a large current from flowing through the standard cell. When the potential drop of the potentiometer and the electromotive force of the unknown cell or standard are nearly balanced, key 3 is pressed to obtain maximum galvanometer sensitivity. When the total resistance of the unknown cell is comparatively small, such as 100 ohms, a potential difference of, say, 0.0001 volt between the potential drop in the slide-wire and the electromotive force of the unknown cell can cause sufficient current to flow through the galvanometer to produce a deflection. However, if the resistance of the unknown cell is, say, 5 megohms, the current through the circuit due to 0.0001 volt is only $0.0001/(5 \times 10^6)$, or 2×10^{-11} amp, which is too small to turn the coil of an ordinary galvanometer. In such a case, it is necessary to use an electrometer or electronic voltmeter.

Special potentiometers have been designed for thermocouple work. These include the Wenner potentiometer[6] and the White potentiometer. Double potentiometers of the White type are particularly useful in calorimetry, where it is desired to measure, practically simultaneously, two temperatures which are appreciably different, without resetting dials.

RECORDING POTENTIOMETERS.   There are now available recording potentiometers which perform the operations of balancing against an applied potential automatically and of standardizing themselves by balancing against the standard cell or Zener diode at regular intervals. The automatic balancing mechanism replaces the galvanometer in the usual potentiometer circuit. This electronic device consists of an amplifier which drives a motor whose direction and speed depend upon the polarity and voltage of the unbalance. The motor moves the contact on the potentiometer slide-wire, and the attached pen makes a trace on the moving chart. The basic circuitry for such instruments has been described by Witherspoon.[7]

Here again damping is required in the potentiometer recorder, it being generally achieved by a resistance-capacitance damping circuit in the input to the amplifier in the instrument. The theory of damping in such systems is described in several places.[8]

The precision of recording potentiometers at any point on the chart is about 0.3 percent of the full-scale voltage. Since so many measurements may be reduced to a measurement of voltage, recording potentiometers can be used in a wide variety of applications. They can be fitted with mercury switches or microswitches which can operate heaters or motors at certain potentials.

More specific information as to the potential required for a full-scale displacement, the time required for the pen to move across the full scale to a new balance point, the rate at which the paper moves under the pen of the instrument, etc., may be obtained from the manufacturers.

STANDARD CELLS.   The electromotive force of the Weston cell in the new absolute system is 1.0186 volts at 20°. This cell is set up in an airtight H-shaped vessel, with platinum wires sealed through the bottoms for connection with the electrodes, as shown in Fig. 145. The positive electrode consists of pure mercury, which is covered by a thick paste of mercurous sulfate and a small quantity of cadmium sulfate. The negative electrode is a cadmium amalgam containing 12.5 percent cadmium. On the top of the solidified amalgam and the mercurous sulfate paste are placed some rather large and clear cadmium sulfate crystals; then the cell is filled with a saturated solution of cadmium sulfate. The ends of the tubes are closed, allowing sufficient air space for thermal expansion. The materials must be thoroughly purified. The temperature coefficient of the cell is small, so that the

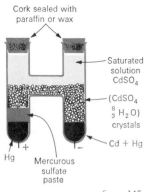

Cork sealed with paraffin or wax

Saturated solution $CdSO_4$

($CdSO_4$ $\frac{8}{3}H_2O$) crystals

Cd + Hg

Hg   Mercurous sulfate paste

figure 145.

Weston normal cell.

electromotive force may be given with accuracy sufficient for most purposes by the expression

$$E_t = E_{20} - 0.0000406(t - 20)$$

In actual laboratory measurements the unsaturated cadmium cell is more commonly used than the Weston cell. This cell contains a solution of cadmium sulfate saturated at 4° and has the advantage that the temperature coefficient (0.00001 volt deg^{-1}) is less than for the saturated cell. The voltage of the unsaturated cadmium cell varies between 1.01856 and 1.01910 int volts.

The following precautions should be taken in using standard cells: (a) The cell should not be exposed to temperatures below 4°C or above 40°C. (b) Abrupt changes of temperature should be avoided because they may produce temporary variations of several hundredths of 1 percent in the electromotive force. (c) All parts of the cell should be at the same temperature. (d) Current in excess of 0.0001 amp should never pass through the cell. (e) The electromotive force of the cell should be redetermined at intervals of a year or two.

REFERENCE ELECTRODES.   The cells whose electromotive forces are to be determined may be considered to be made up of two electrodes, one an indicator and the other a reference electrode. The calomel electrodes and the silver–silver chloride electrode are the most commonly used reference electrodes.

There are three types of calomel electrodes in common use, depending upon the concentration of the potassium chloride solution used: 0.1 N, N, or saturated. The electromotive forces at 25° of the three calomel electrodes are as follows:[9]

$Hg, Hg_2Cl_2(s); KCl (0.1 N)$ $E = 0.3356$
$Hg, Hg_2Cl_2(s); KCl (1.0 N)$ $E = 0.2802$
$Hg, Hg_2Cl_2(s); KCl (sat)$ $E = 0.2444$

Commercially available saturated calomel cells such as that illustrated in Fig. 43 (page 187) are especially convenient for work in the physical chemistry laboratory.

For the silver–silver chloride electrode the electromotive force at 25°C is given by the statement

$Ag(s), AgCl(s); KCl (0.1 N)$ $E = 0.2880$

The calomel electrode may be prepared as follows. The electrode consists of a test tube with a sidearm bent down at right angles and fitted with a ground-glass plug which serves as a salt bridge, the current being carried by the thin liquid film between the wall of the tube and the plug. It is preferable to have the tube made of Pyrex and the plug of soft glass. If the plug is cooled before insertion into the tube, it will become firmly fixed upon subsequent warming to room temperature. This type of junction has a high resistance and requires the use of a high-resistance galvanometer (approximately 100 ohms) in the potentiometer circuit. A junction of this type should be kept immersed in potassium chloride solution when not in use so that the liquid film will not dry out. An alternative method of reducing diffusion at the junction involves constricting the end of the bridge arm to a tip which contains solidified agar. This connecting bridge has a much lower resistance than the type with a ground-glass joint. Better still is a simple connecting salt bridge with sintered glass at the ends.[10]

The glass vessels are cleaned and rinsed thoroughly, using distilled water for the latter operation. The calomel paste is made by grinding calomel in a mortar with purified mercury and potassium chloride solution of the concentrations indicated above (depending on the type of electrode to be used). A few milliliters of redistilled mercury is placed in the tube and is covered with calomel paste to a depth of approximately 1 cm. The tube and the bridge are then filled with potassium chloride solution of the desired concentration, and an electrode consisting of a platinum wire fused to a copper wire, which is sealed in a glass tube mounted in a stopper, is placed in the pool of mercury.

In all precise electromotive-force work, oxygen must be carefully excluded from the cell. This may be done by bubbling purified nitrogen through the solutions in the cell for a period of time. Tank nitrogen is readily freed from small amounts of oxygen by passing over heated copper turnings. The effluent nitrogen is bubbled first through a sample of the solution in a presaturator and then into the solutions in order to avoid excessive evaporation from the latter. Another type of purification system is described by van Brunt.[11]

MEASUREMENT OF ELECTROLYTIC CONDUCTANCE [12]

AC WHEATSTONE BRIDGE. The measurement of electrolytic conductance using the Wheatstone bridge has been described in Experiment 27. Alternating current must be used to prevent electrical polarization of the electrodes, and this introduces a number of problems not present in dc bridge measurements. A simple Wheatstone bridge for measurements of resistance with direct current is illustrated in Fig. 39. Resistance R_3 is adjusted to bring points A and B to the same potential, as indicated by the absence of a galvanometer deflection when the tap switch is closed. If a dc source is used, only pure resistances are involved, and when A and B are at the same potential,

$$\frac{R_1}{R_2} = \frac{R_3}{R_4}$$

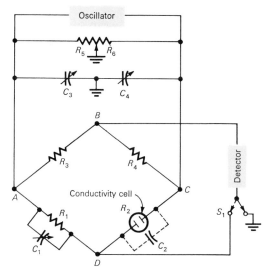

figure 146. Alternating-current bridge circuit.

If an alternating source of current is used, it is necessary to balance the reactances in the circuit as well as the resistances. In the ac bridge circuit illustrated in Fig. 146 the conductance cell presents an admittance which includes a capacitative term as well as a conductance term, because there is an accumulation of charge on each plate as well as a flow of charge through the cell when a potential difference exists between the terminals. The cell is therefore electrically equivalent to a resistance R_2 in parallel with a capacitance C_2. The balance condition for alternating current can be satisfied only by introducing at least one other reactive element. Capacitor C_1 is added for this purpose. For a null output, the resistors and C_1 must be adjusted to make the ac potential at points B and D equal in respect to both phase and amplitude. These dual conditions are met when the equations

$$\frac{R_1}{R_2} = \frac{R_3}{R_4} = \frac{C_2}{C_1}$$

are satisfied.

In practice, adjusting R_1, R_3, R_4, and C_1 leads to a minimum in bridge output but not to a perfect null. The reason is that if D and B are not at ground potential, current flows through the detector or earphones to the ground via distributed capacity. In precise measurements, it is necessary to provide means for adjusting points D and B to ground potential. The most common circuit for this purpose is referred to as a Wagner earthing device. As illustrated in Fig. 146, this device consists of adjustable resistances (R_5 and R_6) and capacitances (C_3 and C_4) between A and ground and between C and ground. After adjusting the bridge to the minimum signal of the detector or earphones, the switch S_1 is closed to the grounded position, and the resistances and capacitances of the Wagner earthing device are adjusted to the minimum signal of the detector. In this way, point B is brought to ground potential. It should be noted that since alternating current is used, the potentials of C and A vary sinusoidally with time, one being above ground potential while the other is below ground potential. Finally, the main-bridge balance is readjusted, with the detector in the original position.

The source of power is usually a vacuum-tube oscillator such as that described on page 634. Such an oscillator may be designed to give a pure-sine-wave current so that the current in one direction exactly offsets that in the other. At audio frequencies a telephone headset may be used as a detector, but to attain the best results, an amplifier must be used, since the bridge current should be maintained at a low value in order to avoid heating effects in the conductivity cell. This is achieved by limiting the voltage input from the oscillator.

Edelson and Fuoss[13] have given the specifications for the construction of a portable audio-frequency conductance bridge. They discuss the use of an oscilloscope for obtaining the balance points.[13,14] This method is more sensitive and indicates the resistance and capacitance balance separately.

The theory and design of ac bridges for measuring the conductance of electrolytic solutions have been discussed by Jones and Josephs[15] and by Shedlovsky.[16]

The construction of bridges for high-precision work (accuracy of 0.02 percent or better) has been discussed by Dike[17] and Luder.[18]

The resistance coils of the Wheatstone bridge must be wound noninductively; i.e., the wire is doubled back in the middle, and the two parts of the wire with current going in opposite directions are side by side. Coils can be constructed in this way so that the difference between dc and 20,000 Hz ac resistances is less than 0.01 percent.

Electrolytic conductance may also be measured with direct current using nonpolarizable electrodes.[19] This method is capable of quite precise results, but it is applicable only to those electrolytes for which nonpolarizable electrodes are available.

CONDUCTIVITY CELLS.   A number of forms of conductivity cells are shown in Fig. 147. The cells are usually constructed of highly insoluble glass, such as Jena 16III or Pyrex, or of quartz. The platinum electrodes should be heavy and well anchored, so that the cell constant will not change when the cell is used frequently. The conductivity cell for a given measurement should be chosen with an appropriate cell constant, so that the resistance will not fall far below 1000 ohms, where excessive polarization difficulties are encountered with the usual apparatus, or above 10,000

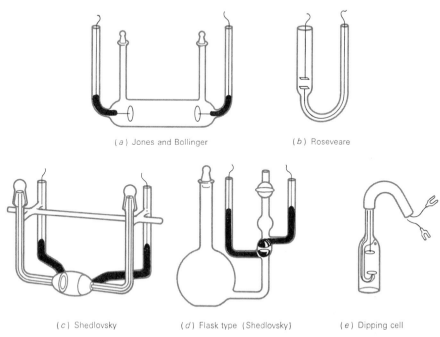

(*a*) Jones and Bollinger   (*b*) Roseveare

(*c*) Shedlovsky   (*d*) Flask type (Shedlovsky)   (*e*) Dipping cell

figure 147. Conductivity cells.

to 30,000 ohms, where errors due to insulation leakage are encountered. For solutions of low conductance, the electrodes should be large and close together. For solutions of high conductance, the electrodes should be smaller and farther apart. Jones and Bollinger[20] have shown that in many cells which have been commonly used, the filling tubes are relatively close to the electrode leads, so that disturbing parasitic currents can flow through capacitance-resistance paths, and these can produce variations in cell constant with resistance. The cell designed by Jones and Bollinger is shown in Fig. 147*a*. The Roseveare cell (Fig. 147*b*), with the corners of the thin platinum plates welded in the glass, is easy to make. In measuring the conductivity of solutions which show any tendency to foam, it is desirable to use a cell with conical electrodes (Fig. 147*c*), through which the electrolyte is flowed into the cell. The flask-type cell shown in Fig. 147*d* is useful for preparing and measuring the conductivities of very dilute solutions without risk of contamination from atmospheric or other impurities.[21] The dipping-type cell (Fig. 147*e*) is not suitable for precise measurements but is often convenient for practical measurements. A Freas-type conductivity cell is illustrated in Fig. 40 (page 172).

Polarization may be practically eliminated by using a pure-sine-wave alternating current of moderate frequency and by coating the electrodes with platinum black. The electrodes can be platinized by immersing them in a solution containing 3 g of platinic chloride and 0.02 g of lead acetate in 100 ml of water and connecting them to two dry cells connected in series. The current is regulated by means of a rheostat, so that only a small amount of gas is evolved. After the electrodes are coated with platinum black, they are removed from the solution and thoroughly washed with distilled water. Any traces of chlorine adsorbed from the plating solution may be removed by continuing the electrolysis, with the same connections, in a dilute solution of sulfuric acid. In precise measurements of electrical conductance, it is especially important to test for the possibility of polarization.[22] Polarization has the effect of increasing the measured resistance and, in general, is less important at higher frequencies.

A cell with platinized electrodes should always be filled with distilled water when stored.

For work in very dilute solutions the platinized surface must be dispensed with because it is so difficult to rinse out the last traces of electrolyte from it. Bright platinum electrodes are used, but since some polarization results, it is necessary to make measurements at several frequencies and to extrapolate the values to infinite frequency.

CONDUCTANCE OF POTASSIUM CHLORIDE SOLUTIONS.   In a very careful and exacting research, Jones and Bradshaw[23] have redetermined the electrical conductance of standard potassium chloride solutions for use in the calibration of conductance cells. The results of the work are summarized in Table 1. The values given in this table do not include the conductance due to water, which must be added and should be less than $\kappa_{H_2O} = 10^{-6}$ ohm^{-1} cm^{-1} in work with dilute solutions. The potassium chloride should be fused in an atmosphere of nitrogen to drive out water, and in the case of salts which are deliquescent, it is necessary to use a Richards bottling apparatus[24] to avoid exposure to air.

Table 1. Specific Conductance of Standard Potassium Chloride Solutions

Grams of potassium chloride per 1000 g of solution (in vacuum)	Specific conductance, ohms^{-1} cm^{-1}		
	0°C	18°C	25°C
71.1352	0.065176_a	0.097838	0.111342
7.41913	0.0071379	0.0111667	0.0128560
0.745263	0.00077364	0.00122052	0.00140877

[a] Lowering of the last figure indicates that it is uncertain.

CONDUCTANCE WATER.   In all conductance measurements made in aqueous solution it is necessary to have very pure water. Distillation in a seasoned glass vessel and condenser with ground-glass joints or with a block-tin condenser can give water with a specific conductance of about 1×10^{-6} ohm^{-1} cm^{-1} if a little potassium permanganate is added to the flask. If such a distillation is carried out in air, the water is saturated with the carbon dioxide of the air (0.04 percent). Some of the dissolved carbon dioxide can be removed to give a higher resistance by bubbling carbon dioxide-free air through the water.

It is interesting to note that Kohlrausch and Holborn[25] reported the preparation of purified water with a specific conductance at 18° of only 0.043×10^{-6} ohm^{-1} cm^{-1}.

Conductance water for laboratory use may be prepared on a large scale by redistilling distilled water and condensing in a block-tin condenser. By condensing the water at relatively high temperatures, the absorption of carbon dioxide is reduced.

MEASUREMENT OF CURRENT AND QUANTITY OF ELECTRICITY

The most direct method for the measurement of quantity of electricity involves the use of a coulometer, as, for example, a silver coulometer, in which silver is plated on a platinum crucible from a silver nitrate solution. Other coulometers have been used in which iodine is liberated from a potassium iodide solution and titrated with standard sodium thiosulfate solution, copper is deposited from an acidified copper sulfate solution, or water is decomposed and the volume of gas evolved is measured.

The most accurate and convenient method for the determination of a steady current is to measure the potential drop across a standard resistance through which the current flows, as illustrated earlier on page 29. The current is calculated by Ohm's law. When accurate resistors are required in the laboratory, calibrated resistors of the National Bureau of Standards type[26] should be used. These resistors are constructed of selected manganin wire and are immersed in oil. The limit of error is 0.01 percent for wattage dissipation up to 0.1 watt and 0.04 percent up to 1.0 watt. Since the limit of error on a good standard resistance is 0.01 percent and the potential may be determined versus a standard cell for which the electromotive force is known to 0.01 percent, the current may be calculated with considerable accuracy. For accurate measurements of current in transference or electrical-heating experiments, it is desirable to have rather constant current. This may be accomplished by means of an electronic current regulator[27] or a commercial constant-current power supply.

MEASUREMENT OF ELECTRICAL ENERGY
IN CALORIMETRY[28]

Calorimetric data are generally given in units of electrical energy. The unit is the absolute joule, the product of absolute volts, absolute amperes, and time in seconds. The 15°-calorie is 4.1840 abs joules.[29] With the voltage E in absolute units, the resistance R in absolute ohms, and the time in seconds, the energy dissipated is E^2t/R absolute joules, or $E^2t/4.1840R$ defined calories. The resistance of the heater may be measured with a Wheatstone bridge, or it may be computed from the voltage drop across the heater, and the current determined by measuring the voltage drop across a standard resistor connected in series with the resistor. Current and voltage drop are maintained at a constant value during the performance of an experiment. A basic wiring diagram is shown in Fig. 148.

The heating coil H is of manganin or other wire having a low temperature coefficient of resistance. It may be wound on mica, insulated between two mica sheets, and encased in a silver or copper sheath or made by winding resistance wire on a threaded tube of anodized aluminum.

The current is supplied by steady storage batteries B in good condition or preferably by an electronically regulated supply. The standard resistance R consists of

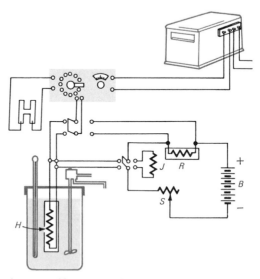

figure 148. Measurement of electrical energy in calorimeter.

uncovered manganin or constantan wire or other alloy having a negligible temperature coefficient of resistance. It is immersed in oil to keep the temperature nearly constant, and it should be proved that the current used in the experiment does not raise the temperature of the wire sufficiently to change its resistance. Any excessive heating effect may be reduced by using wire of larger diameter and greater length. The resistance of R should be chosen so that the potential drop across it may be conveniently measured with the potentiometer.

The potentiometer is used to measure the potential drop across the heater or across the standard resistance. It is not permissible to use a voltmeter, because the voltmeter itself carries some current and acts as a shunt around the resistance which is being measured. A voltmeter may be used, however, in a compensation method, if it is used with a galvanometer.

The upper double-pole double-throw switch connects the potentiometer either to the heating coil or to the standard resistance. It must be of good quality, with good contacts and no electrical leakage across the base.

The current is kept constant by continuous adjustment of the rheostats. It is important to have the contacts of the rheostat in good condition. They should be rubbed with emery paper frequently and coated with a thin film of petroleum jelly. Two rheostats are convenient, one for coarse adjustments and one for fine adjustments.

If plenty of batteries are available, it is well to use a large number and bring the current down with a high resistance. In this way, any slight change in the resistance of the circuit has a slight effect on the current. The circuit is closed by throwing the upper double-pole double-throw switch to the left. Before starting a determination, the lower switch is thrown to the right for several minutes so that the current will flow through a resistance J which is approximately equal to the resistance of the heating coil. In this way the battery reaches a steady condition before the experiment is started. If the switch is thrown immediately to the left from the position of open circuit, the battery voltage drops rather rapidly at first and renders difficult the control of the current at a constant amperage. The calculation is simplified and the accuracy is increased if the current is kept constant throughout the experiment.

The time of passage of the current is determined with a stopwatch or an electric clock or timer. If the time, as measured with a stopwatch, is the least accurate factor, a chronometer may be used, or the time may be increased by decreasing the rate of heating. Stopwatches used for such work should be checked frequently, since they are likely to get out of order.

For precision work an equipotential shield is provided to eliminate stray currents. All the instruments are set on a piece of sheet metal, which is grounded, and under every insulator is placed a grounded metal shield.

MEASUREMENT OF TRANSFERENCE NUMBERS AND IONIC MOBILITIES[30]

The moving-boundary method[31] for the determination of transference numbers and ionic mobilities has largely replaced the earlier Hittorf method. This has happened because the velocity of a moving boundary may be measured considerably more accurately (to ± 0.02 percent) than the change in concentration of an ion in an electrode chamber of the Hittorf apparatus. The moving-boundary method also has the advantage that it may be applied to mixtures of ions, particularly proteins. In the moving-boundary apparatus used in Experiment 27, the anode was made of metallic cadmium, so that the solution following the moving boundary was cadmium chloride. In general, it is desirable to be able to use other salts as following electrolyte, and so there is the problem of forming an initially sharp boundary between the leading electrolyte and the following electrolyte. This is best achieved by the shearing mechanism invented by MacInnes and Brighton.[32] The construction of the modern apparatus is shown in Fig. 149. Heavy glass plates C_1, C_2, C_3, and C_4 are ground so that C_1 and C_2, and C_3 and C_4, fit well together and rotate on each other. The moving-boundary tube A is mounted in C_2 and C_3. The silver–silver chloride electrodes E' and E are attached to C_1 and C_4. Electrode chambers are required to prevent products of the electrode reaction from reaching the moving-boundary cell. If the boundary is to be a rising boundary, it is formed at C_3-C_4 as follows: Electrode vessel E and the tube connecting it to C_4 are filled with the indicator electrolyte; the glass plates C_1 and C_2 are clamped firmly together, and C_3 is rotated with respect to C_4 so that initially the tube A does not connect to electrode E. Tube A and E' are then filled with the leading electrolyte. Electrode E' is shut to the atmosphere, and A is rotated into juxtaposition with the hole in C_4, connecting E. Upon application of the current, the sharp boundary formed by this method moves up tube A. The time required for the boundary to move between graduations on A is determined. Since it is not practical to use coulometers for the measurement of current, it is desirable to hold the current constant by one of the devices mentioned on page 628 and to use a potentiometer and an accurately known resistance to measure the current.

It is necessary to apply a correction for the volume change due to the electrode reaction and to the migration of ions into and out of the region between the moving boundary and the closed electrode.[33]

If neither the indicator nor the leading electrolyte contains colored ions, it is necessary to locate the boundary by the difference in refractive index of the two solutions. In the case of sharp boundaries, this may be done by focusing a lens on the boundary while placing an illuminated slit behind and somewhat below the boundary. If the refractive-index gradients are not sharp, as in the case of moving

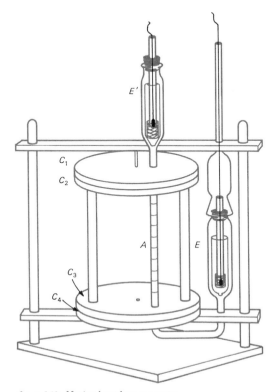

figure 149. Moving-boundary apparatus.

protein boundaries, more complicated optical methods must be used for the determination of boundary velocity. One such method is the schlieren optical system, which is described in Chapter 21.

MEASUREMENT OF DIELECTRIC CONSTANT

Some methods for the measurement of dielectric constant are more suitable for use in dipole-moment determinations, while others are preferred where dielectric loss is involved. Restricting the description to the former, we may classify them as heterodyne-beat methods, resonance methods, and bridge methods. The first two have been discussed to a certain extent in connection with Experiments 34 and 35. An interesting modification of resonance method has been described by Wyman[34] for use when the solutions under investigation have an appreciable electrical conductance.

The bridges for measurement of the dielectric constants of solutions are normally made up of two resistance arms and two capacity arms. By a symmetrical construction of the bridge elements, the inductance is kept to a minimum amount. One of the advantages of the bridge is that if the solution or material being studied has some slight conductance, allowances for it can be made in balancing the bridge and a loss factor determined.

The steady current flowing in a dc circuit depends upon the magnitude of the applied potential and the resistance of the circuit. In the case of alternating current, the equilibrium current is limited by the circuit impedance Z, which depends not only on the circuit resistance R but also on the reactance X, which arises from the capacitance and inductance in the circuit. The reactance, and hence the impedance, is a function of the frequency of the current. The phase relationships between the currents and voltages in the various branches of an ac circuit depend also upon the reactances of the components involved.

The impedance bridge is a four-terminal network of the type shown in Fig. 150a, in which the arms are impedances, which will, in general, consist of some combination of resistance, capacitance, and inductance. The familiar Wheatstone bridge is a particular type of impedance bridge where all four arms are pure resistances; in ac bridge operation this represents a limiting condition only, as indicated below. Two common forms of impedance bridge are shown in Figs. 150b and c, together with a statement of the conditions which are satisfied at bridge balance, i.e., zero voltage across the detector. These conditions are obtained from the requirements that the voltages appearing at the two points across which the detector is connected (measured relative to some common reference point) must be equal both in magnitude and in phase when the bridge is balanced. The frequency of the applied voltage may or may not enter explicitly into the balance equations.

Capacitance measurements with the impedance bridge are usually made by the

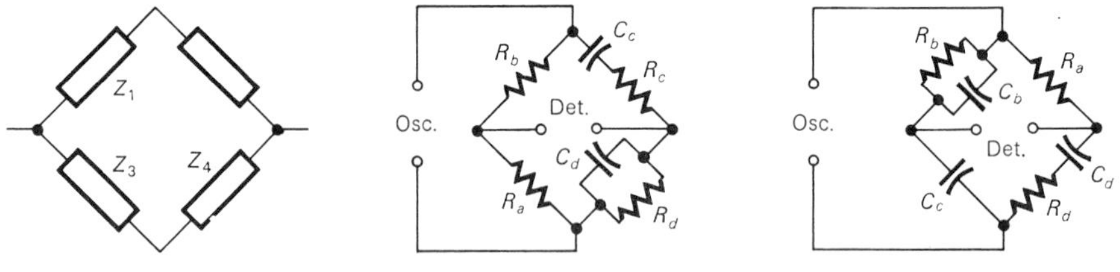

$$4\pi^2 f^2 = \frac{1}{R_d R_c C_d C_c} \qquad \frac{C_d}{C_c} = \frac{R_b}{R_a} - \frac{R_c}{R_d} \qquad C_d = \frac{R_b}{R_a} C_c \qquad R_d = \frac{C_b}{C_c} R_a$$

(a)   (b)   (c)

figure 150. (a) *Generalized representation of impedance bridge;* (b) *Wien impedance bridge;* (c) *Schering impedance bridge.*

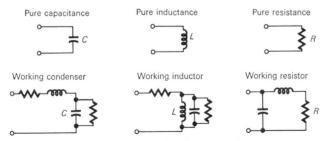

figure 151. Schematic representation of working circuit elements.

substitution method. In the parallel-substitution method, a standard capacitor is connected in the bridge circuit in parallel with the unknown capacitor, and the bridge is balanced. The unknown capacitor is then removed, and the bridge again balanced by resetting the standard capacitor. The change in capacitance of the standard required to reestablish balance is then equal to the capacitance of the unknown plus any changes in lead capacitances or other stray-capacitance effects which may have been involved. The series-substitution method may also be employed. Because of these lead capacities, etc., it is desirable where possible to use a variable-capacitance cell for the measurement of dielectric constants, as described under Experiments 34 and 35, since high accuracy is thereby achieved.

It is customary to think of resistances, capacitances, and inductances as separately realizable entities. In actual practice, however, it is impossible to construct a pure resistance, capacitance, or inductance. Any circuit element is an impedance in which one contribution may predominate but in which all three appear. As the operating frequency increases, this fact becomes more and more important. In Fig. 151 are given the schematic representations of capacitance, inductance, and resistance, showing to a first approximation how each of these circuit elements behaves in an ac circuit.

Careful attention to proper shielding and grounding of the bridge is required in order to prevent the environment, including the operator, from influencing the performance of the bridge. Successful work, particularly at high frequencies, requires a sound understanding of the theory of ac bridge circuits, which is treated comprehensively by Hague[35] and by Hartshorn,[36] and a good background of practical experience in the field.

The power source for the bridge must have adequate frequency stability and power output. At audio frequencies the resistance-capacitance-coupled oscillator mentioned above is a versatile and reliable unit. At radio frequencies, crystal-controlled oscillators or stable variable-frequency oscillators of several types are available. Suitable power sources in practically any frequency range are available from commercial manufacturers.

MEASUREMENTS AT VERY HIGH FREQUENCIES.   At frequencies in excess of 10^8 Hz the difficulties encountered in working with the traditional circuits become very great. There is the frequency range 10^8 to 3×10^9 Hz, in which the measuring techniques are still rather unsatisfactory, and there is the microwave region, 3×10^9 to 6×10^{10} Hz, that portion of the frequency spectrum in which the free-space wavelength corresponding to the frequency of oscillation is comparable in magnitude with that of the laboratory equipment. A large variety of satisfactory equipment has now been developed for use at the microwave frequencies, and the precautions which are required in order to obtain accurate measurements have been set down. These methods of measurement of dielectric constant are classed as being resonator methods or transmission methods; both depend on the transmission properties of a dielectric-filled waveguide or coaxial line. (A resonator can be considered as a length of waveguide short-circuited at both ends.) The technique and equipment employed in such work have been described in a number of places, including text and reference works.[37]

References

1. *Natl. Bur. Std. U.S. Circ.* C459, 1947.
2. J. Strong, "Procedures in Experimental Physics," pp. 328, 590, Prentice-Hall, Inc., Englewood Cliffs, N.J., 1938.
3. V. Z. Williams, *Rev. Sci. Instr.*, **19:** 135 (1948).
4. H. Gershinowitz and E. B. Wilson, Jr., *J. Chem. Phys.*, **6:** 197 (1938); F. S. Mortimer, R. B. Blodgett, and F. Daniels, *J. Am. Chem. Soc.*, **69:** 822 (1947).
5. C. Tanford and S. Wawzonek in A. Weissberger (ed.), "Technique of Organic Chemistry," vol. 1, "Physical Methods in Organic Chemistry," 3d ed., pt. 4, Interscience Publishers, Inc., New York, 1960.
6. L. Behr, *Rev. Sci. Instr.*, **3:** 109 (1932).
7. J. E. Witherspoon, *Instruments*, **25:** 900 (1952).
8. A. Tustin (ed.), "Automatic and Manual Control," pp. 249ff, Butterworth & Co. (Publishers) Ltd., London, 1952.
9. W. J. Hamer, *Trans. Electrochem. Soc.*, **72:** 45 (1937).
10. H. A. Laitinen, *Ind. Eng. Chem. Anal. Ed.*, **13:** 393 (1941).
11. C. van Brunt, *J. Am. Chem. Soc.*, **36:** 1448 (1914).
12. R. A. Robinson and R. H. Stokes, "Electrolyte Solutions," 2d ed., p. 87, Academic Press Inc., New York, 1959; T. Shedlovsky in A. Weissberger (ed.), "Technique of Organic Chemistry," vol. 1, "Physical Methods in Organic Chemistry," 3d ed., pt. 4, Interscience Publishers, Inc., New York, 1960.
13. D. Edelson and R. M. Fuoss, *J. Chem. Educ.*, **27:** 610 (1950).
14. G. Jones, K. J. Mysels, and W. Juda, *J. Am. Chem. Soc.*, **62:** 2919 (1940).
15. G. Jones and R. C. Josephs, *J. Am. Chem. Soc.*, **50:** 1049 (1928).
16. T. Shedlovsky, *J. Am. Chem. Soc.*, **52:** 1793 (1930).
17. P. H. Dike, *Rev. Sci. Instr.*, **2:** 379 (1931).
18. W. F. Luder, *J. Am. Chem. Soc.*, **62:** 89 (1940).

19. L. V. Andrews and W. E. Martin, *J. Am. Chem. Soc.*, **60:** 871 (1938); H. I. Shiff and A. R. Gordon, *J. Chem. Phys.*, **16:** 336 (1948).
20. G. Jones and G. M. Bollinger, *J. Am. Chem. Soc.*, **53:** 1411 (1931).
21. T. Shedlovsky, *J. Am. Chem. Soc.*, **54:** 1411 (1932).
22. G. Jones and G. M. Bollinger, *J. Am. Chem. Soc.*, **57:** 280 (1935).
23. G. Jones and B. C. Bradshaw, *J. Am. Chem. Soc.*, **55:** 1780 (1933).
24. T. W. Richards and H. G. Parker, *Proc. Am. Acad. Arts. Sci.*, **32:** 59 (1896).
25. F. Kohlrausch and L. Holborn, "Leitvermögen der Elektrolyte," 2d ed., Teubner Verlagsgesellschaft, Leipzig, 1916.
26. E. B. Rosa, *J. Res. Natl. Bur. Std. U.S.*, **4:** 121 (1912).
27. D. J. Le Roy and A. R. Gordon, *J. Chem. Phys.*, **6:** 398 (1938); P. Bender and D. R. Lewis, *J. Chem. Educ.*, **24:** 454 (1947).
28. J. M. Sturtevant in A. Weissberger (ed.), "Technique of Organic Chemistry," vol. 1, "Physical Methods in Organic Chemistry," 3d ed., pt. 1, Interscience Publishers, Inc., New York, 1959.
29. H. F. Stimson, *Am. J. Phys.*, **23:** 614 (1955).
30. R. A. Robinson and R. H. Stokes, "Electrolyte Solutions," 2d ed., pp. 43, 102, Academic Press Inc., New York, 1959; M. Spiro in A. Weissberger (ed.), "Technique of Organic Chemistry," vol. 1, "Physical Methods of Organic Chemistry," 3d ed., pt. 4, Interscience Publishers, Inc., New York, 1960.
31. D. A. MacInnes and L. G. Longsworth, *Chem. Rev.*, **11:** 171 (1932).
32. D. A. MacInnes and T. B. Brighton, *J. Am. Chem. Soc.*, **47:** 994 (1925).
33. L. G. Longsworth, *J. Am. Chem. Soc.*, **65:** 1755 (1943); L. G. Longsworth and D. A. MacInnes, *ibid.*, **62:** 705 (1940).
34. J. Wyman, Jr., *Phys. Rev.*, **35:** 623 (1930).
35. B. Hague, "Alternating Current Bridge Methods," 5th ed., Sir Isaac Pitman & Sons, Ltd., London, 1945.
36. L. Hartshorn, "Radiofrequency Measurements by Bridge and Resonance Methods," John Wiley & Sons, Inc., New York, 1941.
37. S. Roberts and A. von Hipple, *J. Appl. Phys.*, **17:** 610 (1946); C. H. Collie, J. B. Hasted, and D. M. Ritson, *Proc. Phys. Soc. London*, **60:** 71 (1948); H. M. Barlow and A. L. Cullen, "Microwave Measurements," Constable & Co., Ltd., London, 1950; C. F. Montgomery (ed.), "Technique of Microwave Measurements," Radiation Laboratory Series, vol. 11, McGraw-Hill Book Company, New York, 1948.

Chapter 25

Spectroscopy

The term *spectroscopy* is applied to the study of the interaction of matter with an electromagnetic field. This interaction produces absorption, emission, and scattering of the radiation, and energy is in general transferred between the field and the system. A study of the strength of the interaction as a function of the frequency of the radiation, or other variables, leads to information about the energy levels of the system and about transitions between them. Spectroscopic techniques can be classified according to the type of information about molecules they yield or according to the experimental techniques used. The latter classification will be used here. Some experimental techniques which are not of the general type described above are also called spectroscopy, e.g., mass spectroscopy (discussed in Chapter 27).

SPECTROMETER CHARACTERISTICS

The essential features of a spectrometer are a source of radiation, a sample container, and a detector. If the source is broadband, producing many frequencies at once, some kind of monochromator or filter must also be included.

Spectroscopic work covers a very wide range of frequencies, as illustrated in Fig. 152. The devices used to generate, transmit, and detect the electromagnetic fields, and to determine the frequency or wavelength, vary from one region of the electromagnetic spectrum to another. Techniques may be classified into three broad categories depending on the relationship between the wavelength λ of the radiation and the characteristic dimensions l of the apparatus.

Circuit methods: $\lambda \gg l$
Transmission-line methods: $\lambda \sim l$
Optical methods: $\lambda \ll l$

An additional category might be added at the short-wavelength end, when λ becomes comparable to, or less than, atomic dimensions, since still different experimental techniques must then be used. The fact that laboratory apparatus is typically 0.1 to 10 cm in size establishes approximate limits on these regions, as indicated in Fig. 152. The energy changes introduced by the frequencies in question are also indicated in the figure, together with the types of energy levels which are commonly investigated in these spectral regions.

Spectrometers are judged on the basis of several concepts, which will be introduced before specific spectrometers are considered.

RESOLUTION.   One important characteristic of a good spectrometer is its ability to separate closely spaced lines corresponding to different absorption processes. The resolution of a spectrometer is measured in terms of the observed width of lines which are known to be inherently narrow or, equivalently, by the minimum separation at which two overlapping lines can be recognized as distinct.

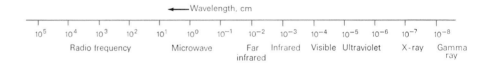

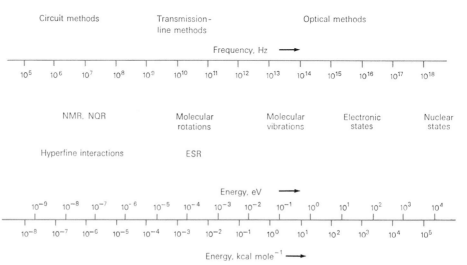

figure 152. Electromagnetic spectrum and associated energies.

All spectral lines in a plot of, say, adsorption versus frequency, appear with a finite line width and a definite line shape. Two common line shapes are the lorentzian and the gaussian,† illustrated in Fig. 153. An ideal spectrometer will not distort the line shape or add to the line width. These properties then give additional information about the system being studied. Unfortunately, in real spectrometers there are often instrumental sources of line width and of line-shape distortion.

ABSORPTION AND DISPERSION.   The most familiar spectroscopic situation deals with the absorption of radiation of a particular frequency by the sample. It is often also possible, and sometimes desirable, to investigate dispersion by determining the frequency dependence of the refractive index, or something closely related to it. These phenomena are intimately related: it is quite characteristic for rapid

† A discussion of these line-shape functions is given in Ref. 1.

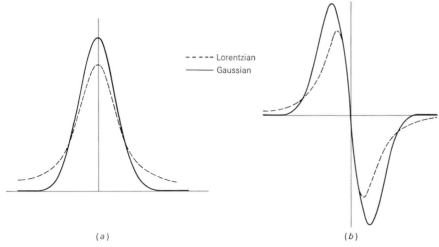

figure 153. *Lorentzian and gaussian line shapes: (a) absorption; (b) derivative of absorption.*

changes in refractive index with frequency to occur in the neighborhood of absorption lines (Fig. 154). Dispersion should not be confused with the first derivative of absorption produced by some experimental techniques (page 293) although the line shapes may appear similar (Fig. 153).

The same molecular processes which produce optical absorption and dispersion are manifested somewhat differently in the case of a substance which has been closely coupled to an electric circuit by being placed between the plates of a capacitor or within a closely wound coil. In the circuit case, absorption of energy by the sample has an effect equivalent to the introduction of resistance into the circuit, while dispersion effectively causes a change in the apparent value of the capacitance or the inductance, as the case may be. In the radio-frequency and microwave regions it is frequently convenient to relate absorption and dispersion to the phase of transmitted or reflected radiation, relative to that of the incident beam.

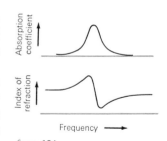

figure 154.

The typical relationship between absorption and dispersion.

SENSITIVITY.   The amount of sample which can effectively interact with radiation in the spectrometer is usually limited. This may be due to limited availability of the material being studied or to essential instrumental restrictions. In addition, the strength of the interaction varies greatly from one line to another, even within one spectral region. The ability of the spectrometer to detect and accurately measure even a very weak interaction is thus critical.

Since electronic signals can readily be amplified, the factor determining spectrometer sensitivity is the signal-to-noise ratio at the detector output. The noise is most often generated in the detector, but the radiation source sometimes produces a significant amount of noise also. The measure of the sensitivity of a spectrometer is the signal-to-noise ratio obtainable for a given absorption coefficient or, equivalently, the minimum detectable absorption coefficient.

Sensitivity can often be improved by an averaging process, since noise is a random function of time, while the signal level at each point in the spectrum is in principle constant. One method of sensitivity enhancement is to scan over the spectral range being investigated many times. The signal intensities are measured at closely spaced points (typically 1024 points over the whole spectrum), and the values at each point are averaged over the many sweeps. This can be done by a device that is essentially a simple-minded digital computer, commonly known as a *computer of average transients* (CAT). A second, different, approach is to reduce the bandwidth of the spectrometer and sweep very slowly over the spectral range, in effect averaging one point at a time by means of a long time constant in the output.

BANDWIDTH AND RESPONSE TIME.   Since the noise power generated in electrical systems is proportional to the bandwidth (page 597), it is advantageous to reduce the bandwidth of the amplification system of a spectrometer to a minimum. One effective method of doing this is to code the information at a particular frequency and to detect only this frequency in a phase detector (page 634).

The response time of the system τ is inversely related to the bandwidth. As the bandwidth is reduced, the response time is inescapably lengthened, and thus the rate at which the spectrum can be swept is reduced. An attempt to sweep too rapidly for the response time of the system will result at first in distortion of line shapes or, if extreme enough, in the complete loss of the signal. A time of the order of $10\,\tau$ should be allowed for passing through a line.

At the very slow sweep rates required by a long time constant, new sources of noise become important. The slow variation of voltages, source intensity, etc., called *drift* are no longer negligible. A practical limit is thus established for the narrow-bandwidth averaging technique. Some problems, such as variation in source intensity, can be partially alleviated by using comparative measurements, in which a portion of the source radiation which does not interact with the sample is compared with that which does.

SATURATION.   It often appears that the sensitivity of a spectrometer could be increased by using more intense radiation. If the sample absorbs a certain fraction of the incident radiation, then the absolute magnitude of the absorption will be greater, and therefore more easily detected, if the incident intensity is increased. This is not always the case, however.

The probability of causing a molecule in state a, with energy E_a, to make a transition to a state b, with a different energy E_b, by applying radiation of frequency

$\nu = |E_b - E_a|/h$ is exactly equal to the probability of causing a molecule in state b to make a transition to state a. The two processes, absorption and stimulated emission, will thus differ in rate in a bulk sample only if there is a population difference between states a and b. If state b, with $E_b > E_a$, has a lower population than state a, then more $a \rightarrow b$ transitions than $b \rightarrow a$ transitions will be induced and there will be a net absorption of energy by the sample from the radiation. An absorption spectrum can in principle be observed in this case, but it also follows that the population of state b will tend to increase relative to that of state a. The spectrum may thus disappear. This effect is known as *saturation*, and it becomes more important as the power in the radiation field is increased. If a spectrum is to be observed under steady-state conditions, some mechanism must exist whereby equilibrium populations tend to be reestablished.

RELAXATION.   Two types of mechanisms exist. There is a finite probability of spontaneous emission from an excited state, returning it to the ground state. This spontaneous emission occurs in the absence of a radiation field, or when a field is present, it adds to the probability of downward transitions. It is frequency-dependent and is highly effective in the infrared and above. It is rather ineffective in the microwave and radio-frequency regions.

The other mechanism involves the transfer of energy in the system from one degree of freedom to another without radiation. Thus energy may be transferred from electronic or nuclear spins, or from molecular rotations, to translational energy. At thermal equilibrium, the populations of all states are determined by their Boltzmann factors. In a nonequilibrium situation, states differing only in their rotation or their spin quantum numbers, for example, may not follow a Boltzmann distribution, or the distribution may be characteristic of a temperature different from the bulk temperature of the sample. In these cases reference is sometimes made to a rotational temperature, a spin temperature, etc. Such temperatures can be infinite, if states of different energy are equally populated, or even negative, if a higher-energy state is more highly populated than one of lower energy. Relaxation processes tend to redistribute energy among the different degrees of freedom, returning them all to equilibrium at a common temperature.

Although saturation can be a problem in spectroscopy, the study of relaxation processes under conditions of partial saturation gives additional information about the system under investigation and is itself an important branch of spectroscopy.

SOURCES AND DETECTORS

Sources of radiation are clearly necessary for spectroscopy, as well as for other applications. They are either broadband, like incandescent lamps; line, like mercury arcs and gas discharge lamps; or monochromatic, like lasers and electronic oscillators. Different sources also vary greatly in the intensity of the radiation they produce.

GAMMA RAYS.   Lying at the high-energy end of the electromagnetic spectrum, γ rays are emitted when atomic nuclei in excited states, often the product of radioactive decay, fall into lower-energy states. They may be emitted at one or a few discrete frequencies or over a frequency range. Sources of γ rays of particularly sharp frequency are used in Mössbauer spectroscopy, and will be discussed later under that heading. Gamma rays are very penetrating and, because of their high energy, can cause severe damage in many materials, including the human body. Gamma-ray sources must always be well shielded and appropriate precautions for handling radioactive sources observed (see Chapter 27).

X-RAYS.   X-rays are produced when one of the inner electrons in a heavy or moderately heavy atom is removed and an outer electron "falls into the hole." In an x-ray tube, electrons which have been accelerated to a high energy by a high voltage impinge upon a block of metal. The geometry of the system and the shape of the metal block concentrate the emission of x-rays in the desired direction. The wavelength of the x-rays is determined by the metal used, rather than by the energy of the electron beam, but of course a certain threshold energy must be exceeded for each specific wavelength produced.

X-rays also represent a radiation hazard, and shielding is necessary. It is particularly important to remember that radiation damage is cumulative and even a small exposure is to be avoided if it is repeated or extends over a long time.

The most important use of x-rays in chemistry is the x-ray diffraction study of crystal structure, which is not considered a spectroscopic technique and will not be treated here.[2,3]

ULTRAVIOLET.   A mercury-vapor arc in quartz is the most convenient source of light in the ultraviolet if the discrete wavelengths it provides are appropriate. The spectrum of the mercury arc is shown in Fig. 155. Several types of mercury arcs are commercially available. The General Electric AH-6 lamp is a very intense water-cooled lamp. Arc lamps involving other materials provide lines at a variety of wavelengths.

The radiation emitted by the hydrogen arc[4] is nearly continuous when the arc is designed so that the atoms recombine rapidly to give molecular hydrogen. It is valuable for obtaining absorption spectra in the ultraviolet at frequencies higher than the range provided by a tungsten lamp.

VISIBLE AND INFRARED.   A tungsten filament bulb is often the most convenient light source in the visible region of the spectrum. In the infrared, a silicon carbide

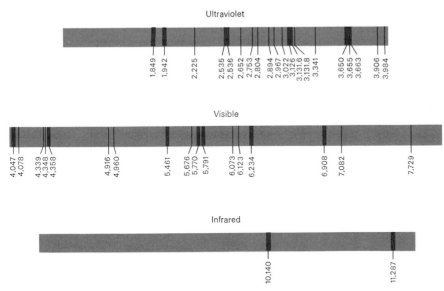

figure 155. Spectrum of the mercury arc. Numbers denote wavelengths in angstroms.

rod (Globar) or a Nernst filament, a spindle of rare-earth oxides, may be used.[5] In either case a solid material is heated electrically, and a continuous wavelength distribution is emitted. The Nernst filament must be preheated by some external means before it becomes conductive. The temperature is then maintained by the current flowing through it. The frequency region of maximum intensity is determined by the temperature of the radiating material, as with a blackbody. The intensity at a given wavelength of an electrically heated incandescent source is very sensitive to the voltage applied. Various constant-voltage transformers and electronically controlled circuits are available to maintain constant intensity.

LASERS.[6]   A novel radiation source known as the *laser* (an acronym for light amplification by stimulated emission of radiation) can no longer be regarded as a laboratory curiosity. Lasers are commercially available at many wavelengths.

The fundamentals of laser operation are most readily seen with reference to Fig. 156. The pumping lamps provide intense radiation at a frequency corresponding to the transition $0 \rightarrow 2$ in the laser material. Nonradiative transitions then rapidly transfer population to state 1, but the lifetime of state 1 is relatively long. With sufficient power in the pumping radiation it is possible to achieve a situation in which the population in state 1 is greater than that of the ground state. When one or a few systems emit spontaneously and fall to the ground state, that radiation will

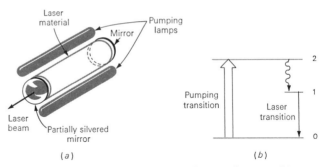

figure 156. Principles of laser operation. (a) Laser configuration; (b) energy levels in the laser material.

stimulate emission in other systems. Because of the population inversion stimulated emission will exceed absorption.

It is a characteristic of stimulated emission that the emitted radiation is in the same direction as the stimulating radiation and in phase with it. In general the radiation will quickly escape from the laser material. Sometimes, however, it will be traveling exactly along the axis of the cylinder. On reaching the end it strikes a carefully aligned optically flat mirror. It is then partially reflected back into the laser material. The reflected beam will continue to bounce back and forth between the two ends, gaining intensity by stimulated emission on each passage. The transmitted portion coming from one end is the useful laser radiation.

Important characteristics of laser radiation, consequences of its method of generation, are the following: it is extremely monochromatic, in a well-defined direction, and for many lasers it is coherent. Laser radiation can be very intense. Although the efficiency with which pumping radiation is converted into laser radiation is never 100 percent and may in some cases be quite small, the monochromaticity of the laser beam and its linearity, which allows it to be efficiently focused onto a small area, mean that lasers provide a far higher intensity of radiation in a definite frequency and spatial region than has previously been available. These properties make lasers very useful. Their principal drawback at present is the difficulty of tuning them.

The most common laser materials are ruby, with a laser wavelength at 6943 Å, and He-Ne, at 6328 Å, but many other wavelengths in the visible and the infrared are available.† Higher frequencies are produced by frequency-doubling techniques: when a high-intensity laser beam of frequency ν is applied to an appropriate material, a substantial fraction of the radiation emitted will have frequency 2ν .

MICROWAVE.   The common laboratory source of microwave radiation is a *klystron oscillator*.[8] A klystron is a special type of vacuum tube in which electrons

---

† A survey of Laser types and their characteristics, as of 1967, is given in Ref 7.

are made to oscillate in a tunable resonant chamber. The oscillating electrons radiate at the frequency of oscillation, which is in the microwave region. Since the frequency of oscillation depends on cavity dimensions, the klystron is tuned by changing the size of the cavity, either mechanically or thermally. Small changes in frequency are made by shifting the voltage applied to one of the electrodes called the *reflector*. Klystrons are essentially monochromatic sources, although in some applications FM noise is significant, as well as the more common AM noise. Voltages in a klystron power supply must be kept very carefully regulated.

Microwave radiation can also be produced by a number of other devices, including backward-wave oscillators. Recently solid-state microwave generators have become available for the lower microwave frequencies.

RADIO FREQUENCY. In the radio-frequency region conventional vacuum-tube or solid-state oscillator circuits and amplifiers can be used to produce monochromatic radiation of any desired intensity. These techniques are discussed in Chapter 28.

MONOCHROMATORS AND FILTERS

If the radiation source is broadband or contains lines of significant intensity at several frequencies, some means must be provided for isolating a particular frequency or narrow frequency band.

Filters are the cheapest and most convenient means of restricting radiation to a narrow range of frequencies in the visible. Although filters absorb some of the desired light, they absorb much more of the light in other parts of the spectrum. They are available in standard sizes and thickness. About 80 different optical filters are described, with code numbers and their transmission as a function of wavelength, by the Corning Glass Works.[9] They include various portions of the visible, and ultraviolet and infrared, both with and without transmission of visible light. Some typical transmission curves are shown in Fig. 157.

Filters can often be made by filling a cell with an appropriate liquid or solu-

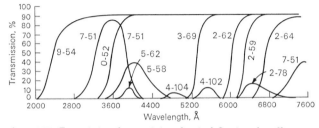

figure 157. Transmission characteristics of typical Corning glass filters.

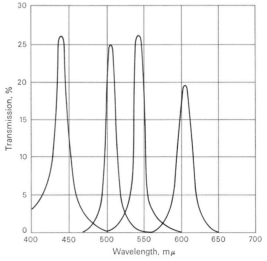

figure 158. Spectral transmission for four interference filters.

tion. A 1 *M* solution of cupric chloride 1 cm thick, for example, transmits most of the light between 4000 and 6000 Å but absorbs the remaining light, and a concentrated solution of iodine in carbon disulfide is opaque to visible light but transmits reasonably well in the infrared. Filters containing organic dyes in gelatin are available commercially† with many transmission characteristics.

Interference filters consist of thin evaporated layers of dielectric material between semitransparent metallic films on glass. A narrow range of wavelengths is transmitted, all others being reflected. Filters with transmission peaks every 50 to 100 Å are available,§ and transmission characteristics of a few of these are illustrated in Fig. 158. Multilayer interference filters are able to transmit as much as 70 percent of light with a half bandwidth of 70 Å.

Filters involving resonance circuits or cavities can be constructed for the radio-frequency and microwave portions of the spectrum (see Chapter 28). Since monochromatic sources are generally available, they are used chiefly in noise-reduction applications or in connection with harmonic generators.

Monochromators are more selective but more expensive than filters. They allow a substantial portion of the spectrum to be swept by simple mechanical adjustments, and are available in various ranges from the far ultraviolet to the far infrared. They are of prism or diffraction-grating type.

† Kodak Wratten Filters, Eastman Kodak Co., Rochester, N.Y.

§ Farrand Optical Co., Inc., New York; Baird Associates, Inc., Cambridge, Mass.; Bausch and Lomb Co., Rochester, N.Y.

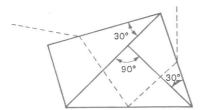

figure 159. *Constant-deviation prism.*

A useful prism is the constant-deviation type, illustrated in Figure 159. It can be regarded as being built up of two 30-60° prisms and a 90° totally reflecting prism. As the prism is rotated, the beam which issues at a given angle with respect to the incident beam changes in frequency. It is isolated by a fixed slit.

A grating is usually a carefully ruled metal film on glass. The ruled lines must be very accurately parallel and evenly spaced. Light reflected at a given angle from various points of the grating will interfere destructively except for certain wavelengths. The wavelength effectively diffracted at a given angle is given by

$$\lambda = \frac{2d}{N} \sin \theta \cos \frac{\delta}{2} \tag{1}$$

where N = an integer, the *order* of the reflection

d = spacing of ruling on the grating

δ = angle at the grating between the incident beam and the diffracted beam of wavelength λ

θ = the angle between the normal to the grating face and the bisector of δ

These angles are illustrated in Fig. 160. The close similarity between this equation and the Bragg equation (page 322) is obvious. The use of a grating is illustrated in the infrared spectrometer discussed on page 553.

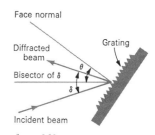

figure 160.

Angles appearing in the grating equation (1).

DETECTORS

Radiation detectors are nearly as varied as sources. In most spectroscopic applications it is important that the detector be sensitive and be a minimal source of noise.

In the γ- and x-ray regions of the spectrum, radiation may be detected photographically or by any of a variety of counters. These counters are discussed in Chapter 27.

Photographic techniques can also be used in the ultraviolet, visible, and near infrared. Except for special applications, however, it is more common in modern spectrometers to use photoelectric detection.

When light of frequency above a certain critical value strikes a metal, electrons are emitted. The cutoff frequencies are lowest for the alkali and alkaline-earth metals. If the metal target is incorporated into a vacuum tube together with a wire positively charged with respect to the photoelectric metal, the photoelectrons will pass to the wire. A current will thus flow when light is incident to the cell. This current can be detected by a galvanometer or amplified electronically. For light of very low intensity, a photomultiplier tube can be used. In a photomultiplier a number of electrodes are arranged in series at appropriate angles. Each electrode is at a more positive potential (typically 100 volts) than the previous one. The first electrode is coated with cesium or another material which ejects electrons when light falls on it. The photoelectrons are attracted to the second electrode, being accelerated by the potential difference. Each electron when it hits ejects several more, which are accelerated toward the third electrode, and so the process continues, producing a substantial current at the final electrode for a very small amount of incident light.

A number of solid-state light-sensitive devices are also available.[10] Light falling on a semiconductor can excite electrons from the nearly filled to the nearly empty band, thus increasing both the number of electrons and of holes available for conduction. A decrease in the resistance of the photoconductive cell results. Cadmium sulfide cells containing a broad-area polycrystalline photoconductive surface combine high sensitivity in the visible with moderate response speed. Lead sulfide cells are also common.

Silicon or germanium p-n alloy junctions can serve as radiation detectors. A single crystal of p-type silicon with a 0.5-μ layer of n-type material diffused into it will convert light directly to electrical energy with no external power required. Silicon photovoltaic cells are used to convert solar energy to electrical energy. Germanium p-n junctions are frequently connected with a reverse bias to function as photoconductive cells, with good sensitivity and response speed well beyond the audio.

Detectors which depend on the heating effect of radiation cover a wide range of frequencies but are particularly useful in the infrared. A fine thermocouple junction, blackened to absorb radiation, can be used. A bolometer contains a thin strip of platinum which is part of a sensitive Wheatstone bridge. Its resistance changes with temperature as the incident radiation heats it. A Golay cell is essentially a sensitive gas thermometer in which pressure changes are observed.

Bolometers are also used in microwave power measurements, but their relatively slow response prevents their use in spectrometers where the useful information appears at a high modulation frequency. It is also difficult to match a bolometer properly to the input of many amplifiers. Crystal diodes, such as the 1N23, are commonly used. They are more sensitive than bolometers but also generate more noise. In addition to Johnson noise (page 615), excess noise is produced in the crystal when direct current flows (as it does whenever the crystal detects microwave power). This excess noise is referred to as *semiconductor noise*. The semiconductor-noise power per unit bandwidth in the vicinity of a frequency f varies roughly as $1/f$ in the audio- and radio-frequency regions.

If the microwave signal could be amplified by a low-noise amplifier prior to detection, the signal-to-noise ratio could be improved. Maser amplifiers have been used in this connection, but they are limited to only a single frequency and low noise performance requires great care in operation. Microwave amplifiers based on traveling-wave tubes may become important in the future, although at present their noise figures are too high for them to be effective in improving signal-to-noise ratios in the very low-level signals encountered in spectroscopy.

Detection at radio frequencies is discussed in the chapter on electronics (page 610).

RADIO-FREQUENCY SPECTROMETERS.   A nucleus with nonzero spin angular momentum has associated with it a magnetic moment.[11,12] The magnetic moment will interact with an applied magnetic field to give different energies to states having different components of nuclear spin in the direction of the applied field. Transitions between these levels are observed in nuclear magnetic resonance (NMR) spectroscopy. If the nuclear spin is greater than $\frac{1}{2}$, there may also be an electric quadrupole moment associated with the nucleus. (This is a consequence of the fact that the charge distribution within the nucleus is not spherically symmetric.) Except in molecules of very high symmetry there will also be a quadrupole moment associated with the electronic charge distribution. These two moments will interact and again produce a splitting of energy levels of different nuclear spin quantum number. Transitions among them are observed in nuclear quadrupole resonance (NQR) spectroscopy.

The energy splittings encountered in NMR for reasonable laboratory magnetic fields, and in NQR, correspond to frequencies in the radio-frequency region. Two radio-frequency spectrometers designed for investigation of nuclear magnetic resonances will be described. One illustrates the crossed-coils method of Bloch, Hansen, and Packard,[13] while the other employs a regenerative oscillator, utilizing a principle introduced by Roberts[14] and employed by Pound and Knight.[15] The latter is very similar to instruments employed for NQR investigations.

The sample to be investigated by NMR, which may be solid, liquid, or gaseous, is placed in a constant uniform magnetic field H_0, taken to define the z direction. The magnetization vector of the sample, defined as the net magnetic-moment vector (or in this case the part arising from the nuclear moments) per unit volume, will lie along the z axis. Transitions can be observed if a radio-frequency field of proper frequency is produced by impressing a voltage across a coil wound around the sample vial. This radio-frequency field is linearly polarized along the coil axis, here taken as the y axis. When a field at the resonance frequency, given by

$$\nu = \frac{|\gamma|}{2\pi} H_0 \tag{2}$$

where γ is the magnetogyric ratio of the nucleus, is applied in this fashion, the magnetization vector becomes tilted slightly and precesses about the z axis at the frequency of the radio-frequency field. The magnetization may therefore be resolved into components, a constant component along the z direction and a component of fixed magnitude rotating in the xy plane. This rotating component, which exists only when the sample is stimulated at a frequency close to resonance, will induce a voltage in any coil with its axis in the xy plane. In the crossed-coils method, a second coil (x coil) is placed with its axis perpendicular to the axis of the y coil, so that the voltage coupled directly from the y coil in the absence of the x coil is minimized. The sample resonance is then detected by observing the very weak voltage which is induced in the x coil by the rotating sample magnetization. In single-coil methods, the reaction of the sample on the y coil itself is observed as a very slight change in its impedance, this change being detected from the effect on the circuit of which it is a part.

A block diagram of the crossed-coils apparatus of the type originally used by Bloch, Hansen, and Packard[13] appears in Fig. 161. A signal generator supplies radio-frequency power to the y coil. The receiver (tunable radio-frequency amplifier and detector) is tuned to the frequency of the signal generator. A sawtooth voltage synchronized with the oscilloscope horizontal sweep is fed to the field modulation coils to produce a repetitive sweep in H_z. The receiver output is observed on the oscilloscope, effectively as a function of H_z.

A communications receiver can be used in this apparatus, though it is necessary for good sensitivity to precede it with a low-noise radio-frequency preamplifier, since the noise which limits sensitivity is generally that produced in the first stage of amplification of the voltage from the x coil. To permit the use of slow rates of sweep, the output for the oscilloscope is taken from the second detector of the receiver, since the gain of the audio amplifier in the receiver drops off rapidly below the audio range.

Since it is not practicable to achieve perfect geometry in the construction and alignment of the coils, the signal from a sample resonance would be swamped by radio-frequency leakage directly from the y coil to the x coil unless arrangements were made to balance this out. A satisfactory means of doing so is provided by two suitably oriented metallic paddles, or one-turn coils, which can be rotated to modify the coupling between the y and x coils. These are adjusted alternately until the leakage voltage is minimized.

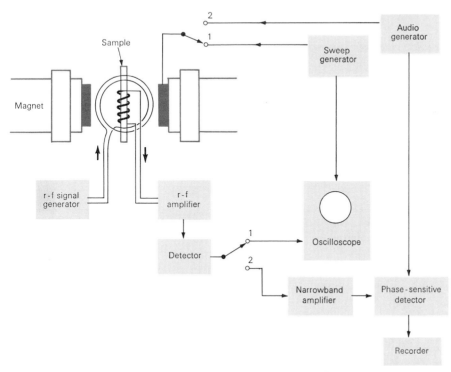

figure 161. Block diagram of a NMR spectrometer similar to that of Bloch, Hansen, and Packard. The apparatus used by Bloembergen, Purcell, and Pound differs from this mainly in that a radio-frequency bridge is employed instead of the crossed-coils arrangement.

In order to obtain a line shape corresponding to simple absorption, rather than a mixture of absorption and dispersion, it is necessary to detect that component of the voltage in the x coil which has the same phase as the voltage at the y coil. Although this objective can be achieved by the use of a radio-frequency phase-sensitive detector, the same result was accomplished in the earlier spectrometers by introducing a reference radio-frequency voltage of suitably adjusted phase into the receiver. Provided this voltage is larger than the signal from any sample resonance, it has the effect of converting an ordinary peak-following detector into a phase-sensitive detector. The inherent limitation in this scheme is that any fluctuations in this reference signal, whether they arise from variations in signal-generator output or from changes in gain of the receiver, can cause troublesome drift in the detector output.

To obtain good sensitivity it is essential to employ a narrow bandwidth and slower rates of sweep than are feasible in the presence of this drift. One very good method of obtaining narrower bandwidths with good base-line stability is the field-modulation method. In this scheme, which corresponds to position 2 of the switches in Fig. 161, the sawtooth sweep is not used. An audio generator supplies a current to the modulation coils to produce a sinusoidal term in H_z. The output from the receiver when the frequency is close to that of a sample resonance is now an audio voltage at the modulation frequency. This voltage is detected by a phase-sensitive detector with reference derived from the modulation oscillator. The output of the phase detector is displayed by a chart recorder while the average value of H_z is changed at a slow and constant rate. The signal, for small modulation amplitude, is proportional to the derivative of the true line shape, as explained in detail on page 293. As an alternative to modulation of the field, an equivalent effect can be produced by frequency-modulating the radio-frequency oscillator of the signal generator.

The modulation frequency must be kept lower than the line width if distortion of the line shape is to be avoided. However, the effect of flicker noise from the amplifier is minimized by using modulation frequencies as high as is feasible within this restriction. Typically, modulation frequencies in the range 25 to 400 Hz are used. This modulation method is used in modern research instruments designed for the study of relatively wide, weak resonances, as are commonly found in solids.

In the design of spectrometers for the observation of the relatively narrow resonances (often below 1 Hz) characteristic of liquids, emphasis has been placed on developing techniques for producing magnetic fields of very high stability and homogeneity. An example of a high-resolution spectrometer is illustrated in Fig. 162. The magnetic field is stabilized to within about 1 part in 10^8 at 14,000 gauss by a circuit which senses changes in the field and corrects for these by driving an appropriate current through a pair of coils mounted on the magnet pole caps and also, at the same time, acting through the regulator circuit of the power supply for the main field coils of the electromagnet. Careful temperature control of the magnet is essential to maintaining good stability. From time to time the magnet current is carried through an empirically determined cycling procedure to restore good homogeneity. Homogeneity can also be improved by passing small carefully controlled currents through specially shaped shim coils mounted on the faces of the magnet pole caps.

Crystal-controlled oscillator circuits are employed in the radio-frequency generator unit, since stability of the spectrometer frequency is as important as that of the magnetic field. The frequencies given in the diagram are those for the 60-MHz generator, which is used for the detection of proton resonances at about 14,000 gauss.

Since line widths of the order of 1 Hz are obtained, clearly the audio-modulation method of detection described above is not appropriate and a different approach is used for the achievement of narrow bandwidths. In principle, a suitably narrow bandwidth and good absorption-line shape could be obtained by use of a sharply tuned

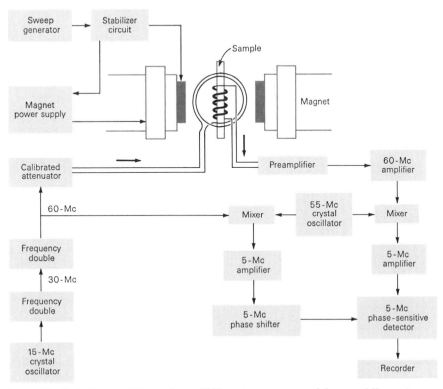

figure 162. Block diagram of high-resolution NMR spectrometer (principal features of Varian Associates model V-4311).

amplifier at 60 MHz followed by a phase-sensitive detector (page 634) at this frequency. At the present state of the art, however, these functions can be performed more satisfactorily by heterodyning (page 208) to 5 MHz and employing a tuned amplifier and phase-sensitive detector at this frequency. The detector output, after filtering, is presented on a chart recorder.

Although the noise bandwidth is really determined by the phase-sensitive-detector output circuit, the bandwidth of the radio-frequency amplifier itself has to be kept reasonably narrow in order to permit a large gain to be obtained without the noise level becoming so high as to swamp the final stages.

The regenerative oscillator spectrometer of Pound and Knight,[15] illustrated in Fig. 163, is particularly well suited to those studies for which it is necessary to be able to change the spectrometer frequency easily or even to search continuously over a wide frequency range. The range covered is 1 to 40 MHz. The sample is placed

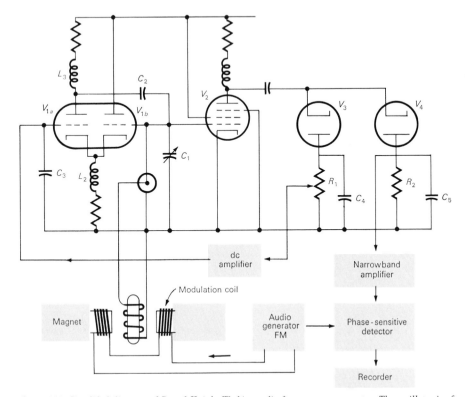

figure 163. Simplified diagram of Pound-Knight-Watkins radio-frequency spectrometer. The oscillator is of the cathode-coupled type. Oscillations in the resonant circuit (page 590) consisting of C_1 and the sample coil are sustained through amplification by V_1 and feedback of power through C_2. Pentode stage V_2 is a radio-frequency amplifier. The amplitude of the oscillations is held approximately constant through the action of diode detector circuit V_3, R_1, C_4 (page 610), which determines the bias of V_{1a}. Modulation of the radio-frequency voltage produced by sample absorption (as in Fig. 161) is detected by V_4 to yield a signal at the modulation frequency. This in turn is amplified and detected to produce a deflection of the recorder pen.

in the coil of the tank circuit of the oscillator. To search for resonances, the oscillator is tuned by means of a synchronous motor which drives the rotor of the variable capacitor C_1 of the tank circuit. When a resonance frequency of the sample is encountered, the absorption of energy by the sample causes a decrease in the Q of the tank circuit, and as a result the amplitude of oscillation is very slightly reduced. Audio modulation of either the field or the spectrometer frequency is used so that good sensitivity can be obtained by a narrowband amplifier, phase-sensitive detector, and chart-recorder system. This type of spectrometer has been widely employed for the study of resonances in solids.

An NQR spectrometer differs from an NMR spectrometer in that no magnetic field is involved. It is also much more difficult than in the NMR case to predict frequencies at which resonance will occur. Single-coil techniques are used, and the frequency is swept over a wide range automatically by means of a motor-driven variable capacitor.

MICROWAVE AND ESR SPECTROMETERS.   The transitions associated with the pure rotational spectra of molecules and those between different states of electron spin projection in laboratory magnetic fields correspond to frequencies in the microwave region. Some of the characteristics and design principles of a microwave spectrometer of the transmission-line type will be considered here.[16,17] An ESR (electron spin resonance) spectrometer of the common bridge type will also be examined.

The purpose of the spectrometer to be described is to display the pure rotational spectra in gaseous samples with sufficient resolution for general molecular structural studies. For such work it is desirable to have an instrument which (*a*) covers a wide frequency range, corresponding to a substantial portion of the microwave region, (*b*) is capable of detecting lines with absorption coefficients† of the order of 10^{-6} to 10^{-9} cm^{-1}, and (*c*) displays these lines with a width of the order of 1 MHz or less.

The peak value of the absorption coefficient varies from one line to another and also depends on the temperature and pressure of the sample. As a rule, the width of the lines is due to collision broadening. The width from this source is, from a simple kinetic-theory calculation, proportional to the pressure (Fig. 164).

---

† The absorption coefficient α is defined by $P = P_0 e^{-\alpha l}$ where P_0 is the microwave power entering a transmission line as a traveling wave and P is the power passing a cross section at a distance l farther along the line. In an empty waveguide, the power absorption is due to ohmic losses associated with wall currents; if the waveguide is filled with a gaseous sample which has an absorption line at the given microwave frequency, a very slight increment is added to α. It is this increment which the spectrometer must detect.

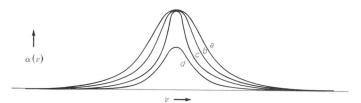

figure 164. Microwave absorption line, plotted against frequency ν, for a series of different pressures, with pressure decreasing in the sequence a, b, c, d. Curves a, b, c illustrate the narrowing of the collision-broadened line which accompanies a decrease in pressure; curve d shows the decrease in peak value which eventually occurs when the pressure becomes so low that saturation is observed at the given microwave power level. Saturation can be avoided by reducing the power level, but the sensitivity becomes poorer at very low power levels.

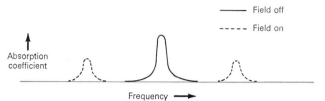

figure 165. Absorption line of a sample in the absence of an electrostatic field and with the field applied. The field-on pattern is merely illustrative since the Stark patterns vary from one line to another.

For the width to be kept below 1 MHz, the sample pressure must ordinarily be of the order of 10^{-2} mm or less. At such low pressures, the microwave power must be held to a reasonably low level, usually below 1 mw, if saturation effects are to be avoided.

As an example of a spectrometer which very satisfactorily meets these requirements, we shall describe a Stark effect modulation spectrometer. The Stark effect (page 276) is the splitting and shifting of spectral lines which occurs when the sample is subjected to an electrostatic field (Fig. 165). The introduction of the principle of Stark effect modulation by Hughes and Wilson[18] provided an elegant and effective means of increasing sensitivity. Figure 166 is a block diagram of a spectrometer of the Hughes-Wilson type. Most research instruments now in use represent elaborations of this basic scheme.

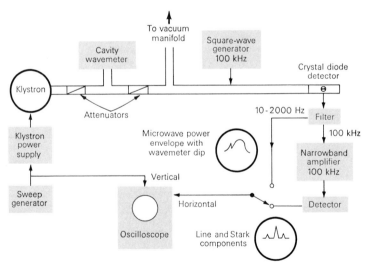

figure 166. Block diagram of Stark modulation microwave spectrograph of the Hughes-Wilson type. Typical frequency range is 10,000 to 14,000 MHz.

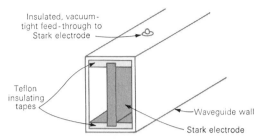

Insulated, vacuum-
tight feed-through to
Stark electrode

Teflon
insulating
tapes

Waveguide wall

Stark electrode

figure 167. Details of Stark cell construction.

The absorption cell consists of a length, perhaps 10 ft, of waveguide closed off at both ends by vacuumtight mica windows. The sample is introduced into the cell from a vacuum manifold, through a few small holes so located in the waveguide wall as to cause the least interference with the propagation of the microwave fields. To permit the application of an electrostatic field, a strip of brass, supported and insulated by two grooved Teflon tapes, is placed inside the sample cell (Fig. 167). This Stark electrode is located in a plane which is perpendicular to the electric component of the microwave field and so does not seriously affect its transmission.

The cavity wavemeter, which is used for approximate† measurement of the microwave frequency, consists of a cylindrically shaped cavity, coupled to the main waveguide through a small hole. The cavity length is determined by the position of a micrometer-driven piston. A sharp dip in the power transmitted past the cavity hole occurs when the frequency is such that a cavity resonance is excited.

After going through the absorption cell, the microwave radiation is detected by a special microwave diode, consisting of a small semiconductor crystal mounted in a coaxial cartridge suitable for admitting power at microwave frequencies. The crystal diode rectifies the microwave currents and produces an output direct current, the magnitude of which depends upon the amount of incident microwave power (page 609).

A sawtooth voltage sweep is introduced at one of the klystron electrodes to produce a corresponding sweep in the microwave frequency, over a range of perhaps 50 MHz. The same sweep voltage is applied also to the horizontal deflection channel of the oscilloscope. The shape of the microwave power envelope arriving at the crystal is somewhat irregular because of the variation in oscillator output and in the microwave-system transmission characteristics over the course of a sweep. The crystal detector output follows this envelope. This can be displayed on the oscilloscope when desired. The wavemeter, when suitably tuned, produces a visible dip in this pattern. If there were a sample absorption line within the region swept, how-

† Cavity wavemeters presently available commercially offer an accuracy of the order of 0.1 percent and a precision of 0.01 percent.

ever, it would not be seen against this relatively uneven background unless it were an unusually strong line.

A zero-based square-wave voltage is applied between the Stark electrode and the waveguide. The frequency f_m of the modulating field produced in this way is relatively low, so that one may think of it as an electrostatic field which is alternately being switched on and off. As the klystron frequency goes through the frequency of a line, the microwave voltage at the crystal is modulated at the frequency of the Stark field; this happens because the sample absorbs at one frequency while the field is off and at different frequencies when the field is on. As a result, at those times during the sweep when the klystron is going through the frequency of a sample absorption line, there exists in the crystal output an ac component, at the frequency f_m, which is due to sample absorption. A filter broadly tuned to this frequency passes this component on to the narrowband amplifier. There is a signal at the point along the sweep where the sample absorbs when the field is off and other smaller signals at points where the sample absorbs when the field is on. The latter are called *Stark components;* their position along the sweep changes when the amplitude of the Stark modulation voltage is changed. The ac voltages due to sample absorption at the filter are very minute and are actually buried in noise voltages often thousands of times larger.

The narrowband amplifier is tuned to the modulation frequency and selectively amplifies the ac voltage from the absorption line, but not the slow variations in microwave power level. It also discriminates against noise generated in the crystal, except for noise voltage components whose frequencies are within the passband of the amplifier. The detector output follows the envelope of the ac voltage from the amplifier.

The limitation on sensitivity with this spectrometer is in most instances the noise generated within the crystal. As this is random noise, the noise power at the output of the narrowband amplifier is proportional to the bandwidth, while the signal power is independent of the bandwidth. Therefore the signal-to-noise ratio improves as the bandwidth is narrowed. The bandwidth of the amplifier must be left broad enough, however, to accommodate drift in the modulation frequency or in the amplifier components which determine the amplifier tuning, for deterioration of signal will obviously occur if the modulation frequency is not well within the amplifier passband. Typically, the bandwidth is of the order of a few percent of f_m.

Because of the $1/f$ noise of the crystal detector (page 616), the modulation frequency is chosen high enough to achieve a significant reduction in crystal noise below that present at the low audio-frequencies. However, if f_m becomes comparable with the line width (expressed in frequency units), the modulation causes the lines to appear broadened. Thus the choice of modulation frequency involves a compromise between sensitivity and resolution. The modulation frequency is usually chosen

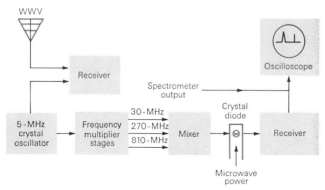

figure 168. Block diagram of microwave frequency-measuring equipment.

in the range 50 to 100 kHz. In many cases, a communications receiver which extends into this range is used as narrowband amplifier and detector.

When it is desired to employ narrower bandwidths than are readily achievable with the arrangement of Fig. 166, the simple detector can be replaced by a phase-sensitive detector and chart recorder.[18] With this arrangement, somewhat lower modulation frequencies are sometimes used, down to about 5 kHz.

Figure 168 is a block diagram of a typical microwave frequency standard used for accurate frequency measurements. The 5-MHz crystal-controlled oscillator provides a stable reference which is monitored against standard frequency transmissions of Station WWV. Various harmonics of this frequency are generated by the multiplier stages, and these, when fed into the microwave crystal, produce a lattice of accurately known frequencies at intervals of 30 MHz in the microwave band of interest. A sample of the klystron oscillator power enters the same crystal and mixes with power from the standard. The receiver is used to detect a beat note (usually the one of lowest frequency) between the klystron and one of the reference harmonics. The output of the receiver is displayed on the oscilloscope along with the output of the spectrometer detection system. When the klystron is swept, the beat note passes so rapidly through the passband of the receiver that only a sharp pip appears on the oscilloscope. Suppose, for example, that with the receiver tuned to a frequency f_r, a beat mark is produced at the instant at which the klystron passes through the frequency of an absorption line. Then the frequency of the line is $30n \pm f_r$ MHz, where n is an integer. The value of n is determined with the aid of a cavity wavemeter, and the choice of sign made by noting the direction in which the position of the beat mark changes when the receiver frequency is increased. The accuracy of a system such as this is inherently of the order of a few kilocycles per second, so that the accuracy of frequency measurement is practically always limited only by the breadth of the lines.

Electronic counters are commercially available to determine frequencies up to 100 MHz or more to accuracies approaching 1 part in 10^8. They may be combined with harmonic generators and mixers to determine frequencies well into the microwave region. One commercial unit extends a counter range to 35 GHz. The accuracy of such a measurement depends on that of the counter time base, generated by a carefully thermostated crystal oscillator. Recalibration against standard frequencies (page 624) is necessary at intervals.

Stark modulation spectrometers are commonly used over the frequency range of 10,000 to 40,000 MHz. At higher frequencies, the construction of the Stark cell presents difficulties as smaller waveguide sizes are used.

An ESR spectrometer[8] differs from a microwave spectrometer for studies of pure rotational spectra because of the fact that the sample must be contained in a volume over which a uniform magnetic field of appreciable strength can be maintained. The waveguide absorption cell is therefore replaced by a resonant cavity. The cavity functions as a resonant element and concentrates microwave energy in a small volume about the sample. Solid, liquid, or gaseous samples may be used.

A typical bridge-type ESR spectrometer is shown in Fig. 169. The output of the klystron oscillator passes first through an isolator. This is a section of waveguide containing a small piece of ferrite material in the steady magnetic field produced by a small permanent magnet. It has the property of transmitting microwave power in one direction with very little loss but absorbing almost completely any microwave radiation moving in the other direction. Its function is to prevent any effect on the amplitude or frequency stability of the klystron of varying reflections of power from other parts of the spectrometer. An attenuator is also introduced to provide control of the microwave power level.

The heart of the spectrometer is the hybrid tee, which functions as a bridge. Power incident from arm E is divided equally between arms 1 and 2. Any power in these arms which is reflected from the cavity or the slide-screw tuner and thus returns to the hybrid tee is recombined in such a way that the power in the H arm is proportional to the *difference* in the returning powers in arms 1 and 2.

The cavity may be either rectangular or cylindrical. Microwave power enters from the waveguide through a small hole, and a standing-wave pattern is built up. Energy is lost by reemission through the coupling hole back into the waveguide and through dissipation in the cavity walls. The Q of the cavity can be defined as the ratio of the power stored in the cavity to that lost, by all means, per cycle. If the resonance condition is fulfilled for a sample in the cavity, energy will be transferred to the sample and the power reflected from the cavity will be less than when conditions are off resonance.

The termination at the end of arm 2 absorbs all the power incident on it. The slide-screw tuner provides for the insertion of a metal rod into the waveguide, from

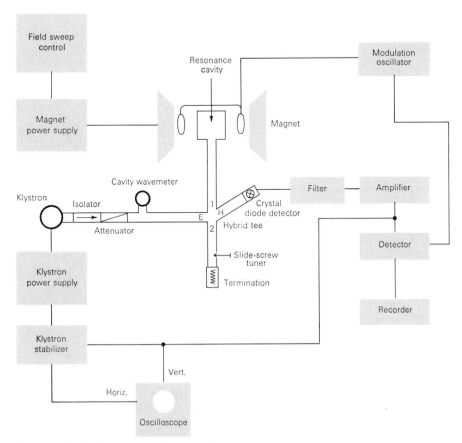

figure 169. Block diagram of homodyne bridge ESR spectrometer.

which power is reflected. The degree of insertion of the rod and its position in the waveguide can be adjusted to duplicate the off-resonance reflection from the cavity and balance the bridge. A small amount of imbalance is deliberately retained so that the crystal detector is biased to operate in an optimum region of its response behavior and provided with a phase reference signal.

The cavity concentrates power effectively at only a few frequencies, corresponding to different standing-wave modes, determined by the cavity dimensions. The ESR spectrometer is therefore operated at fixed frequency while the magnetic field is swept. The signal appearing at the crystal detector as a function of microwave frequency is shown in Fig. 170. The falloff at either side corresponds to the limit of the range of oscillation of the klystron. The sharp dip in the center is due to the

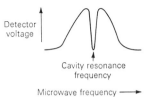

figure 170.

Detector output as a function of microwave frequency (klystron reflector voltage).

cavity. The klystron stabilizer serves to keep the klystron very near the cavity resonance frequency. A small ac voltage is superimposed on the dc voltage supplied to the klystron reflector. This has the effect of frequency-modulating the klystron. A portion of the output of the crystal is fed into a phase-sensitive detector referenced to the modulation frequency. The phase-detector output will be zero if the klystron frequency is exactly at the cavity resonance frequency. For slightly different frequencies, a dc error signal will be produced, with sign dependent on the direction of deviation. This error signal is added to the voltage applied to the klystron reflector in such a way as to return the klystron to the cavity resonance frequency.

It should be noted that this signal is essentially independent of the presence of a sample or the strength of the magnetic field. The magnetic field is also modulated (at a frequency different from that of the klystron modulation), and a signal at the field modulation frequency appears when the field and microwave frequency fulfill the electron spin resonance condition. In order to minimize the effects of detector noise, the modulation frequency should be high, but in addition to the limitation imposed by line width, as in the microwave spectrometer discussed above, there is difficulty in getting magnetic-field modulation at high frequency through the walls of the cavity. The field modulation frequency commonly used is 100 kHz.

Common microwave frequencies are about 9.5 and 35 GHz. Available spectrometers offer sensitivities of $5 \times 10^{10} \, \Delta H$ spins at the lower frequency, where ΔH is the line width in gauss. Sensitivity is greater at higher frequencies, but the allowed sample volume is less.

FIELD-STRENGTH MEASUREMENT.　The NMR phenomenon provides a means of determining magnetic field strengths very accurately. Gauss meters are commonly of the single-coil type since they are tunable over a wide range of frequencies. Samples containing hydrogen or lithium are commonly used. The limiting factor in the measurement is the impossibility of having the NMR probe and the ESR sample simultaneously at the same point in the magnetic field.[19]

For many purposes a determination of field strength based on the Hall effect is sufficient. This effect is illustrated in Fig. 171. When a semiconductor crystal is placed in a magnetic field and a current passed through it in a direction perpendicular to the field, a potential difference will develop in the third mutually perpendicular direction. This is a consequence of the Lorentz force, which tends to accelerate charges moving in a magnetic field in a direction perpendicular to both their motion and to the field. The voltage difference at the crystal faces is

$$E_h = R_h \frac{IH}{t} \tag{3}$$

where I = current
H = magnetic field strength
t = thickness of crystal

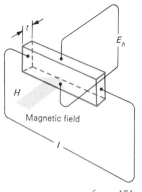

figure 171.

Diagram illustrating the Hall effect.

R_h = Hall coefficient of semiconductor material, inversely proportional to number of carriers per unit volume

The typical Hall effect crystal is quite small and very thin. Hall voltages are likely to be in the microvolt or millivolt region, but the signal-to-noise ratio is excellent.

Either a nuclear resonance or a Hall probe may be used to control the magnetic field, keeping it at a desired value. The NMR technique is similar to that described for klystron-frequency lock with a nuclear resonance taking the place of the cavity resonance. A Hall probe can be made part of a bridge circuit, with an amplified error signal used to correct the magnet current.

INFRARED SPECTROMETER.   To introduce some of the principles encountered in the design of an optical spectrometer, we shall discuss the Baird Associates† infrared spectrophotometer as modified by Evans for high-resolution work. This is one of several examples reported[20] in which commercial infrared instruments have been modified by the introduction of gratings to permit the study of the rotational structure of vibration-rotation bands of simple gaseous molecules. The range of this particular instrument is 2 to 12 μ.

A diagram of the optical system appears in Fig. 172. Because the performance

† Baird Associates, Cambridge, Mass.

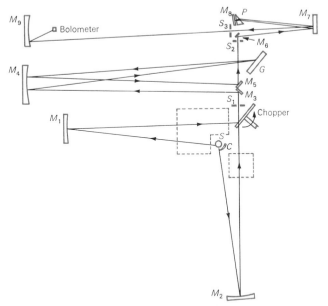

figure 172. Optical system of Baird Associates double-beam infrared spectrophotometer as modified by Evans for high-resolution work.

of the spectrometer is limited by the energy available from the source, the source is kept at as high a temperature as is feasible.

The source S is enclosed in a water-cooled housing, which has two apertures through which beams of infrared radiation emerge. These beams are focused by the concave mirrors M_1 and M_2 onto slit S_1. En route from the mirrors to the slit, the beams pass through two wells (dotted), one of which contains the sample cell and the other the reference cell. Immediately in front of the slit is a semicircular mirror which is rotated at 10 revolutions per second. This half-mirror sends the sample and reference beams alternately through the entrance slit S_1. Near the source is the *comb C,* which intercepts a part of the reference beam and thereby functions as a variable attenuator. The position of the comb is controlled by a mechanism described below.

The portion of the instrument in the optical path between S_1 and S_3 functions as a bandpass filter which selects radiation of a narrow range of wavelengths, which is then focused on the detector by M_9. Slit S_1 is in the focal plane of mirrors M_1 and M_2 and also in that of M_4. The divergent rays which pass through a single point in the aperture of S_1 are directed by plane mirror M_3 toward the concave spherical mirror M_4 and are then collimated by M_4 into a beam of parallel rays which go on to the diffraction grating G. The cross-sectional area of this bundle of rays is determined by the size of M_4, which is made large enough to utilize the full area of the grating. The grating is a carefully ruled film of aluminum on glass. Radiation is reflected from the grating in various directions, according to wavelength. The diffracted rays which reach M_4 are subsequently brought to a focus in the plane of the exit slit S_2. The wavelength which falls at the midpoint of S_2 is given by the diffraction grating Eq (1). Numerical values for the instrument described here are $\delta = 4°$, $d = 0.00667$ mm, $\lambda = (13.33/N) \sin \theta \ \mu$. The spectrum is scanned by rotating the grating to change θ, and hence the wavelength emerging from the center of S_2.

We have thus far considered only the radiation emitted from a single point in S_1 which reaches a single point in S_2. To achieve adequate sensitivity it is necessary of course to employ slit apertures of appreciable size. Rays emerging from different points in S_1 travel in slightly different directions in the collimated beam; similarly, the exit slit S_2 accepts rays reflected from the grating in a small range of angles. Therefore the radiation which eventually leaves through S_2 corresponds not to a single value of θ, but to a small range of angles, $\Delta\theta$. This range, called the *angular slit width,* is determined by the widths of S_1 and S_2. Through Eq. (3), $\Delta\theta$ determines the range of wavelengths $\Delta\lambda$ which actually reaches the detector.

For the present spectrometer, operated in the first order with a typical setting $\theta \approx 30°$, the slits are usually adjusted to give a resolution $\Delta\lambda$ of about 0.5 cm^{-1} at 1500 cm^{-1}.

The optical path between S_2 and S_3 consists of a small prism monochromator —comprising plane mirror M_6, spherical collimating mirror M_7, sodium chloride prism P, and Littrow mirror M_8—which serves only to reject all diffraction orders other than the one desired. For example, if the grating monochromator is set for 1500 cm^{-1} at $N = 1$, then radiation at 3000, 4500, . . . cm^{-1} also passes through S_2. The desired order is selected by rotating the prism-mirror assembly (P and M_8) to direct the radiation of this order onto the slit S_3.

The rays from S_3 are focused onto the bolometer detector by M_9. The bolometer is a thin strip of platinum, about 0.2×2 mm, coated with platinum black to absorb the infrared radiation. The bolometer forms one arm of a balanced dc Wheatstone bridge (Fig. 173). The effect of incident radiation is to alter the temperature, and hence the resistance, of the bolometer and thereby to produce an unbalance in the bridge. To minimize effects of convection currents and to improve sensitivity, the bolometer is mounted in an evacuated housing with a KBr window through which the radiation enters.

Unless the beams which come from the sample and reference paths are of equal intensity, the bolometer is subjected to radiation intensity which is modulated at 10 Hz, and a 10-Hz voltage appears at the output of the bridge. This voltage is amplified and then detected by a phase-sensitive detector. The phase of the 10-Hz voltage, and hence the sign of the detector output, depends upon which of the two beams has the greater intensity. The output of the detector controls a motor which

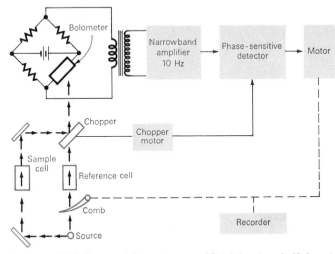

figure 173. Block diagram of electronic system of Baird Associates double-beam infrared recording spectrophotometer. A few parts of the optical system (Fig. 172) have been included to clarify the function of the electronic system.

moves the comb until the 10-Hz signal output from the bolometer bridge is reduced to zero. The pen of a chart recorder is coupled mechanically with the comb. As the spectrum is scanned by slow rotation of the grating, the changes in transmission of the sample are indicated by changes in position of the comb, and hence of the pen, which occur as the servomechanism operates to keep the beams matched. Thus the recorder traces the percent transmission as a function of wavelength.

It is instructive to examine the way in which various features of the design of the optical system and detection system affect the sensitivity and resolution of the spectrometer.† To do this, we shall assume that the resolution is determined by the optical system and that the sensitivity is limited by the noise generated in the detection system.

The response of the bolometer is of the form

$$e = S \, \Delta P \tag{4}$$

where e is the unloaded output-voltage change produced by a change ΔP in power level incident on the bolometer, with the current through the bolometer held constant. The parameter S depends on the design of the bolometer and on the conditions under which it is operated.

If the bolometer is placed in a bridge with the resistances of the other three arms equal to that of the bolometer (matched condition), the output voltage of the unloaded bridge§ is $\frac{1}{2}S \, \Delta P$. For the case of a beam chopped at 10 Hz, this is the peak-to-peak value of a square wave. The effective signal voltage is the sinusoidal component of this at 10 Hz, as only this component is amplified by the narrowband amplifier. The rms value e_s of this component is (page 583)

$$e_s = \frac{\sqrt{2}}{\pi} \left(\tfrac{1}{2}S \, \Delta P \right) = \frac{1}{\sqrt{2} \, \pi} S \, \Delta P$$

The rms noise output of the bridge at open circuit, within the effective bandwidth B of the amplifier-detector system, is (page 617)

$$e_n = (4kTRB)^{1/2} \tag{5}$$

where k = Boltzmann constant

$\quad T$ = absolute temperature of detector

$\quad R$ = output resistance of bridge

For the matched bridge, R is equal to the bolometer resistance. The noise figure F (page 616) of the bolometer and amplifier system is unity if the only noise detected is Johnson noise from the bolometer and bridge and is larger than unity if

---

† The discussion of sensitivity and resolution given here is based in part on an equation derived by Strong (Ref. 21); see also Refs. 20 and 22.

§ Several applications of Thévenin's theorem (page 592) are made in this derivation.

excess noise is generated in the bolometer or bridge or is added in the amplification process. In practice, a noise figure close to unity is achieved in this case. The signal-to-noise ratio at the output of the detection system is

$$\frac{e_s}{e_n} = \frac{S\,\Delta P}{2\pi(2FkTRB)^{\frac{1}{2}}} \tag{6}$$

To calculate ΔP, let us suppose that the sample absorbs a fraction γ of the incident power. Then $\Delta P = \gamma P$, where P is the radiant energy reaching the bolometer per unit time. The power P in watts is given by

$$P = B_\lambda A \frac{h}{f} \Delta\theta \, \Im \, \Delta\lambda \tag{7}$$

where B_λ = brightness of source, watts cm^{-2}-steradian^{-1} per centimeter of wavelength range, at given wavelength

$\quad\quad A$ = cross-sectional area, cm^2, of collimated beam

$\quad\quad h$ = height of monochromator slits S_1, S_2, cm

$\quad\quad f$ = focal length of collimating mirror M_4, cm

$\quad\Delta\theta$ = angular slit width, radians = slit width divided by f

$\quad\quad\Im$ = transmission fraction of optical system of spectrometer

$\quad\Delta\lambda$ = wavelength range passed by monochromator

Hence the signal-to-noise ratio is

$$\frac{e_s}{e_n} = \frac{S\gamma}{2\pi(2FkTRB)^{\frac{1}{2}}} B_\lambda A \frac{h}{f} \Delta\theta \, \Im \, \Delta\lambda \tag{8}$$

The resolution $\Delta\lambda$, assumed to be determined by the grating dispersion $d\theta/d\lambda$, is given by

$$\Delta\lambda = \Delta\theta \left(\frac{d\theta}{d\lambda}\right)^{-1} \tag{9}$$

It is apparent that increasing the slit width increases the sensitivity but impairs the resolution.

A useful figure of merit for a spectrometer of this type under the assumptions specified above is

$$\frac{e_s/e_n}{\gamma(\Delta\lambda)^2} = \frac{S}{2\pi(2FkTRB)^{\frac{1}{2}}} \left[B_\lambda A \frac{h}{f} \, \Im \, \frac{d\theta}{d\lambda} \right] \tag{10}$$

which is obtained by substituting Eq. (9) in (8) and keeping only the purely instrumental parameters on the right-hand side. This equation shows very clearly the nature of the compromise which must be made between sensitivity and resolution.

As a rule, the goal of the designer is to make the right side of Eq. (10) large. The factors in front of the square brackets are determined by the detecting system. The quantity $S/(FR)^{\frac{1}{2}}$ is a figure of merit for the bolometer-bridge-amplifier system and should be made as large as is possible with available detectors. To obtain a high figure of merit, the bolometer is made with a small sensitive area, and then M_9 is so designed as to form an image of S_3 which approximately matches the size of the bolometer. The bandwidth B is chosen to be as narrow as possible, consistent with a tolerable response time. Within the brackets, B_λ is determined by the source temperature, which is limited by the materials of construction of the source; the remaining factors are determined by the optical system. The practical upper limit to A is usually the area of the grating because this is the most expensive optical element in the system, and its cost rises rapidly with its area. Hence it is made as large as practical and the system designed to utilize it fully. With proper choice of cell window material and use of a suitably blazed grating for the wavelength range of interest, the factor $\mathfrak{I}$ can be made satisfactorily close to unity. The grating dispersion $d\theta/d\lambda$ is limited largely by the technical problems associated with the ruling of precision gratings. Practical difficulties of a geometrical nature make it difficult to increase h/f beyond a certain point.

If the cross-sectional area of the collimated beam is to be large enough to fill the area of the grating, the solid angle Ω formed by the rays entering the monochromator through S_1 must not be too small. But Ω equals the solid angle subtended at S_1 by M_1 and M_2. The areas and focal lengths of M_1 and M_2 must be chosen accordingly. Similarly, if all the light emerging from S_2 (in the desired grating order) is to reach the detector, the solid angles subtended by M_7 at S_2 and by M_9 at S_3 must not be made too small.

MÖSSBAUER SPECTROMETERS.[23]

When an atomic nucleus falls from an excited state to its ground state, γ rays are emitted. Part of the energy of the transition may be transformed to kinetic energy of nuclear motion, however, so that a distribution of frequencies of the emitted γ rays results. This does not happen if the atom is tightly bound in a crystal lattice. In that case, the transferred motion would have to appear as motion of the entire crystal. Its mass is so great that the effect is negligible, and γ rays of very sharply defined frequency result. Such a γ ray may be absorbed by another, equivalent nucleus, raising it to an excited state.

The exact spacing of nuclear energy levels depends on the electron distribution about the nucleus. This is the basis of chemical applications of the Mössbauer effect: a γ ray emitted by a given nucleus may not be within the absorption band of an otherwise identical nucleus in a different electronic (therefore chemical) environment. Mössbauer spectroscopy studies the shift in absorption frequency with chemical environment.

To do this it is necessary to have a range of frequencies available, and nuclear emissions are not readily tunable. Frequency shifts of an appropriate magnitude can be produced by the doppler effect. If the source of radiation is moving relative to the detector, there will be a shift in frequency ν of

$$\delta\nu = \nu\frac{v}{c} \tag{11}$$

where v = relative velocity, taken positive when source is approaching detector

   c = velocity of light

Sinusoidal motions are readily produced, but the analysis of the resulting spectra is not simple. It is more convenient for spectral analysis to produce motion in which the velocity of the source relative to the sample changes linearly with time. This corresponds to constant acceleration and to displacement which is parabolic in time. It is, of course, impossible to continue a uniform acceleration indefinitely, and in practice the direction of motion is reversed at intervals producing oscillatory parabolic motion. This motion is produced by first generating an electronic signal of the appropriate waveform and applying it to an electromechanical transducer like a loudspeaker voice coil. An electronic circuit to do this is described by Wertheim.[23]

References

1. L. Petrakis, *J. Chem. Educ.*, **44:** 432 (1967).
2. P. J. Wheatley, "The Determination of Molecular Structure," 2d ed., Oxford University Press, Fair Lawn, N.J., 1968.
3. M. J. Buerger, "Crystal Structure Analysis," John Wiley & Sons, Inc., New York, 1960.
4. R. H. Munch, *J. Am. Chem. Soc.*, **57:** 1863 (1935).
5. C. N. Banwell, "Fundamentals of Molecular Spectroscopy," McGraw-Hill Book Company, New York, 1966.
6. M. Brotherton, "Masers and Lasers: How They Work, What They Do," McGraw-Hill Book Company, New York, 1964; B. A. Lengyel, "Introduction to Laser Physics," John Wiley & Sons, Inc., New York, 1966.
7. T. H. Maiman, *Phys. Today*, **20** (7): 27 (1967).
8. C. P. Poole, Jr., "Electronic Spin Resonance: A Comprehensive Treatise on Experimental Techniques," Interscience Publishers, Inc., New York, 1967.
9. "Glass Color Filters," Corning Glass Works, Corning, N.Y., 1960.
10. "RCA Photocells, Solid State Photosensitive Devices," Radio Corporation of America, Electronic Components and Devices, Harrison, N.J., 1966.
11. J. D. Roberts, "Nuclear Magnetic Resonance," McGraw-Hill Book Company, New York, 1959.
12. J. D. Pople, W. G. Schneider, and H. J. Bernstein, "High-resolution Nuclear Magnetic Resonance," McGraw-Hill Book Company, New York, 1959.
13. F. Bloch, W. W. Hansen, and M. Packard, *Phys. Rev.*, **69:** 127 (1946); **70:** 474 (1946).
14. A. Roberts, *Rev. Sci. Instr.*, **18:** 845 (1947).
15. R. V. Pound and W. D. Knight, *Rev. Sci. Instr.*, **21:** 219 (1950); R. V. Pound, *Progr. Nuclear Phys.*, **2:** 21 (1952).

16. W. R. Gordy, W. V. Smith, and R. F. Trambarulo, "Microwave Spectroscopy," John Wiley & Sons, Inc., New York, 1953.
17. J. J. Townes and A. L. Schawlow, "Microwave Spectroscopy," McGraw-Hill Book Company, New York, 1955.
18. R. H. Hughes and E. B. Wilson, Jr., *Phys. Rev.*, **71:** 562 (1947); K. B. McAfee, Jr., R. H. Hughes, and E. B. Wilson, Jr., *Rev. Sci. Instr.*, **20:** 821 (1949).
19. B. G. Siegal, M. Kaplan and G. K. Fraenkel, *J. Chem. Phys.*, **43:** 4191 (1965).
20. A. R. H. Cole, *J. Opt. Soc. Am.*, **44:** 741 (1954); R. C. Lord and T. K. McCubbin, Jr., *ibid.*, **45:** 441 (1955); M. V. Evans, private communication.
21. J. Strong, *J. Opt. Soc. Am.*, **39:** 320 (1949).
22. P. Jacquininot, *J. Opt. Soc. Am.*, **44:** 761 (1954).
23. G. K. Wertheim, "Mössbauer Effect: Principles and Applications," Academic Press Inc., New York, 1964.

Chapter 26

Photochemistry†

† General references are listed as Ref. 1.

The first practical requirement in photochemical research is a source of light having sufficient intensity to produce a measurable reaction in a reasonable length of time. For quantitative research it is necessary to use nearly monochromatic light and to know the intensity of the light. Perhaps the greatest difficulty in photochemical technique lies in the fact that any means of restricting the light to a narrow range of wavelengths reduces its intensity and makes the measurement of the chemical change difficult.

Apparatus for producing light and radiation, for monochromators and filters for restricting the light to specified wavelengths, and electronic detectors for measuring the light intensity were discussed in Chapter 25. The present chapter is concerned with apparatus and techniques for studying photochemical reactions, which require high intensities of light. The low intensity of light from monochromators may be partially offset by using larger prisms and lenses, long exposures, and capillary arcs of high intensities. Lasers can supply monochromatic light of very high intensity.

TECHNIQUES FOR PRODUCING ATOMS AND RADICALS

Photochemically activated molecules may produce atoms, free radicals, or other activated intermediates, which are important in understanding the mechanism and kinetics of the reaction.

Flash photolysis[2] is a significant development in which photochemical reactions are carried out with light of great intensity, caused by the momentary discharge of electricity accumulated in large condensers. These intermittent flashes between electrodes are focused onto the reacting system, and they are sufficiently intense, for example, to dissociate chlorine gas Cl_2 into colorless Cl atoms. The rate of recombination of the atoms is measured with photoelectric cells which record the amount of light transmitted through the mixture of chlorine molecules and atoms. The same technique is used for studying the photodecomposition of organic molecules and the recombination rate of free radicals and atoms.

Molecules may be rendered highly reactive not only by the absorption of light but by a sudden input of energy in the form of shock waves.[3] A gas or a mixture of gases is placed under extremely high pressures on one side of a diaphragm. When the pressure is increased still further, the diaphragm ruptures and the gas moves into the low-pressure chamber with great velocity, the molecules having kinetic energy equivalent to that which they would have at very high temperatures. Gaseous bromine molecules are broken down into bromine atoms, for example, and the new products and their rate of recombination are recorded by light absorption, using a photomultiplier and oscilloscope.

New information is being obtained concerning free radicals and energy-rich intermediates which are produced by the absorption of light or by exposure to radioactivity. Again, these "hot" radicals can be made by a nuclear transformation which releases enormous energies within a molecule that contains a disintegrating atom. A hot atom or radical can be produced with energy much in excess of the average energy of the surrounding molecules and has not yet come into thermal equilibrium with them. These hot atoms are responsible for interesting phenomena, which throw light on the mechanisms of some reactions. An excellent review is given by Willard.[4]

Important progress in the study of free radicals and hot free radicals is being made by freezing them in solid materials, such as a frozen solution, and measuring the light absorption of these immobilized units.

THERMOPILES

A thermopile is made of a number of couples of unlike metals arranged in series with blackened-metal receivers attached to the junctions. The hot junctions are placed in the path of the light, and the cold junctions at one side in the shadow. Radiations of all wavelengths are absorbed by the black receivers and converted into heat so that the temperature of one set of junctions is increased.

The elements are chosen so as to give a maximum thermoelectric effect and electrical conductance with a minimum of heat conductance between the two junctions. The material should be as thin as possible, to minimize the heat capacity, without being too fragile. Bismuth-silver thermopiles are often used, but copper-constantan and platinum-tellurium elements are satisfactory. Detailed instructions for constructing thermopiles are given by Strong.[5] The theory and practice of thermopile construction have been discussed critically by Leighton and Leighton.[6]

The ordinary linear thermopile is smaller than the reaction cell behind which it is placed, and it is necessary to move the thermopile over the whole area of the transmitted beam in order to obtain an average value. Large-area thermopiles which do their own integrating are convenient for photochemical investigations. The thermocouples, thoroughly insulated with glyptal lacquer, are attached with de Khotinsky cement to the back of a blackened receiver of sheet silver, 10 by 40 mm in area and 0.02 mm in thickness. The cold junctions are attached in a similar manner to another silver sheet of the same size and heat capacity located at the side of the entering light beam. Twenty or more thermocouples of copper and constantan (Advance) wire are connected in series. The number is chosen so as to give the critical damping resistance for the galvanometer. The junctions are soldered with pure tin, using rosin for a flux and removing all excess tin. The thermocouples and receiver are attached to a bakelite frame, which is then set into a rectangular block of aluminum or other metal. Radiation strikes the blackened receiver through a quartz window.

The thermopile is connected directly to the galvanometer, and the deflection is proportional to the current through the galvanometer, which, in turn, is proportional to the voltage generated by the difference in temperature of the junctions. The temperature difference is proportional to the energy of radiation falling on the receivers. The radiation receiver is covered with lampblack, together with a little platinum black to increase the heat conductance. The mixture is suspended in methanol containing a trace of shellac, applied to the receiver, and allowed to evaporate. The black surface is practically nonselective, converting radiation of all wavelengths directly into heat. This is the advantage over photoelectric cells, which are much more sensitive than thermopiles but which respond only in restricted regions of the spectrum.

The thermopile should have a resistance equal to the critical-damping resistance of the galvanometer, so that a quick return to the zero reading is obtained. The galvanometer scale is arranged to slide back and forth so that it may be conveniently set at zero before each thermopile reading in order to avoid any error due to drift. The drift is caused by thermal inequalities in the thermopile circuit produced by unequal or fluctuating room temperature or by air currents. Evacuation not only improves the constancy of the zero point, but it may increase the sensitivity several fold.

CALIBRATION.   The deflection of the thermopile-galvanometer system is sufficient for comparative results, but for investigations connecting the quantity of chemical reaction with the energy absorbed (molecules per erg or per quantum), deflections are converted into absolute units. The quantity of radiation falling on the thermopile is obtained in ergs per second or in watts by calibration with a carbon-filament lamp standardized at the National Bureau of Standards.† The apparatus is shown in Fig. 174. The standard lamp L is connected to storage batteries or other steady source of direct current, and the rheostat R is adjusted until the ammeter M gives an exact reading corresponding to one of the values given in the calibration table accompanying the lamp (for example, 0.4 amp). A black screen A, 1 m square, is set 100 cm back of the lamp, and another one, B, having a square hole 25 cm on a side, is set 25 cm in front of the lamp, with the opening directly in front of it. The thermopile slit is mounted exactly 200 cm from the tip of the lamp. The lamp is rotated so that the two lines etched on either side of the globe are in line with the thermopile, giving the same conditions as used in the standardization. The room must be dark and free from objects that may reflect light on the thermopile S, and the operator must remain at a considerable distance.

The slit is narrower than the thermopile receivers, and its area must be accurately determined. If there are horizontal gaps between the receivers, a correc-

---

† These lamps may be purchased at a nominal cost from the National Bureau of Standards. The calibration is described in Ref. 7.

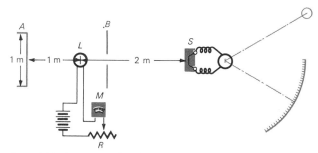

figure 174. Arrangement for calibration of a thermopile.

tion for the area must be made. The effective area is multiplied by the radiation in watts per square millimeter as given on the calibration sheet.

Windows on the thermopile are necessary to prevent fluctuations in output due to temperature changes and air drafts. Glass windows are unsatisfactory if the thermopile is to be calibrated in absolute units, because glass absorbs some of the infrared radiation emitted by the carbon-filament lamp. Polished quartz windows are preferred for this purpose, and a correction is made by finding the small percentage decrease in deflection caused by interposing a second quartz plate in front of the pile. The same procedure can be used for glass windows also, but the correction is larger.

The total radiation E may be calculated from

$$E = \frac{g}{s} \, atr$$

where s = galvanometer deflection with standard lamp

g = galvanometer deflection with monochromator

r = radiation from standard lamp, ergs $\sec^{-1}$ mm^{-2}, under conditions specified by National Bureau of Standards

a = area of slit

t = time of radiating

REACTION CELLS

Flasks or open dishes may be used for qualitative work. In precision work, the cell has front and back plates of polished quartz or glass, and the cell is placed between the exit slit of the monochromator and the thermopile.

Quartz is transparent throughout the whole visible and ultraviolet range down to 2000 Å. Pyrex in 2-mm thickness will transmit 10 percent at 3000 Å, whereas Corex will transmit 20 percent at wavelengths as low as 2750 Å. Vycor glass contains 96 percent silica and is fairly transparent as low as 2500 Å. All these glasses are transparent throughout the longer ultraviolet, visible, and short infrared. Window glass is suitable for visible light.

Glass cells of various sizes and shapes for holding filter solutions or chemically reacting systems may be purchased, or they may be constructed by fusing a Pyrex tube around a closely fitting circular window cut from a polished plate of Pyrex with a revolving brass tube and emery powder. For some reactions, the cells may be made with polished-glass plates cemented to the ends of a glass tube with Tygon cement or other cement which is inert toward the solution used. When requirements for optical precision are not too great, the photoreaction cells may be made conveniently from Lucite or other plastic material, sawed out to the right size and held together tightly with brass screws and a little cement. It is usually desirable to arrange the cell so that it is almost completely filled by the light beam.

Corrections for the light reflected at an interface are necessary in accurate work. The light should strike the windows at right angles, but even under these conditions, about 4 percent of the light is reflected at each glass-air or quartz-air surface. The light reflected at a glass-water or quartz-water surface is practically negligible. The fraction of light reflected at right angles is given by Fresnel's formula

$$\frac{I_r}{I_i} = \left(\frac{\mu - 1}{\mu + 1}\right)^2$$

where I_r, I_i = intensity of reflected and incident light, respectively
μ = ratio of refractive indices of two media

The light entering the inside of the empty cell is greater than that registered on the thermopile receivers, by an amount that depends on the number of quartz-air (or glass-air) surfaces through which the light passes. Sometimes the light reflected from the thermopile window passes back through the cell. The corrections are usually small, and they vary with the particular arrangement of cells and thermostat windows. Usually the corrections can be made to cancel out in gas reactions by placing an empty cell in the path of the light when a zero reading is made. The difference in energies registered on the thermopile gives the energy absorbed. In the case of solutions, the amount of light absorbed is obtained by subtracting the galvanometer reading with solution in the cell from the reading with pure solvent in the cell. It is a great convenience to have two cells exactly alike, either one of which may be slid into the path of the light.

PHOTOGRAPHY [8]

Photography is indispensable for much laboratory work in physical chemistry. It is discussed in the experiments on spectroscopy (38) and Raman effect (41).

When sensitized grains of silver halide in the gelatin are exposed to light, they are activated in such a way that they are more easily reduced to silver by a suitable mild reducing agent. The action of light produces the *latent image*. Nuclei are produced in the silver halide crystals by the light, and when the plate is immersed in a solution of the proper reducing activity (proper reduction potential in the electromotive-force series), each grain containing a nucleus is reduced to silver. This process is called *development*, and the reducing solution is called a *developer*. The silver halide grains that do not contain nuclei are reduced only after a much longer period of development. The production of nuclei for the latent image depends upon the presence of imperfections in the crystal produced by impurities or strains formed during the nucleation and crystallization. The light energy produces electrons and positive holes which are trapped at imperfections in the crystal, giving mobile silver and halogen atoms which start reaction with the developer. A diffusion of the silver and halogen atoms to the surface of the grain is involved.

After development, the plate is *fixed*. In this process the unreduced grains of silver halide are dissolved in sodium thiosulfate ("hypo"), leaving behind the grains which had light-induced nuclei and which were accordingly reduced by development to give black grains of silver. The parts of the plate that received the brightest light when the plate was exposed become the darkest when the plate is developed and fixed, and the finished plate is called a *negative*.

After fixation is complete, the plate is thoroughly washed and dried.

The amount of change produced on the plate by the action of the light is, of course, dependent upon the amount of light energy acting. Obviously, the same amount of light can be admitted through the lens by using a small aperture and long exposure, or a large aperture and short exposure. Better definition and greater depth of focus are obtained by the use of a small aperture.

In a perfect negative, i.e., one which has silver deposits in the various areas proportional in amount to the intensities of light reflected from the corresponding areas of the objects being photographed, the densities of the deposit are proportional to the logarithm of the corresponding exposure. If a series of identical plates are given exposures increasing in geometrical progression (so that their logarithms increase in arithmetical progression), and all plates are subjected to exactly the same process of development, a curve is obtained that shows the density of the developed photographic silver image as a function of the logarithm of the exposure, known as the *characteristic curve* of the film or plate. It has the general form shown in Fig. 175.

If the exposures are such as to bring the photographic plate into the region of the straight line, II, excellent results will be obtained. Good results may often be obtained, however, when the exposure is somewhat less, in the curved part, I, because the failure to give direct proportionality between density and exposure is partly compensated for in the printing of the positive.

The use of an exposure meter is recommended, with proper attention to the

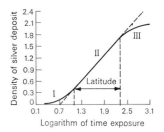

figure 175.

Characteristic curve for the exposure of a photographic plate.

exposure index of the particular type of film or plate used. These indices are supplied by the manufacturer and recorded on the box containing the films.

Emulsions with large grains (silver halide crystals) are faster. They are not suitable for enlargements or lantern slides, and for these a finer-grained, slower emulsion is used. For photographing laboratory material and apparatus and line drawings, high contrast is necessary. Process plates are suitable. For spectrographic work, panchromatic plates sensitive to all wavelengths of visible light are necessary.

In taking photographs, the camera is focused and then the loaded plateholder is inserted in the back of the camera, without changing the position of the camera. The shutter is closed, the diaphragm is adjusted to the desired opening, and the slide is pulled out of the plateholder. In the *f* system, the diaphragm opening represented by *f*/16 means that the diameter of the opening is one-sixteenth of the focal length of the lens. With the setting of *f*/16 and a process plate, an exposure of about 10 sec or less is appropriate, but the exact exposure is obtained from the exposure meter. After exposure, the slide is immediately replaced. It is a convention that the black side of the handle on the slide is always placed outward after the plate has been exposed.

Three trays containing developer, water, and fixing bath are placed near the red lamp in the darkroom. At 20°C, the image appears in 15 to 20 sec, but development should be continued until the details in the shadows are brought out, usually requiring about 3 min. Control of temperature is very important, as a few degrees of change greatly influence the action of the developer. The proper point at which development should be stopped must be learned by experience. The plate is then rinsed to remove alkali and placed in the fixing bath for about 20 min, or at least 5 min after all the halide (white) is apparently removed from the plate. After fixing, the plate is washed for at least half an hour in running water, placed on a rack, and allowed to dry. Upon completion of washing, the surface of the wet plate should be swabbed off (lightly) with wet absorbent cotton before drying. This removes any possible sediment that may have collected on the gelatin surface.

Ready-mixed developers may be purchased, or the developer may be prepared from the formula which accompanies the plate or film.

A suitable formula for process plates or contrast plates or films is prepared by mixing the two following solutions *A* and *B*.

Solution *A*		Solution *B*	
Water	100 ml	Water	100 ml
Hydroquinone	2.5 g	NaOH	4.0 g
Na_2SO_3	2.5 g		
KBr	2.5 g		

The following solutions are recommended for fixing plates or prints:

Solution *A*		Solution *B*	
Water	500 ml	Glacial acetic acid	10 g
Hypo ($Na_2S_2O_3 \cdot 5H_2O$, crude)	125 g	Powdered alum	10.5 g
		Na_2CO_3	10.5 g
		Water	75 ml

The water of solution *A* may be heated to dissolve the hypo quickly, but it must be cooled before solution *B* is added.

The reagents are added in the order given, and when they are fully dissolved, the hardening solution *B* is poured into the thiosulfate, or hypo, solution *A* slowly, with stirring. The hypo must be *fully* dissolved before adding the hardener; otherwise sulfur will be deposited.

A simple photographic emulsion of silver halide and gelatin is usefully sensitive only to the blue, violet, and near ultraviolet. The short-wavelength limit is set by absorption of the gelatin; sensitivity falls rapidly below 2500 Å and is negligible below 1900 Å. The long-wavelength limit is set by the absorption of the silver halide and varies from about 4300 Å for pure silver chloride to 5200 Å for silver bromoiodide.

In order to photograph radiation of wavelength less than 2000 Å, it is necessary to use emulsions very low in gelatin or to sensitize an ordinary emulsion with a fluorescent coating. The low-gelatin emulsions are called *Schumann plates*. Improvements of the Schumann plate are made by Ilford, Ltd., and sold as Q plates.

Ordinary emulsions may be sensitized to the short ultraviolet by coating with a material which, when exposed to ultraviolet, fluoresces with emission of radiation, to which the emulsion is sensitive. This method is convenient and gives good sensitivity, but the resolving power is lowered by spreading of the image. The ethyl ester of dihydrocollidinedicarboxylic acid is suitable for the short-ultraviolet sensitization. It may be obtained from the Eastman Kodak Co. as "ultraviolet sensitizing solution, No. 3177," or the Kodak plates may be purchased with the proper fluorescent coatings.

Sensitization to wavelengths longer than those absorbed by the silver halide is produced by adding to the emulsion special dyes (optical sensitizers) which are absorbed on the silver halide and sensitize it to the light absorbed by the dyed grains. The chemistry of the dyes and the mechanism of their action are described by Mees (Ref. 8, chaps. 23–25) and James and Higgins (Ref. 8, chap. 14).

Dyes of many chemical types may act as long-wavelength sensitizers, but the most important classes are the cyanines and merocyanines. Photographic plates and films sensitive to the green (orthochromatic), to the whole visible spectrum (panchromatic), and to the infrared out to 9000 Å and farther are available.[9]

Applications of photography to specific problems are described in a number of books.[10]

References

1. C. R. Masson, V. Bockelheide, and W. A. Noyes, Jr., in A. Weissberger (ed.), "Technique of Organic Chemistry," vol. 2, "Catalytic, Photochemical, and Electrolytic Reactions," 2d ed., Interscience Publishers, Inc., New York, 1956; L. J. Buttolph in A. Hollaender (ed.), "Radiation Biology," vol. II, chap. 2, McGraw-Hill Book Company, New York, 1955; J. F. Scott and R. L. Sinsheimer in *ibid.*, chap. 4; R. B. Withrow and A. B. Withrow in *ibid.*, vol. III, chap. 3, 1956.
2. M. I. Christie, R. G. W. Norrish, and G. Porter, *Proc. Roy. Soc. London Ser. A,* **216:** 152 (1952).
3. T. Carrington and N. Davidson, *J. Phys. Chem.,* **57:** 418 (1953).
4. J. E. Willard, Radiation Chemistry and Hot Atom Chemistry, *Ann. Rev. Phys. Chem.,* **6:** 141 (1955).
5. J. Strong, "Procedures in Experimental Physics," Prentice-Hall, Inc., Englewood Cliffs, N.J., 1938.
6. P. A. Leighton and W. G. Leighton, *J. Phys. Chem.,* **36:** 1882 (1932).
7. W. W. Coblentz, *Natl. Bur. Std. U.S. Bull.,* **11:** 87 (1915).
8. S. R. Arenson, *J. Chem. Educ.,* **18:** 122 (1941); T. Evans, J. M. Hedges, and J. W. Mitchell, Theory of Photographic Sensitivity, *J. Phot. Sci.,* **3:** 1–11 (1955); T. H. James and G. C. Higgins, "Fundamentals of Photographic Theory," John Wiley & Sons, Inc., New York, 1948; J. E. Mack and M. J. Martin, "The Photographic Process," McGraw-Hill Book Company, New York, 1939; C. E. K. Mees, "The Theory of the Photographic Process," The Macmillan Company, New York, 1942; C. B. Neblette, "Photography: Its Principles and Practice," 5th ed., D. Van Nostrand Company, Inc., Princeton, N.J., 1955.
9. Eastman Kodak Co., "Kodak Photographic Films and Plates for Scientific and Technical Use," 8th ed., Rochester, N.Y., 1960.
10. C. P. Shillaber, "Photomicrography in Theory and Practice," John Wiley & Sons, Inc., New York, 1944; Eastman Kodak Co., "Photomicrography," 14th ed., Rochester, N.Y., 1944; W. Clark, "Photography by Infrared," 2d ed., John Wiley & Sons, Inc., New York, 1946; C. C. Scott, "Photographic Evidence," Vernon Law Book Co., Kansas City, Mo., 1942; T. A. Longmore, "Medical Photography," Focal Press, London, 1944; C. H. S. Tupholme, "Photography in Engineering," Faber & Faber, Ltd, London, 1945; G. E. Mathews and J. I. Crabtree, Photography as a Recording Medium for Scientific Work, *J. Chem. Educ.,* **4:** 9 (1927); Eastman Kodak Co., "Infrared and Ultraviolet Photography," Rochester, N.Y., 1959.

Chapter 27

Nuclear and Radiation Chemistry†

† General references are given as Ref. 1.

AVAILABILITY OF ISOTOPES

For the 104 elements which are now known, over 600 isotopes are known.†
Isotopes of an element have different mass numbers, i.e., total number of neutrons and
protons in the nucleus, but the same atomic number, i.e., number of protons in the
nucleus. The nuclei of most of these isotopes are unstable and decay with the
emission of α particles, β particles, γ rays, or positrons, which can be detected with
instruments discussed in this chapter. The relative concentrations of stable isotopes
may be determined by mass spectrometry, spectroscopy, nuclear magnetic resonance,
and other methods to be mentioned.

Research involving isotopes has been greatly accelerated by the increased
availability of radioisotopes and stable isotopes from the U.S. Atomic Energy
Commission, the Atomic Energy Commission of Canada, and similar agencies of
other nations. A number of radioisotopes which do not occur naturally have been made
available in large quantities by the operation of uranium nuclear reactors which
produce neutrons in high concentrations. A great variety is available at nominal prices
in quantities of fractions of a millicurie to several curies. (A curie is the quantity of a
radioisotope that supplies 3.7×10^{10} disintegrations per second.) Catalogs, price
lists, and application blanks can be obtained from the U.S. Atomic Energy Commission, Isotopes Division, Oak Ridge, Tenn. A prospective purchaser of isotopes must
state that he is equipped to handle them, that his monitoring instruments are properly
calibrated, that he will take adequate precautions in the handling of waste materials,
and that he will accept legal responsibility for any damage from radioactivity.

Many organic and inorganic compounds which have been synthesized with
radioactive elements for use in tracer experiments are commercially available.

Radioisotopes may be formed by (n,α), (n,p), and (n,γ) reactions in uranium
nuclear reactors in which neutrons are absorbed and α particles, protons, or γ rays
are emitted. Production of these isotopes involves the insertion of an element, often
in the form of a chemical compound, contained in a small aluminum container, into
the nuclear reactor for a few days or months. Important examples of isotopes made
in this way are ^{14}C, ^{32}P, ^{82}Br, and ^{59}Fe.

Isotopes may be obtained also by separation from fission products produced in a
nuclear reactor.

Stable isotopes including ^{2}H, ^{10}B, ^{18}O, ^{198}Hg, ^{13}C, ^{15}N, and a number
of metals in concentrated form are available.

---

† Tabulated information on various nuclei is available in Ref. 2. Various supplements have been
issued, and now nuclear-data sheets may be obtained on subscription from the National Academy of Science–
National Research Council, Washington, D.C. In these sheets new data from the literature are cumulated
with old data. A complete list of isotopes and isotopically labeled compounds which are commercially
available is given in Ref. 3

Deuterium oxide is produced on a large scale by the use of exchange reactions carried out between water vapor and liquid water with solid catalysts in a column operation. In the later stages of purification fractional distillation and electrolysis are used. Nitrogen 15 is concentrated by the passage of ammonia gas upward through a packed tower, down which a stream of ammonium chloride solution is passing.[4]

The isotopes of a number of metals have been separated electromagnetically, i.e., by use of large mass spectrometers designed for the purpose, and are available from the U.S. Atomic Energy Commission on a loan basis.

ANALYSIS FOR STABLE ISOTOPES

MASS SPECTROMETRY.[5]   In a mass spectrometer ions in the gas phase are accelerated by an electric field and passed through a magnetic field to separate the ions according to their mass-to-charge ratio. A gas to be studied is ionized by bombarding it with electrons which have been accelerated by a potential difference of 50 to 100 volts. Solids may be used if they are first vaporized with a hot filament. The type of mass spectrometer designed by Nier[6] is illustrated in Fig. 176. The glass tube, containing an ion source at one end and an ion collector at the other,

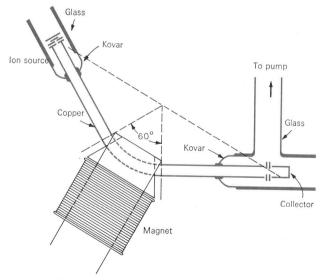

figure 176. Nier mass spectrometer.

is evacuated with a diffusion pump. It must be heated before it is used in order to drive out traces of adsorbed water and gases. The ions to be studied are accelerated by a potential of from 800 to 1000 volts, and the ion paths are bent by the field of an electromagnet (about 3500 gauss). The ion current at the collector is amplified and recorded automatically. Mass spectrometers are finding extensive use for the analysis of organic mixtures as well as isotope ratios.

ANALYSIS FOR RADIOACTIVE ISOTOPES[7]

IONIZATION CHAMBERS. In an ionization chamber a low voltage is applied across two electrodes, usually a wire at a positive potential and a coaxial cylinder at a negative potential. If a source of radiation is brought into the vicinity of the ionization chamber, ions are produced in the gas in the chamber, and these ions are attracted to the electrodes. As the potential difference between the wire and the cylinder is increased, the current increases because more gaseous ions are drawn to the electrodes before they have a chance to recombine. Eventually a saturation current is reached because all the electrons and gas ions produced by the radiation are drawn to the electrodes. This plateau, or saturation, region is shown by BC and $B'C'$ in Fig. 177. The saturation current produced by α particles is greater than that produced by the same number of β particles of the same energy. Because of their greater range, β particles may not expend all their energy within the chamber but may pass on through while the α particles all come to rest in the chamber.

The current produced in an ionization chamber is typically of the order of 10^{-17} amp. Such small currents may be measured using electrometer tubes or a vibrating-reed electrometer. In this latter device a metallic reed serving as one plate of a condenser is driven at a fixed frequency, such as 400 Hz, and an alternating current is obtained from the ionization current. This signal is amplified by an ac amplifier.

Ionization chambers are especially useful for α particles which produce a large specific ionization. Since the range of α particles is low, the sample is usually placed inside the chamber.

PROPORTIONAL COUNTERS. As the voltage between a positive wire and the negative coaxial cylinder is increased above that required to collect all the ions produced by radiation, the current again increases, as shown in Fig. 177. The primary electrons are accelerated sufficiently in the field of the center electrode to produce secondary ionization by their interaction with gas molecules. The electrons must be accelerated sufficiently between collisions for their kinetic energy to become greater than the ionization potential of the gas. The resulting ionization becomes a chain re-

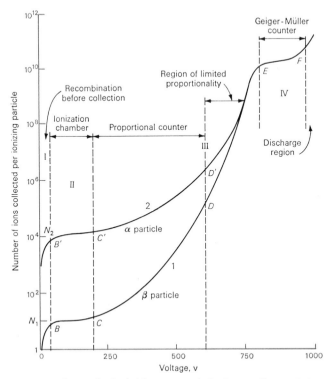

figure 177. Curves of pulse height versus applied voltage to illustrate ioniza-tion, proportional, and Geiger-Müller regions of operation [Adapted from Montgomery and Montgomery, J. Franklin Inst., **231:** 447 (1941), by R. T. Overman and H. M. Clark, in "Radioisotope Techniques," p. 27, McGraw-Hill Book Company, New York, 1960.]

action, and an avalanche of ions is produced by each ionizing particle. As seen from the figure, the amplification achieved in a proportional counter is of the order of 10^4. The amount of charge collected per particle which expends all of its energy in the chamber is proportional to the initial energy of the particle, as in the ionization chamber. Because of the internal amplification it is more easily detected, and less external amplification is needed.

In a proportional counter the discharge due to a single ionizing particle is localized in a small portion of the tube. The discharge stops when all the electrons produced are swept to the collector. Very high counting rates may be used compared with Geiger-Müller counters. Windowless proportional counters in which methane or other counting gas flows continuously through the tube are widely used for determining α and weak β activities.

The voltage required for a proportional counter ranges from several hundred to several thousand volts, depending upon the pressure and ionization potential of the gas used and the geometry of the counter. A very stable high-voltage supply is required for a proportional counter since the number of ions collected per ionizing particle depends markedly upon the voltage. As the voltage is raised above the proportional region, the strict proportionality between amount of ionization produced in the gas by an α or β particle and the amount of charge collected is lost, as shown in Fig. 177.

GEIGER-MÜLLER COUNTERS.   In the Geiger-Müller region *EF*, the electric field strength in the vicinity of the positively charged wire in the center of the coaxial conducting tube becomes so high that a mechanism is developed whereby the discharge always produces the same total current regardless of the initial number of pairs of ions formed by the α or β particles.

In this region an α particle and a β particle give rise to the same number of ions collected at the electrode. Therefore Geiger-Müller counters cannot be used to distinguish between radiations of different energies. In contrast to the proportional region, the discharge covers the entire length of the central wire. There is a danger that multiple pulses may arise from the radiation emitted by the ions as they collect electrons from the wall in becoming neutral atoms again. This is prevented by adding a quenching gas which dissociates in absorbing the deexcitation radiation. As seen in Fig. 177, the gas amplification factor is of the order of 10^6 to 10^9.

One of the problems with Geiger-Müller counters is that although the collection time for electrons is of the order of a microsecond, the collection time for positive ions is of the order of several hundred microseconds. During this time the tube is inoperative, and an ionizing particle coming in this dead time will not be counted. A correction for such coincidence losses may be made[8] if the resolving time of the tube and electronic circuit is known. Much faster disintegration rates may be measured with proportional counters because the positive ions from one pulse do not have to be collected before another pulse can be received. Their resolving times are in the range 0.01 to 0.01 μsec.

Almost every β particle which enters a Geiger-Müller counter is counted because it produces one or more ions in the gas. Only about 1 percent of the gamma rays are counted because their probability of being absorbed in the counter gas, often argon, is low.

SCINTILLATION COUNTERS.[9]   When ionizing radiation is absorbed by matter, electrons are ejected from atoms and molecules, and light may be emitted as electrons return to the vacated energy levels. This radiation may be detected with a photomultiplier tube as illustrated in Fig. 178. In the absorption of a γ ray, a quantum of

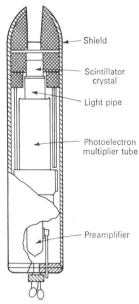

Shield

Scintillator crystal

Light pipe

Photoelectron multiplier tube

Preamplifier

figure 178.
Schematic diagram of probe type of scintillation detector.

energy is transferred to an electron as kinetic energy. This energetic electron loses its excess energy in a series of interactions with atoms and molecules. Thus a single γ-ray photon gives rise to many photons, and the amount of light produced is proportional to the energy of the radiation. Solid scintillation counters are widely used for γ rays because they have a high density and are therefore effective in absorbing the γ rays. Gas-filled counters have a low sensitivity for γ rays since the probability that a γ ray will be absorbed in a small volume of gas is low.

Certain organic or inorganic solids or organic liquids emit light when irradiated and may therefore be used in scintillation counters. Sodium iodide with a trace of thallium as an excitant is widely used for high-efficiency γ-ray detection because of its high density, high light output, and high transparency. Since NaI is hygroscopic, the crystal is encased in a watertight metallic shell. For β counting, anthracene and stilbene crystals are often used. For the detection of weak β radiation, the sample to be measured may be dissolved in a liquid-scintillator solution. The scintillator, such as p-terphenyl, may be dissolved in a solvent, such as benzene or toluene. When ^{14}C and ^3H are counted in this way, the pulses from the low-energy β radiation are so small that they may be in the range of the random noise from the photomultiplier tube. This thermal noise is reduced by cooling the photomultiplier tube.

The luminescence decay period for NaI (activated by Tl) is 3×10^{-7} sec, and the decay periods for the organic phosphors which are used are 10 to 100 times shorter. Thus if the associated electronic equipment does not introduce appreciable delay, high counting rates may be handled without significant coincidence losses.

A scintillation counter may be used in conjunction with a pulse-height analyzer to determine the energy spectrum of γ rays. In a single-channel analyzer only those pulses falling in a certain voltage range are counted. In more complex instruments there are a number of channels, so that pulses in different energy ranges may be counted simultaneously. In addition to the sharp peak in the plot of number of particles versus energy which is due to the ejection of photoelectron by the γ ray, there is also Compton scattering of the γ-ray photons. The Compton scattering results in a broad region of pulse heights at lower energies.

SOLID-STATE DETECTORS. Various crystals have been used for the detection of radiation by coating opposite faces with electrodes and measuring the conductivity by use of special circuits. Since the crystals are insulators, no appreciable current flows in the absence of ionizing radiation. When ionizing radiation is absorbed by the crystal, electrons are raised to the conduction band and positive holes are left in the valence band. Under the influence of an applied electric field, the electrons and holes drift through the crystal, and the current is measured in the external circuit. These detectors have the advantage of (*a*) high stopping power, so that high-energy particles may be detected with small detectors, (*b*) proportionality between absorbed particle energy and pulse height, and (c) fast rise times so that high counting rates may be used.

NEUTRON COUNTERS.   Neutrons may be detected by use of the (n,α) reaction

$$^{10}_{5}B + ^{1}_{0}n \longrightarrow ^{4}_{2}He + ^{7}_{3}Li$$

The α particle and lithium nucleus have a total kinetic energy of 2.78 Mev and produce ionization which may be detected by use of an ionization chamber, proportional counter, or scintillation detector. The ^{10}B is available in its isotopic form and may be placed in the detector as BF_3 or as a thin coating of boron on the inside of the counter chamber.

Neutrons may also be detected by the determination of the proton, γ ray, or fission products produced in (n,p), (n,γ), and $(n,$ fission$)$ reactions. Since the probabilities of different reactions depend in different ways upon the neutron energies, different reactions are useful for determining fast and slow neutrons. Neutrons may also be detected by the radioactivity of fission products in a counter containing uranium. The fissionable material is usually incorporated into the chamber in the form of a thin film or a multiple-plate construction.

MISCELLANEOUS METHODS.   When a particle passes through matter at a velocity which is greater than the speed of light in the medium, the particle is decelerated and radiation is emitted. The first detailed study of the visible light emitted was made by Čerenkov. The Čerenkov radiation is analogous to the bow wave of a ship which is moving faster than the velocity of surface waves or to the shock wave produced by an object moving through air at a supersonic velocity. The radiation may be detected by use of photomultiplier tubes. For the detection of single particles, special techniques are required, since the intensity is low. For example, only 250 photons of visible light is produced per centimeter of path by an electron traveling through Lucite at nearly the speed of light.

Ionizing radiation produces latent images in photographic plates. When the plate is developed, grains of silver appear along the tracks of particles.[10] By study of these paths and the records of various types of collisions, nuclear physicists obtain detailed information about mass, charge, and energy of the particles. Special nuclear emulsions have been developed for this type of work, since ordinary emulsions are not suitable. The general blackening of a photographic plate may be used to measure the integrated intensity of radiation, as in a film badge for radiation safety. The location of radioactivity in a sample may be determined by laying the sample on a photographic plate in the dark (radioautography). For example, the location of radioactive compounds on a filter-paper chromatogram or the location of radioactive phosphorus in a leaf may be determined. If the intensity of radiation is weak, long times may be used.

Ionizing particles may also be detected by use of cloud chambers and bubble chambers. In a bubble chamber the pressure on a liquid is suddenly reduced to a value below the vapor pressure. Under these circumstances ionizing radiation produces nuclei for the formation of bubbles, and these small bubbles are photographed. Bubble chambers containing liquid hydrogen are especially useful for studying the particles produced with high-energy accelerators.

RADIATION CHEMISTRY [11]

Convenient sources of high intensities of γ radiation are available for the study of the chemical changes which occur when γ rays are absorbed by matter. Cobalt 60 may be purchased from commercial companies with specific activities up to 300 curies per gram. These sources require special housings to make them safe and convenient to use.[12] Unseparated fission products from nuclear reactors may also be used as high-intensity-radiation sources. The fission products are simply used in the form of spent fuel elements.[13]

RADIATION SAFETY.[14]   The safe handling of radioactive material and the proper disposal or storage of waste are vital requirements for any laboratory investigations using radioactive isotopes.[15] Many excellent instruments are now on the market for monitoring laboratories and personnel. Photographic film badges, pocket meters, and portable instruments for detecting and measuring α, β, and γ rays are available.

In general, it may be stated that for laboratory work, the exposure of the body to 0.1 roentgen per week is the maximum dosage which is permitted. However, the detailed specification[16] of exposure limits is complicated and depends upon the type of radiation, the part of the body exposed, the possibility of getting the material in the body, etc. The rules and regulations for handling radioactivity are spelled out in Title 10 Part 20 of the Federal Register, entitled Standards for Protection against Radiation, Nov. 17, 1960.

References

1. M. Calvin et al., "Isotopic Carbon," John Wiley & Sons, Inc., New York, 1959; W. D. Claus (ed.), "Radiation Biology and Medicine," Addison-Wesley Publishing Company, Inc., Reading, Mass., 1958; S. Glasstone, "Sourcebook on Atomic Energy," D. Van Nostrand Co., Inc., Princeton, N.J., 1958; G. Friedlander, J. W. Kennedy, and J. M. Miller, "Nuclear and Radiochemistry," 2d ed., John Wiley & Sons, Inc., New York, 1964; A. Hollaender (ed.), "Radiation Biology," vol. I, McGraw-Hill Book Company, New York, 1954; A. C. Wahl and N. A. Bonner, "Radioactivity Applied to Chemistry," John Wiley & Sons, Inc.,

New York, 1951; R. T. Overman and H. M. Clark, "Radioisotope Techniques," McGraw-Hill Book Company, New York, 1960; W. J. Price, "Nuclear Radiation Detection," McGraw-Hill Book Company, New York, 1958; C. C. H. Washtell, "Introduction to Radiation Counters and Detectors," Philosophical Library, Inc., New York, 1960.

2. Nuclear Data, *Natl. Bur. Std. U.S. Circ.* 499, 1950.

3. J. L. Sommerville (ed.), "The Isotope Index," Scientific Equipment Co., Indianapolis, 1967.

4. H. G. Thode and H. C. Urey, *J. Chem. Phys.*, **7:** 34 (1939); C. A. Hutchinson, D. W. Stewart, and H. C. Urey, *ibid.*, **8:** 532 (1940).

5. A. J. B. Robertson, "Mass Spectrometry," John Wiley & Sons, Inc., New York, 1954; R. W. Kiser, "Introduction to Mass Spectrometry and Its Applications," Prentice-Hall, Inc., Englewood Cliffs, N.J., 1965.

6. A. O. Nier, *Rev. Sci. Instr.*, **11:** 212 (1940).

7. A. H. Snell (ed.), "Nuclear Instruments and Their Uses," vol. I, "Ionization Detectors, Scintillators, Cerenkov Counters, Amplifiers: Assay, Dosimetry, Health Physics," John Wiley & Sons, Inc., New York, 1962.

8. Overman and Clark, Ref. 1, p. 55.

9. J. B. Birks, "Theory and Practice of Scintillation Counting," The Macmillan Company, New York, 1964.

10. H. J. Yogada, "Radioactive Measurements with Nuclear Emulsions," John Wiley & Sons, Inc., New York, 1949.

11. P. Ausloos (ed.), "Fundamental Processes in Radiation Chemistry," John Wiley & Sons, Inc., New York, 1968.

12. R. F. Firestone and J. E. Willard, *Rev. Sci. Instr.*, **24:** 904 (1953); H. A. Schwarz and A. O. Allen, *Nucleonics*, **12(2):** 58 (1954); W. S. Eastwood, *ibid.*, **13(1):** 52 (1955).

13. J. W. Loeding, E. J. Petkus, G. Yasui, and W. A. Rogers, *Nucleonics*, **12:** 14 (1954).

14. Safe Handling of Radioactive Isotopes, *Natl. Bur. Std. U.S. Handbook* 42, 1949; Maximum Permissible Amounts of Radioactive Isotopes in the Human Body and Maximum Permissible Concentrations in Air and Water, *Natl. Bur. Std. U.S. Handbook* 52, 1953; Permissible Dose from External Sources of Ionizing Radiation, *Natl. Bur. Std. U.S. Handbook* 59, 1954.

15. W. D. Claus, "Radiation Biology and Medicine," Addison-Wesley Publishing Company, Inc., Reading, Mass., 1958.

16. "Standards for Protection against Radiation," U.S. Atomic Energy Commission, 1958.

Chapter
28

Electronics

The object of this chapter is to provide an introduction to certain principles of electronics at a sufficiently quantitative level to be of practical use to the physical chemist who seeks to understand the capabilities and limitations of electronic circuits well enough to use them intelligently. The coverage is selective rather than broad, with the aim of going deeply enough to enable the reader to utilize the more detailed reference material available in the literature.

AC CIRCUIT THEORY

This section contains a résumé of certain principles of ac circuit theory. While it is not necessary to have a thorough knowledge of this subject in order to assimilate the material in subsequent sections, a good understanding of these principles can be a valuable aid to the physical chemist who uses electronic techniques.

BASIC CIRCUIT ELEMENTS.   The three basic elements of ac networks are the resistor, the capacitor, and the inductor. Each of these is a two-terminal device which can be characterized by a single parameter, R, C, or L. A resistor is a device for which the value of current I (entering one terminal and leaving the other) at a given instant depends only on the voltage or potential difference E existing at that instant between the two terminals. The resistance R is defined by

$$R = \frac{E}{I} \tag{1}$$

A capacitor is a device capable of storing or accumulating charge, so that the charge Q which at a given instant has accumulated on one plate depends only on the voltage E existing between the two terminals at that instant. A charge of equal magnitude and opposite sign exists on the other plate. The capacitance C is defined by

$$C = \frac{Q}{E} \tag{2}$$

An inductor is a device which generates a well-localized magnetic field when a current passes through it, such that the magnetic flux Φ linking the inductor at a given instant depends only on the current through it at that instant. The inductance L is defined by

$$L = \frac{\Phi}{I} \tag{3}$$

Any real physical device, a coil of wire, for example, shows all three of these properties to some degree, but it is possible to construct devices for which one or the other of these properties is so dominant over the other two as to make the idealization a useful one. Thus a coil of wire of high resistance may be primarily resistive, while a coil of very low resistance wire may be primarily inductive, at the frequency at which it is intended to be used.

While the defining equations (2) and (3) are of course of fundamental importance, one is often more interested in the relationship between instantaneous current and voltage for each of these devices. The three relations are

$$E = RI \tag{4}$$

$$\frac{dE}{dt} = \frac{1}{C} I \tag{5}$$

$$E = L \frac{dI}{dt} \tag{6}$$

Equation (4) follows directly from (1); Eq. (5), from (2), since I equals dQ/dt; and Eq. (6), from (3), since E equals $d\Phi/dt$.

PROPERTIES OF SINUSOIDAL WAVEFORMS.   Of all the forms of time dependence encountered for voltages and currents in electronic circuits, the sine-wave, or sinusoidal, form is of preeminent importance:

$$A(t) = A_0 \cos (2\pi ft + \phi) \tag{7}$$

where $A(t)$ = instantaneous value of current or voltage at time t
$\qquad f$ = frequency, Hz
$\qquad \phi$ = phase angle, radians
$\qquad A_0$ = amplitude or peak value
As an alternative to the amplitude or peak value, two other measures are commonly used,

$2A_0$ = peak-to-peak (p-p) value
A_{rms} = root-mean-square (rms) value

with A_{rms} defined by

$$A_{\text{rms}} = \left\{ \frac{1}{t_0} \int_0^{t_0} [A(t)]^2 \, dt \right\}^{\frac{1}{2}} \tag{8}$$

whence

$$A_{\text{rms}} = \frac{1}{\sqrt{2}} A_0 = 0.707 A_0 \tag{9}$$

where t_0 = period of one cycle = $1/f$.

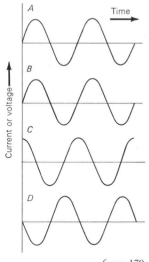

Current or voltage

A   Time

B

C

D

figure 179.

*Phase relationships for
sinusoidal waveforms.*

As a matter of notational convenience, $A(t)$ is often written

$$A(t) = A_0 \cos (\omega t + \phi) \tag{10}$$

where $\omega = 2\pi f =$ circular frequency, in radians per second.

   Several of the possible phase relationships which can exist between two sinusoidal waveforms are illustrated in Fig. 179. Two sinusoids are said to be *in phase* if they pass through corresponding points of their respective cycles at the same instant, as, for example, A and B of Fig. 179. Waveforms A and C are said to be 90° *out of phase*, while A and D are 180° out of phase. Waveform D is said to *lead C* by 90°, while B *lags C* by 90°.

   It is useful to remember that an arbitrary sinusoidal wave, such as $A(t)$ of Eq. (7), can be resolved into a pure sine and a pure cosine term of the same frequency f:

$$A_0 \cos (2\pi ft + \phi) = (A_0 \cos \phi) \cos 2\pi ft - (A_0 \sin \phi) \sin 2\pi ft \tag{11}$$

   Often it is necessary to find the waveform $B(t)$ which results when two sinusoidal waveforms $A'(t)$ and $A''(t)$ of the same frequency are added:

$$B(t) = A'(t) + A''(t) \tag{12}$$

where

$$A'(t) = A_0' \cos (\omega t + \phi') \tag{13}$$
$$A''(t) = A_0'' \cos (\omega t + \phi'') \tag{14}$$

One could imagine plotting the two waveforms and adding the instantaneous values point by point, but this is awkward as a quantitative procedure, and not even particularly helpful as a way of visualizing the resultant. Two efficient methods of obtaining the sum will be described below, but it is worth noting that both are really based on Eq. (11). The waveforms $A'(t)$ and $A''(t)$ can each be resolved into sine and cosine terms; then the sine terms can be added directly, and similarly for the cosine terms. There is a clear analogy here with the composition of two vectors to form a resultant, an analogy which is of considerable help in understanding ac circuit problems.

   Method A: Vector Diagrams. The waveforms $A'(t)$ and $A''(t)$ can each be considered to be generated by rotation of a vector† of constant length in a plane. Thus let two vectors **A'** and **A''** be arranged as in Fig. 180a at $t = 0$. If these rotate counterclockwise with circular frequency ω, the components along the horizontal axis generate, respectively, $A'(t)$ and $A''(t)$. Obviously, the vector sum **B** formed in

   † To avoid the connotation of geometrical significance which sometimes is attached to the term *vector*, the rotating vector which generates a sinusoidal waveform is called in many electrical-engineering texts a *phasor*, or *sinor*.

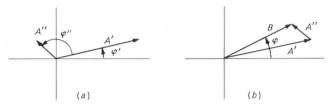

figure 180. Vector diagrams illustrating (a) vectors representing sinusoids
$A'(t)$ and $A''(t)$ at $t = 0$ and (b) addition of two sinusoids to form a
resultant.

the usual way from $\mathbf{A}'$ and $\mathbf{A}''$ similarly generates $B(t) = A'(t) + A''(t)$. The
amplitude and phase angle of $B(t)$ are thus obtained readily, as shown in Fig. 180b.
The amplitudes of the sine and cosine components of $B(t)$ can also easily be found
from this diagram as the projections at $t = 0$ of the vector sum along the vertical
and horizontal axes, respectively.

Method B: Complex Algebra. When it is desired to perform a calculation
analytically rather than graphically, the vector diagrams obviously provide the basis
for setting up the necessary equations, and this procedure is often followed. A more
powerful analytical method, however, is available with the use of complex algebra.
The rules for manipulation of complex quantities and the procedure for graphing
them are summarized in the Appendix.

With each of the vectors of Fig. 180 we can associate a complex number, as
follows:

Component along horizontal axis $\longleftrightarrow$ real part
Component along vertical axis $\longleftrightarrow$ imaginary part

In other words, we regard the diagram of Fig. 180 as existing in a complex plane
and associate with each vector $\mathbf{A}'$, $\mathbf{A}''$, . . . the complex number corresponding to
its terminus. Since there is a one-to-one correspondence between these vectors and
complex numbers, the same symbol can be used for both. We get, for $t = 0$,

$$\mathbf{A}' = A_0' e^{i\phi'} = A_0' \cos \phi' + i A_0' \sin \phi' \tag{15}$$
$$\mathbf{A}'' = A_0'' e^{i\phi''} = A_0'' \cos \phi'' + i A_0'' \sin \phi'' \tag{16}$$

etc. The addition of two vectors then corresponds analytically to the addition of two
complex numbers.

One advantage of complex notation is that the time dependence of rotating
vectors can be represented in a very compact fashion. For example, the rotating
vectors $\mathbf{A}'(t)$ and $\mathbf{A}''(t)$ are represented by the complex numbers

$$\mathbf{A}'(t) = A_0' e^{i(\omega t + \phi')} \tag{17}$$
$$\mathbf{A}''(t) = A_0'' e^{i(\omega t + \phi'')} \tag{18}$$

Thus the sinusoidal waveforms $A'(t)$ and $A''(t)$ of Eqs. (13) and (14) can be thought of as the *horizontal components* of two rotating vectors or, equally well, as the *real parts* of two complex numbers.

These methods become particularly helpful in visualizing the addition of two sinusoidal waveforms of different frequency. Suppose, for example, one wishes to add the waveforms

$$C'(t) = C_0' \cos \omega' t \tag{19}$$

$$C''(t) = C_0'' \cos \omega'' t \tag{20}$$

The corresponding vectors $\mathbf{C}'(t)$ and $\mathbf{C}''(t)$ rotate at different frequencies. The horizontal component of the resultant is not a simple sinusoidal waveform. For example, for $\omega' \approx \omega''$, the resultant could be described approximately as a sinusoidal waveform with slowly changing amplitude (phenomenon of beating).

IMPEDANCE.   If the current

$$I_0 \cos (\omega t + \phi) \tag{21}$$

is passed through a resistor, capacitor, or inductor, the *steady-state* voltage $E(t)$ developed across the device is, from Eqs. (4) to (6), as follows:

For the resistor:

$$E_R(t) = R I_0 \cos (\omega t + \phi) \tag{22}$$

For the capacitor:

$$E_C(t) = \frac{1}{\omega C} I_0 \sin (\omega t + \phi) \tag{23}$$

For the inductor:

$$E_L(t) = -(\omega L) I_0 \sin (\omega t + \phi) \tag{24}$$

These results are illustrated by the vector diagram of Fig. 181. The corresponding equations in complex form are obtained by writing the expression for the current,

$$I(t) = I_0 e^{i(\omega t + \phi)} \tag{25}$$

and relating the complex voltages to $\mathbf{I}(t)$ through the use of Eqs. (4) to (6); the results are

$$\mathbf{E}_R(t) = R\mathbf{I}(t) \tag{26}$$

$$\mathbf{E}_C(t) = \frac{1}{i\omega C} \mathbf{I}(t) \tag{27}$$

$$\mathbf{E}_L(t) = (i\omega L)\mathbf{I}(t) \tag{28}$$

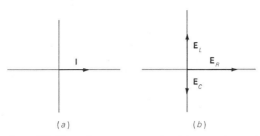

figure 181. Vector diagrams representing (a) assumed current and (b) resulting voltages for resistor, capacitor, and inductor. The vectors are shown for t = 0. They rotate counterclockwise with circular frequency ω. The actual currents and voltages are the components along the horizontal axes.

The reader may verify that Eqs. (26) to (28) are consistent with the vector diagrams of Fig. 181. These equations may be thought of as providing a generalization of Ohm's law to the case of sinusoidal voltages and currents in ac circuits. The great utility of the complex notation can now be appreciated. The relationship between current and voltage takes on a particularly simple and easily remembered form. Furthermore, the equations in complex form are immediately related to the vector diagrams, an important aid to visualization.

Equations (26) to (28) all have the form

$$\mathbf{E}(t) = Z\mathbf{I}(t) \tag{29}$$

where Z is a complex number. The quantity Z which relates $\mathbf{E}(t)$ and $\mathbf{I}(t)$ in this way is called the *impedance*. The reciprocal of Z is a complex number $Y = Z^{-1}$, called the *admittance*, in terms of which Eq. (29) becomes

$$\mathbf{I}(t) = Y\mathbf{E}(t) \tag{30}$$

Admittance and impedance then represent extensions of the concepts of conductance and resistance, respectively. Table 1 gives the admittances and impedances of the basic circuit elements.

An element of a circuit is said to be *linear* if its impedance Z is independent of the amplitude of the current through it and of the voltage across it.

Table 1. Admittances and Impedances of the Basic Circuit Elements

Element	Parameter	Symbol	Impedance	Admittance
Resistor	R	⎓⋀⋀⋀⎓	R	$1/R$
Capacitor	C	⎓⊣⊢⎓	$1/i\omega C$	$i\omega C$
Inductor	L	⎓⊙⊙⊙⎓	$i\omega L$	$1/i\omega L$

IMPEDANCE OF A NETWORK.   If a specified sinusoidal current $I(t)$ passes into a network at one terminal and out at another, some definite steady-state voltage $E(t)$ will exist between these terminals. The impedance Z which relates $I(t)$ and $E(t)$ through Eq. (29) is called the impedance of the network between these terminals.

If two elements with impedances Z_1 and Z_2 are connected in *series*, the equivalent impedance Z_S for the combination is

$$Z_S = Z_1 + Z_2 \qquad \text{series case} \tag{31}$$

If two elements with admittances Y_1 and Y_2 are connected in *parallel,* the equivalent admittance Y_P for the combination is

$$Y_P = Y_1 + Y_2 \qquad \text{parallel case} \tag{32}$$

For networks which can be written in a plane diagram without crossovers, reduction to a single equivalent impedance is always possible by successive application of Eqs. (31) and (32), together with the general relation $Z = Y^{-1}$. For networks not amenable to this approach, the more general methods of mesh and nodal analysis are particularly useful.†

For any given frequency, complex impedances (or admittances) may be plotted in a complex plane and added like vectors. Figure 182 illustrates this *graphical* method of addition for the case of a resistor and capacitor.

The *magnitude* Z_0 of an impedance Z is defined as the magnitude of the complex number Z and may be visualized as the length of the vector from the origin to the point Z in a complex plane. If the current and voltage amplitudes are I_0 and V_0, respectively, it is easily shown that these are related by

$$E_0 = Z_0 I_0 \tag{33}$$

† These methods are treated, for example, in Ref. 1.

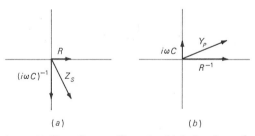

(a)   (b)

figure 182. Vector diagrams illustrating (a) the impedance of a series combination of R and C and (b) the admittance of a parallel combination of R and C.

It may be noted that the rms values are similarly related:

$$E_{\text{rms}} = Z_0 I_{\text{rms}} \tag{34}$$

The *algebraic* procedure for finding the impedance of a network can best be mastered by working out a few examples. Several are given in Table 2. It is instructive to examine the frequency dependence of Z and Z_0 for each of these networks.

Two physically different networks which have the same impedance are called *equivalent networks*. The equivalence may exist only at a single frequency or may hold over a wide range of frequencies. Examples will be encountered below.

Table 2. *Impedances of Several Networks*

Network	Z	Z_0
	$R - i\dfrac{1}{\omega C}$	$\left(R^2 + \dfrac{1}{\omega^2 C^2}\right)^{\frac{1}{2}}$
	$R + i\omega L$	$(R^2 + \omega^2 L^2)^{\frac{1}{2}}$
	$\left(\dfrac{1}{R} + i\omega C\right)^{-1}$	$\left(\dfrac{R^2}{1 + \omega^2 R^2 C^2}\right)^{\frac{1}{2}}$
	$R + i\left(\omega L - \dfrac{1}{\omega C}\right)$	$\left[R^2 + \left(\omega L - \dfrac{1}{\omega C}\right)^2\right]^{\frac{1}{2}}$

PARALLEL RESONANT CIRCUIT.   A very common and important network, the parallel resonant circuit is shown in Fig. 183. The resistance R shown is usually not present in the actual circuit as a separate component but represents the effective resistance of the coil. The quality factor Q of the coil is defined by

$$Q = \frac{\omega L}{R} \tag{35}$$

A coil is said to possess a high Q if $Q \gg 1$. Both Q and R depend on frequency but may often be considered constant over narrow ranges of frequency.

The impedance Z and its magnitude Z_0 may be obtained by a straight-forward calculation; the results are, for $Q \gg 1$,

$$Y = \frac{1}{R + i\omega L} + i\omega C \tag{36}$$

$$Z \approx \omega L Q \frac{1}{1 + iQu} \tag{37}$$

where

$$u = \omega^2 LC - 1 = \frac{\omega^2 - \omega_0^2}{\omega_0^2} \tag{38}$$

$$\omega_0 = \frac{1}{(LC)^{\frac{1}{2}}} \tag{39}$$

The function $Z_0(\omega)$ is plotted in Fig. 183b. It has a peak at $\omega = \omega_0$ and is sharply resonant for $Q \gg 1$. The *bandwidth* $\Delta\omega$ is defined as the separation between the two points at which $Z_0 = Z_{\max}/\sqrt{2}$. From Eq. (37) this is seen to depend on Q:

$$\frac{\Delta\omega}{\omega_0} = \frac{1}{Q} \tag{40}$$

For the high-Q case, Z_0 can be written, when $\omega + \omega_0 \approx 2\omega_0$, as

$$Z_0 \approx \omega L Q \left\{ \frac{1}{1 + 4[(\omega - \omega_0)/\Delta\omega]^2} \right\}^{\frac{1}{2}} \tag{41}$$

which shows clearly how the decrease of Z_0 from its peak value depends on the magnitude of $\omega - \omega_0$ relative to the bandwidth $\Delta\omega$. A useful result to remember is that the impedance at resonance for the high-Q case is $\omega L Q$.

If a voltage is to be injected in series with the coil of Fig. 183a, this same network can usefully be treated as a *series LCR* circuit. The analysis for this case is presented in Experiment 35 (page 222).

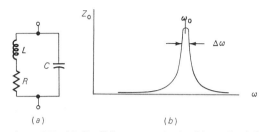

figure 183. (a) Parallel resonant circuit; (b) graph of Z_0 against frequency for parallel resonant circuit.

ACTIVE CIRCUIT ELEMENTS.   The circuit elements considered thus far are capable of dissipating or storing electromagnetic energy but not of generating it. On the other hand, there are circuit elements, such as batteries or vacuum tubes, which act as sources of electromagnetic energy within a network. Such elements are called *active* elements.

An important property of an active element is its *internal impedance*. A simple illustration of this concept is provided by a battery. The voltage measured at the terminals falls below the open-circuit value when current is drawn. To a certain degree of approximation, an actual battery, then, is equivalent to a constant-voltage dc generator (of zero impedance) acting in series with a fictitious resistance R_g, called the internal impedance of the battery. An entirely equivalent representation is that of a constant-current generator (of infinite impedance) acting in parallel with a conductance $G_g = R_g^{-1}$. These are represented in Fig. 184. The circuits b and c are equivalent to a in the sense that for any load impedance connected across the terminals the current through the load and the voltage across it are the same for the two fictitious circuits as for the actual circuit (a).

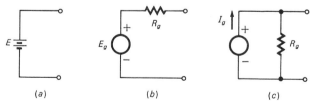

figure 184. Equivalent generators. (a) Actual circuit; (b) equivalent constant-voltage generator with series resistance; (c) equivalent constant-current generator with parallel conductance.

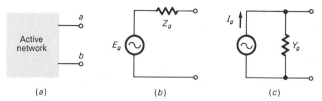

figure 185. (a) *Active network;* (b) *equivalent network by Thévenin's theorem;*
(c) *equivalent network by Norton's theorem.*

THÉVENIN'S THEOREM.   Suppose one has a network (Fig. 185*a*) which contains one or more active elements and which delivers power to a load of impedance Z_L connected across two terminals *a* and *b* of the network. For the purpose of calculating the current through (or voltage across) the load, the active network can be replaced by an equivalent network (Fig. 185*b*), consisting of an idealized voltage generator E_g acting in series with an internal impedance Z_g, where E_g is the voltage produced between the output terminals *a* and *b* when the load is removed and Z_g is the impedance measured between output terminals with load removed and with the active elements of the network replaced by their internal impedances. This is *Thévenin's theorem.* Stated briefly,

E_g = open-circuit voltage of the active network
Z_g = output impedance of the active network

An alternative equivalent circuit (Fig. 185*c*) for the active network is given by *Norton's theorem.* The idealized constant-current generator I_g here acts in parallel with an admittance Y_g, where I_g is the current which would pass from terminal *a* to *b* through a load of zero impedance (short-circuit current) and $Y_g = Z_g^{-1}$ is the output admittance of the active network.

IMPEDANCE MATCHING.   If a variable resistive load R_L is placed across the terminals of the battery of Fig. 184*a*, the power P delivered to the load is

$$P = I_L^2 R_L = \left(\frac{E_g}{R_g + R_L} \right)^2 R_L \tag{42}$$

where I_L is the current through the load. It is easily shown that P is a maximum for $R_L = R_g$.

More generally, if a device of fixed output impedance Z_1 delivers power to a second device of input impedance Z_2, it can be shown that the power delivered to the second device is a maximum for the condition

$$Z_1 = Z_2^* \tag{43}$$

where Z_2^* is the complex conjugate of Z_2. The impedances are said to be *matched* when Eq. (43) holds.

MAGNETIC COUPLING.   Figure 186a shows two circuits which are coupled only through the magnetic flux common to the two coils. A change in current through coil 1 results in a changing magnetic flux which induces a voltage in coil 2. The mutual inductance M is given by

$$M = k\sqrt{L_1 L_2} \tag{44}$$

where L_1 and L_2 are the inductances of the separate coils and k is the coefficient of coupling, which has its maximum value, unity, if all the flux generated by a current in one coil links the other coil. For the purpose of calculating current through the load Z, the network of Fig. 186b is equivalent to that of Fig. 186a.

Two limiting cases will be considered further. For sufficiently *loose* coupling, i.e., such that

$$\omega M \ll |i\omega(L_2 - M) + R_2 + Z| \tag{45}$$

the equivalent secondary circuit reduces to that of Fig. 186c.

For very *close* coupling, that is, $k \approx 1$, and under the additional assumptions,

$$R_1 \ll \omega L_1 \tag{46}$$
$$R_2 \ll \omega L_2 \tag{47}$$
$$Z \ll \omega L_2 \tag{48}$$

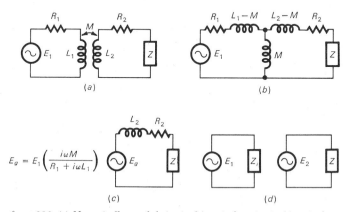

figure 186. (a) Magnetically coupled circuit; (b) equivalent circuit; (c) equivalent circuit for sufficiently loose coupling; (d) equivalent circuit for close coupling, under approximations specified by Eqs. (46) to (48).

the equivalent circuit reduces to Fig. 186*d* and the more familiar equations for transformer coupling are obtained:

$$\frac{E_2}{E_1} = \frac{N_2}{N_1} \tag{49}$$

$$Z_i = \left(\frac{N_1}{N_2}\right)^2 Z \tag{50}$$

where N_1 = number of turns in primary winding
N_2 = number of turns in secondary winding

The point to be especially noticed is that the impedance effectively presented to the voltage source E_1 is the so-called *reflected* impedance Z_i rather than the actual load impedance Z. Thus a closely coupled transformer of proper design [consistent with Eqs. (46) to (48)] can be used to match the impedance of a given load to the impedance of a given voltage generator. This is an efficient and useful arrangement because, to the extent that the above conditions can be met, the match is independent of frequency and is achieved with negligible loss of power.

RESPONSE TO NONSINUSOIDAL EXCITATION.   Since the concept of imped-ance is defined only for sinusoidal waveforms, the above methods of analysis are not directly applicable when nonsinusoidal functions are encountered. Three avenues of approach which are used in treating nonsinusoidal cases will be men-tioned here:

1. Equations (4) to (6) are still applicable and for relatively simple networks often yield the required information quite readily. Thus, for example, application of a step function of voltage to a resistor and capacitor in series leads to the familiar exponential growth of the voltage across the capacitor and the exponential decay of the voltage across the resistor.
2. The methods of Fourier series analysis are applicable if the waveform is periodic. Thus, for example, the zero-based square wave $A(t)$ of Fig. 187 can be expanded in the series[2]

$$A(t) = \tfrac{1}{2}A_0 + \frac{2}{\pi} A_0 \left(\sin \omega_0 t + \tfrac{1}{3}\sin 3\omega_0 t + \tfrac{1}{5}\sin 5\omega_0 t + \cdots\right) \tag{51}$$

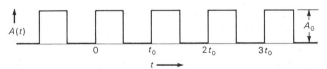

figure 187. Zero-based square wave of period t_0 and peak amplitude A_0.

where ω_0, the fundamental frequency, is $2\pi/t_0$. The impedance of the circuit being known as a function of frequency, one can apply the methods for sinusoidal waveforms separately to each term.

The method of Fourier analysis is also very helpful in a qualitative way. Thus, for example, if a complex voltage waveform is applied to a series RC circuit, one can immediately see that Fourier components at frequencies $f \ll 1/2\pi RC$ will appear principally across the capacitor, while components at frequencies $f \gg 1/2\pi RC$ will appear principally across the resistor. Thus a degree of separation of the Fourier components of the input voltage according to frequency is achieved with this circuit.

3. The method of Fourier integral analysis† and the closely related method of the Laplace transform§ represent very efficient tools for analyzing the response of a circuit to a wide variety of waveforms, both periodic and nonperiodic.

AMPLITUDE MODULATION AND DETECTION. An amplitude-modulated wave may be visualized as a sinusoid with amplitude a function of time. The simplest example is that of pure sine-wave modulation, shown in Fig. 188a.

The waveform

$$A(t) = A_0(1 + m\cos\omega_m t)\cos\omega t \tag{52}$$

† A good introduction to the use of Fourier analysis in ac circuit theory is given in Ref. 2.

§ An elementary introduction to the use of Laplace transforms for the analysis of network response is given in Ref. 3.

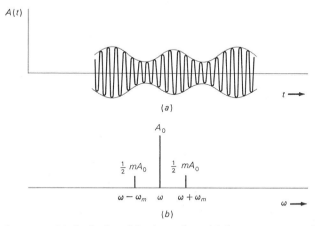

figure 188. (a) Amplitude-modulated waveform; (b) frequency spectrum of waveform bearing pure-sinusoidal modulation at a single modulation frequency ω_m.

is produced when the *carrier* $A_0 \cos \omega t$ is modulated at frequency ω_m. The coefficient m is the *degree of modulation*, ω_m is the *modulation frequency*, and $\cos \omega_m t$ is the *modulation waveform*.

With the use of familiar trigonometric identities, $A(t)$ may be written in the alternative form

$$A(t) = A_0 \cos \omega t + \tfrac{1}{2} m A_0 \cos (\omega + \omega_m)t + \tfrac{1}{2} m A_0 \cos (\omega - \omega_m)t \qquad (53)$$

i.e., as a superposition of three pure sinusoids, a carrier and two *sidebands*. Equation (53) is really an illustration of the Fourier theorem. The frequency spectrum of $A(t)$ is shown in Fig. 188b. It is important to note that $A(t)$ does not contain a component at the modulation frequency ω_m.

For the general case $A(t) = A_0[1 + m(t)] \cos \omega t$ with a more complicated modulation waveform $m(t)$, additional sideband pairs appear. For example, if $m(t)$ is a square wave, then a pair of sidebands is produced for each term in Eq. (51) beyond the first.

Given a modulated waveform, the process of extracting the waveform which corresponds to the envelope (or a reasonable approximation thereof) is called *detection*. A practical detector circuit is described on page 610.

FREQUENCY MODULATION.   A FM waveform is produced if the frequency of a sinusoidal waveform is made to vary with time. For modulation by a pure sinusoid, the instantaneous frequency is

$$\omega = \omega_c + \Delta\omega \cos \omega_m t$$

where ω_c = carrier frequency

$\quad\quad \omega$ = instantaneous frequency of FM waveform

$\quad\quad \Delta\omega$ = peak frequency deviation

$\quad\quad \omega_m$ = modulation frequency

The expression for the FM waveform for this case is

$$A(t) = A_0 \cos \left(\omega_c t + \frac{\Delta\omega}{\omega_m} \sin \omega_m t + \phi \right) \qquad (54)$$

where ϕ is the phase angle. Equation (54) has the form $A_0 \cos \theta(t)$ with $d\theta/dt$ equal to ω. The analysis of $A(t)$ into Fourier components leads to a frequency spectrum which consists of sideband pairs at $(\omega_c \pm \omega_m)$, $(\omega_c \pm 2\omega_m)$, $(\omega_c \pm 3\omega_m)$, ..., as contrasted with a single pair for the corresponding AM case. Another difference between the two types of modulation is in the phase relationships which exist among the sidebands.

MIXING.   The operation of combining two or more waveforms in a nonlinear impedance in order to generate sum and difference frequencies is known as *mixing* or *heterodyning*.

As a simple example, suppose the two voltages

$$E_1 = E_{01} \cos \omega_1 t \tag{55}$$
$$E_2 = E_{02} \cos \omega_2 t \tag{56}$$

are applied in series to a diode (see pages 609 to 611) for which the current-voltage characteristic is represented well enough by the relation

$$I = aE + bE^2 \tag{57}$$

between instantaneous voltage and current. If the substitution

$$E = E_1 + E_2$$

is made in Eq. (57), the resulting current includes, among others, the term

$$bE_{01}E_{02}[\cos (\omega_1 + \omega_2)t + \cos (\omega_1 - \omega_2)t] \tag{58}$$

as the reader may easily show. If this current is made to flow through a resistor, the resulting voltage includes components at the sum and difference frequencies. (There are also components at ω_1, ω_2, $2\omega_1$, and $2\omega_2$.) It is especially to be noticed that the sum and difference frequency terms disappear if b vanishes. It is for this reason that a nonlinear impedance was specified; merely combining E_1 and E_2 in a linear network, however complicated, will not yield any output except at the original frequencies.

HARMONIC GENERATION.   If a pure sinusoidal voltage of frequency ω_1 is applied to a nonlinear impedance, the resulting current contains components at ω_1, $2\omega_1$, $3\omega_1$, ... The process is similar to mixing, the higher harmonics being generated by higher-order terms in the current-voltage characteristic. Practically, this result is achieved by making the amplitude of the driving signal large.

Probably the most common use of harmonic generation is in frequency measurement. Given an oscillator which supplies an accurately known frequency, one may employ harmonic-generating circuits to produce a whole series of higher frequencies of comparable accuracy. The frequency of an unknown signal can then be compared with the nearest of these harmonics by mixing the two voltages and measuring the difference frequency.

BANDWIDTH AND RESPONSE TIME.   The frequency-response characteristics of an amplifier† can be specified by giving the voltage amplitude gain $G(f)$ for a sinusoidal waveform as a function of frequency.

The *half-power bandwidth* $\mathcal{B}$ is defined by

$$\mathcal{B} = f_2 - f_1 \tag{59}$$

† The statements in this section actually apply to any linear network with two input terminals and two output terminals. The most common applications are to filters and amplifiers.

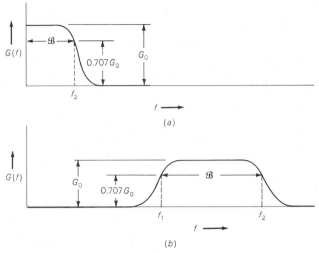

figure 189. Frequency-response characteristics and half-power bandwidths for (a) low-pass amplifier and (b) bandpass amplifier. The function $G(f)$ is the voltage or current gain.

where f_1 and f_2 are the frequencies at which the power gain has dropped to half its peak value, or $G(f)$ to 0.707 times its peak value. The half-power bandwidth is illustrated in Fig. 189*a* for a low-pass amplifier and in Fig. 189*b* for a bandpass amplifier. It is usually easy to determine $\mathscr{B}$ experimentally.

If a step function of voltage (which might be the leading edge of a pulse) is applied at the input of a low-pass amplifier, the resulting output voltage does not reach its final value instantaneously but rather approaches it monotonically or with one or more crossings of the final value. If the output does go beyond the final value momentarily, the maximum amount by which the final value is exceeded is called the *overshoot*. For a *low-pass* amplifier with little or no overshoot, the rise time τ, defined as the time required for the output resulting from a step-function input to rise from 10 to 90 percent of its final value, is related approximately to the bandwidth by[4]

$$\tau \mathscr{B} \approx 0.35 \tag{60}$$

Similarly, if an ac voltage is suddenly applied at the input to a *bandpass* amplifier having little or no overshoot, the rise time, 10 to 90 percent, of the envelope of the resulting output ac voltage approximately satisfies the relation[5]

$$\tau \mathscr{B} \approx 0.7 \tag{61}$$

In general, then, since the response time varies inversely with the bandwidth, an amplifier of wide bandwidth must be employed if fast response is required.

VACUUM TUBES

A vacuum tube consists of a number of electrodes mounted in an envelope of glass or metal, which is usually highly evacuated but which, for specific applications, may contain a suitable gas at low pressure. One electrode, the *cathode*, supplies the electrons required for the operation of the tube. In the high-vacuum tubes and in most gas-filled tubes, the electrons are produced by thermionic emission. The cloud of electrons produced around the heated cathode is termed the *space charge*. The cathode may be of the filamentary type, i.e., a wire heated by the passage of an electric current, or of the indirectly heated type, in which the electron-emitting material is placed on the outside surface of a sleeve which is heated by a separate filament placed within it.

The other electrodes serve to collect the electrons passing through the tube or to control their flow. Most of the collecting is done by one electrode in particular, called the *plate*. The control electrodes are usually referred to as *grids*.

THE DIODE AND RECTIFICATION.   The diode contains two electrodes, a cathode and a plate. If the plate is maintained at a potential positive with respect to the cathode, electrons will be drawn to it from the space charge, where the electron density will be maintained by emission of electrons from the hot cathode. When the plate is negative with respect to the cathode, no electrons will be attracted to it. The diode thus provides a means of controlling the direction of flow of an electric current.

This property is applied in the conversion of alternating current to direct current, a process called *rectification*. The basic circuit employed is shown in Fig. 190a. Let an alternating voltage E_{AB} be impressed across the input terminals. During one half-cycle of the applied voltage, the plate will be positive with respect to the cathode, and current will flow through the tube and through the series load resistor R, producing a potential drop E_0 which will vary during this half-cycle as shown in Fig. 190b. During the other half of the cycle, the plate is negative with respect to the cathode; no current flows through the tube, and the potential drop E_0 across the load resistor R is zero. This circuit is called a *half-wave-rectifier circuit*. The maximum value of E_0 is less than the crest value of E_{AB} because of the potential drop between the cathode and plate.

Both halves of the ac wave are utilized in the full-wave-rectifier circuit of Fig. 190c. During the half-cycle of the applied voltage in which point A is positive and B negative with respect to C, only diode I will conduct, producing a current through R, the load resistor. During the next half-cycle, B will be positive and A negative with respect to C and only diode II will conduct, producing a current which flows through R *in the same direction* as that of the previous half-cycle. The output voltage E_0 across the load resistor then takes the form shown in Fig. 190d. The filtering of this pulsating voltage to obtain a steady dc potential is discussed below.

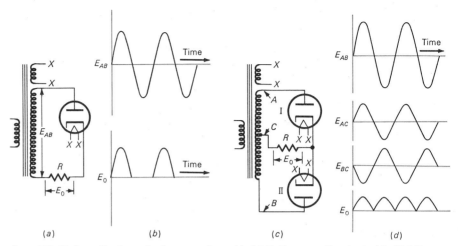

figure 190. Diode rectification and voltage waveforms. (a), (b) Half-wave-rectifier circuit; (c), (d) full-wave-rectifier circuit.

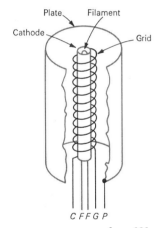

figure 191.

Electrode arrangement of triode.

THE TRIODE AND AMPLIFICATION. The most important applications of vacuum tubes result from the control of the magnitude of the tube current which is made possible by the introduction of additional electrodes. The triode contains, in addition to a cathode and a plate, a third electrode called the *control grid* (or merely *grid*), which ordinarily consists of a wire helix surrounding the cathode and extending its full length, as shown in Fig. 191.

A triode is generally operated with a positive potential E_p applied to the plate and a much smaller negative potential E_c applied to the grid, both potentials being defined conventionally relative to the cathode. The current collected by the plate is denoted by I_p. Figure 192 shows a triode in a simple experimental circuit with variable potentials applied for the study of the relationship of plate current to plate and

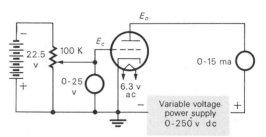

figure 192. Circuit for the determination of characteristic curves for a triode.

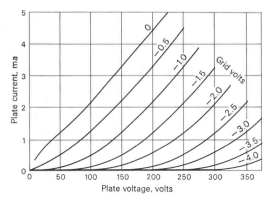

figure 193. Characteristics for 12AX7 triode. I_p versus E_p at constant E_c.

grid potentials. With the grid negative as shown, the current to the grid is negligible in many applications.†

The plate current depends on both plate and grid potentials, the latter having the greater influence because it is so much closer to the cathode and space charge. Thus the grid acts somewhat as a valve which controls the flow of electrons to the plate. The dependence of I_p on the plate and grid potentials is given quantitatively by *characteristic curves*, such as those of Fig. 193.

Amplification is made possible by connecting a suitable load impedance, such as a resistance, in series with the plate of the tube, as shown in Fig. 194. The time-dependent *signal voltage* to be amplified e_s is applied to the grid in series with the fixed *bias voltage* E_{cc}. The latter is ordinarily made large enough in relation to e_s so that the grid is at all times negative with respect to the cathode. The plate current I_p, which varies in response to changes in grid potential, is required to flow through the load resistance R_L, across which is produced a potential drop that changes whenever the plate current changes. By choosing the resistance of R_L to be suitably large, the variation in E_p can be made much greater than the variation in e_s which causes it.

The performance of the tube in amplification can be predicted quantitatively by a graphical procedure based on curves such as those of Fig. 193, or by an analytical *equivalent-circuit* method. The graphical procedure is particularly useful when the signals involved are relatively large. For low-amplitude signals, the equivalent-circuit method is superior. Both are outlined below.

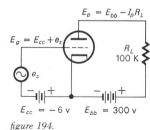

figure 194.

Circuit diagram for single triode-amplifier stage.

† A small grid current may result from the flow of positive ions produced from residual gas in the tube and from the small fraction of the electrons in the space charge which have sufficient kinetic energy to reach the grid despite its negative potential. This current may be a serious source of noise in very sensitive amplifier circuits. There is also an ac grid current due to interelectrode capacitance effects when an ac voltage is applied to the grid.

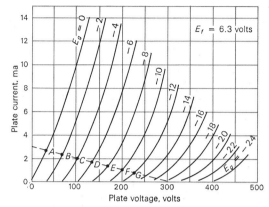

figure 195. Characteristics of a triode with load line, shown dashed, for $R_L = 100$ kilohms, $E_{bb} = 300$ volts.

In either case, the first step is to locate the *operating point* of the tube, i.e., the point on the plate-characteristics graph which corresponds to the condition of the tube in the given circuit when no signal voltage is present. The procedure for finding the operating point is illustrated in Fig. 195. The *load line*, dashed, gives the plate voltage of the tube as a function of the plate current *for the given circuit*. This line, which has the equation $E_p = E_{bb} - I_p R_L$, is constructed by drawing the straight line between the points $(I_p = 0, E_p = E_{bb})$ and $(E_p = 0, I_p = E_{bb}/R_L)$. The operating point may then be found as that point on the load line which corresponds to the actual bias. For the circuit in Fig. 194 the bias is -6.0 volts and the operating point is D. The plate current is then about 1.7 ma, and the plate voltage about 130 volts.

GRAPHICAL ANALYSIS OF TRIODE AMPLIFIER.   When an ac signal voltage e_s is added in series with the fixed bias voltage in the circuit of Fig. 194, the grid voltage varies above and below -6 volts. The behavior of the circuit is then described by a point on the graph of Fig. 195 which moves back and forth along the load line in the vicinity of point D. The relationship of the output waveform to the input is best seen with the aid of a *dynamic transfer characteristic*, shown in Fig. 196. The latter is plotted point by point with data read from the load line of the preceding figure. (Corresponding points on the two graphs are labeled with the same letters.) The graph of Fig. 196 shows directly the plate-current waveform which results for a given grid signal. The output voltage is $I_p R_L$. If the waveform of the output voltage is to be the same as that of the input voltage, i.e., if distortionless amplification is to be obtained, the dynamic transfer characteristic must be a

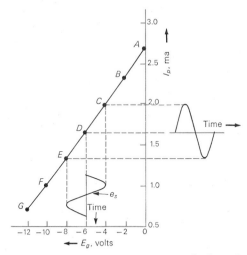

figure 196. Dynamic transfer characteristic for the triode of Fig. 195 with 100-kilohm plate load resistor.

straight line. In the case at hand, the change in grid voltage from -8 to -4 volts produces a change in the voltage across R_L of 65 volts, a sixteenfold voltage amplification.

DYNAMIC CHARACTERISTICS OF TRIODES.   Consider the effect on I_p of making small arbitrary changes in the values of E_g and E_p. Since I_p is determined by E_g and E_p, a small increment dI_p is given by

$$dI_p = \left(\frac{\partial I_p}{\partial E_p}\right)_{E_g} dE_p + \left(\frac{\partial I_p}{\partial E_g}\right)_{E_p} dE_g \tag{62}$$

The dynamic plate resistance and transconductance are parameters of the tube defined by

$$r_p = \text{plate resistance} = \left(\frac{\partial E_p}{\partial I_p}\right)_{E_g}$$

$$g_m = \text{transconductance} = \left(\frac{\partial I_p}{\partial E_g}\right)_{E_p} \tag{63}$$

In terms of r_p and g_m, the expression for dI_p is

$$dI_p = \left(\frac{1}{r_p}\right) dE_p + g_m \, dE_g \tag{64}$$

The relative effectiveness of grid and plate in bringing about changes in tube current is represented by a third parameter,

$$\mu = \text{amplification factor} = -\left(\frac{\partial E_p}{\partial E_g}\right)_{I_p} \tag{65}$$

The relationship of μ to r_p and g_m may be seen by imposing small changes on E_p and E_g, such as to make $dI_p = 0$, and evaluating the right-hand side of Eq. (65) with the aid of Eq. (64). The result is

$$\mu = g_m r_p \tag{66}$$

The instantaneous values of plate voltage, grid voltage, and plate current may be written

$$\begin{aligned} E_g &= E_{og} + e_g \\ E_p &= E_{op} + e_p \\ I_p &= I_{op} + i_p \end{aligned} \tag{67}$$

where E_{og}, E_{op}, I_{op} are the quiescent values corresponding to the given operating point, and e_g, e_p, i_p are time-dependent terms present when an ac input voltage is applied. Provided the time-dependent terms are of small amplitude, so that Eq. (64) holds, the relationship among them is

$$i_p = \left(\frac{1}{r_p}\right) e_p + \left(\frac{\mu}{r_p}\right) e_g \tag{68}$$

GAIN OF A TRIODE AMPLIFIER. With the circuit of Fig. 194, we have $e_i = e_g$ and $e_o = e_p$, where e_i and e_o are the ac components of the input and output voltages, respectively.

The *amplification*, or *gain*, A is defined by

$$A = \frac{e_o}{e_i} \tag{69}$$

Since the output ac voltage arises from the changing plate current through R_L,

$$e_p = -i_p R_L = -\frac{R_L}{r_p} e_p - \frac{\mu R_L}{r_p} e_g \tag{70}$$

$$= -\mu e_g \frac{R_L}{r_p + R_L} \tag{71}$$

$$A = \frac{e_p}{e_g} = -\mu \frac{R_L}{r_p + R_L} \tag{72}$$

The negative result for e_p/e_g corresponds to the fact that an increase in E_g results in an increase in I_p, which in turn causes E_p to decrease. It may be noted that the magnitude of A cannot exceed μ.

EQUIVALENT CIRCUIT. Examination of Eq. (71) shows that e_p is equal to the voltage which would be produced across the load resistance R_L by a hypothetical voltage generator which produced a voltage equal to $-\mu e_g$ and which acted in series with a resistance r_p, as in Fig. 197; i.e., the voltage $-\mu e_g$ is divided in series between r_p and R_L. Thus, within the validity of the gain equation above, the circuit of Fig. 197 is the equivalent for alternating current of the output circuit of the triode amplifier.

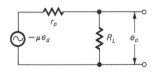

figure 197.

Equivalent circuit for triode amplifier.

The equivalent circuit is found to be of considerable utility in analyzing the performance of triode-amplifier circuits under a variety of conditions. For example, the effect on the gain of changing R_L is easily remembered from Fig. 197. For $R_L \gg r_p$, the gain is approximately μ; for $R_L \ll r_p$, the gain is approximately $g_m R_L$. For fixed input amplitude, the maximum *voltage* output is obtained for $R_L \gg r_p$, the maximum *current* through the load for $R_L \ll r_p$, and the maximum *power* dissipated in the load for $R_L = r_p$.

PRACTICAL TRIODE-AMPLIFIER CIRCUIT. A practical triode-amplifier circuit suitable for the audio-frequency range is shown in Fig. 198. If this is compared with the simpler circuit of Fig. 194, several added features may be noted. Bias is obtained by the use of a cathode resistor R_k. For the case discussed above, the operating point for -6 volts bias would be obtained by making R_k equal to $6/(1.7 \times 10^{-3}) = 3500$ ohms.

The capacitor C_k, which is chosen to have a low impedance to alternating current at the frequencies to be amplified, serves to hold the cathode at ac ground

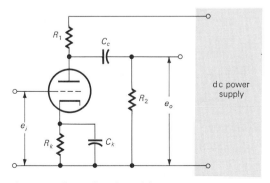

figure 198. Practical triode-amplifier circuit.

potential. If the bypass capacitor C_k were omitted, the periodic variation in plate current produced by e_i would cause also a periodic variation in the potential of the cathode relative to ground, such that the net ac voltage at the true input to the tube, namely between grid and cathode, would be less than e_i. This effect, termed *cathode degeneration*, is largely eliminated by the bypass capacitor, so that the cathode potential remains constant relative to ground and the full ac voltage e_i appears between grid and cathode.

The other important added feature in Fig. 198 is the output coupling network, comprising R_2 and C_c. The coupling capacitor C_c blocks direct current but is chosen to have a low impedance to alternating current in the frequency range to be amplified. The purpose of the output coupling circuit is to eliminate the large dc component of the potential at the plate while passing ac signal voltages with negligible loss.

Observe that with the coupling circuit present, as in Fig. 198, resistors R_1 and R_2 are effectively in parallel in the ac circuit. Thus, in Eq. (72) and in the equivalent circuit of Fig. 197, the effective load resistance R_L to be used is given by

$$R_L = \frac{R_1 R_2}{R_1 + R_2} \tag{73}$$

Equation (72) and the equivalent circuit of Fig. 197 describe the actual behavior of the amplifier circuit of Fig. 198 very satisfactorily over a considerable frequency range, but at very low or very high frequencies the gain drops off. To understand this behavior, the more complete equivalent circuit of Fig. 199 should be considered. Here the cathode bias and coupling capacitors are shown explicitly, and it can be seen that at sufficiently low frequencies the impedance of C_k and C_c will become appreciable and account for a decrease in gain. Also shown explicitly are the interelectrode capacitances and stray capacitances, which account for the reduction in gain at high frequencies.

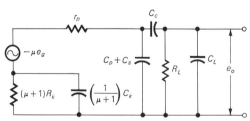

figure 199. More complete equivalent circuit of triode amplifier. C_p = triode output capacitance; C_s = stray wiring capacitance; C_L = capacitance of load.

THE PENTODE.   In addition to causing loss of gain at high frequencies, the inter-electrode capacitances of triodes can also lead to instability at high frequencies because of the resulting interaction between plate and grid circuits. This interaction is very materially reduced by the introduction of additional electrodes in the pentode, or five-electrode tube.

The two additional electrodes of the pentode are grids which are placed between the control grid and the plate. The one nearer the control grid, called the *screen grid*, is maintained at a constant potential positive with respect to the cathode and acts as an electrostatic shield, which may reduce the plate-to-grid capacitance to a value as low as 0.005 pf. Because of the shielding action of the screen grid, the plate voltage has relatively little effect, except at low plate voltages, in determining the plate current, which is principally controlled by the control grid and screen grid potentials. The screen grid also acts as a collector of electrons, but because its effective area is much smaller than that of the plate, this screen grid current is small compared with the plate current under normal operating conditions.

The fifth electrode, the *suppressor grid*, is located between the screen grid and the plate. It is generally operated at cathode potential, and its function is to create a potential distribution between the screen grid and the plate which will cause electrons emitted from the plate (*secondary emission*) to return to the plate.

Characteristic curves for a typical pentode are shown in Fig. 200.

Pentodes typically have much higher values of μ and of r_p than have triodes. With a single pentode-amplifier stage, much higher gain can usually be achieved than with a single triode stage. However, for many applications this advantage is more

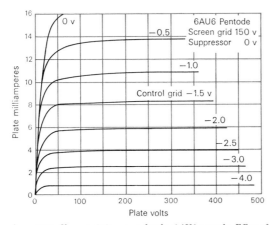

figure 200. Characteristic curves for the 6AU6 pentode. Effect of plate voltage and control-grid bias on plate current for constant screen- and suppressor-grid potentials.

than offset by the even higher gain and added design flexibility offered by twin triodes (two electrically separate, nominally equivalent triodes contained in a single miniature glass tube envelope). Pentodes retain the advantage, however, of much reduced interaction between plate and grid circuits and are therefore particularly useful in radio-frequency amplifier circuits.

SEMICONDUCTOR DEVICES

SEMICONDUCTORS.   The energy levels available to electrons in crystalline solids come in *bands* of closely spaced levels. These bands are separated by *energy gaps,* or forbidden zones. The lowest band of interest here is the *valence band,* which arises from the valence orbitals of the atoms in the material. The next higher band is the *conduction band.* For insulators at normal temperatures the valence band is filled and the conduction band is empty. The gap between them is $\gg kT$, for example, 7 eV for diamond. The electrons are not mobile because they do not have enough energy to reach any unoccupied states. Metals contain a partially filled band. Electrons in such a band are mobile because there are many vacant states of slightly higher energy available. In semiconductors the gap between valence and conduction bands is much less than in insulators, and at finite temperatures there are some vacant states in the valence band and some electrons in the conduction band. Both provide the possibility of conduction.

The semiconductors of greatest technical interest are silicon and germanium, both group IV elements. The valence shells of these atoms contain four electrons, and in a perfect crystal a given atom forms four covalent bonds in a tetrahedral arrangement exactly similar to that of the diamond structure for carbon.

In the pure crystalline state at low temperatures, silicon and germanium are poor conductors. The reason for this is that nearly all the electrons are in the practically filled valence band and hardly any are in the conduction band. If the temperature is raised, an appreciable number of electrons are elevated into the conduction band. At the same time, an equal number of holes is created in the valence band. The resulting material exhibits an increased conductance which is due not only to the motion of *electrons* in the conduction band, but also to the fact that the *holes* in the valence band move when a field is applied. In the pure material, these two species of carriers are equal in number and both make significant contributions to the conductivity.

The addition of small but accurately controlled amounts of specific impurities to a highly purified sample of a semiconductor can have a remarkable effect. Thus, if arsenic is added to germanium, it enters the structure by replacement of germanium atoms. The neutral arsenic atom has five electrons in the valence shell, of which only four are required in bonding to the four neighboring germanium atoms. The fifth electron is quite easily raised to the conduction band. The arsenic is hence called a do*n*or impurity, the additional current carriers introduced by the impurity are *n*egative (electrons), and the resulting semiconductor is referred to as *n*-type germanium.

A different result is obtained by the addition of indium, for example, for which the neutral atom has three electrons in the valence shell. When such an atom enters the germanium lattice by substitution and forms four covalent bonds with neighboring atoms, it must capture an electron from an adjacent germanium atom, thus creating a hole. The indium is called an acce*p*tor impurity, the current carriers introduced by the impurity are *p*ositive (holes), and the resulting semiconductor is classified as *p* type.

The most important applications of semiconductors involve the use of appropriate combinations of *p*- and *n*-type materials in diodes and transistors.

THERMISTORS. Perhaps the simplest semiconductor device is the thermistor, or thermally sensitive resistor. Thermistors are made from mixtures of various metallic oxides sintered together under controlled conditions to yield a ceramic-like material which has a large negative temperature coefficient of resistance, usually of the order of 4 percent $°C^{-1}$ at room temperature. The resistance R at temperature T, to a good approximation, is represented by the relation

$$\log R = \frac{a}{T} + b \tag{74}$$

where a and b are constants characteristic of the thermistor.

The combination of high specific resistance, large temperature coefficient, and good stability makes thermistors particularly suitable for use in temperature control and measurement. They can be sealed in glass for protection from corrosive materials, with the resultant element still small enough to give a very rapid response to a change in the temperature of its environment.

DIODES. When a *p*-type crystal and an *n*-type crystal are joined together, a semiconductor *diode* is formed. The labeling of terminals and the schematic symbol for this device are shown in Fig. 201. Its most important properties are summarized in the current-voltage characteristic, typically of the form shown in Fig. 202. It can be seen that, provided the breakdown voltage is not exceeded, the resistance to current flow in the backward direction is relatively high, while the forward resistance is relatively low. The diode can therefore function as a rectifier, though this is by no means its only use.

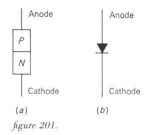

figure 201.

(a) Semiconductor diode; (b) schematic symbol for semiconductor diode.

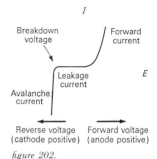

figure 202.

Current-voltage characteristics of p-n junction diode.

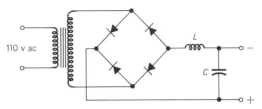

figure 203. Bridge rectifier power supply with semiconductor diodes.

Semiconductor diodes of the power-rectifier type have been developed which are very efficient and reliable, and they are coming into widespread use as rectifiers in dc power supplies. Figure 203 shows a typical bridge rectifier circuit. For either polarity of the ac voltage at the transformer secondary, current can flow in only one direction through the load; it goes through one pair of diodes during one half-cycle and through the other pair during the other half-cycle. As compared with tubes, semiconductors offer the great advantage in this application of low forward resistance, absence of filament power requirements, and indefinite life—provided the maximum ratings are not exceeded. It is very important that the diodes be mounted on heat-dissipating plates, with ample surface area per diode, and that the reverse voltage rating not be exceeded even momentarily. The air circulation provided by a small blower greatly improves the efficiency of cooling.

A related application is that of low-level detection. Although a bridge arrangement of four diodes may be used, the simple circuit of Fig. 204 is often employed. The output is a voltage which follows the envelope of the input ac voltage. The time constant RC is made large compared with the period of the carrier but small enough to follow the component of highest frequency in the modulation envelope. It becomes obvious on consideration that for proper operation of the circuit in this manner, R must be large compared with the forward resistance and small compared with the back resistance of the diode. It is necessary at high frequencies to take into account the shunting effect of the diode-junction capacitance, which effectively appears in parallel with the diode.

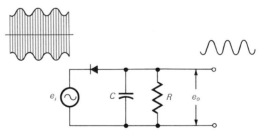

figure 204. Diode detector circuit.

The operation of the diode detector is simplest when the amplitude of the input ac voltage is sufficient to extend into the linear region of the forward-current characteristic (Fig. 202). The diode is then said to act as a *linear detector;* the output voltage varies linearly with the amplitude of the input ac voltage. An input-voltage level of the order of a volt or so is usually sufficient to ensure linearity. If the amplitude of the input voltage is such that the peaks do not reach beyond the bend of the forward-current-characteristic curve, the diode still functions as a detector, though the detection is not linear. In this region, the diode characteristic is approximated by the series expansion

$$I = aE + bE^2 + \cdots \tag{75}$$

If a modulated voltage waveform such as that of Eq. (52) is used, for example, it is not difficult to show that the current contains terms at the frequency ω_m, these terms being generated by the term bE^2. A diode detector operated in this fashion is therefore called a *square-law detector.*

Since the diode presents a highly nonlinear impedance, it is also used for mixing, i.e., combining two ac voltages in such a way as to produce sum or difference frequencies, and for harmonic generation.

Though most types of diodes are ruined if the breakdown voltage is exceeded, the *Zener diode* is designed to be operated in the avalanche region (Fig. 202). Since the voltage across the diode in this region is nearly constant over a rather wide range of currents, the Zener diode can function as a voltage regulator. The regulating capacity is characterized by the dynamic resistance, dE/dI, which depends on the operating point. Zener diodes are available for a number of voltages, ranging from a few volts to over a hundred. The voltage stability in many cases is limited by the variation in breakdown voltage with temperature, ~ 0.1 percent $°C^{-1}$. For use as reference voltage sources, temperature-compensated units are available having temperature coefficients of the order of 0.001 percent $°C^{-1}$ over a wide temperature range and offering a fractional stability of 10^{-5} when operated near 25°C; these units are much more rugged than the conventional Weston standard cell.

When reverse bias (not exceeding the breakdown voltage) is applied to a semiconductor diode, the resistance becomes very high but an appreciable capacitance exists between anode and cathode terminals. Since this capacitance is a function of the voltage, a reverse-biased diode can function as a *voltage-variable capacitor.* Perhaps the most common application is the control of the frequency of an oscillator by means of an applied voltage.

TRANSISTORS. The structure of a *p-n-p* transistor is shown pictorially in Fig. 205. It consists of a very thin layer of *n*-type semiconductor material sandwiched between two *p*-type regions in such a way that two diode-type junctions are formed

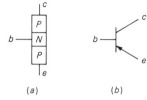

figure 205.

(a) *Structure of p-n-p transistor;* (b) *schematic diagram of p-n-p transistor.*

back to back. The three terminals are designated *emitter, base,* and *collector.* The *n-p-n* transistor is similar but with *n* and *p* interchanged. In the schematic symbol, the emitter terminal is designated with an arrow that shows the direction of easy current flow across the emitter-base junction.

The symmetry which might appear from Fig. 205*a* to exist is destroyed when the usual biasing potentials are applied. For both *n-p-n* and *p-n-p* types, biasing is normally such that the base potential is intermediate between that of the other two electrodes, with the normal emitter-base junction biased in the forward direction and the collector-base junction biased in the reverse direction. The resulting current flow for a *p-n-p* transistor under normal bias conditions may be visualized as follows. Current enters through the emitter lead and leaves via the base and collector leads. The magnitude of the emitter current is then the sum of the magnitudes of the base and collector currents. The current directions are all reversed for the *n-p-n* transistor. This flow scheme is easily remembered from the schematic symbols.

The three circuit configurations employed are designated by specifying which of the three electrodes is common to the input and output circuits (Fig. 206). The generally accepted voltage and current sign conventions are also based upon Fig. 206. For both *n-p-n* and *p-n-p* types, the electrode potentials are measured relative to that of the common electrode, if only one subscript is given. (When no configuration is specified, the common-base arrangement is usually to be assumed.) To avoid ambiguity, symbols with two subscripts are often used, such as E_{ce} for the potential of electrode *c* relative to that of *e*. In any case, the positive direction for current flow is usually *into* the transistor.

Characteristic curves for the common-emitter configuration are given for a typical power transistor in Fig. 207. Note that over a fairly wide range of operating conditions, the collector current is determined mainly by I_b and is relatively insensitive to changes in E_{ce}. The base-collector current amplification factor β for this configuration is defined by

$$\beta = \left(\frac{\partial I_c}{\partial I_b}\right)_{E_{ce}} \tag{76}$$

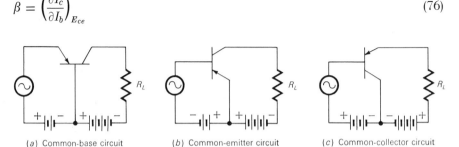

    (*a*) Common-base circuit   (*b*) Common-emitter circuit   (*c*) Common-collector circuit

figure 206. Circuit configurations for p-n-p transistors. The analogous arrangements for n-p-n transistors are the same except for the reversal of all battery polarities.

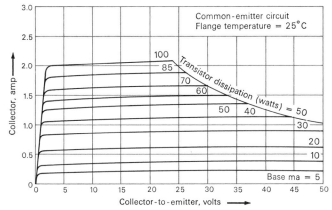

figure 207. Characteristic curves for a power transistor, type 2N1070, common-emitter configuration.

Representative values of β range between 20 and 100. Amplification in this configuration then results from the fact that a small change in base current results in a much larger change in collector current.

One form of equivalent circuit for a transistor, for the common-base configuration, is given in Fig. 208. The parameters r_b, r_e, r_c are dynamic resistance characteristics of the transistor; α is the emitter-collector amplification factor,

$$\alpha = -\left(\frac{\partial I_c}{\partial I_e}\right)_{E_{cb}} = \frac{\beta}{\beta + 1} \tag{77}$$

and r_m is the mutual impedance, $r_m = \alpha r_c - (1 - \alpha)r_b$. These parameters are all to be evaluated at the operating point for the given circuit. The equivalent circuit is included here especially to emphasize a few salient facts about transistors: (a) the input impedance may be relatively low; (b) it depends on the output load impedance; (c) the output impedance depends on the source impedance. In short, the input and output circuits are not nearly as well isolated as for a vacuum-tube circuit. This fact complicates the process of designing transistor circuits.†

Transistors are quite rugged mechanically but are rather easily damaged electrically. The various maximum ratings specified by the manufacturer should be scrupulously respected since there is often little margin left for error. For power transistors, the deratings for higher-temperature operation should not be overlooked, because even if effective cooling is provided, these units are likely to become warm under operating conditions.

† A large collection of carefully designed transistor circuits, "Handbook of Selected Semiconductor Circuits," NAVSHIPS 93484, is available from the Superintendent of Documents, U.S. Government Printing Office. Circuit-design principles are also discussed in this reference.

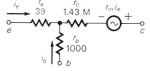

figure 208.

One form of equivalent circuit for a transistor. Numerical values are for Amperex 2N279, for which r_m is 1.38M, resistance in ohms, with $M = 10^6$.

Other disadvantages to be mentioned are the temperature dependence of the characteristic curves and the generation of semiconductor noise, which can in some cases be many times larger than the Johnson noise (page 615).

On the other hand, the small size and low power consumption of transistors offer definite advantages for certain applications. For example, a small amplifier or oscillator, together with batteries sufficient to provide operating power for months, can be housed in a single well-shielded box to achieve a high degree of freedom from hum and from interference due to external electrical influences. Such units can also easily be isolated from ground potential.

Both graphical and analytical methods are available for the analysis of standard transistor circuits. The procedures are recognizably analogous to those outlined above for vacuum-tube circuits but are somewhat more involved because of the greater degree of interaction between input and output circuits.

NOISE

NOISE AS THE ULTIMATE LIMIT TO SENSITIVITY.[2,6]   The ultimate limitation in the accuracy with which an electric current or voltage (or indeed any macroscopic quantity) can be measured, or in the minimum magnitude of voltage or current which can be detected, is set by random fluctuations which are an inherent aspect of the behavior of systems consisting of a large number of (atomic-scale) particles. For example, one may employ an electronic amplifier of high gain to detect or measure a minute unbalance voltage from a Wheatstone bridge. Before the system can give optimum performance, it is necessary to eliminate such factors as poor contacts, effects of variations in power-line voltage, pickup of stray voltages, etc.; with sufficient care, these can all be reduced to a negligible level. But after this has been done, the output of an amplifier of sufficiently high gain will be found to exhibit strictly random fluctuations, which originate from sources located in early stages of the amplifier or in the bridge circuit itself. In this context, such fluctuations are called *noise*.

The general magnitude of noise fluctuations can best be characterized by the mean square noise voltage $\langle E_n^2 \rangle$ or the rms value $\langle E_n^2 \rangle^{\frac{1}{2}}$. It is clear that one cannot expect to detect a signal voltage E_s which is less than the rms voltage of the accompanying noise; nor, for larger signal voltages, can one expect to be able to measure E_s with an error less than the rms noise voltage.

FREQUENCY DISTRIBUTION OF NOISE.   Figure 209 shows several samples of noise observed at the output of an amplifier by means of an oscilloscope. Each trace is a photograph of a single sweep. The source of noise in all cases was the same.

In the first case, the frequency range 1 to 25 kHz was amplified; in the second case, a lowpass filter was added so that only the range 1 to 5 kHz was amplified; in the third case, amplification was restricted to a narrow range, about 50 Hz wide, centered at 5 kHz. For comparison, a 5-kHz sine wave is in Fig. 209*d*. The time scale is the same for all four sweeps, but the oscilloscope amplification was increased progressively for traces *a*, *b*, and *c*.

These observations illustrate the fact that a noise voltage can be considered to consist of a large number of Fourier components (page 594), representing a wide range of frequencies, with some components being passed and others being rejected when the noise is put through a frequency-selective network. Though the mathematical analysis of noise is not simple, this notion can be put on a rigorous basis.

JOHNSON NOISE.   This is the noise, inherent in any resistor, due to thermal agitation of the electrons. The mean square voltage of noise components due to this cause in the frequency range f to $f + \Delta f$ in a resistor of resistance R at temperature T is given by the Nyquist equation,

$$\langle E_n^2 \rangle = 4kTR\,\Delta f \tag{78}$$

In any network, the actual resistor may be replaced by its equivalent: a voltage generator E_n (of zero internal impedance) acting in series with a noiseless resistance R. For 300°K, the value of $4kT$ is 1.65×10^{-20} joule.

SHOT-EFFECT NOISE.   The tube current of a vacuum tube fluctuates on account of the corpuscular nature of electrons. The resulting noise is called *shot-effect* noise. For a triode, the shot-effect component in the plate current in the frequency range f to $f + \Delta f$ is given by

$$\langle I_n^2 \rangle = \left(\frac{0.664}{\sigma}\right) 4k\,T_c\,g_m\,\Delta f \approx 4k\,T_c\,g_m\,\Delta f \tag{79}$$

where σ = a parameter characteristic of a given type of tube (with a value typically between 0.5 and 1.0)

T_c = cathode temperature (about 1000°K for oxide-coated cathodes)

g_m = transconductance of the tube

An equivalent circuit for the tube is obtained by placing a current generator I_n (of infinite internal impedance) in parallel with the tube plate resistance r_p, which is to be considered noiseless. The shot effect is often the principal source of noise in radio-frequency circuits.

FLICKER-EFFECT NOISE.   An additional source of noise in vacuum tubes which is particularly important at audio frequencies is the flicker effect. This noise arises

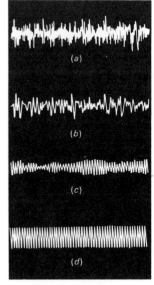

figure 209.

Oscilloscope photographs showing appearance of noise voltage as a function of time. (a) Passband 1 to 25 kHz; (b) passband 1 to 5 kHz; (c) passband 50 Hz wide, centered at 5 kHz; (d) 5-kHz sine wave.

from slow random fluctuations in the rate of emission of electrons from the cathode and is generally ascribed to processes taking place in the oxide coating. The frequency distribution of this noise is such that the resulting mean-square noise current in a small frequency range f to $f + \Delta f$ goes approximately as $f^{-1} \, \Delta f$. At the lower audio frequencies, say below 1 kHz, the flicker effect usually becomes a more serious source of noise than the shot effect.

CURRENT NOISE.   Noise in excess of the Johnson noise is produced when direct current passes through a semiconductor or a carbon-composition resistor. The mean-square noise voltage generated in this way has approximately a $1/f$ frequency dependence. In sensitive amplifiers, particularly for the lower audio frequencies, current noise can be greatly reduced by the use of wire-wound, metal-film, or carbon-film resistors.

NOISE IN TRANSISTORS.   Noise generated in transistors includes Johnson noise and excess noise, the latter with roughly a $1/f$ frequency dependence. Transistors therefore tend to have poor noise characteristics at low audio frequencies. However, improved types are becoming available, and audio amplifiers with reasonably good noise characteristics have been developed.

NOISE FIGURE.   The noise figure F of an amplifier compares the actual noise output obtained under specified conditions with that which would be obtained if no noise were generated within the amplifier itself. For the purpose of defining F, it is assumed that the amplifier input terminals are connected to an external network, or dummy input, which has an impedance matching that of the amplifier input and which generates only Johnson noise. Then F is defined as the ratio of the total noise power output of the amplifier to that due to Johnson noise from the dummy input.

NOISE CALCULATIONS.   Once the noise *sources* have been characterized, as by Eq. (78) or (79), for example, there remains the problem of determining the noise at the *output* of the system. Though it is difficult to make accurate calculations, it is relatively easy to make estimates which are entirely satisfactory for many purposes.

When random noise from several independent sources is combined, one may add the noise powers, the mean-square noise voltages, or the mean-square noise currents. The first-stage gain is often high enough for noise generated in subsequent stages to be negligible.

Consider the case of a linear amplifier. Since the gain is a function of frequency, it is necessary in principle to consider separately the noise in each small frequency range Δf. A means of avoiding this complication in practical calculations is provided by introducing the *bandwidth for noise B*, defined by

$$\int_0^\infty G^2(f)\ df = G_0^2 B \tag{80}$$

where $G(f)$ = voltage or current gain at frequency f, from the location of the noise
source to the output (cf. Fig. 189)

   G_0 = peak value of $G(f)$

Equation (80) furnishes the justification for the following procedure, which is often
used. The noise generator in the equivalent circuit for the source is considered to
have the rms voltage or current corresponding to a frequency range $\Delta f = B$; for
example,

$$\langle E_n^2 \rangle = 4kTRB \qquad \text{Johnson noise source} \tag{81}$$

or

$$\langle I_n^2 \rangle = 4kT_c g_m B \qquad \text{shot-noise source} \tag{82}$$

The subsequent gain for the rms noise is then taken as G_0.

   In many applications, the task of evaluating the integral in Eq. (80) can be
avoided by using the half-power bandwidth $\mathcal{B}$ (page 597) as an approximation for
B. Actually, for the simple LRC resonance curve and for a single-section lowpass
RC filter, $B = (\pi/2)\mathcal{B}$. For the idealized rectangular response curve (gain constant
between frequencies f_1 and f_2, zero outside this range) B exactly equals $\mathcal{B}$.

   A more difficult case, but one of great practical importance, is that of a linear
amplifier followed by a detector, the output of which is filtered and then registered
by a meter, chart recorder, or other indicating device.[†] Let the noise bandwidth of
the amplifier be B_a and that of the detector-filter-indicator system, B_i. For the purpose of estimating signal-to-noise ratios, the system bandwidth for noise B_s now replaces B in Eq. (81) or (82). For either a square-law or linear detector with weak
signal,

$$B_s \approx \sqrt{B_a B_i} \tag{83}$$

For a phase-sensitive detector, or for a linear detector operated with a strong carrier,

$$B_s \approx \tfrac{1}{2} B_i \tag{84}$$

Here the designations *weak* and *strong* pertain to a comparison with the rms noise
level at the detector input. For both B_a and B_i, the approximation $B \approx \mathcal{B}$ may be
used. For a single-section lowpass RC filter, the exact value for B_i is $1/4RC$.
Ordinarily, since $B_i \ll B_a$, much greater noise reduction is obtainable with a phase-
sensitive detector than with a simple square-law or linear detector with weak signal,
for the same output-response time.

   [†] For theoretical treatments of the response of detector circuits to noise, see Refs. 2 and 6.

MEASUREMENTS AND TEST EQUIPMENT

MULTIMETERS.   Widely used in testing electronic equipment, the multimeter consists of a single d'Arsonval meter mounted, with auxiliary networks, in a portable case. The basic meter is a dc microammeter, typically 50 μa full scale. By means of a number of switch-selected shunts, series resistors, batteries, and a rectifier, this single meter serves for measurement of dc or ac voltage, direct current, or resistance. The accuracy is typically 1 or 2 percent of full scale.

While this is a versatile and extremely useful instrument, the user should be aware of its major limitations:

1. When used to measure voltage, the meter presents a certain conductance between its leads, so that connecting the meter between two points in a circuit is equivalent to connecting a resistor between these two points. The voltage actually measured by the meter is therefore not the same as the voltage which existed between the same two points before the meter was connected. The meter resistance can be calculated from the "ohms-per-volt" rating λ, which is usually printed on the face of the meter scale. A typical value is $\lambda = 20,000$ ohms volt^{-1}. The input resistance R_i of the meter is $R_i = \lambda E_f$, where E_f is the *full-scale* voltage for the particular range used.
2. Because of the type of rectifier used, ac measurements cannot reliably be made at frequencies above the audio range with the multimeters now generally available.
3. For ac measurements, the meter deflection is determined by the average rectified voltage; the scale calibration in terms of rms values is correct only for a sinusoidal waveform. The presence of a dc component also causes error.
4. Resistance measurements may be quite meaningless if the circuit contains voltage sources.

ELECTRONIC VOLTMETERS.   The chief limitation of the d'Arsonval meter—that of drawing power from the circuit under test—can be practically eliminated by the use of transistor or tube circuits. Figure 210*a* shows a simple voltmeter circuit which illustrates how this result may be accomplished. With the input leads shorted, the potentiometer is adjusted to bring the meter to zero deflection. When the dc voltage to be measured is applied to the input terminals, the circuit becomes unbalanced and a deflection of the meter results. The use of two twin-triode tubes in a balanced arrangement helps to reduce drift caused by variations in E_{bb} or in filament potential, since to a first approximation these factors affect both triode circuits in the same way and therefore do not alter the meter deflection. The input resistance is high: the upper limit to R is set by the requirement that grid current produce only a negligible error, and the grid current be quite small since in this circuit

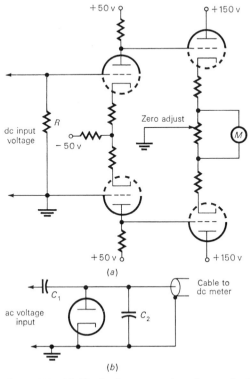

figure 210. (a) *Simple dc vacuum-tube voltmeter circuit;*
(b) *diode probe used for ac voltage measurements.*

the grids are a volt or so negative relative to the cathodes. With this type of circuit, input impedances of the order of 10 to 100 megohms are achieved in modern vacuum-tube voltmeters. The accuracy is typically 1 to 3 percent of full scale.

Measurements of high accuracy (3 to 6 digits) can be conveniently made with digital voltmeters.

Alternating-current voltage measurements are made possible with the addition of a simple diode rectifier (Fig. 210b). The diode load circuit is R and C_2. Capacitor C_1 blocks direct current. To avoid error from the use of long leads at high frequencies, the diode circuit is mounted in a small cylindrical housing connected through a cable to the meter itself. Thus the diode can be brought quite close to the circuit point at which the voltage is to be measured. In this way, ac voltage measurements can be made reliably well into the UHF region (hundreds of megahertz). The input impedance for alternating current for a typical instrument is equivalent to several megohms (dependent on frequency) in parallel with 1.5 pf.

POTENTIOMETRIC INSTRUMENTS.   Modern potentiometric instruments are a far cry from the slide-wire-on-a-meterstick type and offer accuracy par excellence for measurement of dc voltages. The use of a potentiometer should always be considered when an accuracy of better than about 1 percent is desired. The balancing time is of the order of tenths of a second for electronically balanced types, and of course appreciably longer and quite dependent on circumstances for manually operated types. The more accurate instruments in the former class offer five-figure digital readout.

THE CATHODE-RAY OSCILLOSCOPE.   The essential element of the cathode-ray oscilloscope is the oscilloscope tube, whose structure is shown in Fig. 211. It contains an electron gun, which produces a beam of electrons that is directed upon a fluorescent screen at the opposite end of the tube, and deflecting plates, to which voltages may be applied to displace the beam from its equilibrium position. As the beam moves, its path is traced out on the fluorescent screen, where it may be observed or photographed. The electron gun consists of a thermionic emitting cathode, a grid, and two anodes. The grid controls the electron density of the beam and to a lesser extent influences the focusing of the beam by the two anodes which accelerate the electrons toward the screen. The final velocity with which the electrons leave the gun depends upon the potential of the second anode. The intensity of the image is changed by changing the grid voltage, while adjustment of the beam focus is accomplished by changing the potential of the focusing anode. The diaphragms in the electrodes help to keep the beam sharp.

The two sets of deflecting plates in the tube are arranged so that one causes a horizontal deflection of the beam, and hence of the fluorescent spot produced on the screen, and the other a vertical deflection. Because appreciable voltages must be impressed across the plates to produce a large displacement of the beam, amplifiers are provided in the instrument to amplify the actual input voltages. When alternat-

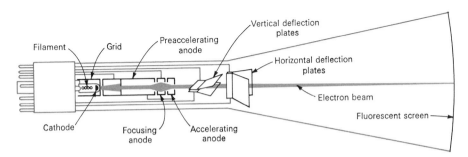

figure 211. Electrode arrangement of cathode-ray oscilloscope tube.

ing voltages are applied simultaneously to the two plates, the position of the spot at any time represents the vector sum of the instantaneous values of the voltages applied to the two sets of plates. Because the screen is phosphorescent as well as fluorescent, the motion of the spot produces a line pattern on the screen, the nature of which depends on the waveshapes, amplitudes, and phase relationship of the two voltages.

A major application of the cathode-ray oscilloscope is in showing the appearance of a voltage as a function of the time. For this purpose the beam is periodically swept horizontally across the screen at a uniform rate to a point of maximum displacement, from which it returns practically instantaneously to its zero position; the frequency of this action is termed the *sweep frequency*, and the special type of voltage wave applied to the horizontal plates to produce it is provided by a special variable-frequency oscillator incorporated in the instrument. The unknown voltage is applied to the vertical-deflection plates. If the frequency of the sweep voltage is adjusted to equality with that of the unknown, the pattern on the screen will represent one cycle of the unknown voltage.

The oscilloscope is also useful for the comparison of the frequencies of two voltages. The two voltages are applied simultaneously to the two sets of deflection plates. If the ratio of the frequencies is a rational number, a closed pattern called a *Lissajous figure* will result, and from its form the frequency ratio can be determined. An unknown frequency is determined by comparing it with the output of a calibrated variable-frequency oscillator, the frequency of which is adjusted until a simple Lissajous pattern is obtained. In Fig. 212 is shown the pattern for a frequency ratio of 3:2.

The relative phase of two sinusoidal voltages of the same frequency can also be determined. The two signals are applied to the two sets of deflection plates. The vertical and horizontal amplifiers are adjusted as required to give equal deflections for the two signals, if the latter are not equal in amplitude. If the two voltages are

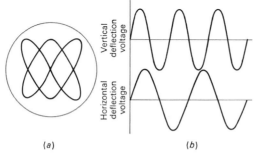

(a)   (b)

figure 212. Typical Lissajous pattern in frequency comparison with the cathode-ray oscilloscope.

in phase, the pattern will be a straight line, making an angle of 45° with the horizontal. If the voltages are 90° out of phase, the pattern is a circle. For other phase angles the pattern is an ellipse.

The input impedance, deflection sensitivity, and frequency-response (or rise-time) characteristics of an oscilloscope depend entirely on the signal amplifier used. In a typical case, the input impedance may correspond to 2 megohms in parallel with 20 pf. This impedance is high enough so that dc and audio-frequency circuits are usually not much affected by connection of the oscilloscope leads; however, the effect of the oscilloscope input capacity may be considerable when added to a radio-frequency circuit.

Perhaps the single most common source of confusion for the novice using an oscilloscope is the occurrence of pickup. The term *pickup* refers to a voltage which appears in the circuit of interest (in this case, the oscilloscope leads and input circuit) as a result of coupling with stray magnetic or electrostatic fields. The ubiquitous power-line fields are a frequent source of pickup, which is then called *hum pickup*. It is imperative that shielded leads be used if reliable indications are to be obtained for circuits with even moderately high impedances, say above a few thousand ohms.

MEASUREMENT OF RESISTANCE, CAPACITANCE, AND INDUCTANCE.   For many purposes, satisfactory measurements of resistance, capacitance, or inductance can be made with an impedance bridge (page 232) or with a suitable UHF equivalent employing transmission-line techniques. It is important to bear in mind the frequency at which the measurement is made, since the effective values of R, L, and C are somewhat dependent on frequency. In general, components handbooks or manufacturer's literature should be consulted with regard to the frequency dependence of these parameters for particular types of components.†

Capacitors and resistors with the size and accuracy required for most applications are available, but it is often desired to wind a coil either to give a certain value of inductance or to resonate with a given capacitance at a specified frequency. This is best done by trial, with available coil-winding data or equations used as a preliminary guide. A trial coil can be connected across a capacitor, and a grid-dip meter used to find the resonant frequency of the resulting circuit. (The principle of the method is essentially the same as that of the resonance method explained on pages 226 and 227.) The value of L can be calculated from the resonant frequency if C is known. Most physical circuits are actually multiply resonant, so it is wise to make a rough calculation based on the winding data to guard against being led far astray by spurious resonances.

---

† As a rough general rule, the resistance of composition-type and film-type resistors and the capacitance of air, mica, or ceramic capacitors are approximately constant from direct current at least through the radio-frequency region and in some cases even into the UHF region. On the other hand, the effective resistance and inductance of coils and of wire-wound resistors may change considerably over a decade of frequency.

MEASUREMENT OF FREQUENCY. Three important methods for the measurement of frequency will be described, in the order of increasing accuracy.

Resonance Method. The signal of unknown frequency is coupled into a resonant *LC* circuit, and the capacitance adjusted to produce resonance with the signal. Grid-dip meters are usually designed to be usable in this way, the dial being calibrated directly in frequency units. This method requires an appreciable amount of signal power; an accuracy of several percent can be expected.

Heterodyne Frequency Meter. This instrument contains a variable-frequency, calibrated oscillator, the output of which is mixed with the signal and adjusted to zero beat. The mixing may take place in the frequency meter or in an external circuit, as may be convenient. A crystal-controlled oscillator and harmonic generator are usually included in the meter to provide a series of closely spaced reference points at which the dial calibration may be checked. The crystal oscillator, in turn, may be adjusted to zero beat with station WWV or a local standard. An accuracy of the order of 0.05 percent is readily achieved in the measurement of an unknown frequency.

Counter. The use of a gated event counter for measurement of frequency is illustrated in the simplified block diagram of Fig. 213. The time-base generator resets the counter electronically to zero and then provides a rectangular pulse, which opens the signal gate for a very precisely controlled period of time, one of the five intervals shown being selected by a switch. During the open period, the input signal passes through to the counter, and the number of cycles is registered. After a brief waiting period, determined by a circuit within the time-base generator, the cycle is repeated. The frequency of the signal is indicated by the number of counts received during the open period, the decimal-point position depending on the interval time selected.

The accuracy of measurement is determined largely by the accuracy of the oscillator which controls the timing of the time-base generator. Commercially available frequency counters offer an accuracy of 5 parts in 10^9, plus or minus one count.

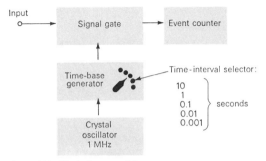

figure 213. Block diagram of frequency counter.

The *period* of the incoming signal may be measured by having the signal control the gate and counting the time-base pulses. The *time interval* between two incoming pulses can be measured by having the first pulse open the gate and the second close it, the time-base pulses being passed and counted during the open period.

TIME AND FREQUENCY STANDARDS.   Broadcasts from the U.S. National Bureau of Standards shortwave radio stations WWV (Fort Collins, Colorado) and WWVH (Hawaii) provide standard carrier frequency transmissions with various standard audio tones and time signals which are valuable for calibration purposes.† The principal transmissions are at 5, 10, and 15 MHz.

The radio and audio frequencies as transmitted are held stable to within a few parts in 10^{11} (in the case of WWV) or 10^{10} (in the case of WWVH). However, changes in the height of the reflecting layer of the ionosphere produce a doppler-effect error which limits the accuracy of received signals to about 3×10^{-7}. The doppler error can be averaged out by extending the measurements over a sufficiently long period of time.

Low-frequency standard signals at carrier frequencies of 60 and 20 kHz are transmitted by the National Bureau of Standards Stations WWVB and WWVL, respectively, both located at Fort Collins. These transmissions are by ground wave and so avoid the doppler error. The frequencies are stable to 2 parts in 10^{11}.

MISCELLANEOUS ELECTRONIC CIRCUITS

FULL-WAVE-RECTIFIER POWER SUPPLY.   A typical circuit employed to provide dc power for vacuum-tube circuits is given in Fig. 214a. It is based on the full-wave-rectifier circuit of Fig. 190c, the output of which is a pulsating voltage equivalent to a steady dc voltage plus an alternating component called the *ripple voltage*. In this case, the main component of the ripple voltage is at 120 Hz. For the reduction of the ripple component in the output voltage, a two-section filter is employed in the circuit of Fig. 214a.

The operation of the filter can be understood by considering that each section consists of an inductor in series with a capacitor as in Fig. 214b. The input voltage E_i has an ac component superimposed on a dc component. The inductor, or *choke*, offers a high impedance to alternating current, while the capacitor offers a low impedance to alternating current; hence only a small fraction of the ac component appears across the capacitor. On the other hand, most of the dc input voltage appears

† The complete schedule of signals transmitted is given in NBS Standard Frequency and Time Services, *Natl. Bur. Std. U.S. Misc. Publ.* 236, 1967.

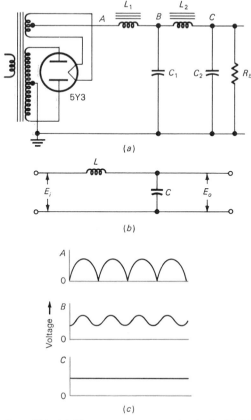

figure 214. (a) Direct-current power-supply circuit; (b) LC
filter section shown separately with input E_i and E_o indicated;
(c) voltage waveforms observed between points A, B, C, and
ground: A-output of rectifier; B-output of first filter stage; C-
output of first filter stage; C-output of second filter stage.

across the load, though there is some loss due to the resistance of the choke-coil
windings. In the less critical applications, only one filter stage is necessary.

The filter shown is of the *choke-input* type. If the first choke is omitted, there
remains a *capacitor-input* filter. The latter yields a higher output voltage, since the
capacitor C_1 then becomes charged nearly to the peak rather than the average value
of the input voltage. The ripple amplitude also increases. In tube manuals, graphs
are often included for rectifier tubes showing the dc output voltage to be expected
for a given ac input voltage, current drain, and type of filter.

The chokes should always be placed in the lead that is *not* grounded, which is usually (but not necessarily) the positive lead. There is considerable capacitance between the secondary of the power transformer and its primary, one side of which is grounded. This capacitance to ground will bypass some ac ripple around the chokes if they are put in the grounded side of the filter. In order for the filter to function effectively, the current through it must not fall below a minimum value. The bleeder resistor R_b provides this minimum current flow and also serves to permit the condensers to discharge rapidly when the power supply is turned off, a safety factor of importance.

The tube illustrated is called a full-wave rectifier and corresponds to two diodes in a single envelope with a common cathode. The output voltage at point C depends upon the secondary voltage of the transformer and upon the current furnished by the rectifier, which determines the potential drops across the tube and the chokes. The output voltage of the circuit will then vary with fluctuations in the line voltage across the transformer primary and with the current drawn by the load. This undesirable feature may be eliminated by the addition of a voltage-regulating section.

AC VOLTAGE REGULATORS.   The ordinary 110-volt ac line voltage is subject to fluctuations arising primarily from the changing load on the line. Regulating units of several kinds are available for providing a constant-voltage output at the line frequency. In the selection of an ac voltage regulator, one should consider the question of stability against variations in load as well as that of stability against fluctuations in line voltage. Electronically controlled regulators are the most versatile, as they provide good regulation against wide variations in either load or line, have a fairly rapid response, and give a sine-wave output with relatively little distortion.

DC VOLTAGE REGULATOR.   For sensitive circuits, it is usually necessary to provide a more stable dc voltage than can be obtained with the circuit of Fig. 214. The usual solution is to add an electronic regulator, illustrated in Fig. 215. The triode V_1, interposed between the unregulated input voltage and the load, functions as an electronically controlled resistance, which is caused to vary in such a way as to minimize changes in the output voltage E_o, whether due to changes in input voltage or to changes in load.

The operation of the circuit is as follows. Resistors R_1 and R_2, which are chosen for good stability, act as a potential divider to provide a voltage E_s, which is a fixed fraction of E_o. Then E_s is compared with the stable reference voltage E_r, supplied by a battery, voltage regulator tube (page 628), or Zener diode (page 611). The differential amplifier circuit amplifies the difference $E_s - E_r$ and causes the grid potential of V_1 to change in a direction opposite to that of any change in E_s.

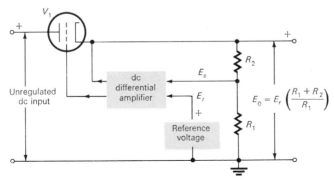

figure 215. The principle of a typical electronic regulator for a dc power supply.

Therefore, if E_o should increase, the grid of V_1 is driven in a negative direction by the differential amplier, the effect being to increase the resistance offered by V_1 to the load current and thus to oppose the original change in E_o. In essence, then, the regulating circuit functions to keep E_s equal to E_r, so that the output voltage is held at the value

$$E_o = E_r \frac{R_1 + R_2}{R_1}$$

The response is rapid enough to minimize hum and other ac disturbances in the audio-frequency range, as well as to correct for slower variations in output. The quality of the regulation varies considerably with the type of circuit used and the load current drawn, but typically both hum and variations in E_o due to changes in line voltage and load are reduced to the millivolt level. The limiting factor is often the stability of the reference voltage or of the first tube of the dc amplifier.

It is worth remembering that this circuit tends to maintain the voltage constant at the sampling network R_1, R_2 rather than at the load. If the load is some distance from the supply, serious deterioration in performance can result from the resistance in the leads or from stray voltages picked up. The remedy for this is to use the technique of *remote sensing*, in which two additional leads, adequately shielded, are used between the supply and load, these leads being so arranged that they do not carry the load current, but merely sample the voltage across the load and carry this information to the differential amplifier.

The output voltage can be made continuously variable over a substantial range by introducing a potentiometer into the sampling network so that the ratio $(R_1 + R_2)/R_1$ becomes adjustable.

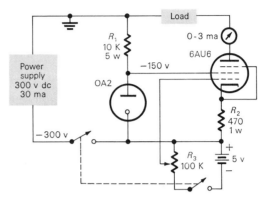

figure 216. Circuit diagram for constant-current supply.

CONSTANT-CURRENT SUPPLY.　When it is desired to maintain a constant dc current through a changing resistive load, as in Experiment 59, the circuit of Fig. 216 may be employed. The plate current of the pentode tube is nearly independent of the plate voltage over a fairly wide range of voltages when the control-grid and screen-grid potentials are held constant (cf. Fig. 200). If the load resistance increases, the potential drop across it increases and the plate voltage of the tube decreases, but the resulting change in plate current is small. The current stability is further improved by the presence of the cathode resistor R_2. The plate current is controlled by adjusting the control-grid bias of the tube, a battery being used to obtain a very stable bias voltage.

Constancy of screen-grid voltage is assured by use of the voltage-regulator tube OA2. This type of tube contains a cold cathode of large area, an anode of small area, and a gas, such as argon, at low pressure. It is found that such a tube exhibits a practically constant potential drop from cathode to anode provided the current lies within limits specified by the manufacturer. If the voltage across the series combination of R_1 and the tube changes, the current through the tube will change, but the potential drop across it will not. The total change in voltage then appears across R_1. The resistance of R_1 must be large enough to restrict the maximum current through the VR tube to a safe value, or the voltage-regulating property of the tube will be lost.

CATHODE FOLLOWER.　Figure 217a shows the actual circuit diagram and equivalent circuit for a cathode follower, a device widely used in modern electronic equipment. The operation of the circuit is easily understood qualitatively: if the input signal e_i causes the grid potential to increase positively, the tube current I_p increases; as a result, the cathode potential $I_p R_k$ also increases, the cathode thus

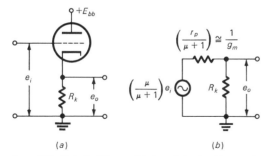

figure 217. Cathode follower. (a) Actual circuit; (b) equivalent circuit.

tending to "follow" the grid. The voltage gain e_o/e_i may be seen from the equivalent circuit, Fig. 217b, to be

$$\frac{e_o}{e_i} = \frac{\mu R_k}{(\mu + 1)R_k + r_p} \approx \frac{R_k}{R_k + (r_p/\mu)} \tag{85}$$

the approximation holding for $\mu \gg 1$. The ratio e_o/e_i is somewhat less than unity, but the power gain is large, as the cathode follower has a relatively high input impedance and low output impedance, the latter typically of the order of a few hundred ohms. Its use is illustrated in Fig. 218. If a load of, say, 10,000 ohms were to be connected at A, the gain of the first stage would drop severely; if the same load were connected at B, the overall gain would drop only slightly. Thus the tube functions to supply the current required by a relatively low impedance load.

AMPLIFIER WITH FEEDBACK. The term *feedback* refers to the process of deriving from the output of an amplifier stage a voltage, called the *feedback voltage,*

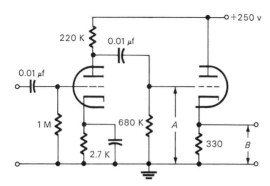

figure 218. Triode amplifier and cathode follower.

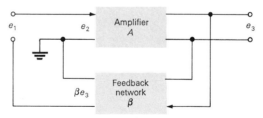

figure 219. Amplifier with feedback.

and applying it in series with the input signal of the same or an earlier stage. If the phase of the voltage fed back is such as to aid the signal to which it is added, the feedback is called *positive,* or *regenerative;* if the phase is such as to oppose the signal, the feedback is called *negative, degenerative,* or *inverse.*

The block diagram of an amplifier with feedback is illustrated in Fig. 219. For simplicity, we shall assume that there is no phase shift in either the amplifier or the feedback network, other than a possible sign change (180° shift). The signal input e_1 is applied between the terminals shown. The total input to the amplifier, however, is $e_2 = e_1 + \beta e_3$, that is, the sum of the signal plus the feedback voltage βe_3. The amplifier itself has gain A, meaning e_3/e_2. The feedback network provides an output which is a certain fraction β of the amplifier output e_3. Then the definition of A leads to

$$(e_1 + \beta e_3)A = e_3 \tag{86}$$

whence the overall gain G, from signal input terminals to output, is

$$G = \frac{e_3}{e_1} = \frac{A}{1 - \beta A} \tag{87}$$

For the case of negative feedback (βA negative), the gain G is less than A in magnitude; with regenerative feedback (βA positive), the gain G is greater than A in magnitude. For $\beta A \approx 1$, the amplifier becomes quite unstable and may break into oscillation.

In some applications, the feedback network produces a phase shift dependent on frequency. In this case, the quantities β and G in Eq. (87) become complex.

Three examples of the application of the principle of feedback follow: (*a*) the use of negative feedback to improve linearity of an amplifier and stability of gain, (*b*) the use of frequency-selective feedback to create a narrowband amplifier, and (*c*) the use of regenerative feedback to convert an amplifier into an oscillator.

AMPLIFIERS WITH NEGATIVE FEEDBACK.   Inverse feedback is used when it is desired to sacrifice gain in order to reduce distortion in the output waveform or to improve the constancy of gain. The use of inverse feedback for this purpose is illustrated in the circuit of Fig. 220. The input signal e_i is applied in series with the fraction† $\beta = R_2/(R_1 + R_2)$ of the output voltage e_o of the amplifier stage. Because the plate voltage is decreased by an increase in grid voltage, the output voltage e_o is $180°$ out of phase with the input voltage and hence the feedback is negative. The feedback voltage βe_o is then $180°$ out of phase with e_i, so that the signal amplification is reduced.

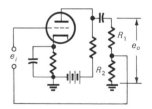

figure 220.

Amplifier stage illustrating inverse feedback.

To see how the advantages mentioned above are obtained, one may apply Eq. (87) to this case. Here βA is negative. For $|\beta A| \gg 1$, the overall signal gain approaches $-(1/\beta)$, and thus is determined mainly by the feedback network, and becomes relatively insensitive to the tube characteristics. For example, for $\beta = \frac{1}{4}$ and $A = -80$, the value of G is -3.81; now if A changes by 10 percent, perhaps by replacement of the tube, G changes by only 0.5 percent.

Another way of introducing negative feedback is to omit the cathode bypass capacitor. Since the alternating current through the cathode resistor is the same as that through the plate load impedance, the ac cathode voltage becomes proportional to the ac output voltage. Furthermore, this ac cathode voltage is effectively subtracted from the input signal because the actual input to the tube is the difference between the grid and cathode voltages. The equivalent circuit of Fig. 197 provides the basis for the quantitative treatment of this case.

NARROWBAND (FREQUENCY-SELECTIVE) AUDIO AMPLIFIER. The term narrowband amplifier refers to a bandpass amplifier of relatively narrow bandwidth. Figure 221*a* is the diagram of a narrowband amplifier which employs feedback as a means of achieving frequency selectivity. The twin-T-network, Fig. 221*b*, exhibits a null output at a frequency $\omega_0 = 2\pi f_0$, satisfying the condition $\omega_0^2 = 1/RC$. The feedback is degenerative, so that the overall amplifier gain is very low $(G \ll 1)$, except for frequencies close to the null frequency f_0 for the twin T. The resulting amplifier response curve is similar to that of a parallel resonant circuit peaked at f_0. With the circuit values shown, the gain at the peak frequency is 2, and the value of Q, the ratio of f_0 to the half-power bandwidth Δf, is 15. The unused half of one twin-triode tube can be used as an amplifier if additional gain is needed.

This circuit can be used throughout the audio range. As to the twin-T component sizes, C should be large enough to minimize effects of variations in stray capacitance and R large enough to avoid overloading the cathode follower. A practical realization of the twin T is shown as Fig. 221*c*.

---

† The output impedance of the triode circuit, if not small compared with R_1, must be taken into account in calculating β. If the impedance of the coupling capacitor is negligible, the output impedance (pages 592 and 605 to 606) of the triode circuit is effectively in series with R_1.

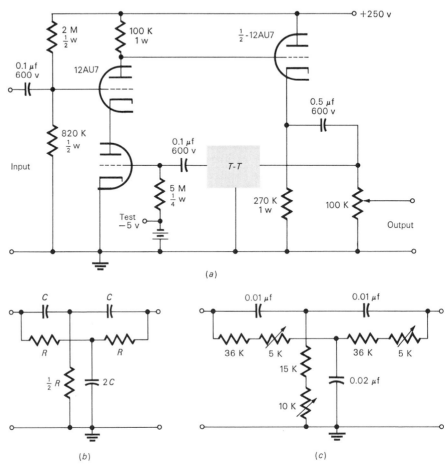

figure 221. Frequency-selective amplifier. (a) Circuit diagram of amplifier. [Adapted from circuit of H. Fleisher in G. E. Valley, Jr., and H. Wallman (eds.), "Vacuum Tube Amplifiers," McGraw-Hill Book Company, New York, 1948.] *(b) Twin-T network. (c) Practical twin-T network, tunable from 380 to 420 Hz.*

VACUUM-TUBE OSCILLATOR.   An oscillator circuit is one which converts dc power into ac power. The use of feedback with an amplifier to achieve this end is illustrated by the diagram of Fig. 222. Here the entire grid excitation for the amplifier is derived by feeding back part of the output into the grid circuit. Note that in this case the amplitude and phase of the feedback voltage are dependent on frequency.

The circuit of Fig. 216 falls in the class of feedback circuits illustrated by Fig. 213, but with the signal input terminals shorted together since there is now no

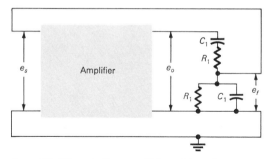

figure 222. Schematic diagram of RC-*coupled oscillator.*

external signal. If the oscillations are to be self-sustaining, two conditions† must be met at the frequency of oscillation:

1. For Eq. (86) to be satisfied with no external excitation ($e_1 = 0$), a necessary condition is

$$A\beta = 1 \tag{88}$$

2. The total phase shift in the amplifier and feedback network must be $0°$.

On the assumption that the amplifier phase shift is negligible, condition 2 is met when the feedback-network phase shift is zero. For the network shown, this happens at only one frequency,[1] $f = 1/2\pi R_1 C_1$. This condition therefore determines the frequency of oscillation. Though other types of feedback networks are used for oscillators, all have in common this property, namely, a phase shift strongly dependent on frequency, this being essential for good frequency stability.

For the network shown, the value of β is $\frac{1}{3}$ when the phase shift is zero, so that Eq. (88) is satisfied for an amplifier gain of 3. In practical terms, the amplifier would be designed to have a small-amplitude gain of somewhat more than 3. The amplitude of oscillations, then, will build up until the nonlinearity in tube characteristics, or some other factor, reduces the gain to a value exactly satisfying Eq. (88). Thus, through the dependence of A on amplitude, Eq. (88) actually determines implicitly the amplitude of oscillations.

An excellent practical audio-oscillator circuit which employs these principles is shown in Fig. 223. Here degenerative feedback (network R_3, R_4, R_5, and the lamp V_3) is used to stabilize the gain at approximately the value 3. Good amplitude stability is provided by the use of lamp V_3 in the feedback network; if the amplitude increases, the increased ac power dissipated in the lamp causes the temperature of its filament to increase. The accompanying increase in resistance alters the feed-

† Both conditions are contained in Eq. (88) if A and β are considered in general to be complex quantities.

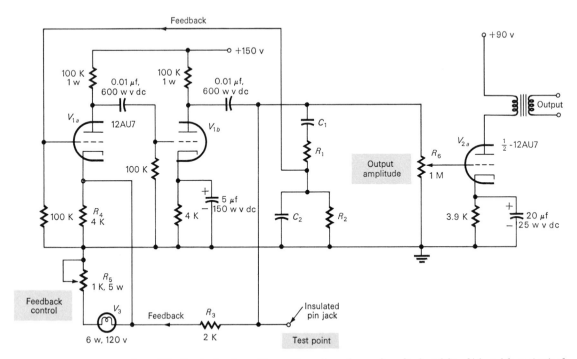

figure 223. Circuit of audio oscillator with good waveform and amplitude stability. [Adapted from circuit of F. E. Terman, R. R. Buss, W. R. Hewlett, and F. C. Cahill, Proc. IRE, **27**: 649 (1939).] *Values of R_1, R_2, C_1, C_2 are chosen for the desired frequency of oscillation. The degenerative feedback control R_5 is adjusted to achieve good amplitude stability consistent with a satisfactory waveform as observed with an oscilloscope connected at the test point. The amplitude is normally about 1 volt rms at this point.*

back ratio in such a way as to decrease the gain and thus to oppose the change in amplitude. A better waveform is obtained in this way than if the amplitude is stabilized through nonlinearity in the tube dynamic transfer characteristic.

PHASE-SENSITIVE DETECTOR. The phase-sensitive detector, illustrated schematically in Fig. 224, has two inputs, a signal input and a reference input. The latter is usually a sine wave,

$$E_r = E_0 \cos 2\pi f_0 t \tag{89}$$

while the signal input is a voltage of the same frequency,

$$E_s = E_0' \cos (2\pi f_0 t + \phi_s) = E_1 \cos 2\pi f_0 t + E_2 \sin 2\pi f_0 t \tag{90}$$

often accompanied by noise.† It is the function of the phase-sensitive detector to

† Several examples of systems which include phase-sensitive detectors occur in Chapter 25.

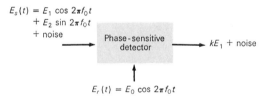

figure 224. *Diagram illustrating the function of a phase-sensitive detector.*

produce an output dc voltage proportional to the amplitude $E_1 = E_0' \cos \phi_s$ of the *in-phase* component of the signal, while responding as little as possible to the quadrature (90°) component and to noise. Of course, it does necessarily respond to noise components with frequencies sufficiently close to f_0. Many practical phase-sensitive detector circuits also respond to noise components close to integral multiples of f_0, but this is seldom a disadvantage because the bandwidth of the preceding amplifier can easily be made narrow enough to eliminate noise components at harmonics of f_0.

Two practical phase-sensitive detector circuits are shown in Fig. 225. In the upper circuit, the reference signal drives a switching relay which feeds the signal alternately to two different points. For the signal component in phase with the reference voltage, the positive half-cycle is always applied to one side and the negative half-cycle to the other. The lowpass filter smooths out the ac components, and there

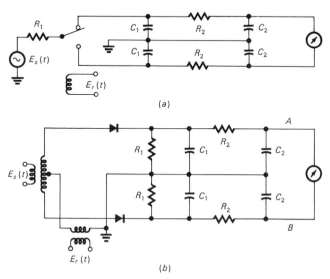

figure 225. *Practical phase-sensitive detector circuits. (a) Relay type; (b) diode type.*

results a dc voltage which produces a meter deflection. However, for the sine component of the signal, the switching times are such that no dc component results in the output, since in this case the switching occurs at the peaks of the signal voltage. For a noise component of frequency f_n, close to f_0, the switching occurs at different points on successive cycles; the resulting output contains no dc component but does contain ac components, some of which are at the frequency $|f_n - f_0|$, which may be low enough ($\lesssim 1/2\pi RC$) to appear at the output of the low-pass filter. The totality of noise components of frequencies close to f_0 in the input therefore produces random low-frequency fluctuations in the output.

With the use of high-speed relays, particularly solid-state types, this and like circuits are useful in the low audio range. The switching can be accomplished electronically with a tube or transistor circuit, up to frequencies of about 200 kHz.

Another type of phase-sensitive detector is that of Fig. 225b. To examine its operation, consider first the case in which there is no signal input. Then the reference input, applied through a transformer, is fed to a split load comprising two ordinary diode detector circuits (page 610) in parallel. The output of each is a dc voltage close to the peak value of the input waveform. If the diodes and other components are suitably matched, the same dc voltage is developed at point A as at B, and the net voltage difference between these output terminals is zero. The reference voltage is chosen to be rather large, say 5 to 15 volts rms. Now let the signal voltage be applied. For proper operation of the detector, this must be much smaller than the reference voltage, perhaps less than a volt. If the signal voltage has the same frequency and phase as the reference, the total voltage (vector sum) applied to one diode circuit is increased while that applied to the other is decreased, and a dc voltage unbalance develops at the output terminals A and B. A signal voltage 90° out of phase with the reference, however, does not produce an output because the vector addition of a small quadrature signal to the large reference voltage produces only a negligible change in amplitude, and hence does not unbalance the two diode circuits. Again, noise components at frequencies f_n close to f_0 produce ac components in the output at frequencies $|f_n - f_0|$.

SAFETY PRECAUTIONS

It is particularly important that the safety factor be kept firmly in mind in working on or with electronic circuits, since the hazards involved in any electrical work where high voltages may be present are considerable. A current of as little as 15 ma has been known to be fatal, and although such shocks usually are produced by higher-voltage sources, the ordinary 110-volt ac line voltage has been known to be sufficient. The hazard is especially great when the skin is moist, particularly as a result of perspiration.

Direct-current voltages are also very dangerous. A particular source of trouble arises when, owning to faulty circuit design, no path has been provided for the discharge of high-voltage condensers after the main power switch has been turned off. A good-quality condenser will retain its charge for quite a time if the only discharge path is by slow leakage of current between the terminals. A bleeder resistor across the condenser may be employed to eliminate this difficulty, but it is well to bear in mind that the time constant for discharging may be several minutes, and also that bleeder resistors may fail, or somehow become disconnected. The only safe course is to develop the habit of ensuring the discharge of high-voltage condensers, before working around them, by momentarily placing a screwdriver or other reliable short across the terminals.

It is desirable to employ double-pole switches in ac or dc power lines. One side of the 110-volt ac line is usually at ground potential; if a nonpolarized plug is used, with a single-pole switch in the line, there is an even chance that one terminal of the load will be live when the main power switch is off, unless, of course, a transformer is used between line and load.

Alterations of circuits should be carried out only with the primary power source disconnected. In the measurement of voltages, etc., properly insulated test leads must be used.

References

1. R. W. Landee, D. C. Davis, and A. P. Albrecht, "Electronic Designers' Handbook," sec. 6, p. 48, McGraw-Hill Book Company, New York, 1957.
2. S. Goldman, "Frequency Analysis, Modulation, and Noise," McGraw-Hill Book Company, New York, 1948.
3. S. Fich and J. L. Potter, "Theory of A-C Circuits," Prentice-Hall, Inc., Englewood Cliffs, N.J., 1958.
4. R. M. Walker and H. Wallman in G. E. Valley, Jr., and H. Wallman (eds.), "Vacuum Tube Amplifiers," chap. 2, McGraw-Hill Book Company, New York, 1948.
5. H. Wallman in G. E. Valley, Jr., and H. Wallman (eds.), "Vacuum Tube Amplifiers," chap. 7, McGraw-Hill Book Company, New York, 1948.
6. A. Van der Ziel, "Noise," Prentice-Hall, Inc., Englewood Cliffs, N.J., 1954.

Chapter
29

Purification
of
Materials

In many experiments, the factor which limits the accuracy of the results is the purity of the materials used rather than the refinement of the measurements. For example, nothing is gained by determining the refractive index of a liquid to five decimal places if it contains an impurity in amount sufficient to alter the refractive index in the third decimal place.

The need for chemicals of both high purity and established purity is great, and it extends to all branches of science. The U.S. National Bureau of Standards has been active in this field, and it now supplies some materials of high purity and provides other services such as standards of purity and descriptions of methods of purification. It is hoped that these services will be expanded. Many of the chemical companies supply chemicals of specified purity, and a committee of the American Chemical Society establishes standards of purity. The Johnson-Mathey Co. specializes in inorganic chemicals of high purity as determined spectroscopically. Some pure hydrocarbons are available from the Phillips Petroleum Company as "research grade" products whose purity has been established by the freezing-point method.

The degree of purity required depends on the material to be investigated, the use which is to be made of it, and the nature of the impurities. The different classes of chemical substances include (a) elements, including selected isotopes; (b) organic compounds, including hydrocarbons and derivatives such as alcohols, amines, and biologically important materials such as carbohydrates and proteins; (c) inorganic materials, including halides, oxides, acids, salts; semiconductors and phosphors; and (d) single crystals.

The level of attainable purity varies greatly. For example, 90 percent purity is sufficient for some types of experiments, but 99.999 is required for some metals and organic substances and "nine nines" for certain components in semiconductors.

There are many techniques for purifying materials, including the following:

Absorption	Fractional distillation
Adsorption	Gas-liquid chromatography
Countercurrent distribution	Precipitation
Crystallization	Sedimentation
Distillation	Single-crystal growth
Electrolysis	Sublimation
Electrophoresis	Zone melting
Extraction	

The following are some of the means which can be used as criteria of purity and sometimes as quantitative measures of the amounts of impurities:

Absorption spectra	Emission spectra
Bioassays	Freezing points
Boiling points	Heats of reaction and phase change
Electrical conductance	Infrared absorption

Magnetic properties	Stoichiometry
Neutron activation	Viscosity
Radioactivity	X-ray spectra
Refractive index	

METHODS.   Considerable technical knowledge and laboratory skill are required for the proper purification of materials for precision measurements. A knowledge of the origin of the starting material is important, since it will suggest the identity of the probable impurities and thus influence the procedure adopted. For example, benzene from petroleum sources invariably contains thiophene and other sulfur compounds. Since these contaminants are more rapidly sulfonated than benzene, they can be removed by shaking the benzene repeatedly with small portions of concentrated sulfuric acid. Another example is provided by commercial "absolute" alcohol, which usually contains traces of benzene introduced in the removal of water from the alcohol by an azeotropic-distillation step (see below).

Purification procedures in general involve both chemical and physical processes. The chemical steps are specifically characteristic of the compounds involved. The physical procedures utilized will be selected most often from among the following processes.

Crystallization. A very useful method of purification available is a series of fractional-crystallization steps. The impurities present must not form solid solutions with the compound being purified. When materials with low freezing points are treated, precautions must be taken to protect them from condensing atmospheric moisture.

Fractional Distillation. This is probably the most common procedure used for the purification of liquids. A very efficient distillation column is required when the boiling points of the impurities are close to that of the major constituent. Azeotropic solutions, because of their constant boiling points at constant pressure, have often been mistaken for pure components. Fractional distillation is sometimes routinely and unimaginatively used in many cases when a better result could be obtained by fractional crystallization or other methods.

Azeotropic Distillation. Here advantage is taken of the formation of an azeotropic mixture involving an impurity to facilitate purification by fractional distillation. In the production of commercial absolute alcohol, benzene is added to the 95 percent azeotrope of ethanol and water obtained by ordinary distillation. A ternary azeotrope of water, ethanol, and benzene can then be fractionated out to remove the water present. Further distillation removes the benzene in a binary azeotrope with ethanol and leaves essentially anhydrous ethanol contaminated with traces of benzene.

Adsorption. The selectivity shown in adsorption processes can result in effective separations which are extremely difficult to obtain by other methods.

The slight differences in the rates of absorption and desorption of two different substances, multiplied many times in the passage of fluid along a packed column with large surface areas, are used not only to identify and estimate concentrations but also to effect separations. Solutes in solution may be separated as illustrated in Experiment 48, and vapors may be separated also. Great progress has been made in vapor-phase chromatography.[1-4] In some experiments a stream of helium or other carrier gas forces the vapors through a column of diatomaceous earth covered with a nonvolatile oil. The rates at which various components travel through the column depend upon the distributions between the gas phase and the oil solution on the solid packing.

Drying. It will commonly be found that the adequate elimination of water from a sample constitutes one of the most difficult problems in the whole purification process. The method of drying employed is determined primarily by the chemical properties of the material. If the product is not used immediately, care must be taken to prevent its recontamination by absorption of atmospheric moisture. Because of its low molecular weight, polar character, and chemical reactivity, relatively small amounts of water can be very troublesome. Heating at elevated temperatures until further heating gives no decrease in weight is often used for drying solids. Frequently a liquid can be freed from water by shaking with a drying agent such as anhydrous calcium chloride or phosphorous pentoxide. Both solids and liquids are often dried by keeping them in a desiccator, for long periods of time, over sulfuric acid or other drying agent.

Zone Refining.[5] The requirements for very high purity of solids used in transistors and similar electronic instruments have led to the perfection of purification by zone melting. A long tube of the frozen solid is melted at one end by a narrow, movable electrical heating unit which melts the material in a narrow disk. The heating coil is moved slowly along the tube, and the melted zone which contains the impurities also moves along the tube, collecting more impurities as it goes. In this way the impurities are displaced to one end. The process is repeated several times.

CRITERIA OF PURITY.   One of the best criteria available for organic compounds is the constancy of the freezing point or melting point throughout the phase transition.[6-8] If a liquid is impure, the impurities will become concentrated in the liquid phase as the solid separates out; the freezing point thus is gradually lowered. If the liquid phase can be treated as an ideal solution, an assumption of adequate validity in many cases, and if no solid solutions are formed, the amount of impurity can be calculated with fair accuracy from the shape of the freezing or melting curve. Since high sensitivity is required in the temperature measurements, a platinum resistance thermometer or multiple-junction thermocouple is used.

Comparison of the normal boiling point, refractive index, etc., with the accepted values for the compound concerned is often used to estimate its purity. A valuable reference tabulation of physical constants of organic compounds has been prepared by Timmermans.[9] Unfortunately, the reference data available are often of inadequate accuracy; the objective evaluation of purity furnished by the freezing-point method is hence much to be preferred in critical cases.

It is important to know the identity as well as the mole fraction of impurity present, in order to judge its effect on the measurements to be made. In addition, direct determination of the important contaminants may be possible through standard analytical procedures.

TYPICAL PURIFICATIONS.   Special techniques are given here as examples for purifying a few common chemicals.

Recommendations concerning the purification of many organic liquids are given in "Organic Solvents."[10] Archibald[11] describes methods for the preparation of pure inorganic compounds, and Farkas and Melville[12] have specified methods for the preparation, purification, and analysis of a number of gases.

Water. Ordinary distilled water is sufficiently pure for most work in physical chemistry, but for some applications, such as conductance measurements, it is necessary to use specially redistilled water.

Steam is generated from a dilute alkaline permanganate solution in a quartz or heavily tinned copper boiler and is partially condensed in a quartz or block-tin condenser. The escaping steam carries off gases evolved by the boiling liquid and prevents exposure of the condensate to the laboratory air. The condensate is collected in a quartz or tinned copper reservoir under air which has been treated to remove carbon dioxide, ammonia, etc. Polyethylene bottles can also be used for storage of water to prevent its contamination by impurities dissolved from glass bottles.

Benzene. Reagent-grade benzene is treated with concentrated sulfuric acid until it gives a negative test for thiophene with isatin. It is then washed repeatedly with water and dried first with calcium chloride and then with sodium. Fractional distillation results first in the elimination of residual water as the binary azeotrope; the product can then be collected. If extreme purity is required, slow fractional crystallization may be employed.

Hydrocarbons. A large number of hydrocarbons have been purified, and their physical properties studied, as part of the work of Research Project 44 of the American Petroleum Institute. The methods employed have been described in a series of publications.[13]

Sodium Chloride. Since sodium chloride has a small temperature coefficient of solubility, it cannot be easily purified by crystallization. A saturated solution of sodium chloride is treated with HCl gas to throw out pure salt. The gas is introduced through an inverted funnel, because a small tube is soon plugged up with the crystals. Rubber connections are attacked by hydrochloric acid, so the connections and the generating bottles are of glass with Tygon tubing connections. The gas is generated by adding concentrated hydrochloric acid, drop by drop, to concentrated sulfuric acid, while shaking to avoid the formation of two layers, which might lead to an explosion.

The precipitated sodium chloride is packed into a funnel, rinsed with a minimum amount of water, and fused in a platinum dish at red heat.

Sodium Hydroxide. For most titrations with alkali, it is necessary to have the alkali free from carbonate to obtain a sharp end point. High-grade commercial sodium hydroxide may be obtained which ordinarily needs no further purification. Sodium hydroxide solution free from carbonate is readily prepared from a saturated stock solution. The carbonate is thrown out as an insoluble precipitate by the high concentration of sodium hydroxide which exists in a saturated solution. The clear supernatant solution is drawn off with a siphon and diluted with carbon dioxide—free water to the desired concentration, at room temperature. The saturated solution is about 15 *M*, and it is kept in a bottle the inside of which has been covered with paraffin.

References

1. H. W. Patton, I. S. Lewis, and W. I. Kaye, *Anal. Chem.,* **27:** 170 (1955).
2. R. Stock and C. B. F. Rice, "Chromatographic Methods," 2d ed., Chapman & Hall Ltd., London, 1967.
3. E. Lederer and M. Lederer, "Chromatography," American Elsevier Publishing Company, New York, 1957.
4. J. G. Kirchner, "Thin Layer Chromatography," in A. Weissberger (ed.), "Technique of Organic Chemistry," vol. 12, E. S. Perry and A. Weissberger (eds.), Interscience Publishers, Inc., New York, 1967.
5. W. J. Pfann, "Zone Melting," 2d ed., John Wiley & Sons, Inc., New York, 1966.
6. W. P. White, *J. Phys. Chem.,* **24:** 393 (1920)
7. E. L. Skau, *J. Phys. Chem.,* **37:** 609 (1933).
8. J. M. Sturtevant in A. Weissberger (ed.), "Technique of Organic Chemistry," vol. 1, "Physical Methods of Organic Chemistry," 3d ed., pt. 1, chap. 10, Interscience Publishers, Inc., New York, 1959.
9. J. Timmermans, "Physico-chemical Constants of Pure Organic Compounds," Elsevier Press, Inc., Houston, Texas, 1950.
10. J. A. Riddick and E. E. Toops, in A. Weissberger (ed.), "Technique of Organic Chemistry," vol. 7, A. Weissberger and E. S. Proskauer (eds.), "Organic Solvents," Interscience Publishers, Inc., New York, 1955.
11. E. H. Archibald, "The Preparation of Pure Inorganic Substances," John Wiley & Sons, Inc., New York, 1932.
12. A. Farkas, H. W. Melville, and B. G. Gowenlock, "Experimental Methods in Gas Reactions," 2d ed., The Macmillan Company, New York, 1964.
13. A. R. Glasgow, Jr., A. J. Streiff, and F. D. Rossini, *J. Res. Natl. Bur. Std. U.S.,* **35:** 355–373 (1945); A. J. Streiff, F. D. Rossini, et al., *ibid.,* **41:** 323 (1948).

APPENDIX

ALGEBRA OF COMPLEX NUMBERS [1,2]

The rules of complex algebra will be stated briefly. A complex number c may always be written in the form

$$c = a + ib \tag{A1}$$

where a and b are real numbers and i is the pure imaginary unit defined by

$$i^2 = -1 \tag{A2}$$

The rules for manipulation of complex numbers are the same as the usual rules of algebra for real numbers, with i treated purely as an algebraic symbol, except for use of Eq. (A2). Thus, for example, the sum of two complex numbers

$$c_1 = a_1 + ib_1 \tag{A3}$$
$$c_2 = a_2 + ib_2 \tag{A4}$$

is

$$c_1 + c_2 = (a_1 + a_2) + i(b_1 + b_2) \tag{A5}$$

The product is

$$c_1 c_2 = (a_1 a_2 - b_1 b_2) + i(a_1 b_2 + a_2 b_1) \tag{A6}$$

and the quotient c_1/c_2 is

$$\frac{c_1}{c_2} = \frac{a_1 + ib_1}{a_2 + ib_2} = \frac{(a_1 + ib_1)(a_2 - ib_2)}{a_2^2 + b_2^2}$$

$$= \frac{a_1 a_2 + b_1 b_2}{a_2^2 + b_2^2} + i \frac{a_2 b_1 - a_1 b_2}{a_2^2 + b_2^2} \tag{A7}$$

In all cases, the resulting complex number has been written explicitly in terms of its real and imaginary parts.

A very useful formula of complex algebra is the Euler equation,

$$e^{ix} = \cos x + i \sin x \tag{A8}$$

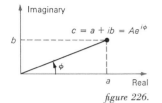

figure 226.

Graph of complex numbers in a plane. The point indicated for the general complex number c = a + ib = Ae^{iφ} has rectangular coordinates (a,b) and polar coordinates (A,φ).

Clearly, any complex number $a + ib$ can be written alternatively in the form $Ae^{i\phi}$, where A and ϕ are real, since a solution always exists for the equations

$$Ae^{i\phi} = A\cos\phi + i(A\sin\phi) = a + ib \tag{A9}$$

Complex quantities may be plotted in a plane as shown in Fig. 226, the abscissa and ordinate being the real and imaginary parts, respectively. The graph provides a very helpful means for visualizing the implications of Eq. (A9): A and ϕ are the polar coordinates of the point representing a complex number, while a and b are the rectangular coordinates of the same point. Multiplication by a real number moves a point directly away from the origin; multiplication by $e^{i\phi}$ rotates a point through an angle ϕ counterclockwise.

The complex conjugate c^* of a complex number c is obtained by replacing i by $-i$ throughout. Thus (c^*c) is a real number, the square root of which is called the *magnitude*, or *absolute value*, of c, often represented by the symbol $|c|$. Clearly, $|c| = A$.

VECTORS[1,2]

Physical quantities which have both magnitude and direction, such as the velocity of a particle or the electric field at a point in space, are described by *vectors*. A physical vector is equivalent to an ordered set of three real numbers:

$$\mathbf{v} = (v_x, v_y, v_z) \tag{A10}$$

These numbers are the components of the vector and are the projections of it on the x, y, and z axes of the coordinate system. (Other choices of the components are sometimes made, but there will always be three independent numbers.) The magnitude or length of the vector is given by $|v| = (v_x^2 + v_y^2 + v_z^2)^{\frac{1}{2}}$.

A vector can be multiplied by a constant by multiplying each component by that constant.

$$a\mathbf{v} = (av_x, av_y, av_z) \tag{A11}$$

Two vectors can be added or subtracted by adding or subtracting corresponding components.

$$\mathbf{u} \pm \mathbf{v} = (u_x \pm v_x,\ u_y \pm v_y,\ u_z \pm v_z) \tag{A12}$$

Vectors can be multiplied in two different ways. One produces a scalar, or ordinary number, as the product, and is called the *scalar*, *inner*, or *dot product*. It is

$$\mathbf{u} \cdot \mathbf{v} = u_x v_x + u_y v_y + u_z v_z$$
$$= |u|\,|v|\cos\theta \tag{A13}$$

where θ is the angle between **u** and **v**. The other way of multiplying vectors produces another vector as the result, and is called the *vector* or *cross product*.

$$\mathbf{u} \times \mathbf{v} = (u_y v_z - u_z v_y, \, u_z v_x - u_x v_z, \, u_x v_y - u_y v_x) \tag{A14}$$

The vector $\mathbf{u} \times \mathbf{v}$ is perpendicular to the plane determined by **u** and **v**, and its magnitude is a maximum when **u** and **v** are perpendicular

$$|\mathbf{u} \times \mathbf{v}| = |u| \, |v| \sin \theta \tag{A15}$$

Note that

$$\mathbf{v} \times \mathbf{u} = -\mathbf{u} \times \mathbf{v}$$

The vector triple product

$$\mathbf{u} \cdot (\mathbf{v} \times \mathbf{w}) = (\mathbf{u} \times \mathbf{v}) \cdot \mathbf{w} \tag{A16}$$

is a scalar quantity equal to the volume of the parallelepiped determined by **u**, **v**, and **w**. Its value is unchanged by a cyclic permutation of **u**, **v**, and **w** but changes sign if any two of them are interchanged. The vector product of three vectors depends on the order in which the multiplications are performed

$$(\mathbf{u} \times \mathbf{v}) \times \mathbf{w} = (\mathbf{w} \cdot \mathbf{u})\mathbf{v} - (\mathbf{w} \cdot \mathbf{v})\mathbf{u}$$
$$\mathbf{u} \times (\mathbf{v} \times \mathbf{w}) = (\mathbf{u} \cdot \mathbf{w})\mathbf{v} - (\mathbf{u} \cdot \mathbf{v})\mathbf{w} \tag{A17}$$

Vector quantities like the electric field or the velocity of a small volume element of a fluid may vary from one point in space to another, both in magnitude and in direction. A vector is thus obtained with each component a function of position

$$\mathbf{v}(x,y,z) = [v_x(x,y,z), v_y(x,y,z), v_z(x,y,z)] \tag{A18}$$

Such a quantity is called a *vector field*.

Certain combinations of the derivatives of a vector field have particular significance in physics. An ordinary, or scalar, function of position can be obtained from a vector field as

$$\text{div } \mathbf{v} = \boldsymbol{\nabla} \cdot \mathbf{v} = \frac{\partial v_x}{\partial x} + \frac{\partial v_y}{\partial y} + \frac{\partial v_z}{\partial z} \tag{A19}$$

It is called the divergence of **v**. The symbol $\boldsymbol{\nabla}$, del, is commonly used. It can be formally considered as a vector with components $(\partial/\partial x, \, \partial/\partial y, \, \partial/\partial z)$, and Eq. (A19) is then interpreted as a scalar product in the usual sense. A vector field resulting from **v** is

$$\text{curl } \mathbf{v} = \boldsymbol{\nabla} \times \mathbf{v} = \left(\frac{\partial v_y}{\partial z} - \frac{\partial v_z}{\partial y}, \, \frac{\partial v_z}{\partial x} - \frac{\partial v_x}{\partial z}, \, \frac{\partial v_x}{\partial y} - \frac{\partial v_y}{\partial x} \right) \tag{A20}$$

It is also possible to construct a vector field from a scalar function of position, $f(x,y,z)$. It is called the *gradient* of f, and is given by

$$\text{grad } f = \mathbf{\nabla}f = \left(\frac{\partial f}{\partial x}, \frac{\partial f}{\partial y}, \frac{\partial f}{\partial z}\right) \tag{A21}$$

Multiple derivatives can also be important. One such combination is

$$\text{grad div } f = \nabla^2 f = \frac{\partial^2 f}{\partial x^2} + \frac{\partial^2 f}{\partial y^2} + \frac{\partial^2 f}{\partial z^2} \tag{A22}$$

also called the laplacian of f. (In older books and in many European books Δ is used in place of ∇^2.) The corresponding expression for a vector is

$$\nabla^2 \mathbf{v} = (\nabla^2 v_x, \nabla^2 v_y, \nabla^2 v_z) \tag{A23}$$

Other combinations of derivatives can often be simplified. Some identities which can be verified by direct substitution are

$$\text{div curl } \mathbf{v} = \mathbf{\nabla} \cdot (\mathbf{\nabla} \times \mathbf{v}) \equiv 0 \tag{A24}$$

$$\text{curl curl } \mathbf{v} = \mathbf{\nabla} \times (\mathbf{\nabla} \times \mathbf{v})$$
$$= \text{grad div } \mathbf{v} - \nabla^2\mathbf{v} \tag{A25}$$

$$\text{curl grad } f = \mathbf{\nabla} \times \mathbf{\nabla}f \equiv 0 \tag{A26}$$

References

1. W. Kaplan, "Advanced Calculus," Addison-Wesley Publishing Company, Inc., Reading, Mass., 1952.
2. H. Margenau and G. M. Murphy, "The Mathematics of Physics and Chemistry," 2d ed., vol. I, D. Van Nostrand Company, Inc., Princeton, N.J., 1956.

DEFINITIONS OF BASIC UNITS

THE INTERNATIONAL SYSTEM.   The International System of Units was adopted by the eleventh General Conference on Weights and Measures in 1960. The abbreviation for this system is SI in all languages. The basic units of this system are defined as follows.

Meter. The meter is the length equal to 1,650,763.73 (exactly) wavelengths under vacuum of the radiation corresponding to the transition between the energy levels $2p_{10}$ and $5d_5$ of the pure nuclide ^{86}Kr.

Kilogram. The kilogram is the mass of the international prototype kilogram, which is in the custody of the International Bureau of Weights and Measures at Sèvres, France.

Second. The second is still formally defined as the fraction 1/31,556,925.9747 (exactly) of the tropical year for 1900 January 0 at 12 h ephemeris time. In October, 1964, however, the twelfth General Conference on Weights and Measures recommended change to a unit based on an atomic radiation frequency and designated for temporary use the value 9,192,631,770 sec^{-1} (exactly) for the frequency of the transition when undisturbed by external fields between the hyperfine levels $F = 4$, $M_F = 0$ and $F = 3$, $M_F = 0$ of the fundamental state $^2S_{\frac{1}{2}}$ of an atom of the pure nuclide ^{133}Cs.

Ampere. The ampere is that constant current which if maintained in two parallel rectilinear conductors of infinite length and of negligible circular cross section at a distance apart of 1 m under vacuum would produce a force between the conductors equal to 2×10^{-7} newton per meter of length.

THE CGS SYSTEM.   The centimeter, gram, and second are defined with reference to the above standards. Thus, 1 cm $= 10^{-3}$ m, 1 g $= 10^{-3}$ kg, and the second is the same in both systems. In the cgs system, however, no new standard unit is introduced for electrical and magnetic quantities; instead units for these are defined in terms of the centimeter, gram, and second. Two sets of cgs electromagnetic units are used, the *cgs-esu system,* based on the esu unit of charge defined through Coulomb's law, and the *cgs-emu system,* based on the emu unit of current defined through Ampère's law.

PHYSICAL-CHEMICAL CONSTANTS†

The following tables give the values of the fundamental constants recommended by the committee on fundamental constants of the National Academy of Sciences–National Research Council. The values of the basic constants are based on the analysis by Cohen and DuMond of experimental data.

INTERNATIONAL ELECTRICAL UNITS.   These units are now considered obsolete but were widely used for calibrations by standards laboratories prior to 1948. The "mean international" units are those estimated by the ninth General Conference on Weights and Measures in 1948. The "U.S. international units" were estimated by the U.S. National Bureau of Standards as applying to calibrations performed by them prior to 1948.

1 mean international ohm = 1.00049 ohms
1 mean international volt = 1.00034 volts
1 U.S. international ohm = 1.000495 ohms
1 U.S. international volt = 1.000330 volts

† F. D. Rossini, "Values of the Fundamental Constants for Chemistry," *Pure Appl. Chem.,* **9**: 453 (1964); Handbook for Authors of Papers in the Journals of the American Chemical Society, pp. 90–100, American Chemical Society, Washington, D.C. 20036, 1967.

Values of the Defined Constants

Constant	Abbreviation or Symbol	Value (exact, by definition)
Unified atomic mass unit	u	1/12 the mass of an atom of ^{12}C
Mole	mol	The amount of a substance, of specified chemical formula, containing the same number of formula units (molecules, atoms, ions, electrons, or other entities) as there are atoms in 12 g (exactly) of the pure nuclide ^{12}C
Standard acceleration of gravity in free fall	g	980.665 cm sec^{-2}
Normal atmosphere, pressure	atm	1,013,250 dynes cm^{-2}
Absolute temperature of the triple point of water[a]	T_{tp}	273.16°K
Thermochemical calorie	cal	4.184 joules
International steam calorie	cal$_{IT}$	4.1868 joules
Inch	in.	2.5400 cm
Pound, avoirdupois	lb	453.59237 g

[a] The difference between the temperature of the triple point of water and the so-called ice point (temperature of equilibrium of solid and liquid water saturated with air at 1 atmosphere) is accurately known: T (triple point) $-$ T (ice point) $= 0.0100 \pm 0.0001$°K.

Recommended Values of the Basic Constants

Constant	Symbol	Value	Estimated uncertainty[a]	Unit
Velocity of light in vacuo	c	2.997925×10^{10}	3	cm sec^{-1}
Avogadro's number	N	6.02252×10^{23}	28	molecules mole^{-1}
Faraday constant	F	9.64870×10^{4}	16	coulomb equiv^{-1}
		2.30609×10^{4}	4	cal volt^{-1} equiv^{-1}
Planck's constant	h	6.6256×10^{-27}	5	erg sec
Pressure-volume product for 1 mole of gas at 0°C and pressure	$(PV)_{0°C}^{P=0}$	2.27106×10^{3}	12	joules mole^{-1}
		2.24136×10	12	cm^3-atm mole^{-1}

[a] Based on three standard deviations applied to last digits of preceding column.
SOURCE: J. W. M. DuMond and E. R. Cohen, *Natl. Bur. Std. U. S. Tech. Bull.* 47, p. 10, 1963.

Recommended Values of the Derived Constants

Constant	Symbol and relation	Value	Estimated uncertainty[a]	Unit
Elementary charge	$e = \dfrac{F}{N}$	4.80298×10^{-10}	20	$cm^{\frac{3}{2}} \, g^{\frac{1}{2}} \, sec^{-1}$ (esu)
Gas constant	$R = \dfrac{(PV)^{P=0}_{0°C}}{T_{0°C}}$	8.31433	44	joules $deg^{-1} \, mole^{-1}$
		1.98717	11	cal $deg^{-1} \, mole^{-1}$
Boltzmann constant	$k = \dfrac{R}{N}$	1.38054×10^{-16}	9	erg $deg^{-1} \, molecule^{-1}$
Electron rest mass	m_e	9.1091×10^{-28}	4	g
Proton rest mass	m_p	1.67252×10^{-24}	8	g
Rydberg constant	$\mathcal{R}_\infty$	1.0973731×10^{5}	3	cm^{-1}
Bohr radius	a_0	5.29167×10^{-9}	7	cm
Gyromagnetic ratio of proton	γ	2.67519×10^{4}	2	rad $sec^{-1} \, gauss^{-1}$
Bohr magneton	μ_B, or β	9.2732×10^{-21}	6	erg $gauss^{-1}$
Nuclear magneton	μ_N	5.0505×10^{-24}	4	erg $gauss^{-1}$

[a] Based on three standard deviations applied to the last digits of preceding column.

Special Units

Unit	Abbreviation	Value
Angstrom	Å	10^{-10} m
Micron	μ	10^{-6} m
Liter	l	$10^{-3} \, m^3$
Bar	bar	10^{-5} newton-m^{-2}
Torr	torr	$101325/760$ newton-m^{-2}
Millimeter of mercury	mm Hg	$13.5951 \times 980.665 \times 10^{-2}$ newton-m^{-2}
Degree Celsius	°C	$T(°K) - 273.15$

REDUCTION OF BAROMETER READINGS ON A BRASS SCALE TO 0°

$$P_0 = P - P \frac{\alpha t - \beta(t - t_s)}{1 + \alpha t}$$

$P_0 = $ barometer reading reduced to 0°

$P = $ observed barometer reading

$t = $ temperature of barometer, °C

$\alpha = 0.0001818 =$ mean cubical coefficient of expansion of mercury between 0 and 35°

$\beta =$ linear coefficient of expansion of scale material, 18.4×10^{-6} for brass

$t_s =$ temperature at which scale was calibrated, normally 20°C

Correction to Be Subtracted from Barometer Readings

t, °C	720 mm	730 mm	740 mm	750 mm	760 mm	770 mm	780 mm
15	1.76	1.78	1.81	1.83	1.86	1.88	1.91
16	1.88	1.90	1.93	1.96	1.98	2.01	2.03
17	1.99	2.02	2.05	2.08	2.10	2.13	2.16
18	2.11	2.14	2.17	2.20	2.23	2.26	2.29
19	2.23	2.26	2.29	2.32	2.35	2.38	2.41
20	2.34	2.38	2.41	2.44	2.47	2.51	2.54
21	2.46	2.50	2.53	2.56	2.60	2.63	2.67
22	2.58	2.61	2.65	2.69	2.72	2.76	2.79
23	2.69	2.73	2.77	2.81	2.84	2.88	2.92
24	2.81	2.85	2.89	2.93	2.97	3.01	3.05
25	2.93	2.97	3.01	3.05	3.09	3.13	3.17
26	3.04	3.09	3.13	3.17	3.21	3.26	3.30
27	3.16	3.20	3.25	3.29	3.34	3.38	3.42
28	3.28	3.32	3.37	3.41	3.46	3.51	3.55
29	3.39	3.44	3.49	3.54	3.58	3.63	3.68
30	3.51	3.56	3.61	3.66	3.71	3.75	3.80

VAPOR PRESSURE OF WATER

t, °C	P, mm Hg	t, °C	P, mm Hg	t, °C	P, mm Hg	t, °C	P, mm Hg
0	4.6	26	25.2	40	55.3	90	526.0
5	6.5	27	26.8	45	71.9	95	634.0
10	9.2	28	28.3	50	92.5	96	657.7
15	12.8	29	30.1	55	118.1	97	682.1
20	17.5	30	31.8	60	149.5	98	707.3
21	18.7	31	33.7	65	187.6	99	733.2
22	19.8	32	35.7	70	233.8	100	760.0
23	21.0	33	37.7	75	289.3	101	787.6
24	22.4	34	39.9	80	355.5	102	816.0
25	23.8	35	42.2	85	433.8	103	845.3

DENSITY OF WATER IN GRAMS PER CUBIC CENTIMETER

Degrees	0	0.1	0.2	0.3	0.4	0.5	0.6	0.7	0.8	0.9
0	0.999841	847	854	860	866	872	878	884	889	895
1	0.999900	905	909	914	918	923	927	930	934	938
2	0.999941	944	947	950	953	955	958	960	962	964
3	0.999965	967	968	969	970	971	972	972	973	973
4	0.999973	973	973	972	972	972	970	969	968	966
5	0.999965	963	961	959	957	955	952	950	947	944
6	0.999941	938	935	931	927	924	920	916	911	907
7	0.999902	898	893	888	883	877	872	866	861	855
8	0.999849	843	837	830	824	817	810	803	796	789
9	0.999781	774	766	758	751	742	734	726	717	709
10	0.999700	691	682	673	664	654	645	635	625	615
11	0.999605	595	585	574	564	553	542	531	520	509
12	0.999489	486	475	463	451	439	427	415	402	390
13	0.999377	364	352	339	326	312	299	285	272	258
14	0.999244	230	216	202	188	173	159	144	129	114
15	0.999099	084	069	054	038	023	007	*991	*975	*959
16	0.998943	926	910	893	877	860	843	826	809	792
17	0.998744	757	739	722	704	686	668	650	632	613
18	0.998595	576	558	539	520	501	482	463	444	424
19	0.998405	385	365	345	325	305	285	265	244	224
20	0.998203	183	162	141	120	099	078	056	035	013
21	0.997992	970	948	926	904	882	860	837	815	792
22	0.997770	747	724	701	678	655	632	608	585	561
23	0.997538	514	490	466	442	418	394	369	345	320
24	0.997296	271	246	221	196	171	146	120	095	069
25	0.997044	018	*992	*967	*941	*914	*888	*862	*836	*809
26	0.996783	756	729	703	676	649	621	594	567	540
27	0.996512	485	457	429	401	373	345	317	289	261
28	0.996232	204	175	147	118	089	060	031	002	*973
29	0.995994	914	885	855	826	796	766	736	706	676
30	0.995646	616	586	555	525	494	464	433	402	371

* Where values are designated with an asterisk, the first three figures are to be supplied from the zero column in the next lower row.

CONCENTRATION SCALES

No one concentration scale is well suited to all purposes; the different scales possess advantages or disadvantages as situations change. The mole-fraction scale is useful in theoretical work because it is symmetrical with respect to all components and because several measurable properties of an "ideal" solution are proportional to mole fraction. When transport phenomena are being studied, a volume-based scale such as molarity must be employed. It is also used in analytical chemistry. The molality scale, on the other hand, is useful in dealing with equilibrium phenomena.

In macromolecular chemistry one often works with substances of unknown molecular weight, and the scales grams per milliliter and per 100 ml, weight percent, and volume fraction find application. The weight percent scale (along with mole fraction and molality) is independent of temperature and pressure, but it cannot be used in the description of flow processes.

Some definitions for six familiar concentration scales are listed below, where g = gram, M = molecular weight of components 1, 2, 3, etc., from j to q, and n = number of moles:

1. Mole fraction

$$X_i = \frac{g_i/M_i}{\sum\limits_{j=1}^{q} g_j/M_j} = \frac{n_i}{n_1 + n_2 + \cdots + n_q}$$

2. Molality (moles component i per 1000 g component 1)

$$m_i = \frac{1000 g_i}{g_1 M_i} = \frac{1000 n_i}{n_1 M_1}$$

3. Molarity (moles of i per liter of solution having density ρ)

$$C_i = \frac{1000 \rho g_i/M_i}{\sum\limits_{j=1}^{q} g_j} = \frac{1000 \rho n_i}{n_1 M_1 + n_2 M_2 + \cdots + n_q M_q}$$

4. Grams c of component i per 100 ml of solution

$$c_i = \frac{100 \rho g_i}{\sum\limits_{j=1}^{q} g_j} = \frac{100 \rho n_i M_i}{n_1 M_1 + n_2 M_2 + \cdots + n_q M_q}$$

5. Weight percent of component i

$$w_i = \frac{100g_i}{\sum\limits_{j=1}^{q} g_j}$$

6. Volume fraction ϕ of component i ($\bar{v}$ = partial specific volume in solution)

$$\phi_i = \frac{\rho\bar{v}_i g_i}{\sum\limits_{j=1}^{q} g_j} = \frac{c_i\bar{v}_i}{100}$$

COLOR-CODE CONVENTIONS FOR RESISTORS

Color	Significant figure	Decimal multiplier	Tolerance, percent
Black	0	1	
Brown	1	10	1
Red	2	100	2
Orange	3	1,000	3
Yellow	4	10,000	4
Green	5	100,000	5
Blue	6	1,000,000	6
Violet	7	10,000,000	7
Gray	8	100,000,000	8
White	9	1,000,000,000	9
Gold		0.1	±5
Silver		0.01	±10
No color			±20

Resistors

Axial-lead type

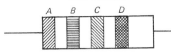

A. First significant figure, resistance in ohms
B. Second significant figure
C. Decimal multiplier
D. Tolerance

SCHEMATIC SYMBOLS FOR ELECTRONIC-CIRCUIT COMPONENTS

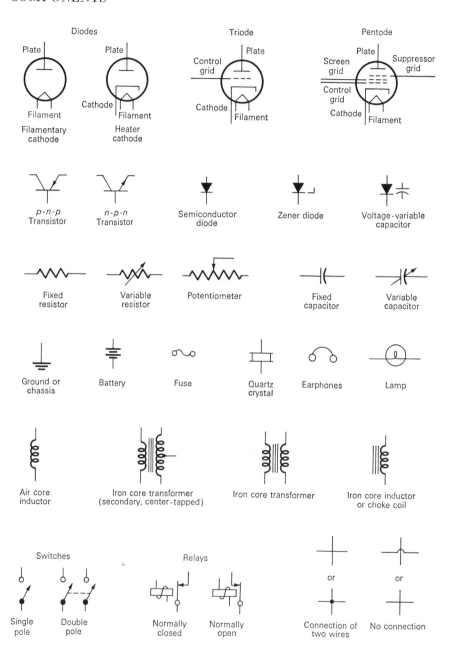

COMMON ABBREVIATIONS OF ELECTRONICS

Units

a = ampere
Hz = hertz
f = farad
h = henry
v = volt
w = watt
Ω = ohm

Unit Prefixes

p = pico· $= 10^{-12}$
n = nano· $= 10^{-9}$
μ = micro· $= 10^{-6}$
m = milli· $= 10^{-3}$
k = kilo· $= 10^{3}$
M = mega· $= 10^{6}$
G = giga· $= 10^{9}$
T = tera· $= 10^{12}$

Miscellaneous

ac = alternating current, i.e., time-dependent, with average value zero
dB = decibel [measure of relative power: $dB = 10 \log (P_2/P_1)$]
dc = direct current, i.e., time-independent
p-p = peak-to-peak
rms = root-mean-square (page 583)

Examples

1 pf = 1 picofarad = 10^{-12} farad
1 MHz = 1 megahertz = 10^6 Hz
1 kMHz = 1 GHz = 10^9 Hz
1 kΩ = 1 kilohm = 10^3 ohms

For *resistances*, the symbol Ω is usually omitted from schematic diagrams, it being understood that the value is in ohms. Thus 100 k means 10^5 ohms.

For *capacitors*, the unit is often omitted from schematic diagrams, with the understanding that the unit is μf for values less than 1 and pf for values greater than 1.

For *frequencies*, the "per second" was often omitted in technical writing until recently, megacycles per second then being abbreviated as Mc, or sometimes as mc. The preferred frequency unit is now the hertz (1 Hz = 1 cycle/sec).

Index